RADAR

E-2C "Hawkeye" Airborne Early Warning and Control System (AWACS).
(Photograph courtesy of U.S. Naval Reserve)

RADAR

PRINCIPLES, TECHNOLOGY, APPLICATIONS

BYRON EDDE

For book and bookstore information

http://www.prenhall.com
gopher to gopher.prenhall.com

Prentice Hall PTR
Upper Saddle River, NJ 07458

Library of Congress Cataloging-in-Publication Data

Edde, Byron
  Radar : principles, technology, applications / Byron Edde
     p. cm.
  Includes index.
  ISBN 0-13-752346-7
  1. Radar.   I. Title.
TK6575.E33
621.3848—dc20   92-10330
                                                  CIP

Editorial/production supervision and
    interior design: bookworks
Cover design: Ben Santora
Manufacturing buyer: Susan Brunke

 © 1993 Prentice Hall PTR
Prentice-Hall, Inc.
Simon & Schuster / A Viacom Company
Upper Saddle River, New Jersey 07458

*All Rights Reserved. No part of this book
may be reproduced in any form or
by any means without permission in writing
from the publisher*

Printed in the United States of America
10 9 8 7 6 5

ISBN 0-13-752346-7

Prentice-Hall International (UK) Limited, *London*
Prentice-Hall of Australia Pty. Limited, *Sydney*
Prentice-Hall Canada, Inc., *Toronto*
Prentice-Hall Hispanoamericana, S.A., *Mexico*
Prentice-Hall of India Private Limited, *New Delhi*
Prentice-Hall of Japan, Inc., *Tokyo*
Simon & Schuster Asia Pte. Ltd., *Singapore*
Editora Prentice-Hall do Brasil, Ltda., *Rio de Janeiro*

CONTENTS

Preface xi

Chapter 1:   **RADAR FUNDAMENTALS**
- 1-1. Introduction 1
- 1-2. Radar Principles 3
- 1-3. Target Information Extraction 8
- 1-4. The Radar Equation: An Introduction 29
- 1-5. Signals and Signal Processing: An Introduction 38
- 1-6. Types of Radars and Radar Functions 44
- 1-7. Radar Compared to Other Electromagnetic Sensors 64
- 1-8. A Short History of Radar 64
- 1-9. Where to Go for More Information 69

Chapter 2:   **THE RADAR SYSTEM: Functions and Parameters**
- 2-1. Coherence 71
- 2-2. Frequency Generation 72
- 2-3. Transmitter Functions and Parameters 77
- 2-4. Waveform Spectra and Bandwidths 81
- 2-5. Antenna Principles, Functions, and Parameters 95
- 2-6. Receiver Description and Functions 105
- 2-7. Signal Processing Functions and Parameters 123
- 2-8. System Stability Requirements 129
- 2-9. Further Information 133

Chapter 3:   **THE RADAR EQUATION**
- 3-1. Radar Equation Introduction 135
- 3-2. Summary of the Radar Equation 136
- 3-3. Point Targets in Noise 140
- 3-4. Losses in the Radar Equation 147
- 3-5. Signal Processing and the Radar Equation 155
- 3-6. Radar Equation with Pulse Compression 157
- 3-7. The Radar Equation for Search Radars 159
- 3-8. The Radar Equation for Tracking Radars 161
- 3-9. The Radar Equation for CW and Pulse Doppler Radars 162
- 3-10. Radar Equation for Area Targets and Clutter 164
- 3-11. Radar Equation for Volume Targets and Clutter 168
- 3-12. Radar Equation for Self-Protection Jamming 170
- 3-13. Radar Equation for Standoff Jamming 173
- 3-14. Radar Equation in Active Augmentation 175
- 3-15. Bistatic Radar Equation, including Missile Illumination 176
- 3-16. Beacon Equation 178
- 3-17. Further Information 182

Chapter 4: TARGETS AND INTERFERING SIGNALS
- 4–1. Radar Cross-Section (RCS) Definition and Fundamentals 184
- 4–2. RCS Fluctuations 186
- 4–3. Target Fluctuation Models 191
- 4–4. Radar Cross-Section of Fundamental Shapes 194
- 4–5. Radar Cross-Section of Complex Objects 201
- 4–6. Glint 203
- 4–7. Bistatic Radar Cross-Section 208
- 4–8. Target Doppler Spectra 209
- 4–9. Polarization Scattering Matrix 211
- 4–10. Interfering Signals — Noise 213
- 4–11. Interfering Signals — Surface Clutter 216
- 4–12. Airborne Clutter: Weather, Chaff, Insects, and Birds 230
- 4–13. Moving clutter 237
- 4–14. Further Information 248

Chapter 5: TARGET ECHO INFORMATION EXTRACTION
Part 1 — Detection
- 5-1. Detection Introduction 250
- 5–2. Detection in Noise 253
- 5–3. Signal Integration and Target Fluctuations 256
- 5–4. M of N Detection 268
- 5–5. Threshold-Setting Concept — CFAR 271
- 5–6. Further Information 277

Chapter 6: TARGET ECHO INFORMATION EXTRACTION PART II:
Sampling, PRF Classes, Range, Doppler, Height-Finding
- 6–1. Sampling 278
- 6–2. Direct and I/Q Sampling 289
- 6–3. Ranging 293
- 6–4. Target Velocity (Doppler Shift) 302
- 6-5. Range and Velocity with CW and Pulse Doppler Waveforms ... 315
- 6–6. Evaluation of the PRF Classes 323
- 6–7. Radar Height-Finding 326
- 6–8. Further Information 329

Chapter 7: TRACKING RADAR
- 7–1. Tracking Introduction 331
- 7–2. Tracking System Properties and Parameters 333
- 7–3. Conical Scan Angle Tracking 335
- 7–4. Lobing Angle Tracking 341
- 7–5. Amplitude-Comparison Monopulse Angle Tracking 343
- 7–6. Phase-Comparison Monopulse Angle Tracking 348
- 7-7. Wideband Monopulse Tracking 350
- 7–8. Track-While-Scan Principles 351
- 7–9. Range Tracking 352

- 7–10. Velocity Tracking 355
- 7–11. Tracking Accuracy 356
- 7–12. Tracking Servos 364
- 7–13. Track Acquisition 369
- 7–14. Track Anomalies 370
- 7–15. Further Information 374

Chapter 8: RADAR TRANSMITTERS AND MICROWAVE COMPONENTS
- 8–1. Transmitter Introduction and Functions 376
- 8–2. Klystron Transmitters 378
- 8–3. Traveling Wave Tube (TWT) Transmitters 381
- 8–4. Crossed-Field Amplifier (CFA) Transmitters 388
- 8–5. Magnetron Transmitters 391
- 8–6. Solid-State Transmitters 394
- 8–7. Modulators 396
- 8–8. High-Voltage Power Supplies 401
- 8–9. Transmitter Vacuum and Cooling Systems 402
- 8–10. Transmitter Monitoring and Testing 403
- 8–11. Microwave Components 406
- 8–12. Waveguide 410
- 8–13. Further Information 414

Chapter 9: RADAR ANTENNAS
- 9–1. Antenna Principles 416
- 9–2. Arrays of Discrete Elements - Principles 422
- 9–3. Radar Antenna Configurations 427
- 9–4. Sidelobe Suppression Techniques 428
- 9-5. Reflector Antennas 431
- 9–6. Lens Antennas 440
- 9–7. Mechanically Scanned Arrays 441
- 9–8. Electronically Phase-Steered Arrays 450
- 9–9. Frequency-Steered Arrays 461
- 9–10. Antenna Pedestals and Data Pick-Off Devices 464
- 9-11. Further Information 469

Chapter 10: RECEIVERS AND DISPLAYS
- 10–1. Receiver Description and Functions 471
- 10–2. Superheterodyne Principles 473
- 10–3. Radio-Frequency (RF) Processor 475
- 10–4. Mixers 481
- 10–5. Local Oscillators 484
- 10–6. Intermediate Frequency Amplifiers 489
- 10–7. Demodulators 492
- 10–8. Radar Receiver Examples 494
- 10–9. Receiver Testing and Monitoring 498

10-10. Radar Displays 502
10-11. Further Information 506

Chapter 11: RADAR SIGNAL PROCESSING I: Introduction and Background

11-1. Introduction 507
11-2. System Fundamentals and Definitions as Applied to Radars ... 510
11-3. Signal Integration 510
11-4. Correlation 522
11-5. Convolution 525
11-6. Spectrum Analysis 528
11-7. Fast Algorithms: FFT, Fast Convolution, Fast Correlation 536
11-8. Processing Errors and Windows 539
11-9. Windows and Resolution 549
11-10. Recovery from Samples - Interpolation 552
11-11. Synthesis of Complex Data from Magnitude-Only 555
11-12. Digital Filter Fundamentals 555
11-13. Further Information 562

Chapter 12: RADAR SIGNAL PROCESSING II: Moving Target Indicators and Doppler Processing

12-1. Doppler and Moving Target Indicator (MTI) Fundamentals 564
12-2. MTI Principles and Methods 566
12-3. Blind Doppler Shifts and PRF Stagger 571
12-4. De-Staggering and Processing 577
12-5. MTI and MTD with Moving Radars and Moving Clutter 578
12-6. Limitations on the Improvement Factor 582
12-7. CW, High PRF, and Medium PRF Doppler Processing 586
12-8. MTI Testing 589
12-9. Further Information 592

Chapter 13: HIGH RESOLUTION RADAR

13-1. Resolution Review 593
13-2. Pulse Compression Fundamentals 595
13-3. Evaluation of Waveforms for Range and Doppler Resolution .. 598
13-3. Analog Pulse Compression 607
13-5. Digital Pulse Compression 611
13-6. High Cross-Range Resolution Introduction 616
13-7. Doppler Beam-Sharpening (DBS) 618
13-8. Sidelooking Synthetic Aperture Radar 627
13-9. Synthetic Aperture With Rotating Objects 640
13-10. Further Information 651

Chapter 14: SPECIAL RADAR TOPICS

14-1. Radar Data Processing Introduction 653
14-2. Coordinate Systems and Netting 654

14-3. Coordinate Conversions 656
14-4. Secondary Surveillance Radar (SSR) 663
14-5. Radiation Hazard 673
14-6. Radomes 677
14-7. Radar Electronic Warfare (EW) Introduction 682
14-8. Further Information 689

Appendices
    A. Unit Conversions 691
    B. Transmitter Safety 693
    C. AN/ Equipment Designation 695
    D. Greek Alphabet and Metric Prefixes 698
    E. Decibels 699

Index 705

PREFACE

The use of radar has grown to the point where it affects us virtually every day, whether or not we realize it. Our society would be very different without it. The worldwide air traffic control system is based on radar. Defense systems rely heavily on radar to locate friendly, unknown, and hostile aircraft, ships, and other vehicles. Ships and boats use it to navigate and to avoid running into one another. The earth itself is monitored from space-based radars. Law enforcement officials use it to measure the speed of cars and trucks.

This book is about *modern* radar. It was written to help readers learn the subject, from overall concepts to detailed block diagrams to the algorithms of signal processing. It will be useful to readers at many levels, ranging from radar novices to professionals who specialize in one facet of radar but need information on the other aspects of the subject. Many readers will also find useful reference material.

To better serve a wide range of readers, discussions of advanced topics have been separated from those of principles. For example, Chapter 1 describes radar systems from a conceptual level. Later chapters provide in-depth discussions of the same topics and cover materials more appropriate to advanced readers.

The book is the outgrowth of courses taught by the author over the past 12 years to radar professionals, graduate engineers with no prior radar knowledge, technicians, technical managers, and general-interest audiences. The materials have been tested in the classroom and in the field, with over a thousand people having used and commented on them.

The book is divided into five main subject areas. The first covers the topics needed by all persons who deal with radars and radar-based systems. The others give in-depth information about the various parts of radar systems.

The first four chapters cover radar principles. In Ch. 1, the radar is described as a sensor and from the systems level. Chapter 2 describes the radar from the standpoint of its system functions and parameters. Chapter 3 covers the radar equation. Chapter 4 deals with radar cross-section, targets, and interfering signals. For short courses, Chapter 1 contains material everyone associated with radars and radar-based systems should know. Chapters 2 through 4 contain additional information needed by those requiring more detailed knowledge.

Chapters 5 and 6 treat the recovery of information from signal echoes. Chapter 5 describes the detection of targets in interference. Chapter 6 covers sampling, pulse repetition frequency classes, and the recovery of target range, Doppler shift, and height.

Radar technology is presented in Chs. 7 through 10. Chapter 7 covers automatic tracking, tracking errors, and tracking algorithms. Chapter 8 describes radar transmitters, modulators, and microwave components. Chapter 9 covers radar antennas and arrays. Chapter 10 describes radar receivers and displays.

Chapters 11 through 13 describe the processing of radar signals. Chapter 11 covers the general principles and theory of signal processing, including techniques for correcting processing errors. Chapter 12 describes Doppler processing. Chapter 13 is about high resolution processing — pulse compression and synthetic aperture.

Chapter 14 covers special topics. The data processing unit includes descriptions of radar coordinate systems, coordinate conversions, and data-linking of radars. Secondary surveillance radar, including identification, friend or foe — IFF — and the air traffic control radar beacon system — ATCRBS — are covered here. Radomes are described in this section, as are microwave radiation hazard, the principles of radar electronic warfare, and low probability of intercept radars.

Background information is included, where appropriate, to assist less experienced readers. If a concept arises that the author feels many readers may not understand, he includes an explanation. This is done because of his observation that people who use these materials come from widely diverse backgrounds and may not have knowledge of what some would believe to be traditional electrical engineering lore (even electrical engineers with strong digital backgrounds may have forgotten much of it). Although little prior knowledge is assumed, the author has tried to compartmentalize background information so that experienced readers can easily skip it.

Many people helped in the creation of this book, and I would like to acknowledge their contributions. Thanks to Charley Shelton, Dottye Spencer, and the late Jim Rumsey for the early opportunity to put these materials into practice. Thanks to Dean Mensa for his insight and encouragement. Thanks to Karen Gettman and Maura Vill from Prentice Hall for pushing and encouraging me to get this project going and keep it moving. Thanks to Karen Fortgang of *bookworks* for her guidance and patience during the production stages. Thanks especially to Martha Edde, my wife, who encouraged me to write the book and who tolerated the time and expense necessary to put it together.

Most of all, thanks to the students, without whom this project would not have happened. They tolerated early and incomplete drafts; they suggested changes and additions; they caught errors. Their enthusiasm made it almost impossible to quit. They actually read the materials, for which I am eternally grateful.

Byron Edde
Camarillo, California

RADAR

1

RADAR FUNDAMENTALS

1-1. Introduction

RADAR (*RA*dio *D*etection *A*nd *R*anging) is a method of using electromagnetic waves to remote-sense the position, velocity, and identifying characteristics of targets. This is accomplished by illuminating a volume of space with electromagnetic energy and sensing the energy reflected by objects in that space. Radar as we know it today dates from the 1930s and usually, but not always, uses energy in the microwave frequency bands (roughly 0.5 to 100 GHz). It is used for many sensing tasks, including but not limited to:

- Detecting and locating ships and land features for ship collision avoidance
- Navigating aircraft and ships in bad weather or at night
- Detecting, locating, and identifying aircraft for air traffic control
- Measuring altitude above the surface for aircraft and missile navigation
- Detecting and locating severe weather for ground, ship, and aviation safety and comfort
- Detecting, locating, and classifying aircraft, ships, and spacecraft for defense purposes
- Giving early warning of hostile aircraft and spacecraft while they are hundreds or thousands of miles away
- Mapping land and sea areas from aircraft and spacecraft
- Locating and imaging ground objects for navigation and targeting
- Detecting ground moving vehicles, such as tanks, for defense purposes
- Controlling weapons, such as guns and missiles
- Measuring distance and velocity for spacecraft navigation and docking
- Precisely measuring the position and/or velocity of objects in space for instrumentation purposes
- Precisely measuring distances for land surveying
- Detecting and measuring objects under the ground's surface
- Measuring motor vehicle velocities for safety, automatic control, and law enforcement

This first chapter is an overview of radar plus some necessary background information. Where appropriate, equations show the relationships between the radar and target parameters and the information in the echo. These relationships are stated without proof and without regard to special conditions. More formal presentations and developments are found in later chapters.

Figure 1-1 shows four radars of three different types. The topmost system is the AN/SPS-10 surface search radar. Its mission is to locate objects on the water for collision avoidance and ship defense, and to locate land masses for navigation. See Appendix C for the AN/ system designation scheme.

Figure 1-1. Shipboard Radars. (Photograph courtesy of ITT Gilfillan)

The large square antenna is part of the AN/SPS-48 air search radar (ASR), which continuously scans or searches a volume of space to find aircraft. It is a three-dimensional radar in that it can discriminate targets in range, horizontal (azimuth) angle, and vertical (elevation) angle. Signal processing helps separate echoes from low-flying aircraft from those caused by the surrounding water and terrain. These interfering echoes from the radar's surroundings are called clutter: ground or land clutter when from land, sea clutter when from lakes and oceans, and weather clutter when from water in clouds and precipitation.

The antenna for this particular radar is an array, which is a large antenna made up of numerous smaller antennas, called elements.

The structure across the top of the AN/SPS-48 antenna is the secondary surveillance radar (SSR) antenna, which is used to identify and give other information about aircraft. This antenna is connected to an interrogator, separate from the radar itself, which transmits a coded signal to a transponder in the target. The transponder, found in various forms in most civilian and military aircraft, replies with its own coded signal providing specific information about the aircraft. In its military identification form, it is called Identification, Friend or Foe, or IFF. When used for civilian and military air traffic control, it is known as the Air Traffic Control Radar Beacon System (ATCRBS). The presently available modes allow the aircraft to identify itself and give its altitude. The new Mode S will allow selective interrogation and data linking. See Ch. 14 for more information.

The two AN/SPG-55 systems are tracking radars. Their mission is to follow single targets to accurately determine their positions and velocities. It also provides the transmitter function for various semi-active radar homing (SARH) missiles. These missiles do not have a radar transmitter and can only listen for echoes caused by the radar's illumination. The large circular structure on each system is the main radar antenna. The other antennas perform missile illumination and guidance, electronic warfare (EW), and auxiliary functions.

Fighter and strike aircraft do not have the space or lifting capability to have separate radars for each function required. The radar in Fig. 1-2 is the AN/APG-70, a multi-mode radar found in the F-15C/D/E Eagle. It combines air search, surface search, tracking, gun control, ground mapping and targeting, navigation, missile illumination, and other functions into a single radar.

Figure 1-2. AN/APG-70 Multimode Airborne Radar. (Photograph courtesy of Hughes Aircraft Company, Radar Systems Group)

1-2. Radar Principles

The function of radar is to detect and locate targets and to report information about them. The majority of radars illuminate targets by transmitting short bursts of illumination energy and listen for echoes with the transmitter silent. These are known as pulsed radars. Most radar functions can also be accomplished by transmitting a continuous wave (CW) and listening for echoes while the transmitter is radiating. See Ch. 7 for waveforms.

Figure 1-3 is a simplified block diagram of a pulsed radar. It is monostatic in that its transmitter and receiver are in the same location. This particular system shares a single antenna between transmit and receive; dual antenna systems with transmit and receive in the same location are also monostatic. The functions of the major blocks are described next.

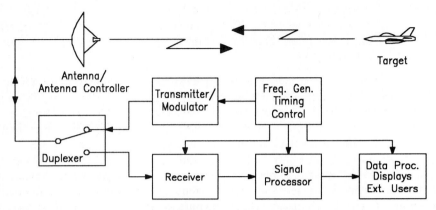

Figure 1-3. Radar Simplified Block Diagram.

— **Frequency generation, timing, and control:** This block generates the frequencies and synchronization signals required by the system. It determines when the transmitter fires and how other system functions relate to the time of transmission. It exercises direct control over the system's parameters. Most of the other blocks in the system interact with it.

— **Transmitter:** The transmitter generates the radio signal which is used to illuminate the target and from which the echo is derived. Transmitters can be either amplifying types, where the waveform is generated at low levels and amplified, or they can be power oscillators, where the illumination signal is generated at the required energy level.

— **Modulator:** This function controls the transmitter in pulsed systems, turning it on and off to form the pulse. In CW radars, it provides the modulation used by some to find target range (Ch. 7). Pulsed modulators are of two types. High level modulators supply the main DC operating power for the transmitter. This may be, for example, a negative pulse of 40,000 volts and 50 amperes, or more. Low-level modulators supply a control voltage for a grid in the transmitting tube. The control voltage is a DC pulse of a few hundred volts potential and a few amperes of current.

— **Duplexer:** The duplexer switches the antenna of monostatic single-antenna systems between the transmitter and receiver, allowing the antenna to be shared by both functions. Since the switching action must take place within nanoseconds, the switch is electronic.

— **Antenna:** The antenna concentrates the illumination signal into a narrow beam radiated in a single preferred direction, intercepts the target echo signals from this same preferred direction, and matches the system impedances to those of the propagation medium. The antenna can be steered, so that the radar can search or track in many directions.

— **Antenna controller:** The antenna controller positions the antenna beam to the desired azimuth and elevation angles and reports these angles to the system controller and data processor. In some systems, it also determines the antenna beam shape and/or polarization. With mechanically steered antennas, the controller serves an electromechanical or electrohydromechanical positioner. With electronically steered antennas, such as phased arrays, it computes the control data necessary to steer the transmit and receive beams.

— **Receiver**: The receiver amplifies the echo signals to a level sufficient for later system components, such as the signal processor. It filters incoming signals to remove out-of-band interference (channel-select filtering) and to optimize the response in a specific type of interference (matched filtering).

— **Signal processor**: This function processes the target echoes and interfering signals to increase the target echo signal level and suppress the interference, thereby increasing the signal-to-interference ratio. It performs the detection function, making the decision as to whether or not a target is present, and recovers information about targets such as range and Doppler shift.

— **Data processor**: The data processor stores and processes the location of detected targets. In some search radars it smooths the data and extrapolates the targets' positions in the track-while-scan function. In some track radars it functions as the track servo, processing angular errors into signals which control the antenna's motion. Some radars furnish target data to other locations through data links, in a process called netting. In these, the data processor converts target position data to a coordinate system which is understandable to all systems in the net. At the receiving system, the data processor converts the net data to local coordinates which are understandable to the local system.

— **Displays**: The displays put the information in a form usable to radar operators and others, such as air traffic controllers and weapon system operators and supervisors.

Figure 1-4 and the following explanation show the process by which a typical pulsed radar detects and locates a target.

The frequency generation and timing system periodically causes the transmitter to generate a burst, or pulse, of illumination electromagnetic energy. The pulse's parameters vary with the mission and type of the radar and the mode in which it is operating. It may have a power level from milliwatts to megawatts. Its width may be nanoseconds to milliseconds. The rate at which pulses are generated (the pulse repetition frequency, or PRF) may be a few pulses per second to hundreds of thousands of pulses per second. At the same time the transmission occurs, a timer is started, shown on the left side of Fig. 1-4.

The illumination energy is delivered by the duplexer to the antenna, where it is concentrated by the antenna's *gain* into a beam pointed in the direction the radar is examining.

The concentrated energy propagates through the atmosphere at a speed near that of light in free space. This forward propagated wave has electric and magnetic fields in the ratio 120π (≈ 377) Ohms, which is the characteristic impedance of the atmosphere or free space.

If the wave encounters an object with propagation characteristics (impedance) different from those of the medium, a fraction of the energy hitting the object reflects. The reflection can be in any or many directions, but the only echo energy which is of use to the radar is that propagating back toward it, called backscatter. There is also backscatter energy from the ground or sea over which the beam propagates and from atmospheric conditions through which it propagates. This clutter interferes with the detection of targets.

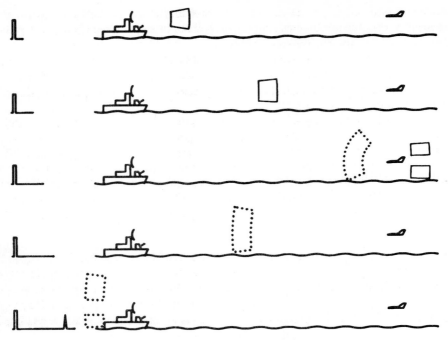

Figure 1-4. Pulse Radar Concept.

The backscatter energy propagates to the radar, where a fraction of it is intercepted by the antenna. The amount of power intercepted from the direction of the beam is determined by the antenna's area. The antenna also intercepts interfering signals, such as clutter and intentional interference, which compete with the target echoes.

The captured target echoes and interference are directed by the duplexer to the receiver, where they are amplified and filtered. The receiver also unavoidably generates random interference of its own, called noise. Like other interference, this noise competes with the desired target echoes.

Target echoes and all the accompanying interference then go to the signal processor, where they are treated in such a way as to enhance target echoes and suppress interference. This interference suppression is not perfect, and interference residue appears with target echoes at the output of the signal processor.

The processed signals and interference residues are compared to a reference level, usually a voltage, called the detection threshold. If the composite signal exceeds the threshold, a detection is declared. If the detection is caused by a desired target, it identifies the target's presence. If caused by interference, the detection is a false alarm, and indicates the presence of a nonexistent target. The consequences of a false alarm can range from minor, such as a lighted spot on a display cathode ray tube, to severe, such as a gun automatically firing on target detection. The threshold must be set carefully so that the probability of a false alarm occurring is small. If a desired target signal is too small to cross the threshold, the target is not detected. It will be shown that of the two detection errors, the false alarm usually has the more severe consequences.

Detection is described statistically. The probability that a target echo will cross the threshold is the probability of detection. The probability of interference alone crossing the threshold is the probability of false alarm.

When a detection occurs, the timer is strobed (sampled) to measure the object's round-trip propagation time, and hence its range. The antenna position encoders are also strobed to find where the beam was pointed at the time of detection. This is assumed to be the target's angular location.

Motion between the target and the radar causes the echo frequency to be slightly different from that of the illumination. In systems so equipped, this Doppler shift is detected and/or measured. The moving target indicator (MTI) function in some radars detects whether or not a Doppler shift is present, but does not measure its value. In other systems, a moving target detector (MTD) identifies the Doppler shift as having one of a few values. Other radars and modes, CW, pulse Doppler (PD), and medium PRF (MPRF), measure and report the Doppler shift value and thus the velocity of the target. Section 1-7 further identifies different radar types.

Interfering signals: As stated earlier, the radar receives several forms of interference, which complicate the detection and target measurement process. If they are large enough, they can mask altogether the desired target echoes. They can also cause the measured target parameters to be in error. The five basic kinds of interference are shown in Fig. 1-5 and described below. One of signal processing's roles is to suppress these interfering signals.

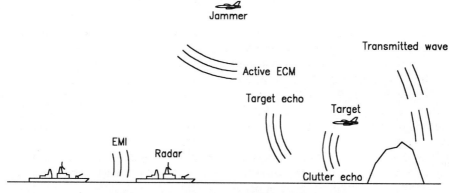

Figure 1-5. Radar Signals and Interference.

— *Noise*, caused by the random motion of electrically charged particles which occurs at all temperatures above absolute zero, is unavoidably generated in the radar's receiver, with small amounts also from the antenna and transmission lines, and from external sources, principally the sun. The randomness of noise compared to the orderliness of target echoes allows the noise to be suppressed.

— *Clutter* is unwanted signal echo from sea, land, and weather. It is a real echo signal which is usually suppressed based on its Doppler shift being different from that of desired targets.

– *Electronic countermeasures (ECM)*, or jamming is intentional interference generated in an attempt to disrupt detection of target echoes. ECM may behave in the radar system as though it were high-level noise (noise jamming), as a rough approximation of a target or multiple targets, or as target(s) which are essential clones of legitimate target echoes. Noise jamming is suppressed as though it were noise. The ability to suppress other kinds of jamming depends on how much they differ from real target echoes.

– *Electromagnetic interference (EMI)* is accidental interference from friendly sources, such as other radars, communication systems, and friendly jammers. EMI is suppressed primarily by prevention. Signal processing may reduce its effect, depending on how different it is from legitimate target signals.

– *Spillover* occurs mainly in continuous wave (CW) radars, and is caused by operating the transmitter and receiver simultaneously. It is the leakage from the transmitter into the receiver. A related form of interference, caused by internal reflections within the radar itself, is known as internal clutter.

1-3. Target Information Extraction

If the target echo signals after signal processing are sufficiently greater in amplitude than the interfering signals, information about the targets can be extracted from them. Target parameters can then be measured. The following information can be derived from radar echoes, recognizing that all radars do not measure all of the listed parameters.

1-3-1. Detection: Detection determines whether or not a target is present. It is accomplished by comparing signal plus interference to a threshold (Fig. 1-6). Table 1-1 shows the four conditions of detection. If a target is present and a detection occurs, the result is correct. If no target is present and no detection occurs, the result is also correct. The detection errors are (a) when a target is present but not detected, and (b) when no target is present but a detection of interference occurs. Detection is covered in Ch. 5.

Table 1-1. Detection Principle.

Target?	Detection?	Result
No	No	Correct
Yes	Yes	Correct
Yes	No	Error
No	Yes	Error (False Alarm)

Note in Fig. 1-6 that the only way one can discriminate with certainty between target echoes and interference is to be told which is which. After signal processing, the two have exactly the same pulse shape, and the only way to discriminate is by amplitude. This is a basic problem in radar detection, since weak targets and strong interference residue can easily be confused with one another. For this reason, detection can only be described by probabilities. The display shown in Fig. 1-6, signal amplitude versus time, is called an A-Scope.

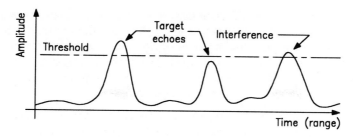

Figure 1-6. Detection Concept.

1-3-2. Target position locating: Target position is described in spherical coordinates (Fig. 1-7). The principal axes are range (R), azimuth angle (θ_{AZ}), and elevation angle (θ_{EL}). The references and origin are described below and the methods of information recovery follow.

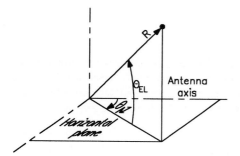

Figure 1-7. Spherical Coordinates.

Range is a target's distance from the radar, regardless of direction. The process of measuring range is known as ranging.

Local horizontal is, for land-based radars, the plane passing through the antenna's center of radiation and perpendicular to the earth's radius passing through the same point. For shipboard and airborne radars, it is the plane containing the vehicle's roll and pitch axis. Some shipboard antennas are stabilized in roll and pitch and thus have the same local horizontal definition as land-based systems. Others are stabilized in roll only, with the local horizontal being a plane through the roll axis such that its intersection with the sea surface is a line perpendicular to the roll axis.

Local vertical is a line perpendicular to local horizontal through the radar's space reference, which is usually the center of radiation of the antenna.

Azimuth angle is the antenna beam's angle on the local horizontal plane from some reference, shown in Fig. 1-8 for various radar platforms. In land-based radars, the azimuth reference is usually true north, with some air traffic control radars using local magnetic north. Shipboard radars are usually referenced to ship's head, which is a line parallel to the ship's roll axis. Zero azimuth is toward the bow of the ship. Airborne radars reference the roll axis on the local horizontal, the same as ships, with zero being in the direction of flight.

Elevation angle is the angle between the radar antenna beam's axis and the local horizontal.

1-3. Target Information Extraction

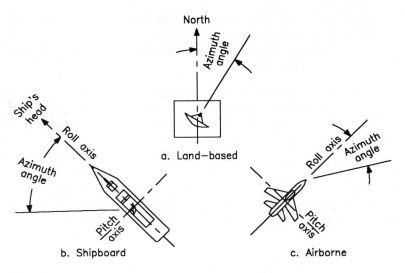

Figure 1-8. Azimuth Angle References.

1-3-3. Ranging: Ranging is accomplished by measuring the time delay between the radar's transmission and the detection of the target's signal echo (Fig. 1-9). Time is usually measured from the center of the transmit pulse to the center of the received echo (centroid ranging), but occasionally from the leading edge of the transmit signal to the leading edge of the echo (leading-edge ranging).

$$R = \frac{c\,T_p}{2} \quad \text{(meters)} \tag{1-1}$$

R = range to target (meters)
T_p = round-trip propagation time (seconds)
c = the propagation velocity (nominally 3.0×10^8 m/s)

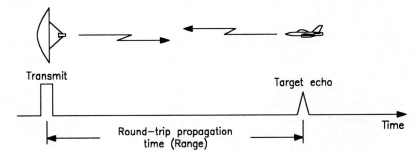

Figure 1-9. Radar Ranging Concept.

Example 1-1: A target echo is detected 129.30 μs after the radar transmits. How far from the radar is the target in nautical miles?

Solution: Equation 1-1 yields target range in whatever units result from the propagation velocity multiplied by the propagation time. In the example, the range is 19,400 m (19.40 km). Since there are exactly 1852 meters in a nautical mile (Appendix A), the range is 10.47 nmi.

Range ambiguity: Ambiguities occur when multiple target positions produce the same reported information. They are caused by basing measured information on incorrect references. Range ambiguities occur when received echoes are attributed to the wrong transmit pulses. In Fig. 1-10, E_1 is the target echo caused by transmit pulse T_1, E_2 is caused by T_2, and so forth. Most systems initially assume that the propagation time to a received echo starts at the latest transmit pulse. Thus if there are no intervening transmissions between a transmit pulse and its echo, the initial range report is correct (Fig 1-10a). This is called a range zone one target. *If it is known that the target is in zone one*, the system ranges unambiguously. *Any target for which the range zone is not known is ambiguous.*

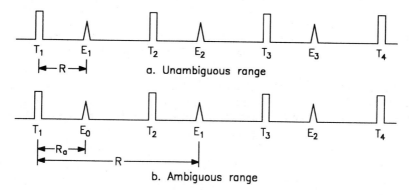

Figure 1-10. Range Ambiguity.

The target in the Fig. 1-10b has a propagation time greater than the time between transmitted pulses. This is a range zone two target. Most radars initially report this target's range as R_a (apparent range), instead of R (true range). The process of determining true range from apparent is called deghosting, and is covered in Ch. 6.

Example 1-2: A radar has a pulse repetition frequency (PRF) of 1250 pulses per second (pps). What is the maximum range which targets can have if they are to be in the first range zone?

Solution: A zone 1 target can have a propagation time no greater than the time between transmit pulses (actually the time between pulses less the pulse width, but this will be ignored for now). The time between pulses at 1250 pps is 1/1250 s or 800 μs. Equation 1-1 gives this range as 120,000 m (120 km or 64.8 nmi).

Range granularity and quantization: Signals, data, and displays are treated digitally in almost all modern radars, from simple small-boat radars to the most sophisticated multi-mode systems. The data is thus granular, having only discrete possible values. Granularity is defined as the size of the minimum data output change. Radars which handle target positions digitally can only report a target's range as discrete values, such as 100 m, 200 m,

300 m, and so forth. If the target is actually at 321.7 m, its position is reported as 300 m. All dimensions in which these radars report target information are granular.

Granularity is sometimes confused with resolution, but they are totally different concepts. Resolution, discussed later in this section, is the ability to separately detect multiple targets; granularity is simply the smallest possible output data change.

In systems having discrete data outputs, all dimensions (range, azimuth, elevation, and Doppler) are treated not as continuous quantities but in small discrete increments called bins. Range bins are discrete quantities of time. Each bin contains the smallest time division available to the radar for signal gathering, range data output, and display. Multiple signals within the same range bin are sensed as their vector sum.

Search radar range is almost always treated as a sequence of bins, whose widths are approximately the same as the radar's range resolution. The first bin occurs immediately after the transmitter output, and bins occur continuously until the radar's maximum instrumented range is reached (Fig. 1-11). A particular bin usually occurs at the same time after each transmitter output. An exception is found in some airborne radar ground mapping modes, where the bins are stabilized to fixed ground positions (Ch. 14).

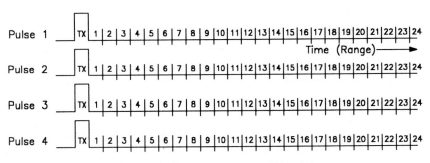

Figure 1-11. Range Bins — Search Radar.

Tracking radars also have range bins, often called range gates, shown in Fig. 1-12. Instead of the bins being in discrete, fixed locations, they move continuously or in very small steps in response to the range track. In many tracking radars, a single gate is used to assure that the angle track errors come only from a single target (gate C in Fig. 1-12). In others,

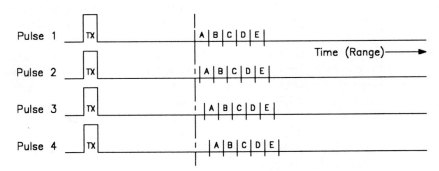

Figure 1-12. Range Bins — Tracking Radar.

Chap. 1. Radar Fundamentals

several contiguous gates move in unison. The target being tracked is in the center bin, called the track gate, and the remaining bins are guard bins or guard gates. Guard gates are for track initiation, to sense separations from targets being tracked, such as missile launches, and to counter certain forms of deceptive ECM.

1-3-4. Angular position: The azimuth and elevation positions of targets are described by the angles between the target's location and some reference. They are measured by determining the antenna's pointing angles at the time a signal detection is made, on the presumption that signals detected always originate from the direction of the antenna's beam at the time of detection (Fig. 1-13). Under certain conditions, this may not be true, and false angular position reporting may result. More accurate results are obtained by reading antenna position at the center of a detection sequence (see centroid detection in Ch. 7).

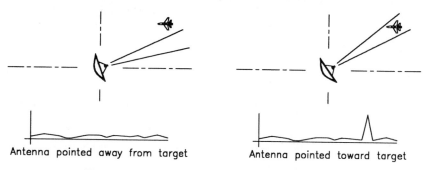

Figure 1-13. Angular Target Location Concept.

Azimuth and elevation bins are data storage locations representing the smallest changes sensed in the respective dimensions.

True ambiguity in angles is rare and occurs only in cases where the radar's antenna is an array with very wide element spacings. These antennas produce multiple main lobes called grating lobes. Apparent ambiguities result when echoes are received through the antenna's sidelobes (Ch. 4).

1-3-5. Velocity measurement and discrimination: Target radial velocity is extracted from the Doppler frequency shift between the transmitted signal and the received echo. Doppler shift is the frequency shift of the echo signal caused by the target's motion with respect to the radar. It is the electromagnetic equivalent of the acoustic effect heard when a vehicle sounding its horn is driven past an observer. As the vehicle approaches, the observer hears the horn at a higher than actual pitch; as it moves away, the pitch is heard lower than actual (Fig. 1-14). The Doppler shift is used both to measure the velocity of targets and to resolve targets occurring at the same time but moving at different velocities. The latter use is the primary method of discriminating moving targets from clutter.

Following are the pertinent relationships regarding Doppler, to be derived in Ch. 6. By definition the Doppler shift is the difference between the frequencies of the received and transmitted waves. A positive Doppler shift is from an inbound target and a negative shift from a target outbound from the radar.

$$f_d \equiv f_R - f_T \text{ (Hertz)} \qquad (1\text{-}2)$$

f_d = the Doppler shift (positive Doppler is from targets approaching the radar; negative is from targets moving away from the radar — Hertz or 1/seconds)

f_R = the frequency of the target echo (Hertz)

f_T = the transmitted frequency (Hertz)

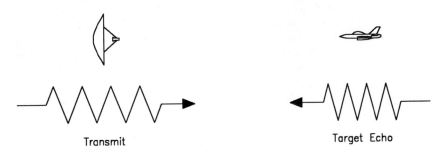

Figure 1-14. Doppler Shift.

If a target's radial velocity with respect to the radar is much less than the velocity of propagation, Eq. 1-3 describes the Doppler shift. If this condition is not met, an exact relationship is found in Ch. 6.

$$f_d \approx 2 f_T \frac{v_R}{c} \quad \text{(Hertz)} \tag{1-3}$$

v_R = the radial velocity difference between target and radar (see below)

c = the velocity of propagation

Only the radial component of velocity (Figs. 1-15 and 1-16) contributes to the Doppler shift. Figure 1-17 shows the role of radial velocity in the detection of targets in clutter. To a fixed radar, no target Doppler shift occurs when the velocity is tangential to the radar's antenna

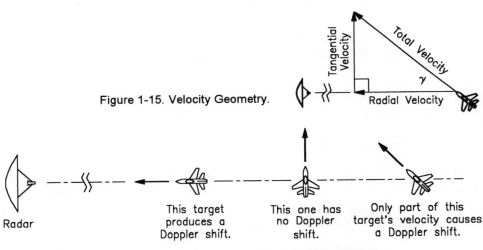

Figure 1-15. Velocity Geometry.

Figure 1-16. Doppler Shift as a Result of Radial Motion.

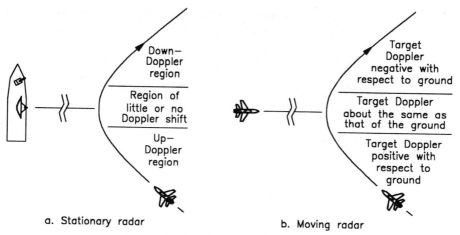

a. Stationary radar b. Moving radar

Figure 1-17. Doppler Shift Conditions.

axis. Thus in the region of little or no Doppler the target may not be resolvable from clutter. Where the radar is moving, both the clutter and the target have Doppler shift. In the region where the target's velocity is primarily perpendicular to the axis of the radar, its Doppler shift is about the same as that of the clutter and discrimination between them may not occur.

Accounting for the difference between target velocity and radial velocity requires Eq. 1-3 to be modified.

$$f_d \approx 2 f_T \frac{v}{c} \cos \gamma \text{ (Hertz)} \qquad (1\text{-}4)$$

v = the velocity difference between target and radar (m/s)

γ = the total angle (in three dimensions) between the target/radar velocity vector and the axis of the radar antenna.

$$\cos \gamma = \cos \gamma_H \cos \gamma_V \qquad (1\text{-}5)$$

γ_H = the horizontal angle between the radar's axis and the target's velocity vector (Fig. 1-18)

γ_V = the vertical angle between the radar's axis and the target's velocity vector.

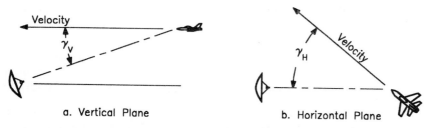

a. Vertical Plane b. Horizontal Plane

Figure 1-18. Geometric Relationships in the Doppler Shift.

1-3. Target Information Extraction

Figure 1-19 shows a block diagram of a simple Doppler radar of the type used for traffic law enforcement. The transmitter is a continuous-wave (CW) oscillator operating on a single frequency at a power level of 1 to 100 mW. Three of the frequencies allocated for this function are 10.525 GHz, 24.150 GHz, and 34.36 GHz. The transmitter feeds an antenna whose gain is on the order of a few hundred. The CW echo from the target is directed to one input port of the mixer by the microwave circulator. At the mixer's second input port is a sample of the transmitted signal. The mixer produces a signal whose frequency is the difference between the frequencies of the two inputs, which is the Doppler shift. If the Doppler signal amplitude exceeds a preset threshold, it is gated to a counter where the Doppler cycles are counted for a certain time. The count is scaled and displayed as the target's radial velocity. No correction is made for the angle between the target's velocity vector and the axis of the radar antenna — this type of radar assumes the radial velocity equals the total closing velocity. The error causes a lower-than-actual velocity to be indicated.

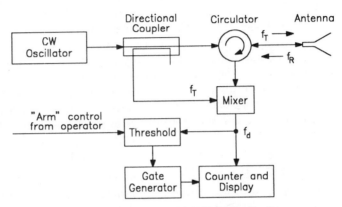

Figure 1-19. A Simple CW Doppler Radar.

Example 1-3: An automobile is moving toward a stationary police radar at 65 statute miles per hour (smph — 5280' mi). The car's velocity vector is coincident with the axis of the radar. The radar transmits on a frequency of 24.150 GHz, in the so-called K-Band. What is the frequency of the received echo signal and the Doppler shift?

Solution: Equations 1-3 and 1-4 give meaningful results only if vehicle and propagation velocities are in the same units. Since velocity of propagation is usually in meters per second, the vehicle's velocity must be converted to 29.06 m/s (0.4470 m/s per smph — Appendix A).

Using the numbers for this example, Eq. 1-3 gives a Doppler shift of 4,678 Hz. By definition, the Doppler shift from a target approaching the radar is *positive*, meaning that the receive frequency is higher than that transmitted. From Eq. 1-2, the received frequency is therefore 24.150,004,678 GHz.

Note: The radar's exact transmitted frequency may differ slightly from that specified. Since Doppler shift is directly proportional to transmitted frequency, it will change for a mistuned transmitter. Large frequency errors cause large velocity errors; if the above radar were mistuned 2 GHz higher than specified, a 65 smph vehicle would produce 5066 Hz of Doppler shift. Since the system is calibrated so that 4678 Hz Doppler indicates 65 smph, this vehicle would register 70.4 smph. There are reports of occasional mistuning of these radars to prevent radar detectors from sensing them, causing erroneous measurements.

The Doppler shift shown in Fig. 1-14 is exaggerated to make the point, and although illustrative, is somewhat misleading. The Doppler frequency shift for endo-atmospheric targets is usually so small that the received signal frequency over reasonable pulse widths is, for all practical purposes, the same as that of the transmit signal. To sense the Doppler shift, one therefore cannot look at a single echo pulse, but must observe several successive echoes and sense the phase change between them. This is illustrated later in Sec. 1-5.

Example 1-4: Consider a 10.0-GHz radar transmitting a 0.5-μs pulse with a PRF of 250,000 pps illuminating a target with a radial velocity of Mach 2 closing. Find the Doppler shift, the frequency change across one pulse, the signal phase change between echo pulses, and the phase change across 2048 echo pulses.

Solution: The Doppler shift is 45,070 Hz (Eq. 1-3 – Mach 1 at sea level is about 338 m/s). Across one pulse, this Doppler shift goes through 0.0223 cycle, or 8.1° of phase, which is not sufficient for even a rough estimate of the Doppler shift (spectrum analysis theory says that for a 0.5-μs measurement time, frequencies must be at least 2 MHz apart to be resolved – shown later in this section). The time between pulses is 4.0 μs, which is about 65° of phase at 45,070 Hz. 2048 pulses occupy 8.192 ms, during which time 45,070 Hz goes through about 369 cycles or 132,920° of phase. Over this time, the signal is easily analyzed.

Figure 1-14 is also misleading in that for most purposes, the echo pulse width is the same as that of the transmitted pulse. If the target can be considered a point in space (a point target), there are the same number of cycles in the echo as in the transmission. If the receive frequency is higher than the transmit frequency (closing target), the receive pulse is shorter than the transmit pulse. The Mach 2 target of Ex. 1-4 produces 8.1° of phase across the 0.5-μs 10-GHz pulse. At 10 GHz, 8.1° of phase takes 2.25 ps, compressing the pulse from 0.50000000 μs to 0.49999775 μs. This change is not significant, particularly since after receiver matched filtering, the ability to measure the pulse width to within 25% or so is lost (Ch. 2).

Doppler ambiguities: Frequency ambiguities occur in all systems where the signal is sampled for digital processing, including pulsed radars and digitally-processed CW radars. It will be shown in Ch. 2 that pulsed transmit signals produce spectra which are discrete, composed of a number of individual frequencies rather than containing all frequencies. If the pulses are periodic (have uniform spacing between them), the individual spectral lines, representing individual sinusoids, are uniformly spaced and are separated from one another by the PRF. The received spectrum is of the same form as that of the illumination, except that it is offset in frequency from the illumination spectrum by the Doppler shift. *Doppler ambiguities happen when it is unknown which transmit spectral line caused a particular receive spectral line.* They can occur when the Doppler shift is greater than half the frequency difference between transmit spectral lines, violating the Nyquist sampling criterion (Sec. 6-1). Figure 1-20 illustrates.

1-3. Target Information Extraction

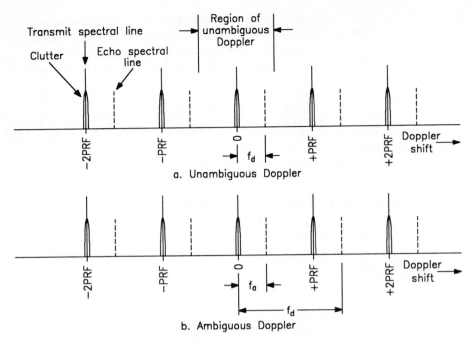

Figure 1-20. Doppler Ambiguity.

The taller solid lines represent the illumination spectrum and are separated by the PRF. The wider features associated with the transmit lines are the spectrum of stationary clutter as viewed from a stationary radar. There are copies of the clutter spectrum, called aliases, associated with each transmit spectral line (Sec. 6-1). The shorter dashed lines are the target echo spectrum. The target in Fig 1-20a has a Doppler shift which is less than half the spacing between illumination spectral lines. This target is unambiguous in Doppler, *if it is known that the spacing shown is correct.*

In Fig. 1-20b, the target's Doppler shift is greater than half the PRF. A spectrum analysis will return the *apparent* Doppler shift f_a, the shift from an echo spectral line to the nearest illumination line. This target is therefore ambiguous in Doppler, and a Doppler deghosting process must be carried out to associate the echo spectrum with the appropriate illumination lines (Ch. 6).

Example 1-5: If a radar operates at a PRF of 1250 pps and has an illumination frequency of 3.00 GHz, find the maximum target radial velocity giving an unambiguous Doppler shift.

Solution: From the discussion, the maximum Doppler shift giving unambiguous results is one-half the PRF, or 625 Hz. Equation 1-3 solved for radial velocity gives a velocity of 31.25 m/s, or 60.7 Kt (about the slowest a single-engine light aircraft can fly).

At this point, the magnitudes of Doppler shifts encountered by practical radars need to be considered. It was shown that a vehicle with a radial velocity of 65 smph (29.06 Kt) illuminated by a transmitter at 24.150 GHz produces approximately 4,680 Hz of Doppler shift. An aircraft moving at 400 Kt radial velocity illuminated at 10.0 GHz has a Doppler shift of 13,724 Hz. A 10.0 GHz radar aboard an aircraft closing with a missile at Mach 5.0 will see a Doppler shift of about 112.7 KHz, which is close to the design Doppler shifts for these systems.

Despite the fact that Doppler shifts in the above ranges are handled comfortably, other rates can provide useful information. For example suppose a 6-m span target is placed on a rotating positioner, illuminated by a 10.0-GHz radar, and rotated about its center at a rate of 0.2 degree per second. If the target is oriented perpendicular to the radar's axis, the extremities of the target have a radial velocity with respect to the radar of 0.0105 m/s. This is about 1/3 wavelength per second, and produces a Doppler shift of 0.70 Hz. This Doppler shift, properly processed, is every bit as useful as that produced by the 400-Kt aircraft. Systems which analyze target scattering, called diagnostic radars, utilize parameters similar to these. At the other extreme are Doppler shifts of laser radars (LADARs); an aircraft closing on a red-light laser radar (wavelength of 700 nm) at 400 Kt will produce a Doppler shift of about 586 MHz.

Doppler granularity: Doppler bins are discrete bandwidths into which target echoes are processed. They result from using either analog filter banks or digital spectrum analysis for the analysis of Doppler signals. Figure 1-21 shows a typical Doppler bin configuration, with two methods of obtaining the spectrum. The analog filter bank finds only the magnitude of spectral components. The digitally processed discrete Fourier transform (DFT) also finds direction (positive frequency is from an inbound target, negative from outbound). In this example, the bins are 100 Hz wide. Details are provided in Chs. 7, 11, and 12.

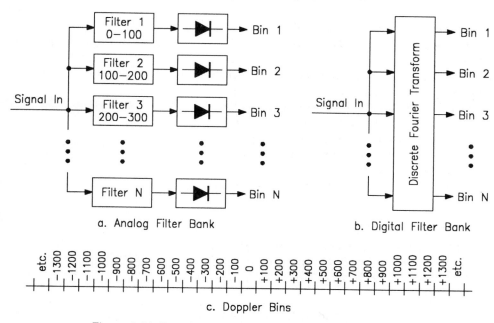

Figure 1-21. Doppler Spectrum Analysis and Doppler Bins.

1-3. Target Information Extraction

1-3-6. PRF classes: Based on range and Doppler ambiguity criteria, pulsed radars are divided into three classes, according to their sample rate, or PRF (Ch. 6).

— *Low PRF* (LPRF) radars and modes wait until echoes from the last transmitted pulse arrive before pulsing again, preventing range ambiguities. At their low rates, Doppler is badly undersampled. *Low PRF is therefore unambiguous in range but highly ambiguous in Doppler.*

— *High PRF* (HPRF, also known as Pulse Doppler) radars and modes meet the Nyquist sampling criterion for the Doppler shifts of all targets the radar is designed to measure. However, little time exists between pulses for ranging. *High PRF is unambiguous in Doppler but highly ambiguous in range.*

— *Medium PRF* (MPRF) radars and modes sample at rates too fast for echoes from long ranges to return before the transmitter pulses again, but too slow to Nyquist sample the Doppler shifts of all targets. *Medium PRF is thus ambiguous in both range and Doppler.* It is, however, not as ambiguous in range as high PRF, nor is it as ambiguous in Doppler as low PRF (it is said to be moderately ambiguous in both range and Doppler). If the ambiguities can be resolved by deghosting, this is a useful PRF class, and gives the best opportunity to simultaneously recover both the range and the Doppler shift of targets. It is used extensively in airborne multimode radars.

1-3-7. Measurement accuracy: Accuracy is a measure of how far the reported target parameter is from the actual parameter, shown in Fig 1-22 for range. Accuracy is normally expressed in absolute terms, such as meters (range), degrees (angular position), and Hertz (Doppler shift). Accuracy is explored further in Ch. 7.

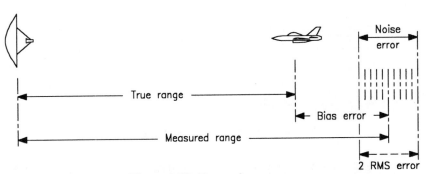

Figure 1-22. Range Accuracy.

Two classes of errors occur. Continuous (but not necessarily constant) offsets from the true value are called bias errors. Their causes are mis-calibrations, such as an erroneous time bases (range) and encoder misalignments, and servo lags, resulting when tracking systems do not keep up with target motions.

The other errors are noise errors, which are random target parameter uncertainties caused by noise and other interfering signals contaminating the target echo. Noise errors are expressed as the standard deviation value in absolute terms (e.g., meters for range).

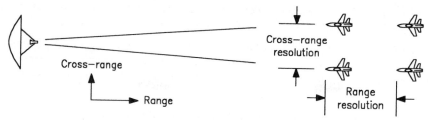

Figure 1-23. Resolution in Range and Cross-Range.

1-3-8. Resolution: Resolution is the ability to separately detect multiple targets or multiple features on the same target, as opposed to reporting multiple targets as a single detection. Targets are resolved in four dimensions, although not necessarily by all radars: range, horizontal (azimuth) cross-range, vertical (elevation) cross-range, and Doppler shift. The range and cross-range resolution dimensions are shown in Fig. 1-23.

A plan position indicator (PPI) display of the same targets is shown in Fig. 1-24. A PPI displays a horizontal map of the radar's environment, with azimuth angle around the periphery of the scope and range as radial distance from the radar's location. In this example display, the radar is in the center.

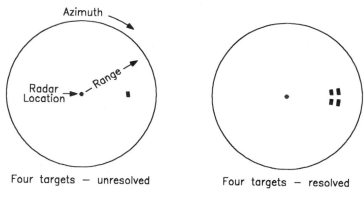

Figure 1-24. PPI Display of Resolution.

Range resolution: Range resolution is the ability to separate multiple targets at the same angular position, but at different ranges. Range resolution is a function of the radar's RF (radio frequency) signal bandwidth, with wide bandwidths allowing targets closely spaced in range being resolved. *To be resolved in range, the basic criterion is that targets must be separated by at least the range equivalent of the width of the processed echo pulse.*

Figure 1-25 shows the effect of pulse width on range resolution. The waveforms represent the echoes as a function of time (range). The pulses are shown after reception and signal processing. Targets, if they are to be resolved, must be separated by a larger range with the long pulse than with the short. A handy way of estimating range and range resolution is to remember that *one microsecond propagation time corresponds to a two-way range of 150 meters.*

1-3. Target Information Extraction

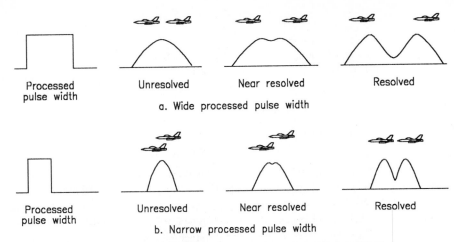

Figure 1-25. Range Resolution and Processed Pulse Width.

$$\Delta R \approx \frac{c\,\tau_c}{2} \quad \text{(meters)} \tag{1-6}$$

ΔR = the range resolution (meters)
c = the velocity of propagation (meters per second)
τ_c = the processed target pulse width (seconds)

Without pulse compression (defined later in this section), the processed pulse width is approximately equal to that of the transmitted pulse. With compression, the processed pulse is narrower than that of the transmitted pulse.

The effective bandwidth of any pulsed wave is approximately the reciprocal of the pulse processed width. Thus range resolution can also be described in bandwidth terms (Eq. 1-8). Time/frequency relationships are described in Ch. 2.

$$B \approx 1/\tau_c \tag{1-7}$$

B = the transmitted matched bandwidth (Hertz)

$$\Delta R \approx \frac{c}{2B} \quad \text{(meters)} \tag{1-8}$$

A common misconception about radar waves is that the shape of the echo pulse can somehow be used to resolve targets. Actually, little or no information is gained from the shape of the echo pulse. Signals from targets and interfering signals are applied to a filter in the receiver called the matched filter (Ch 2). This filter passes a large fraction of the signal energy while restricting the amount of noise energy passed, thus producing the maximum possible signal-to-noise ratio. In doing so, it restricts the bandwidth of signals passed through it sufficiently that pulse shape, along with whatever information it held (which in most cases would still not have been sufficient to tell anything useful about targets), is effectively lost. There is no way to distinguish noise from target echo signal based on pulse shape alone, nor is there a practical way to resolve multiple otherwise unresolved scatterers by pulse width or shape. Representations of signals before and after matched filtering are shown in Fig. 1-26. See Chs. 2, 11, and 13 for more details.

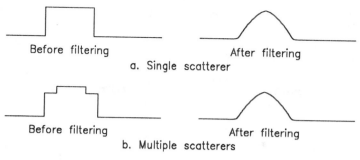

Before filtering   After filtering
a. Single scatterer

Before filtering   After filtering
b. Multiple scatterers

Figure 1-26. Resolution and Pulse Shape.

Example 1-6: A shipboard radar has 0.6-μs transmitted and processed pulse widths (no pulse compression). Two small boats at the same azimuth angle are separated in range by 400 ft. Will the radar detect the boats as two separate targets or as one unresolved target?

Solution: Again, the equations give meaningful results only if the units are consistent. Pulse width in seconds and velocity of propagation in meters per second gives resolution in meters. From Eq. 1-6, the targets must be separated in range by 90 m (295 ft) to be resolved. Since the targets are 400 ft apart, they are resolved and show as two separate blips (spots) on the display.

Enhanced range resolution: A paradox in the design of radar waveforms is that it is desirable to simultaneously have wide transmitted pulses and wide transmitted bandwidths, since wide pulses are good for target detection (Ch. 4) and wide bandwidths are good for range resolution. Without pulse compression, the two desired conditions cannot be realized simultaneously (Eq. 1-8, realizing that without compression the transmitted and processed pulse widths are the same). Pulse compression waveforms containing some form of modulation within the pulse, rather than a simple sinusoid, and their bandwidths are greater than the reciprocal of the transmitted pulse width. Thus with pulse compression waveforms, both conditions can be met simultaneously.

Consider Fig. 1-27. A wide pulse is modulated to create a wide bandwidth and then transmitted. The echo waveform is separated into parts based on the modulation, and each

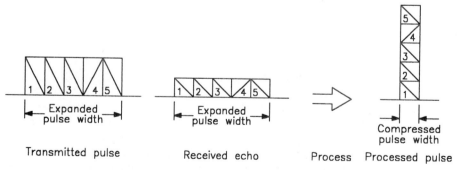

Figure 1-27. Pulse Compression Principle.

1-3. Target Information Extraction

separate part delayed such that they are made simultaneous in time. The component parts, which occur in time sequence in the echo, are summed at the same time, producing a narrower pulse of higher amplitude. Thus the detection capability of the wider transmitted pulse, called the expanded pulse is available at the same time as the range-resolving capability of the narrower processed pulse, called the compressed pulse. See Ch. 13.

Angular and cross-range resolution: Cross-range is the linear dimension perpendicular to the axis of the antenna, specified as azimuth (horizontal) or elevation (vertical) cross-range. Cross-range resolution gives the radar the ability to separate multiple targets at the same range. Resolution in cross-range is determined by the antenna's effective beamwidth, with narrow antenna beams resolving more closely spaced targets. *The criterion for cross-range resolution is that targets at the same range separated by more than the antenna beamwidth are resolved. Those separated by less than a beamwidth are not.*

Figure 1-28 shows the principle of cross-range resolving with antenna beams, in this case with a scanning search radar. The upper portion of the figure shows the relationship between the antenna beams and the targets. The lower portion shows the relative amplitudes of numerous consecutive echoes from the targets as the antenna scans by them. Figure 1-29 shows the geometric development leading to Eqs. 1-9 and 1-10.

$$\Delta X \approx R\, \theta_3 \ \text{(meters)} \tag{1-9}$$

$$\Delta X \approx R\, \theta_3\, [\pi / 180] \ \text{(meters)} \tag{1-10}$$

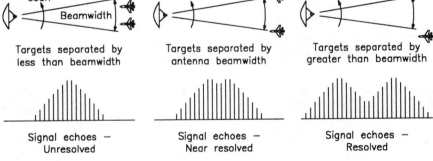

Figure 1-28. Cross-Range Resolution with Antenna Beams.

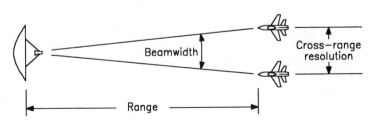

Figure 1-29. Cross-Range Resolution Geometry.

ΔX = the cross-range resolution (meters)
R = target range (meters)
θ_3 = the antenna beamwidth (radians in Eq. 1-9, degrees in Eq. 1-10)

Example 1-7: Many tactical radars have antenna beamwidths of about 3°. How far apart in cross-range must two targets at the same range of 24.5 nmi (45,374 m) be in order to be resolved?

Solution: As with the other examples, this one must also have consistency of units. Equations 1-9 and 1-10 yield cross-range resolution in the same unit as range. By Eq. 1-10, the targets must be separated by at least 1.3 nmi (2,400 m) to be resolved.

It will be shown in Ch. 2 that the beamwidth of an antenna is related to its length and the wavelength of the electromagnetic wave, which is the length of one cycle of the sinusoidal wave as it propagates through the propagation medium (usually the atmosphere). Wavelength is explained later in this chapter.

$$\theta_3 = \lambda / D_{eff} \quad \text{(radians)} \quad (1\text{-}11)$$

$$\theta_3 = (\lambda / D_{eff})(180/\pi) \quad \text{(degrees)} \quad (1\text{-}12)$$

λ = the signal's wavelength (meters)

D_{eff} = the effective length (meters) of the antenna in the direction the beamwidth is being calculated (azimuth or elevation). The effective size of an antenna is typically about 0.7 times its actual size

Using the antenna's dimensions, the cross-range relationship becomes:

$$\Delta X \approx R \lambda / D_{eff} \text{ (meters if } \lambda, D_{eff} \text{ are meters)} \quad (1\text{-}13)$$

Example 1-8: A radar antenna is 12 ft in diameter and the signal wavelength is 5.2 cm. Find how far apart in cross-range two targets at a range of 20 nmi must be to be resolved.

Solution: To make any sense, the length dimensions must be in the same units. The antenna's diameter is thus 3.66 m, the wavelength is 0.052 m, and the range is 37,040 m. The antenna's effective diameter is about 0.7 times its actual diameter (2.56 m). Equation 1-12 gives the beamwidth as 1.16°. Equation 1-10 gives the cross-range resolution at 37,040 m as about 750 m. Alternatively, Eq. 1-13 can be used to find the answer directly without finding the antenna's beamwidth.

Enhanced cross-range resolution: As with range resolution, techniques exist for enhancing cross-range resolution beyond that available with reasonably-sized real antennas. These processes depend on sequentially viewing targets from several different locations with a small real antenna (Fig. 1-30). In the example, the antenna is moved to position 1, the transmitter pulsed, and the signal stored. It is then moved to positions 2, 3, and 4 and the process repeated. After observing the target space from the four positions, the four signals are processed in such a way as to make the effective antenna length the distance it moved. By Eq. 1-13, the cross-range resolution is thus much better than if the antenna were used in only one position.

Systems and modes which resolve with the real antenna's beam are known as real aperture radars (RAR). Those which develop enhanced cross-range resolution by moving

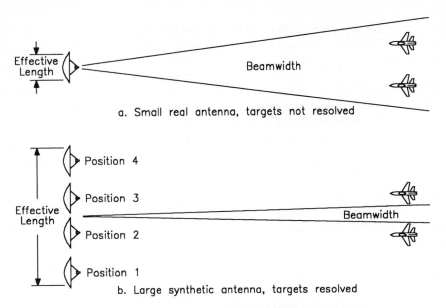

Figure 1-30. Enhanced Cross-Range Resolution (SAR) Concept.

the antenna are known as synthetic aperture radars (SAR). SAR techniques are particularly well adapted to situations where the radar moves rapidly and the targets are stationary, such as an airborne radar viewing the targets on the ground. They also apply where the radar is stationary and the targets move rapidly past. This latter technique is known as inverse synthetic aperture radar (ISAR), and is sometimes used to analyze formations of aircraft from ground-based or shipboard radars to determine how many aircraft are in the formation. This is the raid discrimination function. ISAR is also commonly used in diagnostic radars to analyze the scattering of targets to reduce their radar reflectivity. SAR is covered in Ch. 13.

Example 1-9: An airborne radar stores echoes from targets on the ground as the aircraft travels 280 ft. The signals are then synthetic aperture processed. The antenna is 0.914 m in diameter; the wavelength is 3.0 cm. Find the cross-range resolution at 10 nmi.

Solution: As before, the dimensions must be consistent. The distance the antenna travels is 85.3 m, the wavelength is 0.030 m, and the range is 18,520 m. The diameter of the physical antenna is irrelevant to resolution, but does affect signal-to-noise ratio (Chs. 3 and 13). The effective length of the synthetic antenna is about 60 m (70% of the distance traveled). Equation 1-13 gives the cross-range resolution as 9.3 m, which is the same as it would be if the target were observed once by an antenna which was 85.3 m in diameter.

Note: Under some circumstances, synthetic antennas are effectively twice as long as the real antennas with which they are compared. In other words, the 85-m synthetic antenna may produce the same resolution as a real antenna 170 m long. A full discussion is deferred to Ch. 13.

Doppler resolution: Doppler resolution is the ability to separate targets at the same range, azimuth, and elevation, moving at different radial velocities. It is particularly useful in identifying targets by separating net target motion (called airframe Doppler) from spectral components caused by rotating pieces of the target (jet engine modulations — JEM). *The criterion for Doppler resolution is that the Doppler frequencies must differ by at least one cycle over the time of observation* (Fig. 1-31). Doppler resolution is thus a function of the time over which signal is gathered for processing. Long data gathering times (called the look- or dwell-time) result in smaller Doppler differences being resolved.

$$\Delta f_d \approx 1/T_d \text{ (Hertz)} \tag{1-14}$$

Δf_d = the Doppler resolution (Hertz)

T_d = the look- or dwell-time (seconds)

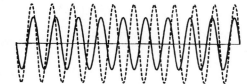

Figure 1-31. Doppler Resolution Concept.

Note that range processing requires short-time looks, so that the target does not move much during the measurement. Doppler processing requires long target look times to gather as many Doppler cycles as possible. One of the design challenges in radar is to achieve good ranging and range resolution, and at the same time achieve good Doppler extraction and Doppler resolution. See Ch. 13.

Example 1-10: An aircraft is chasing a low-flying target, looking down into the ground clutter (a so-called look-down situation). The Doppler shift of the target is 13,500 Hz. The Doppler shift of the clutter to the moving radar is 13,480 Hz. The interfering clutter is at the same range, azimuth, and elevation as the target, and thus the target cannot be discriminated based on location. What is the minimum look time if the target and clutter are to be resolved in Doppler?

Solution: The Doppler shift difference which must be resolved is 13,500 − 13,480 = 20 Hz. By Eq. 1-14, the required look time is 1/20 s. In 1/20 s, the 13,500-Hz shift undergoes 675 cycles and the 13,480-Hz shift 674 cycles, and the criterion of *one cycle of difference* is met.

1-3-9. Signature: A target's radar signature is the composite of all information about it extracted by the radar, used in an identification attempt. Signature includes:

— *Radar cross-section* is the target's relative reflecting "size" and is extracted by simultaneously measuring the received echo amplitude and target range. These two factors, along with a knowledge of the radar's design parameters, allow calculation of RCS (Sec. 1-7 and Chs. 3 and 4).

— The *Doppler spectrum* of a target is an important component of signature, particularly that of airborne targets. It can be used to fix the type and number of engines and the relative RCS of the engines versus the airframe (Ch. 5).

1-3. Target Information Extraction

Figures 1-32 and 1-33 show target Doppler spectra. In Fig. 1-32, the radar is stationary and the main beam clutter (the clutter seen by the antenna's main beam) has a Doppler shift f_{mbc} of zero. Figure 1-33 shows the Doppler spectrum seen by a moving radar, where the main beam clutter has a large net Doppler shift. The moving target is shown to be closing on the radar, since it has a higher Doppler shift than the clutter, assuming the radar antenna is pointed forward.

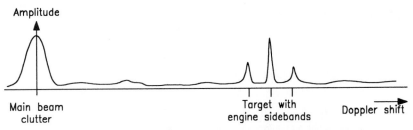

Figure 1-32. Doppler Spectrum Example with a Stationary Radar.

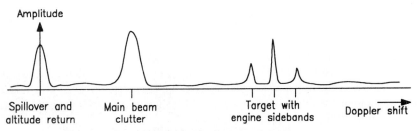

Figure 1-33. Doppler Spectrum Example — Moving Radar.

Both figures show jet engine modulation, manifested by the engine sidebands, surrounding the airframe Doppler. The return at zero Doppler shift in Fig. 1-33 is from an antenna sidelobe illuminating the ground beneath the radar is called the altitude, or normal return (Fig. 1-34).

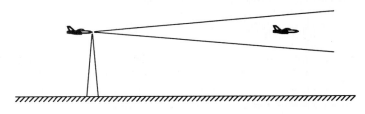

Figure 1-34. Altitude Return (Normal Return).

1-3-10. Imaging: If sufficient resolution is present in both range and cross-range, it may be possible to construct a radar "map" of the target, resolving its individual scatterers. Sufficient resolution for imaging is found in several types of radars and radar modes. Some of these systems use a form of synthetic aperture called Doppler beam sharpening (DBS) to develop high resolution ground maps for navigation and targeting. Other tactical radars image targets to identify them.

A further use for imaging is for the analysis of radar targets to determine the mechanism by which they reflect, with the goal of identifying and reducing the target's radar cross-section (diagnostic radars, such as the Scientific Atlanta 2084 RCS Analyzer). The image of Fig. 1-35 was produced by this type radar. Imaging is covered in Ch. 13.

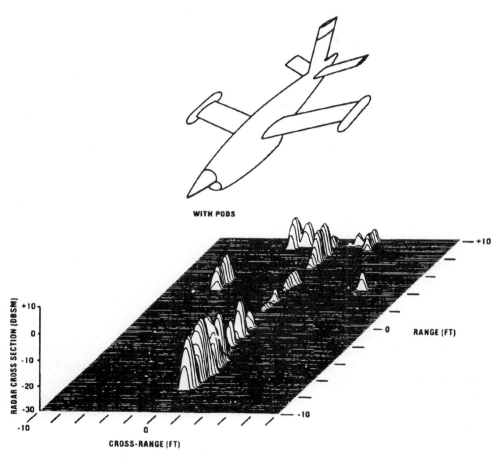

Figure 1-35. Radar Image of Target Drone. (Drawing courtesy of D. L. Mensa; U. S. Government work not protected by U.S. copyright)

1-4. The Radar Equation: An Introduction

In this section, the process by which the radar illuminates the target, receives the echo, and processes it to detect a target's presence is discussed. Methods are presented to find the amount of signal returned to the radar as a function of the radar's parameters, the target's parameters, and the target's range. To better illustrate each step, a numeric example with a hypothetical radar and target is used throughout the discussion. The radar parameters are covered in detail in Ch. 2 and the radar equation is discussed further in Ch. 3.

Example 1-11: The specifications for the radar to be used in this example are typical of those found in long range ship and land-based air search radars. Uses of this type radar include air traffic control and early warning of approaching aircraft.

Transmit power: 3,000,000 Watts (3 MW)
Antenna gain: 4,500
Antenna effective aperture: 20 m^2
Transmit frequency: 1.27 GHz (in the L-Band)
Transmit pulse width: 2.5 μs
Pulse repetition frequency: 350 pps
Receiver noise factor: 2.5

The target which this radar will attempt to detect has the following specifications:

Radar cross-section: 10 m^2
Range from radar: 300 nmi

For each step in the process of illuminating the target and receiving its echo, the values which would be present for this radar and target will be explained.

Transmit: A transmitter generates radio frequency (RF) energy, in this case a pulse 2.5 μs long. At the time of transmission, a timer is started.

Example 1-11a: The example radar would typically emit a pulse of about one to five μs width, and would fire 350 times per second (the PRF). During the time the transmit pulse is on, its power is 3 MW (this is the transmitter's *peak power*). Integrated over the waveform period, the *average power* is 2,625 W (Ch. 2).

Antenna gain: The radio energy from the transmitter is concentrated in a preferred direction by the antenna. The degree of concentration is the gain of the antenna. By definition, an lossless antenna which radiates power equally in all directions is called isotropic, has a gain of unity, and its beam pattern is a sphere. If the same transmitter were connected to an antenna with a gain of 100, the available power would be distributed within a beam which occupies 1/100 of a sphere, and the power density within the beam would be 100 times the power density for the same transmitter connected to an isotropic antenna (approximately — see Eq. 1-15).

Consider the antennas in Fig. 1-36 as connected to identical transmitters and radiating to two identical receivers at the same range. The receiver listening to the isotropic antenna detects a certain amount of RF power. If the receiver listening to the high gain antenna is in that antenna's main beam, it receives more power by a factor of the transmitting antenna's gain. This power must come from somewhere, and in directions other than the transmit main beam, the high gain antenna produces *less* power than the isotropic. In the direction of the beam, the transmitter's power is effectively magnified by the antenna gain.

Using a high gain transmitting antenna gives the same received power with a less powerful transmitter than is required using an isotropic antenna. A transmitter connected to an antenna with a gain of 100 can achieve the same result with 1/100 the power required if an isotropic antenna were used. A receiver would sense the same power, as long as the

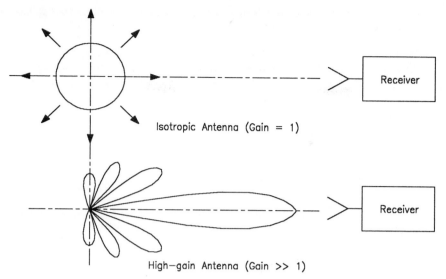

Figure 1-36. Antenna Gain Principle.

ranges were the same, and could not discern whether it was receiving the higher power transmitter and isotropic antenna or the lower power transmitter and high gain antenna.

$$G \approx \frac{\text{Solid angle in a sphere}}{\text{Solid angle in the antenna's beam}} \qquad (1\text{-}15)$$

G = the antenna's gain (dimensionless)*

*The definition given above is correct if power density of the transmitter/antenna combination is compared to the power density from the same transmitter with an isotropic antenna. If the solid angle of a spherical beam (isotropic) is compared to the solid angle of the antenna beam, antenna *directivity*, and not *gain* is defined. For large microwave antennas, directivity and gain are approximate equals.

The power effectively radiated by the transmitter/antenna combination in the direction of the main beam is called the effective radiated power or ERP. It is the product of the power delivered to the transmit antenna and the gain of that antenna. It is the power which would have to be radiated isotropically to produce the same effect as the given transmitter and antenna, in the direction in which the gain is specified.

$$ERP = P_T G_T \qquad (1\text{-}16)$$

ERP = the radar's effective radiated power (Watts)
P_T = the transmit power delivered to the antenna (Watts)
G_T = the gain of the radar's transmit antenna

Example 1-11b: The *ERP* of the example radar is the transmitter power of 3,000,000 W times the antenna gain of 4,500, or 13,500,000,000 W. To produce

1-4. The Radar Equation: An Introduction

the same power in a given receiver at a given range, an isotropic antenna would have to be connected to a 13,500,000,000-W transmitter.

Forward propagation: Power emitted from the transmitter and its antenna propagates away from the antenna at slightly less than the speed of light. For practical purposes, its velocity is approximately 300,000,000 m/s. The ratio of the electric and magnetic field strengths gives a wave "impedance" of 120π (377) Ω. Table 1-2 gives some electromagnetic propagation data.

Table 1-2. Propagation Data.

Propagation constant in free space	1.000000
Velocity in free space	299,792,000 m/s
Propagation constant at sea level	0.999681
Velocity at sea level	299,696,000 m/s
Approximate velocity	300,000,000 m/s
Impedance in free space	120π ($\approx$ 377) Ω
Impedance at sea level	$\approx$ 377 Ω

The power density of the forward signal is the power per unit area of beam cross-section — W/m². Visualize a plane perpendicular to the axis of the antenna at distance R from the radar. Power density is the amount of power falling on each unit area of the plane (Fig. 1-37). It equals the power transmitted divided by the area of the beam at the target's range. The area of the beam is the area of a sphere of radius equal to the range from radar to target, divided by the transmitting antenna's gain, and is inversely proportional to the square of range.

$$P/A_F = \frac{P_T}{4\pi R_T^2 / G_T} \qquad (1\text{-}17)$$

P/A_F = the forward (illumination) power density at the target (Watts per square meter)

R_T = range from radar transmitter to target (meters)

$4\pi R_T^2$ = the surface area of a sphere of radius R_T

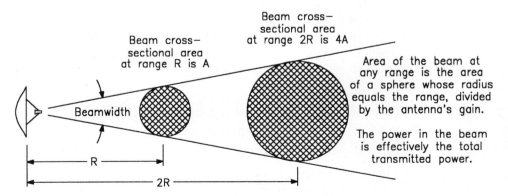

Figure 1-37. Power Density Concept.

Example 1-11c: 300 nmi equals 555,600 m (1852 m/nmi). A sphere of radius 555,600 m has a surface area of 3.879×10^{12} m^2. The antenna beam illuminates 1/4500 of that area, or 862,000,000 m^2. The total transmitted power is 3,000,000 W. Therefore the power per unit area in the beam at the target's range is 3,000,000 W divided by 862,000,000 m^2, or 0.00348 W/m^2.

Target reflection: If the electromagnetic energy propagating through the medium encounters an impedance of other than 120π Ω, a portion of the energy reflects, just as it would if traveling on a transmission line it encountered an improper termination. The directionality of the reflection is determined by the characteristics of the target. For example, a sphere reflects equally in all directions. A flat plate reflects the majority of the energy in a direction determined by the orientation of the plate. That portion of the reflected energy that propagates in the direction of the radar's receiving antenna is called backscatter. *It is the only reflected energy which matters to the radar.*

The power reflected from the target is proportional to the illumination power density and to the reflection characteristics of the target. These characteristics, including the directionality of the reflection, are summarized in the radar cross-section (RCS) of the target.

$$P_{Tgt} = P/A_F \, \sigma \tag{1-18}$$

$$P_{Tgt} = \frac{P_T G_T}{4\pi R_T^2} \sigma \tag{1-19}$$

P_{Tgt} = the effective power reflected by the target in the direction of the radar (Watts)

σ = the target's radar cross-section (square meters)

The concept of radar cross-section is shown in Fig. 1-38 and discussed in Ch. 4. The target is modeled as a "scoop" connected to a radiator which scatters power equally in all

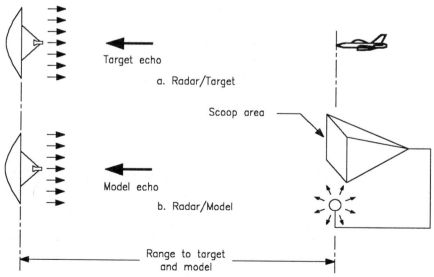

Figure 1-38. Model of Radar Cross-Section.

1-4. The Radar Equation: An Introduction

directions (isotropic radiator). The model is set at the same range as the target and the scoop area varied until the power received by the radar from the model is the same as that from the target. When equality exists, the area of the scoop equals the radar cross-section of the target. Most targets are not isotropic radiators, but they behave as such *in the direction of the radar*.

Figure 1-39 shows three targets presenting equal projected areas to the radar. The sphere scatters power isotropically and returns a small fraction to the radar. It has a small RCS. A circular flat plate mirroring power in the direction of the radar has a much larger RCS, because the power it reflects is directed toward the radar. The same flat plate oriented so that the bulk of its echo power is away from the radar has the smallest RCS of them all.

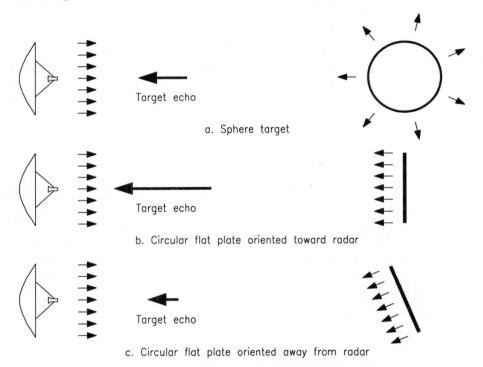

Figure 1-39. RCS Dependence on Shape and Orientation.

The above discussion suggests ways to avoid detection by radars, and three rules for *low observable (stealthy)* targets can be formulated.

− First, the target can be constructed so that its electromagnetic impedance of 377 Ω will absorb rather than reflect illumination energy.
− Second, the target can be constructed so that reflected energy is concentrated into a few very narrow beams. The probability that one of these high RCS "beams" is directed at the radar at the same time the radar is looking in the target's direction is very small. This is one tenet of the faceted design of the F-117A.
− Third, targets can be constructed to diffuse the energy reflected in the direction of the radar as much as possible.

— A fourth concept, discussed in Ch. 4, is to design the target so that reflected energy has a different polarization than that of the illumination. Many radars are insensitive to echoes cross-polarized to the illumination.

Example 1-11d: The illumination power density at the target is 0.00348 W/m². The target has an RCS of 10 m². The equivalent scoop (Fig. 1-38) therefore captures 0.0348 W of illumination power (0.00348 W/m² times 10 m²) and reradiates it. *In the direction of the radar*, the re-radiation can be considered to be isotropic, although it probably is not *actually* isotropic.

Backscatter propagation: The energy reflected from the target propagates away from it at the propagation velocity. The power density in the backscattered wave at the radar is the power effectively isotropically radiated by the target divided by the surface of a sphere of radius equal to the range from target to radar. There is no gain factor in backscatter propagation since the target is treated as though it were isotropic. In a monostatic radar, the range from the target to the radar's receiving antenna equals the range from radar's transmitting antenna to target.

$$P/A_B = \frac{P_T G_T \sigma}{4\pi R_T^2} \frac{1}{4\pi R_R^2} \qquad (1\text{-}20)$$

P/A_B = the backscatter power density at the radar's receiving antenna (Watts per square meter)

R_R = the range from the target to the radar's receive antenna (meters)

Example 1-11e: The target now acts as a transmitter with an equivalent isotropically radiated power (*only in the direction of the radar*) of 0.0348 W. This wave's power density at the radar is this reflected power spread over the surface of a sphere with radius equal to the range from target to radar (555,600 m). The sphere has a surface area of 3.88 x 10^{12} m², and the power density is 8.97 x 10^{-15} W/m².

Echo signal and interference capture: The echo power which propagates back to the radar is captured by the effective area of the receiving antenna. Two factors determine the total power captured: the power density of the echo signal at the antenna, and the effective area of the antenna. *The effective area of a typical radar antenna is about one-half of its actual area.* Figure 1-40 shows the capture by the receiving antenna.

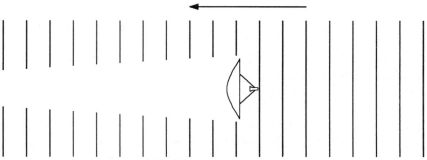

Figure 1-40. Receive Antenna Capture of Echo Energy.

1-4. The Radar Equation: An Introduction

$$P_R = P/A_B A_E \tag{1-21}$$

P_R = the echo power received from the target (Watts)

A_E = the effective capture area of the receive antenna (square meters)

Example 1-11f: The example's receiving antenna has an effective area of 20 m² (which means that it probably has an actual projected area of about 40 m²). The power density of the echo at the receive antenna is 8.97×10^{-15} W/m², from which 20 m² of aperture captures about 1.79×10^{-13} W.

Combining Eqs. 1-12 through 1-21 indicates that the echo power received is proportional to the power transmitted, the transmit antenna gain, the target radar cross-section, the effective area of the receive antenna, and is inversely proportional to the fourth power of range, assuming the range from the transmitter to the target and from the target to the receiver are equal (almost always true).

$$P_R = \frac{P_T G_T \sigma A_E}{(4\pi)^2 R^4} \tag{1-22}$$

Equation 1-22 is a simplified version of the radar equation. It ignores losses in the radar system and propagation path and neglects the fact that many echoes from the same target are usually processed together. It does give a good estimate of the echo power to be expected from a single hit on a target.

For a given radar, the echo power received is a function of the radar's parameters, the RCS of the target, range, and losses. An intuitively useful relationship can be formed by combining some of these quantities, as shown in Eqs. 1-23 and 1-24.

$$P_R = \frac{K_R \sigma}{R^4 L_A} \tag{1-23}$$

$$K_R = \frac{P_T G_T A_E}{(4\pi)^2 L_S} \tag{1-24}$$

L_S = losses in the radar system (dimensionless)

L_A = losses in the propagation path (dimensionless)

K_R represents the radar's parameters (Watt-square meters)

Antenna effective area and gain are proportional to one another (Ch. 2). After translating effective area to gain terms, Eq. 1-24 becomes:

$$K_R = \frac{P_T G_T^2 \lambda^2}{(4\pi)^3 L_S} \tag{1-25}$$

Example 1-11g: Find the "K_R" of the radar used throughout this section. Find the signal echo power received from a 0.1 m² target at 95 nmi. Ignore losses (L_S and L_A are unity)

Solution: From Eq. 1-24, K_R is 1.71×10^9 Watt-square meters. Equation 1-23 gives the received power from a 0.1 m² target at 95 nmi (175,940 m) as 1.78×10^{-13} W (the same as from a 10-m² target at 300 nmi).

The antenna captures interfering signals in addition to the target echoes. This interference can be echoes from the surrounding environment such as sea, land, and weather clutter. Some clutter is captured by the antenna's main beam and some by sidelobes. The antenna's gain and effective area in the direction of sidelobes is much less than that in the direction of the main beam (Ch. 4). If jamming is present, it too is captured by the antenna, much through sidelobes. EMI usually enters the system through the antenna, sometimes through sidelobes. EMI can also enter through power lines, ineffective equipment grounds, and direct capture by equipment and internal transmission lines.

Reception: The captured target and interfering energy is amplified, filtered, and demodulated in the radar's receiver. The receiver increases the amplitude of the signal to a usable level and rejects as much interference as possible.

Thermal noise is mainly generated in the radar itself, usually in the first stages of the receiver. The purpose of the receiver's matched filter, discussed in Ch. 2, is to reduce the effect of thermal noise and noise jamming. The thermal noise power, effectively at the receiver's input, is obtained from Eq. 1-26. Noise is discussed further in Ch. 4.

$$P_N = K T_0 B F \qquad (1\text{-}26)$$

P_N = the noise power at the *input* to the receiver
K = Boltzmann's constant (1.38×10^{-23} W-s/°K or J/°K)
T_0 = 290°K
B = the noise bandwidth of the system
F = the noise factor (if in decibels, it is the noise figure)

Example 1-11h: In Ch. 2, it will be shown that the noise bandwidth is approximately the reciprocal of the processed pulse width, giving a noise bandwidth for our example of 400,000 Hz. The noise factor is 2.5, meaning the receiver produces 2.5 times the noise of an ideal receiver at 290°K. The noise at the receiver's input, from Eq. 1-26, is 4.0×10^{-15} W.

The signal-to-interference ratio determines whether or not sufficient signal is present to detect the target. It is the ratio of signal echo power to interfering power. The signal-to-noise ratio is given in Eq. 1-27.

$$S/N = P_R / P_N \qquad (1\text{-}27)$$

S/N = the power signal-to-noise ratio
P_R = the signal echo power at the input to the receiver
P_N = the noise power at the input to the receiver

Example 1-11i: The signal-to-noise ratio in the example, from Eqs. 1-26 and 1-27, is 44.5 (16.5 dB). In Ch. 5, it will be shown that this signal-to-noise ratio is marginally adequate for reliable target detection without an inordinately high false alarm rate.

1-5. Signals and Signal Processing: An Introduction

Signals recovered by the radar are composites of target echoes and interference. Signal processing's role is (1) to enhance the target echoes and suppress all other signals, and (2) to extract information about the target's behavior, including its position, velocity, and signature. Signal processing exploits differences between the components of the composite signal. Before introducing signal processing, some definitions and typical processes must be examined.

Signal definition: Signals are electrical quantities (voltages and currents, mainly) which convey information. They result from several phenomena. First are the echoes of the illumination wave from desired targets. Second are the echoes of the illumination wave from undesired targets, such as clutter. Third are signals produced within the radar itself, noise and spillover. Fourth are signals transmitted by hostile sources with the intent of interfering with target echoes — ECM, also known as jamming. Fifth are signals generated from friendly sources, but which interfere with the radar's detection of targets — EMI.

Properties of signals: Signals are described by several properties. A typical signal in a pulsed radar is shown in Fig. 1-41, along with an internally generated reference wave called the COHO (COHerent Oscillator). See Ch. 2.

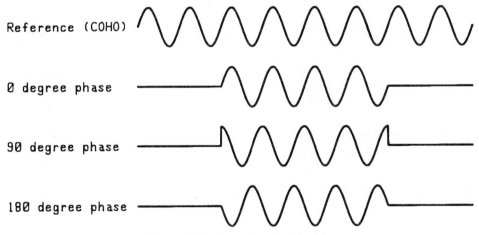

Figure 1-41. Pulsed Radar Signals.

Amplitude, in either voltage or power units, is the quantity of signal present and determines the signal's capacity to overcome interference. The relative amplitude of the target echo compared to that of the interfering signals, in power units, is the signal-to-interference ratio. One principal goal of signal processing is to improve the signal-to-interference ratio.

$$S/I = P_R / P_I \qquad (1\text{-}28)$$

S/I = the signal-to-interference ratio

P_R = the target echo component of the composite signal (Watts)

P_I = the interfering component of the composite signal (Watts)

Frequency is the rate at which the signal vector rotates, and is determined within the radar and by the target's motion (the Doppler shift). The signal is translated from the echo frequency to the COHO frequency (Ch. 2) before it is treated extensively.

Phase is the time offset between the signal and an internal reference, generally the transmitted wave, and is used to find the Doppler shift. In practice, signals are translated to the COHO's frequency and the COHO becomes the phase reference. Two of the three signals in Fig. 1-41 are offset from the COHO. This difference, in radians or degrees, is the signal phase. One cycle of the signal or COHO is 2π rad or $360°$. *Phase only exists as a comparison and a reference must be present. The usual references are the transmitted wave at propagation frequencies and the COHO at translated internal frequencies.*

It is difficult to visualize signal parameters, particularly phase, when they are presented as sinusoidal waves. Thus a different depiction, the phasor, is often used as a "shorthand" way of describing sinusoids. In Fig. 1-42, two signals and a reference are shown both as sinusoids and as phasors. The amplitude and phase are much easier to see in phasor form, with the length of the phasor representing amplitude and its angle representing phase. Zero phase is horizontal to the right, and *the reference is by definition zero phase*. Positive phase angles are counterclockwise from zero. Phasors can be manipulated as though they were vectors.

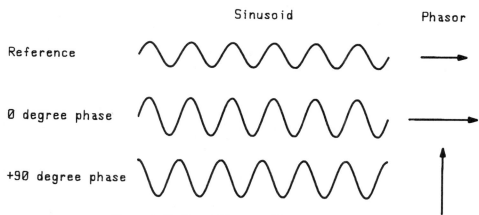

Figure 1-42. Signal Phase and Phasor Notation.

Radars seldom deal with just one echo from a target; they usually illuminate the target for several pulses and collect a series of echoes, as shown in Fig. 1-43. The top row in the figure gives a phasor representation of eight consecutive samples of noise, where both the amplitude and phase are random. The second row is eight consecutive echoes (hits) from a target, caused by eight consecutive zero-phase illumination pulses. The phase is the same for each hit, indicating that the target's range is constant. The third row shows echoes from a target moving radially to/from the radar. Note that the phases change in an orderly pattern.

Signals are composites of target echoes and interference. Figure 1-44 shows eight samples of a typical composite with two components, an echo from a moving target and

1-5. Signals and Signal Processing: An Introduction

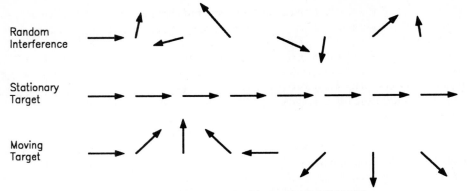

Figure 1-43. Signals from Different Types of Targets.

stationary clutter. The samples shown are from a single range bin and a signal composite exists for each range bin. From each range bin, the radar sees a single signal whose phasor is the sum of the phasors of all components in the resolution cell. Because they are in the same bin, the components are unresolvable in range. The radar sees only the vector sum of the components. Signal processing tries to separate them.

It is extremely important to realize that in pulsed radars signals are almost always processed on a range-bin basis. Signal from range bin 17 are processed only with subsequent samples from range bin 17. Signals in bins 16 and 18 are treated in entirely separate processes. If, for example, the signal process is to sum eight consecutive signal samples (called signal integration), one sum must be performed for each range bin; if there are 256 bins, 256 sums of eight samples must be performed each time the radar pulses eight times (this eight-pulse group, in this case, is known as a look, or dwell).

Signal processing fundamentals: Signal processing attempts to separate composites into their components by exploiting differences between the components. For example, noise is random in both amplitude and phase, whereas signal echoes are orderly, and eight samples of 1.0-V noise sum to a smaller value than eight samples of 1.0-V echoes from a stationary target. Likewise, stationary targets viewed by stationary radars have constant phases, whereas the phases of echoes from moving targets change in an orderly fashion.

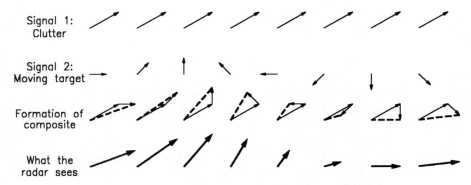

Figure 1-44. Typical Composite Signal from a Single Range Bin.

Consider Fig. 1-45, showing examples of methods used to separate signal components. Figure 1-45a shows a composite of a moving target, clutter, and noise in the time domain (as it would be seen with an oscilloscope). In this view, it is impossible to discriminate the moving target echo from the noise and clutter. In the frequency domain (as viewed on a spectrum analyzer — Fig. 1-45bcd), the different components of the signal are clearly visible. One important type of signal processing is spectrum analysis, converting from a time-view of the composites to a frequency-view.

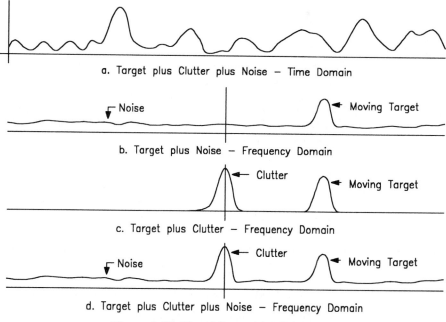

Figure 1-45. Signal Processing Principle.

The signal processing principles shown in Fig. 1-45 are summarized below and treated in detail in Chs. 11 and 12.

— If several samples of a composite of a target echo plus random interference (noise) are summed, or integrated, the orderly target component adds to a larger value than the random interference.

— Orderly target echo signals are enhanced in the presence of random interference by processes which concentrates the orderly components into one (or a few) bins, while the random interference is spread equally over all bins. If 64 samples of the composite echo were sorted into 64 equal Doppler frequency bins, a target component with a constant Doppler shift would concentrate into one of the bins. Noise, containing all frequencies in equal amounts, would spread equally among all bins. Thus in the bin containing the target, the signal-to-noise ratio would be about 64 times its value before the concentration process. This technique, useful with noise and noise jamming, is usually realized as a spectrum analysis.

1-5. Signals and Signal Processing: An Introduction

– Echoes from moving targets will, after spectrum analysis, concentrate in different Doppler bins than will stationary clutter. If the bins containing the clutter are discarded, the signal-to-clutter ratio is improved.

– Illumination with a wave which is encoded generates echoes having the same coding. The properly encoded echoes (targets and clutter) differ from interferers which do not have the proper code. By the process of correlation, the properly encoded signal components are enhanced with respect to the improperly encoded interferers. This process is useful in the presence of noise and some types of ECM.

Very important: To repeat, the composite signal samples which are processed together arrive at the radar simultaneously in time. Only one range bin at a time is analyzed. For example, a signal process might consist of 64 samples of range bin 31. Range bin 32 is analyzed separately – signal in this range bin does not participate in the bin 31 analysis. This is a key concept in radar signal processing.

Signal processing parameters: The effectiveness of signal processing in separating signals from interference is described by a parameter called process gain. It is defined as the ratio of the signal-to-interference ratio out of the process and the signal-to-interference ratio into the process. The process gain of effective signal processing is greater than unity. If not, it shouldn't be used – a piece of wire has a process gain of one.

$$G_p \equiv \frac{S/I_o}{S/I_i} \qquad (1\text{-}29)$$

G_p = the process gain (dimensionless)

S/I_o = the signal-to-interference ratio out of the signal process (dimensionless power ratio)

S/I_i = the input signal-to-interference ratio (dimensionless power ratio)

The process gain of a particular process may be high in the presence of one type of interference – clutter, for example – and low to another type – such as noise. Signal processes must be customized to the type of interference present, and many radars have multiple fundamentally different processes available.

In the special case of clutter, a process gain to clutter is defined, called the MTI improvement factor, or MTI-I. Some authors call it simply the improvement factor, or I.

$$MTI\text{-}I \equiv \frac{S/C_o}{S/C_i} \qquad (1\text{-}30)$$

$MTI\text{-}I$ = the process gain to clutter (dimensionless)

S/C_o = the signal-to-clutter ratio out of the signal process (dimensionless power ratio)

S/C_i = the input signal-to-clutter ratio (dimensionless power ratio)

Effectiveness of signal processing: A radar's purpose is to supply information about targets. In this section, factors which measure the effectiveness of the radar's signal processing at doing its job are introduced, along with a number of evaluation criteria which are common to all radars.

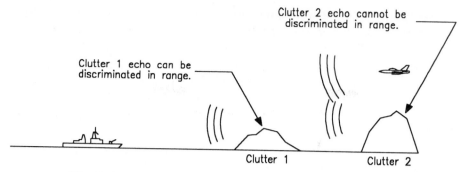

Figure 1-46. Clutter Discrimination with Range and Doppler.

— **Ability to detect targets:** A radar's first task is to detect the presence of targets in interference. The primary factors determining whether or not detections occur are the radar's parameters, the target's radar cross-section and RCS variations or fluctuations (Ch. 4), and the level and character of the interference. The radar parameters include the transmitter's *energy* output (not power, in most cases), the antenna's gain and effective capture area, the receiver's sensitivity, and the signal processor's process gain.

— **Clutter discrimination:** This is a measure of the radar's ability to separate desired targets from echoes from land, sea, and weather. The critical interfering clutter is that which occurs simultaneously in time with the desired target echo. Clutter occurring at times other than the target echo can be separated by time discrimination, called range gating (Fig. 1-46). Signal processing may enhance target echoes enough with respect to clutter to allow the target to be detected. Clutter discrimination is accomplished with two processes: resolution and Doppler analysis.

Resolution limits the amount of clutter present simultaneously with the target echo (Fig. 1-47). Land and sea clutter RCS is proportional to the *area* of clutter within the radar's resolution cell, and they are known as area clutter (Ch. 4). As long as the target is fully enclosed by the resolution cell and all other signal-producing factors are the same, decreasing the resolution cell size reduces the clutter RCS without affecting target RCS. Therefore smaller resolution cells produce higher signal-to-clutter ratios at the input to the signal processor. Figure 1-47 shows the ground area illuminated by two resolution levels. In both cases the target is fully contained within the cell (point target) and has the same RCS. With good resolution, less clutter competes with the target echo.

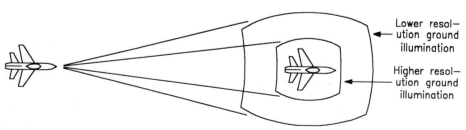

Figure 1-47. Resolution and Clutter Discrimination.

1-5. Signals and Signal Processing: An Introduction

Doppler, or spectrum analysis is then used to separate echo signals by their velocities. Targets moving at different velocities from the clutter are thus discriminated (Figs. 1-32 and 1-33). Two other forms of Doppler clutter processing, moving target indication (MTI) and moving target detection (MTD), identify targets with Doppler shifts different from those of clutter. MTD radars can estimate the velocity of targets.

— **Countermeasures immunity:** Electronic countermeasures attempts to hide targets by presenting interfering signals of greater amplitude than the target echo. Many processes in radars aim specifically to discriminate between target echoes and ECM. Most are based on four principles.

Burn through the jamming by applying enough illumination energy to the target so that the signal echo contains enough energy to overcome the jamming energy.

Use low sidelobe antennas to reduce the amount of jamming which enters the radar through antenna sidelobes.

Use receiver filtering to suppress jamming outside the echo signals' bandpass.

Signal process to reduce the types of jamming which are fundamentally different from the target echoes. Often the difference is in phase coherence from pulse to pulse (Fig. 1-43). Sensing signal phase and processing to enhance the patterns from legitimate targets suppresses some forms of jamming. Other jamming types are indistinguishable from target echoes and must be discriminated by other means.

— **EMI immunity:** Electromagnetic interference (EMI) sources include communication transmitters, other radars, and friendly jammers. EMI can also come from within the radar itself, including leakage from the transmitter and receiver oscillators, which is sometimes known as self-clutter or internal clutter. The primary defense against EMI is prevention, including controlling the interfering emissions, shielding the radar from these emissions, and signal processing to discriminate between target echoes and EMI. Signal processing for EMI depends on the character of the interference, and is similar to processing for ECM.

1-6. Types of Radars and Radar Functions

There are many different radar missions, types of radars, and radar modes. Radar's basic task is to sense objects remotely and to measure these targets' parameters, including location, velocity, and signature. The way a particular radar is configured to carry out its specific task is determined by the specific measurements it is to make, the physical environment in which it must function, and the interferers it is expected to encounter. Following are some of the numerous radar groupings.

1-6-1. Primary and secondary radars: Primary radars (Fig. 1-48a) are those where the radar's transmitter illuminates the target and the echo from the illumination is used to extract information. To primary radars, targets can be cooperative, neutral, or hostile. Secondary radars (Fig. 1-48b) are those where the primary radar's illumination or a separate interrogation signal triggers an active response by the target. It can only be used with cooperative targets. The interrogator transmits a signal, in many cases coded, which propagates to a transponder, or beacon in the target. The transponder, upon detecting the

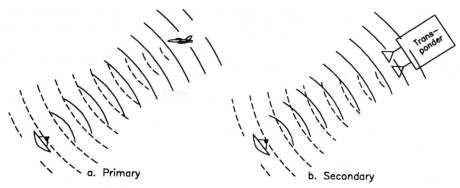

Figure 1-48. Primary and Secondary Radar.

proper interrogation, transmits a reply, which is usually coded, back to the radar. The secondary radar detects this active reply signal, not the echo from the illumination. The reply signal is often of a different frequency than the interrogation, and may even be in a different frequency band. Transponders/beacons are sometimes called RACONs, for *RA*dar bea*CON*s.

1-6-2. Monostatic/Bistatic radars: Radars are classified by the physical relationship between the transmit and receive antennas. Those systems where the antenna is shared by transmitter and receiver, or where separate transmit and receive antennas are in essentially the same location are monostatic (see Fig. 1-3).

Where the transmit and receive antennas are in significantly different location, the radar is bistatic (Fig. 1-49). The angle at the target between the direction to the transmitter and that to the receiver is the bistatic angle. Bistatic radars are often used with continuous wave (CW) or frequency-modulated continuous-wave (FMCW) waveforms, since this configuration produces less spillover interference.

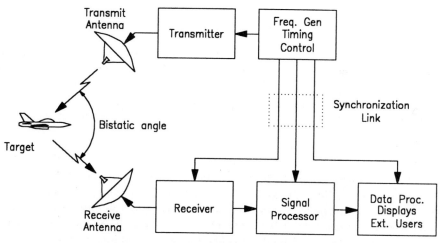

Figure 1-49. Bistatic Radar Simplified Block Diagram.

1-6. Types of Radars and Radar Functions

The bistatic configuration is also used to protect the vehicle carrying the receiver, which is also usually the weapons carrier. Since this threat vehicle does not transmit, the enemy cannot use its illumination to detect (intercept) its presence and locate it. At the same time the transmitter can be placed out of reach of defensive weapons which home on electromagnetic emissions (anti-radiation missiles, or ARMs).

Also, low observable targets can in some cases be detected more easily with bistatic radars (Sec. 4-7).

The functions of the blocks in a bistatic radar are the same as those in a monostatic system, with the major difference being the absence of the duplexer. Synchronization between subsystems may be more difficult if the distances between them are great.

A form of bistatic radar called semi-active radar homing (SARH) is used in some guided missiles Fig. 1-50). The launching platform houses the radar transmitter (called the illuminator) and the missile contains the receiver. The target echo is received by an antenna in the front of the missile, and energy from the illuminator is received by an antenna in the missile's tail (the rear-reference antenna). A comparison of the two signals gives the Doppler shift and the missile/target closure rate (Sec. 6-4). Other information, such as range from missile to target, may or may not be data-linked to the missile. Without data links, SARH systems do not measure target range and must develop guidance solutions from other information (Sec. 6-4).

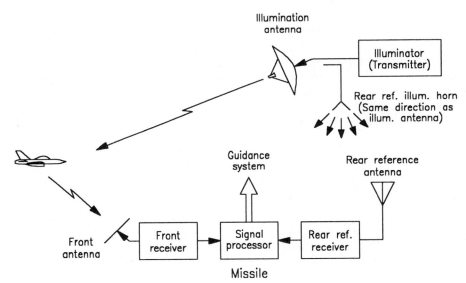

Figure 1-50. Semi-Active Radar Homing Simplified Block Diagram.

1-6-3. Classification by primary radar mission: Radars generally are of one of the types given below. Note that a single radar set may at different times fall into more than one of these classes. For example, it is common for military airborne radars to have search, tracking, and illumination (SARH) modes. These modes may be operated separately, or interleaved, which is cycling the radar quickly through several modes. Some specialized applications are not listed here.

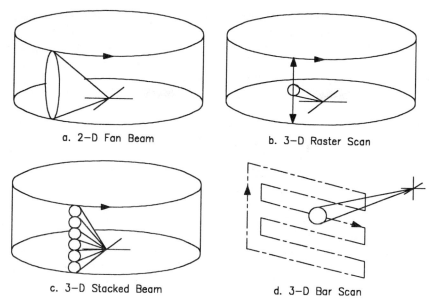

a. 2-D Fan Beam
b. 3-D Raster Scan
c. 3-D Stacked Beam
d. 3-D Bar Scan

Figure 1-51. Search Radar Scan Patterns.

Search radars and modes: These types continuously scan a volume of space without dwelling preferentially at any location. They are used to detect targets and find their range, angular location, and sometimes velocity. The positional and velocity information from a search radar is usually less accurate than that from radars which dwell for long periods on a single target. Typical antenna scan patterns are shown in Fig. 1-51.

— *Surface search* radars and modes look for targets, usually boats or land vehicles, on water or land surfaces.

— *Air search* radars and modes scan the sky for aircraft and missiles.

— *Two-dimensional (2-D) search radars* usually search in azimuth and range, but provide no direct elevation data (Fig. 1-51a). Azimuth beams are narrow for good resolution and elevation beams are wide for coverage of a large volume. Two-dimensional (2-D) radars are commonly found in surface and air search applications. Elevation beams are made wide to prevent the beam lifting off the water when a boat rolls, in the case of surface search (Fig. 1-52), and to provide coverage of large volumes of sky in the case of air search. Figure 1-53 shows the elevation coverage of a typical air search radar.

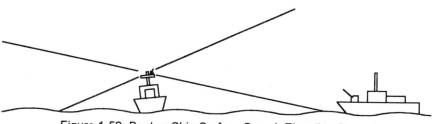

Figure 1-52. Boat or Ship Surface Search Elevation Beam.

1-6. Types of Radars and Radar Functions

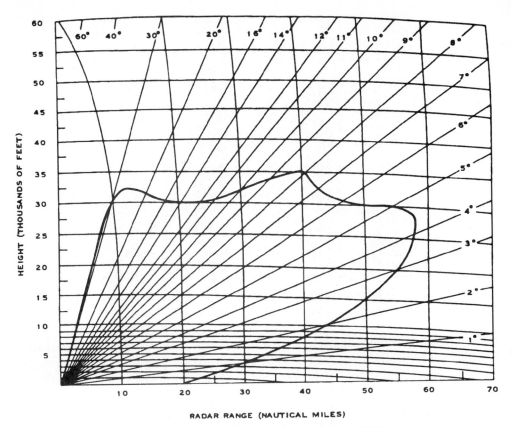

Figure 1-53. Coverage Diagram for the ASR-7.
(Diagram courtesy of the Federal Aviation Administration)

Figure 1-54 shows the antenna of the ASR-9 medium range 2-D air search radar, used principally for air traffic control in the vicinity of airports. Its antenna is characteristic of 2-D systems, being larger horizontally than vertically, which produces narrow azimuth and wide elevation beams (Ch. 3). This antenna has a gain of 33.5 dB and an effective aperture of slightly more than two square meters. The large rectangular antenna on top is for the secondary surveillance radar.

A few 2-D air search radars scan in elevation and range, using wide azimuth beams for coverage. These are the height-finding radars. A comparison of antenna shapes between 2-D radars is given in Fig. 1-55.

— *Three-dimensional (3-D) search radars* provide resolution in range, azimuth, and elevation, and are primarily for air search. Because of the requirement to search in an extra dimension, their scan rates are inherently slower than 2-D systems (Fig. 1-51b). Many 3-D radars, however, use multiple channels to search several elevation angles simultaneously, speeding up the scan. An example is the stacked-beam system of Fig. 1-51c. Figure 1-56 is the U.S. Air Force's AN/TPS-43 stacked-beam radar. Note the multiple horns feeding the reflector. Each horn produces a beam at a different elevation angle. Many modern stacked-beam systems use array antennas (Fig. 1-1 and Ch. 9).

Figure 1-54. Airport Surveillance Radar Model 9 (ASR-9) 2-D Search Radar. (Photograph courtesy of Westinghouse Electric Corporation)

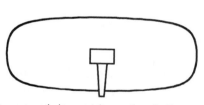
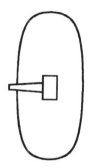

a. Antenna shape for 2—D search, with narrow azimuth beam and wide elevation beam

b. Antenna shape for 2—D height—finding, with wide azimuth beam and narrow elevation beam

Figure 1-55. Comparison of 2-D Antenna Shapes.

Tracking radars and modes: Rather than continuously scanning a volume of space, tracking radars (covered in Ch. 7) dwell on individual targets and "follow" their motion in azimuth, elevation, range, and Doppler (not all radars track in all dimensions). Because of the increased target dwell time, tracking radars and modes can locate objects in space more accurately than can search. Most tracking radars/modes, however, can follow only a single target, since the antenna must be kept continuously pointed at that target. These are known as single-target track (STT) radars and modes.

A few radars can track multiple targets simultaneously. Invariably, an electronically steered array antenna (a phased array) is used so that beam positions can be moved quickly

Figure 1-56. AN/TPS-43 3-D Air Search Radar. (Photograph courtesy of Westinghouse Electric Corporation)

from one target to another. Examples include the U.S Navy's AN/SPY-1 AEGIS radar, COBRA JUDY, Patriot surface-to-air weapon system, and MIR and MOTR instrumentation tracking systems

Tracking radars are classed by how the angle tracking errors are developed (Fig. 1-57). The principal methods are conical scan, lobing, and monopulse. These and other methods are discussed further in Ch. 7.

Conical scan angle tracking (Fig. 1-57a) is accomplished by squinting (displacing slightly off-axis) the antenna's beam axis and then rotating the squinted beam around the antenna's axis. Observing the relative echo amplitude at each of the scan positions, the system determines where in relation to the antenna's axis the target is located, since the echo is strongest with the scanned beam pointing closest to the target. If the transmit beam is on-axis and only the receive beam is squinted and scanned, the system is said to conical scan on receive only, or COSRO. Most older tracking radars were conical scan, scanning both transmit and receive antenna beams. A modern example is the AN/SPN-44 Automated Carrier Landing System radar (Ch. 7).

In a lobing system, the squinted beam is discretely moved to four or more positions around the axis instead of being continuously scanned (Fig. 1-57b). The target location with respect to the axis determines the relative echo amplitude in the beam positions, generating the track errors. If transmit is on-axis and only the receive beam is squinted, the process is lobe on receive only, or LORO. Modern lobing systems move the beam electronically and are LORO. The AN/APG-66 in the F-16 is an example.

Both conical scan and lobing require several pulses and at least one full cycle of beam positions to generate a set of track errors (azimuth and elevation). Their primary advantage is that the antenna has only one receive output and only one receiver is required.

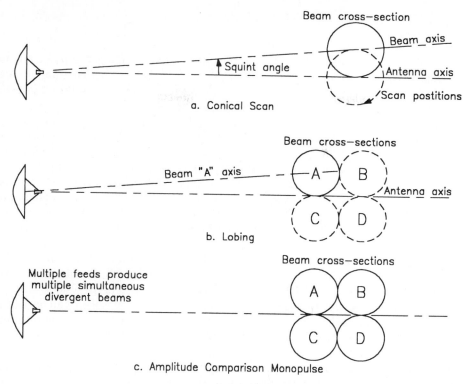

Figure 1-57. Angle Track Methods.

Amplitude-comparison monopulse tracking (Fig. 1-57c) is similar to LORO in that squinted beams are placed in discrete positions and the transmit beam is on-axis. The difference is that the squinted receive beams are generated simultaneously. The relative signal amplitudes in the beams are compared and a full set of angle errors is generated for each echo received. Monopulse has three antenna receive outputs: sum (the normal received signal), azimuth error, and elevation error. This requires three receiver channels. Most modern tracking systems develop their errors using amplitude comparison monopulse. Another form of tracking, phase monopulse, is discussed in Ch. 7.

Track-while-scan: Multiple targets can be followed simultaneously by search radars using track-while-scan, or TWS. As the searching radar passes by each target of interest, its position is reported to a tracking program in the radar's data processor. After several consecutive reports, the data processor can smooth the targets' positions and predict each target's location for the next scan. Because the position updates are separated widely in time — usually several seconds — track-while-scan is not, strictly speaking, tracking. The accuracy of this method is lower than true tracking systems, where the sample rate is high enough to follow target maneuvering. It does, however, allow many targets to be kept track of simultaneously by a single search radar.

Track-while-scan systems develop the targets' positions by centroid detection, wherein the target's signal amplitude is recorded as the antenna scans across it and the centroid of this amplitude pattern is computed. Using centroid detection, target positions are developed with accuracies considerably better than the antenna's beamwidth (Ch. 7).

1-6-4. Classification by frequency band: Radars, although by name implying the use of the radio frequency portion of the electromagnetic spectrum, can function in any spectral region. The low frequency limit for practical radars is set by antenna size (low frequencies require large antennas for appreciable gain), spectral utilization (wide bandwidths are not available at low frequencies), and target characteristics (targets much smaller than the illuminating wavelength have very small RCS). Few radars are found in the lower radio bands, except in cases where special propagation and target characteristics dictate their use.

The upper frequency limit is imposed by the availability of high power electromagnetic generators, by the tiny antenna beamwidths which result from large apertures at short wavelengths, and by the attenuation of high frequency signals by the atmosphere and weather.

Operational radars can, however, be found in all bands. At the lower end, over-the-horizon radars use the band from about 6 to 30 MHz, and are currently in place within the United States and other countries. The U.S. Air Force AN/FPS-118 OTH-B (Over-The-Horizon-Backscatter) radar, the U.S. Navy AN/TPS-71 ROTHR (Relocatable Over-The-Horizon Radar), and the Soviet "Woodpecker" are examples of operational systems. The two principle advantages to this frequency band are that the earth's ionosphere can be used to reflect signals beyond the horizon, and many targets are resonant in this band, causing high RCS. Present HF radars are all bistatic, with the transmitter and receiver separated by up to hundreds of miles.

At the upper frequency end of the spectrum, EHF (extremely high frequency — 30 to 300 GHz), infrared, and light radars (LADARs) are used for special-purpose applications. Their primary features are very high Doppler shifts, very good resolution, targets with high RCS, and very high weather attenuation (light-band radars can't see through fog any better than we can).

Wavelength: Wavelength is an important property of any wave. It is the length of one cycle of a wave as it propagates through a the medium (Fig. 1-58).

Wavelength is related to frequency as

$$\lambda_0 = c / f \qquad (1\text{-}31)$$

$$\lambda_g = v_p / f \qquad (1\text{-}32)$$

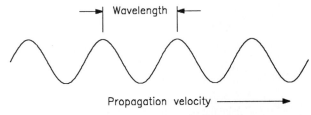

Figure 1-58. Wavelength Concept.

λ_0 = the wavelength in free space (meters)
λ_g = the wavelength in a medium (meters)
c = the free space velocity of propagation (meters per second)
v_p = the medium's propagation velocity (meters per second)

Frequency allocation treaties: As with other services in the electromagnetic spectrum, radars are by international agreement confined to certain frequency bands. Radars usually employ very high power transmitters and are confined to prevent interference with other services. The International Telecommunication Union (ITU) administers this portion of the EM spectrum.

Band designation systems: Several alphabetic band designation systems exist, and two are in common usage. The historic letter-coded bands were developed before and during World War II. The newer sequential letter bands system is somewhat supplanting the traditional scheme, but most radar engineers still use the earlier designation. The historic system is used throughout this book.

Tables 1-3 and 1-4 and Fig. 1-59 show the frequency band designations. The dark bands in the figure are the bands allocated by the ITU for radiolocation (radar). The letters immediately above the bands are their historic designations, and the letters farther above are the new designations.

Table 1-3. Historic Radar Frequency Band Designations

Historic Band Designation	Frequency (GHz)	New Band Designation (GHz)	ITU Radiolocation Assignments (GHz)
LF	< .003	A	
HF	.003-.03	A	
VHF	.03-.3	A < .25; B > .25	.137-.144 & .216-.225
UHF (Incl. P-Band)	.3-1.0	B < .5; C > .5	.420-.540 & .890-.940
L-Band	1.0-2.0	D	1.215-1.400
S-Band	2.0-4.0	E < 3.0; F > 3.0	2.30-2.55 & 2.7-3.7
C-Band	4.0-8.0	G < 6.0; H > 6.0	5.255-5.925
X-Band	8.0-12.5	I < 10.0; J > 10.0	8.5-10.7
K_u-Band	12.5-18.0	J	13.4-14.4 & 15.7-17.7
K-Band	18.0-26.5	J < 20.0; K > 20.0	23.0-24.25
K_a-Band (A-Band)	26.5-40.0	K	33.4-36.0
Q-Band*	33.0-50.0	L	
U-Band*	40.0-60.0	L	
V-Band*	50.0-75.0	L < 60.0; M > 60.0	
E-Band*	60.0-90.0	M	
W-Band*	75.0-110.0	M	
F-Band*	90.0-140.0		
D-Band*	110.0-170.0		
G-Band*	140.0-220.0		

* Denotes an industry-accepted waveguide band — not a standard frequency band.

1-6. Types of Radars and Radar Functions

Table 1-4. New Frequency Band Designations

Band Designation	Frequency (GHz)	Channel Width (GHz)
A-Band	0-.25	.025
B-Band	.25-.50	.025
C-Band	.50-1.0	.05
D-Band	1.0-2.0	.10
E-Band	2.0-3.0	.10
F-Band	3.0-4.0	.10
G-Band	4.0-6.0	.20
H-Band	6.0-8.0	.20
I-Band	8.0-10.0	.20
J-Band	10.0-20.0	1.0
K-Band	20.0-40.0	2.0

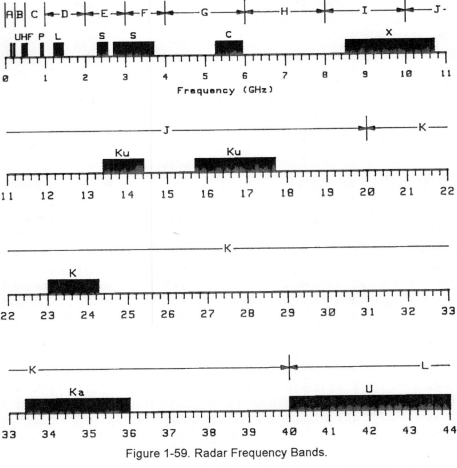

Figure 1-59. Radar Frequency Bands.

54   Chap. 1. Radar Fundamentals

Radar performance and frequency bands: Radar parameters and performance are affected by the frequency band in which they operate. Likewise, the band chosen for a particular radar is determined to a great extent by the radar's mission and specifications. The following discussion is intended to show how the choice of frequency band determines radar performance and how the mission of a radar affects the choice of frequency.

— **Bandwidth:** The bandwidth of the transmitted signal determines, among other things, the range resolution and frequency agility capabilities of the radar. Large bandwidths are available only in the higher frequency bands.

— **Antenna:** Monostatic radars use a single antenna for both transmit and receive. *Gain* is the important transmit parameter, and *effective area* is the critical parameter for receive. Gain and effective area are tied together for all antennas by the following relationship:

$$G = \frac{4\pi A_E}{\lambda^2} \quad \text{(dimensionless)} \qquad (1\text{-}33)$$

G = antenna gain
A_E = effective area (square meters)
λ = wavelength (meters)

Note that for a given antenna gain, low-frequency antennas are larger (have more receive aperture) than high frequency. Search radars must examine a volume of space in search of targets. The time required to examine this volume is the scan time and is set by, among other things, the beamwidths of the antenna. To prevent scan times from becoming impractically long, search beamwidths cannot be too narrow. Beamwidth and gain are related, in that high gains result from narrow beams and *vice versa* (Ch. 9). Thus scan time considerations force an upper limit of about 20,000 (43 dB) or so on the gain of search antennas. Low frequencies are therefore favored for long-range search applications, because of the larger effective area associated with a given gain, allowing more effective capture of echoes.

— **Transmitter:** In general, more radio frequency power can be produced at low frequency than at high. The physical size of radio frequency devices is a linear function of wavelength. Thus at low frequencies (long wavelengths), the device elements are larger, allowing them to handle more voltage and current, and thus higher power. Low frequencies are therefore favored for long-range radars from the standpoint of the transmitter.

— **Receiver:** Traditionally, it has been possible to design less noisy and thus more sensitive receivers at low frequencies than at high, giving low frequencies the advantage for long range radars. Modern microwave technology has somewhat reduced this advantage. On the other hand, it is possible to design higher bandwidth receivers at higher frequencies. From the standpoint of the receiver, there is no clear choice between high and low frequencies.

— **Propagation:** Except for special cases (over-the-horizon, for example) signal propagation clearly favors low microwave frequencies over all other bands. There is less atmospheric signal absorption low frequencies. In precipitation, the low frequency advantage is dramatic (Ch. 3). A given raindrop has over three orders of magnitude more scattering cross-section at X-Band (10 GHz) than at L-Band (1.3 GHz), producing far more clutter and signal scattering (attenuation) at the higher frequency. Conversely, there is more atmospheric and galactic noise in the lower bands, particularly at frequencies lower than microwave.

– **Targets**: If the illumination wavelength is short with respect to the target extent, the target tends to behave as an array of optical reflectors, the overall RCS of which can fluctuate over several orders of magnitude. These fluctuating targets can be difficult to detect and track (Ch. 4). To wavelengths which are comparable to the target extent, resonance effects cause RCS to be sensitive to frequency. Very large RCS is possible, however, and the fluctuations tend to be much less severe than at shorter (optical) wavelengths. If the wavelength is long compared to target extent, targets are Rayleigh scatterers, and have small, non-fluctuating RCS. Each region (optical, resonant, and scattering) has pluses and minuses, and the choice of frequency depends on the radar's mission.

– **Summary**: From the above discussion, it is seen that the performance of various parts of a radar system are markedly affected by the frequency band in which it operates. In general, the longer the range at which the radar must detect targets, the lower the frequency of the radar, down to the VHF bands. High frequencies are favored in applications where long-range detection is of secondary consideration, where high range resolution is required, or where physical constraints dictate small antennas (aircraft and missiles).

Radar types and their frequency bands: Following are examples of the types of radars to be found in the various frequency bands:

– **Lower frequencies (to 30 MHz or so)**: Radars in these bands are those which use the ionosphere as a reflector to view events beyond the horizon, including the U.S. OTH-B and ROTHR radars and the Soviet "Woodpecker."

– **VHF and UHF (30 MHz to 1 GHz)**: Very long-range early warning radars are found in these bands, along with systems to detect and classify spacecraft. Since propagation is primarily line-of-sight, only high altitude targets can be seen at very long ranges.

– **L-Band (D-Band in the new scheme)**: Long-range military and air traffic control search radars are found in this band. It offers a good compromise between antenna size and low weather attenuation of signals. Because of antenna sizes required, most L-Band radars are ground or ship based, such as the FAA's Air Route Surveillance Radar (ARSR) series and the Navy's TAS search radar. A few airborne and spaceborne L-Band radars exist.

– **S-Band (E/F-Band)**: Medium-range ground-based and shipboard search radars use the S-Band. Included are the Airport Surveillance Radar (ASR) series of air traffic control radars, and the Navy's AEGIS multifunction phased array radar (AN/SPY-1). It is also the band of the AN/APY-1 and AN/APY-2 airborne early warning radars, found in various models of the E-3A Airborne Warning And Control System (AWACS) aircraft.

An interesting recent addition to this band is the National Weather Service's next-generation Doppler weather radar (NEXRAD). Even though the RCS of raindrops at S-Band is much smaller than in the presently used C-Band, the superior weather penetration capabilities of this band make it desirable for this mission. The small RCS of the targets is compensated for with high transmitter power and a large antenna.

– **C-Band (G-Band)**: This compromise band allows moderate ranges with reduced antenna size. It is used for search and fire control radars, plus many metric instrumentation radars. Many weather detection radars are also in this band, both ground based (National Weather Service) and in larger aircraft.

— **X-Band (I/J-Band)**: This band is extensively used in applications where antenna size is limited but the extreme atmospheric and weather attenuation of higher bands cannot be tolerated. Most military airborne multimode radars are in this band, as are the missiles associated with them. It is used for small boat radars, and for weather radars for smaller aircraft.

— K_u, K, K_a, **and higher bands (J-, K-, and L-Bands)**: Because of severe weather attenuation, these bands are, at least in the atmosphere, limited to short range systems. High-gain antennas are small, but their apertures are also small. Signals in some parts of the bands are relatively secure from intercept because of very high atmospheric attenuation. Examples include short-range terrain avoidance and terrain following radars, airport surface detection equipment (ASDE) radars, police speed-measuring radars, and other specialized short-range applications. These bands can be used in space-based radars, where atmospheric and weather attenuation is not a factor.

— **Infra-red and visible light bands**: Weather and atmospheric attenuation are the major problems in these bands. Another problem is that for antennas of non-microscopic size, the beamwidths are extremely narrow, making target acquisition difficult. If the beams are widened, the effective capture area becomes very small. The major applications of these bands is in laser rangefinders and optical targeting systems, including many of the so-called "smart" munitions.

1-6-5. Classification by waveform and sample rate: Radars and modes are classed by the types of waveforms they transmit and, in the case of pulsed systems, the rate at which they transmit them. Taking waveforms first, there are four principal types (see Sec. 2-4 for further information on waveforms).

— **Unmodulated CW**: CW radars transmit and receive unmodulated sinusoidal waves. The transmitter and receiver operate simultaneously, and the receiver must amplify weak target echoes in the presence of a strong transmit signal (spillover). Except for very low power systems, spillover control dictates the use of separate transmit and receive antennas. CW is a high-energy waveform and is capable of very long range applications, if spillover is not a problem. CW radars primarily measure the target echo's *Doppler shift* (and therefore its radial velocity) and *angular position*. Range cannot be measured directly without some form of time-variant modulation. (The proximity fuzes of World War II estimated target range by the amplitude of the received echo but their range accuracy was poor.) CW radars are used mainly for velocity search, velocity track, and missile illumination.

— **Modulated CW**: Introducing time-variant modulations on the CW wave adds the capability to directly find target range. Section 6-5 discusses information recovery using modulated CW waves.

— **Gated CW pulse**: The gated CW wave, commonly used in less sophisticated pulsed radars, is an unmodulated sinusoid gated into short bursts. If the PRF is low, the primary measurement from this wave is range and angular position. If Doppler is sensed at low PRF, it usually is to indicate a moving target and not to measure target velocity. If the PRF is high, this becomes a Doppler-measuring waveform with characteristics similar to CW (see high PRF later in this section).

The gated CW wave presents a design paradox. For good detection of long-range targets, the illumination energy needs to be high, for which pulses need to be wide. Good range resolution, on the other hand, requires narrow pulses. The tradeoff yields radars which have either good range resolution but short-range capability or long-range capability with poor range resolution.

— **Complex waveform pulse**: This class of waves is pulsed like gated CW, but the wave within the pulse is modulated to increase the waveform bandwidth. Modulated pulsed waveforms support pulse compression and are capable of simultaneously having high energy (wide pulses) and good range resolution (wide bandwidths). They are finding increased use in more sophisticated radars.

Another benefit of modulating the wave within each pulse is that the wave is more difficult to intercept and to jam. Interception of a radar's illumination signal is a function primarily of peak transmit power, while detection of targets by the radar is a function of waveform energy. This class of waves can achieve high energy with modest peak power levels (wide pulses), and at the same time give good range resolution. They are difficult to jam because the system correlates the received signals with the modulation transmitted. If the proper modulation is not present on an interfering signal, it is attenuated in the signal processor.

1-6-6. Classification by PRF class: The performance of a pulsed radar is also affected by the rate at which the pulses are generated (the pulse repetition frequency or PRF). Three distinct sample rate (PRF) classes, introduced in an earlier section, are recognized. An evaluation of the suitability of each class to different tasks is found in Sec. 6-6.

— Low PRF (LPRF) is mainly for ranging. Range is unambiguous and Doppler is highly ambiguous. This PRF class is found in radars where the true target Doppler shift does not need to be known. MTI and MTD radars usually are of this class.

— High PRF (HPRF), which is also known as pulse Doppler (PD) is mainly for measuring Doppler shifts. It is unambiguous in Doppler and highly ambiguous in range. It is essentially a CW wave which is chopped to allow the transmitter and receiver to operate at different times, letting a single antenna serve both. The RF signal can be an unmodulated sinusoid, whereby this wave performs similarly to unmodulated CW. When ranging is desired, the carrier wave is often modulated, with performance similar to modulated CW.

The difference between the Doppler measuring high PRF and the Doppler sensing low PRF is shown in Fig. 1-60 and discussed in Ch. 6.

— Medium PRF (MPRF) is ambiguous in both range and Doppler, but not so severely that true range and true Doppler of multiple targets cannot be found. It is the only PRF class which allows simultaneous recovery of range and Doppler for large numbers of targets (Ch. 6).

1-6-7. Classification by specific mission and mode: In this section, a large number of specific radar missions are introduced. The descriptions are organized by the location of the radar and the location of the intended target. This list is not meant to be comprehensive, and other applications are discussed throughout the book.

Surface-to-surface missions: These radar missions primarily involve ships and boats searching for other ships and boats. They are relatively short range and, in the main, do not use Doppler.

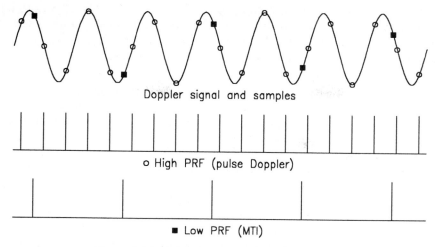

Figure 1-60. Pulse Doppler and MTI Sampling.

Surface search: This is commonly a short- to medium-range search function with the goals of avoiding collisions between boats and ships, and assisting navigation in bad weather. It is 2-D search (azimuth and range) and its range is limited by the line-of-sight restriction of the antenna height and the earth's curvature. Typical maximum ranges are less than 50 miles. It is usually low PRF and Doppler recovery is seldom involved.

ASDE: Airport surface detection equipment has the specialized surface search task of keeping track of aircraft on the ground at airports. It is used by ground controllers in bad weather. It is classed as low PRF, although because of its limited instrumented range, the PRF may be many thousand of pulses per second.

Surface-to-air missions: These are land-based and shipboard radars dealing with aircraft and spacecraft targets. They almost always use the Doppler shift to separate targets from clutter. Many detect but do not measure Doppler shift (MTI or MTD). Others measure target Doppler (CW or PD).

Air search: This mission is usually the medium- (50 miles and above) to long-range (300 miles or so) search for aircraft, both for military and air traffic control purposes. It is commonly low PRF and Doppler is almost always used to separate moving targets from clutter. Secondary radars (identification friend or foe — IFF — and air traffic control radar beacon system — ATCRBS) are commonly co-located with these primary radars.

Early warning: These are long-range radars used to detect and provide information on aircraft and spacecraft, primarily for military purposes. Those designed to detect aircraft have maximum ranges of 300+ miles and those used for spacecraft can sometimes detect targets to 5,000 or more miles. Doppler and pulse compression are extensively used, as are other forms of signal processing. They can be low, high, or medium PRF. They have very high energy transmitters and large antennas. They tend to be in the lower microwave frequency bands.

1-6. Types of Radars and Radar Functions

Spacecraft detecting and cataloging: These radars are similar to those in the early warning category and in some cases are the same radars. They function at very long ranges (to thousands of miles) and are used to maintain catalogs of earth-orbit spacecraft and debris.

Fire control: These systems are used to control surface-to-air weapons, such as the Navy's Standard missile, the Army's Patriot, and guns. The missile-control radars are often high PRF (some are CW) and perform extensive Doppler processing. Low PRF is commonly used for controlling guns. With missiles, there is often an associated illumination function, and sometimes data links are present. They are primarily short- to medium-range tracking radars, but may have an associated air search radar.

Metric instrumentation: These specialized tracking radars locate objects in the atmosphere and in space with extreme accuracy. They are essentially large microwave surveying devices. They are usually either low or medium PRF. They often function as secondary radars, tracking beacons in their targets. Some are phased arrays and can track multiple targets simultaneously.

Weather radar: These low and medium PRF search radars detect and measure the radar cross-section of atmospheric phenomena, primarily to forecast weather and to warn of severe weather. Some search for precipitation and others look for the atmospheric density changes caused by wind shears. They may or may not use Doppler processing. Those that do can report more information about the weather they observe.

Over-The-Horizon (OTH): These specialized long-range search radars operate in the HF bands (3 to 30 MHz). They use the earth's ionosphere as a reflector so they can see over the horizon to search for ships, aircraft, and cruise missiles. They have very high energy transmitters and often employ modulated CW waveforms. Both U.S. types, (AN/FPS-118 Over-The-Horizon-Backscatter OTH-B and AN/TPS-71 Relocatable Over-The-Horizon Radar ROTHR), are bistatic.

Air-to-air missions: Air combat makes extensive use of the multi-mode radars in modern aircraft. These modes are said to be either look-up, if the targets are high enough to be clear of clutter, or look-down, where the radar sees clutter simultaneously with the target. Look-down modes are also capable of looking up, but the converse is not true. Following are some of the many modes used:

Velocity search: This is the longest range search mode in most multi-mode airborne radars. It is look-down and high PRF. It looks for targets which are moving radially to the radar, primarily in nose aspect (the radar and target flying toward one another). It is primarily a Doppler mode and range is often not measured. Search is in both azimuth and elevation, although the long ranges restrict the possible elevation angles for targets.

Range-while-search: This is a medium range look-down search mode, which find target range as well as Doppler. It can be high PRF and use a modulated pulse Doppler wave, or medium PRF.

Track-while-scan: This medium- or high-PRF mode is similar to range-while-search, except that a limited number of targets are accommodated and track files on these targets are maintained in the radar's data processor. These files are used to identify threats, control weapons, and to initiate single-target tracks.

Track: Present day multi-mode airborne radars use mechanically scanned antennas. Thus tracking is done on one target at a time (single-target track — STT). This mode is usually either high or medium PRF. A primary function is to keep the antenna pointed at a target for missile illumination.

Illumination: This mode is used to illuminate the target for SARH missile, such as Sparrow (AIM-7). Some missiles require CW illumination and this mode must be interleaved with a track mode, whose job it is to keep the radar antenna pointed at the target being illuminated. Some missiles can use pulsed illumination (high PRF), and illumination is a by-product of the high PRF track which points the radar antenna.

Raid discrimination: In this mode, enhanced cross-range resolution is developed with Doppler beam-sharpening (DBS — a form of synthetic aperture radar — SAR). See the high resolution air-to-surface missions below, Sec. 1-3, and Ch. 13 for a description of synthetic aperture radar.

Gun director: In this tracking mode, the radar controls a cockpit display which tells the pilot how to point his or her aircraft so that a gun can be brought to bear on a target. Since the bullets must hit the target to be effective, this mode requires highly accurate tracks. It often employs pulse-to-pulse frequency agility to suppress target-induced errors. See Chs. 4 and 7 for discussions of target glint, target-induced tracking errors, and their suppression.

Airborne weather radar: This search mode is used to avoid severe weather. The radar primarily measures precipitation rates. High-rate rainfall is associated with severe turbulence and weather. Radars dedicated to this task may or may not sense the precipitation's Doppler shift. Those which do give a better picture of the weather situation viewed.

Air-to-surface missions: Airborne radars of this type view the surface of land or water to find objects on the surface, to navigate, and to target weapons to features such as buildings and bridges.

Real-beam ground mapping: This search mode scans a wide area of the ground under the radar and is used in modern systems mainly for navigation. It can be low or medium PRF. It usually does not use Doppler and does no special signal processing.

Doppler Beam-Sharpening (DBS) mapping: This search function provides a high resolution map of a limited-extent ground surface for navigation and weapons targeting. It takes advantage of the geometry of the search to allow resolution of multiple objects lying within a single antenna beam position and a single range bin. As shown in Fig. 1-61, these normally unresolvable scatterers may have slightly different radial velocities and may be separable in Doppler. DBS is usually medium PRF. Chapter 13 covers all aspects of synthetic aperture.

Synthetic Aperture Radar (SAR) mapping: This mode is similar to DBS in that it provides high resolution ground maps and images of large targets, such as ships.

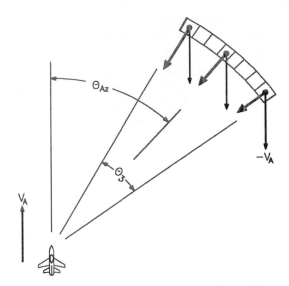

Figure 1-61. DBS Principle. Note the slightly different radial velocities of the three scatterers.

The difference is in the processing. SAR processing can be focused and resolutions obtainable are considerably better than with DBS. It is used for mapping, reconnaissance, and weapons targeting.

Terrain Following (TF) and Terrain Avoidance (TA): These modes allow an aircraft or missile to navigate around and over obstacles while flying low to the ground to avoid detection by other radars. The two modes are similar and several different definitions exist. Using Stimson's definitions [3], terrain following functions in the vertical, guiding the vehicle over obstacles. Terrain avoidance adds the horizontal dimension and allows flight paths over and around obstacles, optimized to avoid detection by enemy radars. These are short-range pulsed modes, either low or medium PRF. Doppler may or may not be used. If available, low energy waveforms are used to avoid interception by the enemy. Some radars dedicated to these tasks are circularly polarized to minimize echoes from precipitation (Chs. 4 and 9). The terrain information thus derived is either displayed as guidance information to a pilot or is coupled into the autopilot, which flies the aircraft or missile.

Ground moving target search: The objective of this search function is to detect from airborne platforms vehicles moving on the ground. The clutter levels are high and discrimination between clutter and a small RCS target moving at almost the same velocity is difficult. Primary processing is of Doppler shift. This mode is short range and has low energy requirements. Medium and high PRF are used, with high PRF having the advantage in the Doppler dimension and medium PRF being advantageous in range. It is discussed in Ch. 12.

Sea surface search: In this function, the radar searches for objects on the surface of the ocean, such as boats, submarines, and periscopes. The desired targets have much lower RCS than the clutter and do not move rapidly, if at all. Doppler clutter suppression is therefore normally not available. The technique commonly used is to have the radar dwell on the surface for a long time, during which the sea clutter becomes very noiselike (Ch. 4). Signals are then processed as though the interference were noise. This is a short- to medium-range mode and has moderate energy requirements. It is often done with a dedicated radar, such as the AN/APS-116 in the Navy's S-3A antisubmarine aircraft.

Radar altimeter: This task uses a dedicated radar which measures the distance from the radar to the ground. The antenna beam is continuously pointed downward and does not scan, unless to stabilize it as the aircraft or missile rolls and pitches. The RCS of the ground viewed vertically (a high-angle area target — Ch. 3) is very large and signal decreases as the square of range, rather than the fourth power as it does with point targets. Thus the energy requirements are low. Most of these radars use a modulated CW waveform.

Doppler update: In this mode, the antenna beam is pointed to a few discrete locations on the ground and the Doppler shift measured. This information, after correction for geometry, updates the velocities calculated by the aircraft's inertial navigation system (INS), improving navigation accuracy.

Multi-mode systems: Many radars, particularly military systems where space and/or cost considerations dictate a single antenna, incorporate several modes in a single system, including search, track, and track-while-scan. These multi-mode systems are mostly either airborne or have phased array antennas. An example is the AN/APG-70 of Fig. 1-2. It is common for these radars to have surface search modes, look-up (out of clutter) air search modes, look-down (into clutter) air search modes, single target track (STT) modes, spotlight multi-target track modes (where the radar dwells on one target, then switches to another, and so on), synthetic aperture ground mapping and target discrimination modes, and most of the other modes attributed to the airborne systems listed above.

Search modes in these radars are generally 3-D, but whereas dedicated 3-D search systems have stacked-beam capability, multi-mode scan patterns are usually sequential in elevation, such as the raster scan of Fig. 1-51b. Radars mounted in the noses of aircraft commonly search in a pattern similar to the bar scan of Fig. 1-51d. Track-while-scan is common in their air search modes.

Track modes in multi-mode radars include single target track for accurately finding a single target's position and for controlling weapon systems during attack phases. Spotlight multi-target tracking allows accurate tracking and weapon control for several targets.

Synthetic aperture ground mapping mode allows high resolution ground maps to be formed for navigation and targeting.

Multi-mode systems are usually much more complex (and expensive) than dedicated single-task radars, but may be less so than the sum of the many radars a single one replaces. Because neither nose-mounted airborne nor phased array antennas can scan 360° in azimuth, scan limits are also different from most dedicated land-based and shipboard systems. Whereas the latter two types usually scan horizontally through 360°, multi-mode systems in most cases can scan only a sector. Phased array systems cover the full azimuth circle by implementing multiple antennas or by mechanically steering a single antenna.

1-7. Radar Compared to Other Electromagnetic Sensors

Radar is just one of several electromagnetic (EM) sensor types used to detect, locate, and analyze remote objects. The others include passive infrared (IR), active (self-illuminating) IR, and passive visual-light sensors. The following discussion compares the various EM sensors.

Self-illumination, as done by radars and active IR systems, has the advantage of being able to directly measure range and velocity by comparing the echo signal to the illumination. Self-illumination also allows the sensor to function in the dark and with cold targets. Passive passive sensors have problems with both situations.

Self-illuminating sensors, on the other hand, can be intercepted and located by enemy detection systems. Radio-frequency active sensors, with their narrow bandwidths and predictable frequencies, are at a particular disadvantage here. It has been said that activating a radar in a combat situation against a sophisticated enemy is simply an invitation to be shot at, plus providing the guidance signal for the enemy's weapons.

Because of their longer wavelengths and predictable frequencies, radars are the easiest of the EM sensors to jam.

The long wavelengths (low frequencies) used by radars have the advantage of penetrating clouds, smoke, and precipitation far better than shorter wavelength passive or active IR and visual sensors. Because of their long wavelengths, however, radars have poorer resolution (ability to separate multiple closely-spaced objects).

Radars are the largest, heaviest, and most expensive of the EM sensors, due mainly to their long wavelength and need for an illuminator (transmitter). Active IR and visual illuminators can be much smaller than those operating at radio frequencies.

Despite their disadvantages, radars can detect and measure targets in ways no other EM sensors can, thanks primarily to their long wavelengths and self-illumination. This is particularly true at night, in bad weather, and with non-emitting (cold) targets.

1-8. A Short History of Radar

The principles by which radars function were known from the very early days of research into radio phenomena. Several attempts were made in the 1920s and early 1930s to devise useful radars, primarily to assist in ship collision avoidance. It was not, however, until just prior to the World War II that it became practically useful and attained the prominence it now holds for early warning, air traffic control, weapon control, ship and aircraft collision avoidance, and weather sensing. A summary of the milestones in the development of modern radars follows (adapted from [1] and [9]).

Between 1886 and the 1920s, the principles by which radar operates were formulated and preliminary techniques derived. In 1886-1888, Heinrich Hertz demonstrated the generation, reception, and scattering of electromagnetic waves in the radio bands. In 1903-04, Christian Hulsmeyer developed and patented a primitive form of collision avoidance radar for ships. In 1922 M.G. Marconi suggested, in his acceptance speech for the IRE (now IEEE) Medal of Honor, an angle-only radar for ship collision avoidance. In 1922, A.H. Taylor and L.C. Young of the Naval Research Laboratory (NRL) demonstrated a bistatic radar for guarding harbors. Most of these early radars were CW.

The mid-1920s saw the introduction of pulsed radars. In 1925 the first short-pulse echo from the ionosphere was observed on a cathode ray tube by G. Breit and M. Tuve of Johns Hopkins University. The first reported detection of an aircraft target was made in 1930 by L.A. Hyland of NRL. He later became president of Hughes Aircraft. This was a CW measurement. During 1934, the first photo of a short-pulse echo from an aircraft was made by R.M. Page of the U.S. Naval Research Laboratory. NRL was destined to play an important role in the development of radar, a role which continues today. The first demonstrations of short-pulse range measurement of aircraft targets were made in 1935 by British and German scientists.

Operational radar systems came about in the mid-1930s. The first, the Chain Home system, was installed in Britain in 1937. It was designed by Sir Robert Watson-Watt and played a critical role in the Battle of Britain, pinpointing the location of German raids and allowing the Royal Air Force to concentrate its forces in repelling these raids rather than having to search for enemy aircraft by patrolling. Churchill said that three things played decisive roles in winning the Battle of Britain: the RAF pilots, the Spitfire fighter, and radar.

The United States installed its first operational shipboard radar, the XAF, on the battleship USS *New York*. It had a surface search (for ships) range of 12 miles and an air search range of 85 miles.

The single most important step in bringing about microwave radars was the development of the *magnetron* (Ch. 9). It was completed by a joint British/U.S. program in 1939, and made possible high power microwave radars, with their attendant relatively small antennas.

By 1941, the United States had produced about 100 SCR-270/271 (Signal Corps Radio models 270 and 271 – Fig. 1-62) early warning radars. On the morning of December 7, 1941, one of these radars, located on Oahu, detected a large number of aircraft approaching Pearl

Figure 1-62. SCR-270 Radar. (Reprinted from Barton [5]. © 1984 IEEE; used with permission)

1-8. A Short History of Radar

Harbor. The information was misinterpreted and no action was taken. This surely must have been the original, but far from the last, C^3I (Command, Control, Communication, Intelligence) breakdown involving radar. These radars were also used extensively during the Korean conflict. Since their operating frequencies are in the present FM broadcast band, they are no longer in use.

World War II changed the world in many ways, including the establishment of radar as an indispensable tool for remote sensing of the enemy and the directing of weapons toward that enemy. One area involved the development of tracking radars to more accurately locate enemy aircraft. An early tracker, the SCR-268, is shown in Fig. 1-63. It developed track errors, which were presented to operators, who in turn manually positioned the antenna.

Figure 1-63. SCR-268 Tracking Radar. (Reprinted from Barton [5]. © 1984 IEEE; used with permission)

Probably the most significant tracking radar of the era, though, was the U.S. SCR-584 (Fig. 1-64). Through primitive but effective computers, it was accurate enough to direct a battery of four anti-aircraft guns. In the later stages of the war, when the V1 "buzz bomb" (actually an early cruise missile) was directed against Britain, the SCR-584 and its guns played a pivotal role. As stated on the PBS Nova program "Echoes of War," on one of the last days of the V1 attacks, of 104 missiles detected, only four made it through the radar/gun shields.

Airborne radars also played a large role in the war. They proved particularly useful in finding Axis submarines and in guiding bombing raids over Europe in bad weather and at night. An unusual early air-to-air system was the "Lichtenstein" radar developed by the Germans for use on the Bf-110 fighter. This VHF radar had a dipole array mounted externally on the aircraft which reportedly slowed its top speed by 150 mph [3]. The development by the Allies of the magnetron allowed American and British radars to operate in the microwave bands and to produce small, light-weight airborne radars. Much of the World War II effort was directed to this end. One result was the first aircraft basically designed around its radar — the P-61 Black Widow.

Figure 1-64. SCR-584 Fire Control Radar. (Reprinted from Barton [5]. © 1984 IEEE; used with permission)

More recent developments: Before and during World War II, several laboratories were established to develop radar, the most famous being the Radiation Laboratory (Rad Lab) at the Massachusetts Institute of Technology. Virtually all of the techniques used in modern radars were identified in this massive effort, although many of these discoveries were not implemented until recently because the technological advances required lagged far behind the theoretical. Some recent improvements include:

— **Microwave electronically steered antennas:** The development of microwave electronically steered antennas ("phased arrays") allowed such advances as instantaneous beam steering, true multi-target tracking, and such ECCM advantages as adaptive nulling. Several complex radar systems have been built based on this principle, including the AN/SPY-1 multifunction shipboard tactical radar, the AN/MPQ-53 Patriot air defense radar, the AN/FPS-108 COBRA DANE long range radar, the AN/FPS-115 PAVE PAWS solid-state early warning radar, and the MIR (multi-target instrumentation radar) and MOTR (multi-object tracking radar) instrumentation radars. See Ch. 9.

— **Phase-stable and gridded microwave transmitter amplifiers:** This development, which resulted in the gridded traveling wave tube amplifier found in many modern radars, was necessary before Doppler signal processing could be developed to make possible detection of moving targets in returning clutter many orders of magnitude more echo power than the targets. This ability to detect moving targets in strong clutter led to the look-down (into clutter), shoot-down (into clutter) capabilities of many modern air-to-air and surface-to-air missiles. See Ch. 8.

— **Digital computer signal processing:** The use of the digital computer has given new dimensions to radar signal processing, making possible good look-down capabilities and such advances as synthetic aperture radars (SAR), in which angular resolution cells can be made many times smaller than the physical width of the antenna beam. Chapters 11 through 13 address this topic.

— **Pulse compression:** Pulse compression, the ability to transmit a wide pulse and signal process the echoes into narrow pulses, has solved the dilemma facing radar designers as a result of the need for wide transmit pulses (for high energy) to detect distant targets, and at the same time resolve closely spaced targets in range (requiring narrow pulses). Modern radars are no longer saddled with the compromises in this area which once drove radar design. See Ch. 13.

— **Fast Fourier Transform (FFT) algorithm:** The development of a computationally efficient way of doing Fourier transforms revolutionized radar signal processing by allowing very rapid signal spectrum analysis to be performed. Before its development by Cooley and Tukey in the mid-1960s, over one million complex multiplications were required to do a 1024 point complex Fourier transform on a digital computer. Using the FFT algorithm, the number of complex multiplications to do the same transform is reduced to approximately 5,000, a reduction by a factor of 200. As we shall see later, nowadays signal processing engineers will go to great lengths to organize a problem so it can be solved with Fourier transforms, and radar signal processing computer designs are optimized to perform this calculation. The FFT and its radar applications are covered in Chs. 11, 12, and 13.

— **Synthetic aperture radar (SAR):** SAR is a technique by which small real antennas can be made to behave like large antennas, simply by moving them to different positions and extracting target information from each antenna location. The data are then processed together, as though coming from a very large real antenna. SAR makes possible cross-range resolutions which could not be achieved with real antennas. Terrain mapping, target identification, and target analysis are some of the uses of synthetic aperture radar. Synthetic aperture radar techniques are found in Ch. 13.

— **Sophisticated smoothing and prediction algorithms:** The development of very sophisticated and effective ways of smoothing data derived from widely spaced samples and predicting future data behavior has made possible very smooth and accurate tracks on radar targets. Their implementation has also resulted in the ability of a computer attached to a search radar to "track" many targets simultaneously, even though samples of the targets' positions may occur seconds apart. Several algorithms have been developed to do this task, the most famous of being the Kalman filter. Tracking algorithms are introduced in Ch. 8.

— **Low-observable targets:** The development of low radar-observable ("stealthy") targets has made the radar's task much more difficult. As we shall see in Chs. 3 and 4, the radar has a strong advantage over targets which compete with internal noise. In this case, the target's echo signal-to-noise ratio increases as the inverse fourth power of range as the target and radar move nearer one another. In order to cut the radar's detection range by a factor of ten, the target's radar cross-section must be reduced by a factor of 10^4 or 10,000.

When the competing interference is clutter or self-protection jamming, much of the advantage shifts to the target. The signal-to-clutter ratio varies as the inverse first power of range, and the signal-to-self-protection-jamming ratio varies as the inverse second power of range. In order to cut detection range in clutter by a factor of ten, RCS must be reduced by only a factor of ten. In self-protection jamming, a detection range decrease by a factor of ten requires that RCS be reduced by a factor of 100. This is also covered in Chs. 3 and 4. Fig. 1-65 shows the U.S. Air Force B-2, or "Stealth" low observable bomber.

Figure 1-65. B-2 Low-Observable Bomber. (Photograph courtesy of Northrop Corporation)

1-9. Where to go for more information

Obviously no one book can cover all aspects of radar. The last time this was attempted, the result was the MIT Rad-Lab series of over fifty volumes and tens of thousands of pages. And that covered only part of the subject through the mid-1950s.

The goal of this book is to introduce and describe radar principles and techniques, without necessarily providing all the details and data needed to implement them. It is *not* a goal to provide an extensive radar bibliography. On the other hand, if interest in the details of a particular subject is sparked, it wouldn't do to leave the reader without a clue as to where to find more information. To these aims, the last section of each chapter will include a *short* list of those books and papers which (in the author's opinion) are the best sources of further information on the topics covered. If these references provide the needed information, well and good. If not, *they* have the extensive bibliographies necessary to point readers toward the critical detail. Specific works cited in the text are also included in these sections.

General references: The following works are appropriate general references for persons requiring greater depth and breadth of coverage than is found here. The following two references are fundamental to a radar engineer's library.

[1] M. I. Skolnik (ed.), *Radar Handbook,* 2nd ed., New York: McGraw-Hill, 1990.

The first edition of the *Radar Handbook* is in some areas more comprehensive than the second, although obviously dated in its treatment of some subjects. Some people may need a copy of each edition.

[2] M. I. Skolnik (ed.), *Radar Handbook,* 1st ed., New York: McGraw-Hill, 1970.

A less rigorous introduction to airborne radars is found in the following book, which also may be of interest to persons dealing with other radar types. The illustrations are outstanding.

[3] G. W. Stimson, *Introduction to Airborne Radar*, El Segundo, CA: Hughes Aircraft Company, Radar Systems Group, 1983.

An interesting treatment of the early days of radar is found in the autobiography of Sir Robert Watson-Watt, the developer of Britain's Chain Home early warning system of World War II. Another good review of early radars is found in a special edition of the *IEEE Transactions on Microwave Theory and Techniques*.

[4] Sir Robert Watson-Watt, *The Pulse of Radar*, New York: Dial Press, 1959.

[5] D.K. Barton, "A Half-Century of Radar," *IEEE Transactions on Microwave Theory and Techniques*, vol. MTT-32, no. 9 (September, 1984).

A comprehensive source of open-literature information on radar systems world-wide is *Jane's Radar and Electronic Warfare Systems*, published biennially. It contains system data plus the address and public information contact person for the manufacturer.

[6] B. Blake (ed.), *Jane's Radar and Electronic Warfare Systems*, 2nd ed., Coulsdon, Surry, U.K.: Jane's Defence Data, 1990.

The Institute of Electrical and Electronics Engineers (IEEE) periodically compiles indices of radar articles published in a wide variety of technical and scientific journals. The latest, covering 1977 through 1984, was published in the *Proceedings of the IEEE 1985 International Radar Conference*, and repeated in the *IEEE Transactions on Aerospace and Electronic Systems (AES)*.

[7] C.H. Gager, "Cumulative Index on Radar Systems for 1977-1984," *IEEE Transactions on Aerospace and Electronic Systems*, vol. 25, no. 4, July 1989.

The previous index, through 1976, is available as IEEE publication JH-4675-5 (IEEE, 345 East 47th Street, New York, NY 10017).

Understanding radar terminology is a large part of understanding the subject. The IEEE publishes a standard on radar terminology. It has as its stated goal "promoting clarity and consistency in the use of radar terminology."

[8] IEEE Standard 686-1982, *IEEE Standard Radar Definitions*, IEEE, 345 East 47th Street, New York, NY 10017.

Two books by E. Brookner give much information on radar techniques and have compilations of data on existing radars.

[9] E. Brookner, *Radar Technology*, Norwood MA: Artech House, 1977.

[10] E. Brookner, *Aspects of Modern Radar*, Norwood MA: Artech House, 1988.

Annually, the IEEE sponsors the National Radar Conference (during alternate years, it is the International Radar Conference), usually in the spring. The proceedings are a valuable source of recent information.

Radar information is found in some IEEE transaction, particularly Aerospace and Electronic Systems, Antennas and Propagation, and Microwave Theory and Techniques.

Target and radar measurements are a major topic of the Antenna Measurement Techniques Association (AMTA), an industry group.

2

THE RADAR SYSTEM:
Functions and Parameters

Introduction

The types of information that can be recovered by a radar depend on the operating parameters of the radar, the targets, and the propagation medium. This chapter examines the radar system and its parameters, including the following:
— Coherence and frequency generation
— Transmit parameters
— Transmitted waveforms, spectra, and bandwidths
— Antenna parameters
— Receiver parameters
— Signal processing parameters
— System stability requirements

2-1. Coherence

Coherence deals with the radar's treatment of signal phase. There are three levels of coherence in radars: coherent, coherent-on-receive, and non-coherent (sometimes called incoherent). Signal phase is the measurement made to recover echo Doppler; thus only coherent and coherent-on-receive systems can recover Doppler shift.

Coherent radars are those in which the phase of the illumination signal is derived from stable internal sources and is constant and predictable. A coherent system "knows" the phase of each illumination pulse prior to transmission and can compare the phase of any echo to it. This class of radars can detect or measure the Doppler shift of any target. Since the illumination signal comes from a master oscillator and the transmitter is an amplifier, this type is sometimes called a master oscillator power amplifier (MOPA) radar (Fig. 2-1).

Coherent-on-receive, sometimes called quasi-coherent, systems have unpredictable illumination phases which are measured and "remembered" for use as the internal reference. The system phase reference is always with respect to the latest illumination pulse,

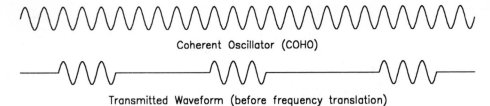

Figure 2-1. Coherent Phase Relationships.

and each time the transmitter fires, previous references are "forgotten." Thus coherent-on-receive systems cannot recover Doppler at medium or high PRF. This type of coherence is used in radars having magnetron transmitters. See Fig. 2-2.

Non-coherent radars ignore the illumination and received signal phases and are thus not suited to detecting or measuring the Doppler shift.

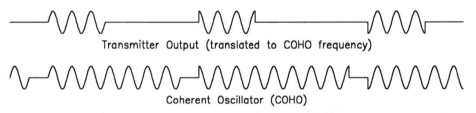

Figure 2-2. Coherent-On-Receive Phase Relationships.

2-2. Frequency Generation

The methods of generating the sinusoids required for operation of the radar depend on the level of coherence. At each level, many frequency generation schemes are possible, with only a sampling described here.

Frequency generation in coherent radars: Figure 2-3 and Table 2-1 show the basic frequency relationships within a coherent radar. In this implementation, two oscillators, the STALO (*STA*ble *L*ocal *O*scillator) and the COHO (*COH*erent *O*scillator) interact to form all the required sinusoids. The STALO operates on a frequency which is the difference between the radar's illumination frequency and the superheterodyne receiver's intermediate frequency (IF). The COHO operates at the intermediate frequency. In multiple-conversion superheterodyne systems, there is more than one local oscillator and the COHO operates at the frequency of the last IF (Ch. 10). In some implementations, a single frequency standard at frequency f_0 is used to synthesize the COHO and STALO.

The single-sideband modulator combines the STALO (at frequency f_S) and COHO (f_C) sinusoids to produce a third sinusoid which is at the sum of the STALO and COHO frequencies. This signal is pulsed and amplified and becomes the illumination wave (f_T).

The target echo frequency is that of the illumination plus the Doppler shift ($f_R = f_T + f_d$). This signal and the STALO are applied to a mixer, which produces a copy of the received echo, but is shifted down in frequency by the frequency of the STALO ($f_T + f_d - f_S$). Since

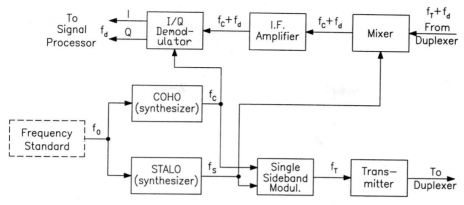

Figure 2-3. Frequency Generation in a Coherent Radar.

the transmit frequency is the sum of the STALO and COHO frequencies, the output of the mixer is $f_S + f_C + f_d - f_S$, or $f_C + f_d$. This signal is applied to the IF amplifier.

After processing in the IF amplifier, the echo signal is applied to the I/Q demodulator, where the COHO frequency is subtracted, leaving $f_C + f_d - f_C$, leaving a signal which changes at the Doppler shift (f_d) rate.

Table 2-1. Radar System Frequency Summary

Signal:	Symbol:	Equivalence:
STALO	f_S	
COHO	f_C	
Illumination	f_T	$f_S + f_C$ (usually)
Doppler shift	f_d	$f_R - f_T$
Target echo received	f_R	$f_T + f_d$
Echo out of receiver mixer	$f_I = f_C + f_d$	$f_T + f_d - f_S$
Echo out of I/Q demodulator	f_d	$f_C + f_d - f_C$

An alternative method of frequency generation is presented in Fig. 2-4. Here, a single frequency standard (the system master oscillator, or SMO) feeds three synthesizers which generate the COHO, STALO, and transmit frequencies. Although not discrete oscillators, the names COHO and STALO apply. The three signals are phase-locked through the SMO.

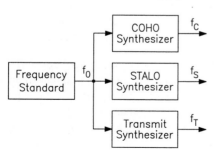

Figure 2-4. Alternate Frequency Generation Method.

2-2. Frequency Generation

Example 2-1: A radar operates with a STALO at 5.535 GHz and a COHO at 320 MHz. The radar tracks a target which is moving radially outbound at 14,733 smph. Assume that all the frequencies given are exact, the single sideband modulator produces the upper sideband (the sum frequency), that the receive mixer produces a frequency which is the difference between its two inputs, and that the velocity of propagation is 300,000,000 m/s. Find the following frequencies:

    The transmit (illumination) frequency

    The received echo's frequency

    The frequency out of the RF signal processor

    The frequency out of the mixer

    The frequency out of the IF amplifier

    The rate of change of the signal out of the I/Q demodulator

In addition, find a second receiver input frequency which will produce *exactly* the same output from the I/Q demodulator.

Solution: Much of the information needed for a solution is found in Table 2-1. The transmit frequency is the sum of the STALO and the COHO frequencies, 5.855,000,000 GHz (assumed exact)

The radial velocity is 6586.2 m/s, and the Doppler shift (Eq. 1-3) is 257,080 Hz. Since the target is outbound, the Doppler shift is negative and the frequency received is less than that transmitted. The received frequency is 5.854,742,920 GHz.

The RF signal processor does not change the frequency of the signal upon which it acts. Therefore it output frequency is 5.854,742,920 GHz.

The frequency of the mixer output is the difference between its inputs from the RF signal processor and from the STALO, in that order. The mixer output is therefore at 319.742,920 MHz.

The IF amplifier does not change the signal's frequency, and its output is at 319.742,920 MHz.

The output of the I/Q demodulator changes at the difference frequency between the IF amplifier output and the COHO. This frequency is $-257,080$ Hz, or the Doppler shift.

Visualize an interfering signal whose frequency is the transmit frequency less twice the COHO frequency plus the Doppler shift (as though from an inbound target). In this example, its frequency would be 5.215,257,080 GHz. It will produce exactly the same I/Q output as the target in our example. It is called the image frequency and can be a serious source of interference in some radars, particularly as a jamming signal. It will be further explored in Ch. 10.

Frequency generation in coherent-on-receive radars: Some radars have transmitters which are power oscillators (magnetrons). In these systems, the transmit frequency and phase are determined by the magnetron rather than by the STALO and COHO. If this type radar is to recover Doppler information, its COHO must, like any other coherent system, be phase-locked to the illumination. Instead of having the COHO/STALO determine the illumination frequency and phase, the STALO frequency is set by the illumination and

COHO's frequencies, and the COHO phase is locked to each transmitted pulse. The frequency relationships in Table 2-1 hold, with the understanding that the illumination frequency is free-running and the STALO and COHO are derived from it.

Figure 2-5 shows the difference between the two systems. In Fig. 2-5a, the transmitter is coherent, transmitting on the same phase with each pulse. The echo is from a target whose motion causes a phase change of +45° for each pulse. If the sample rate (PRF) is high enough for unambiguous Doppler measurement (a pulse Doppler system/mode), the target is closing.

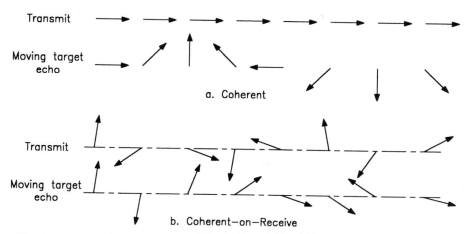

Figure 2-5. Phase Relationships in Coherent and Coherent-On-Receive Systems.

Figure 2-5b gives the same phasors for a coherent-on-receive system. The transmitted phase is now random, and the receive echo's phase appears so also. A close inspection shows that the relationship between the echo and illumination phases is exactly as with the coherent system. It is the *difference* between the echo and illumination phases that determines the Doppler shift.

Figure 2-6 shows the frequency generation in a coherent-on-receive radar. There are now two internal oscillators, both locked to the transmitter. They are, as in coherent systems, the STALO, which functions as the receiver's local oscillator, and the COHO, which runs at the receiver's intermediate frequency and forms the demodulation reference. The STALO is servoed by the automatic frequency control (AFC) circuit so that its frequency is exactly the difference between those of the transmitter and COHO. The COHO is phase-locked to each illumination pulse in the COHO lock circuit (Ch. 10).

The net result is coherence, but with limitations. The process of locking the COHO and STALO produces less accurate phases than using very stable oscillators. Also, the transmitter generates each pulse on a different unpredictable phase, and once the transmitter fires, the COHO is relocked and the phase of the previous pulse is lost. This precludes extracting Doppler from echoes in any PRF class other than low. This is a crucial limitation in modern systems.

2-2. Frequency Generation

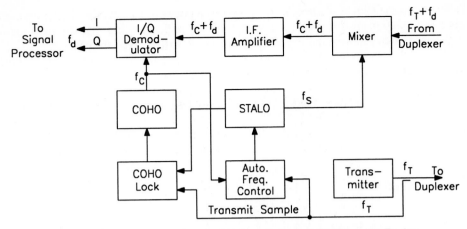

Figure 2-6. Frequency Generation in a Coherent-On-Receive Radar.

Frequency generation in non-coherent radars: Radars which do not recover Doppler have no need for phase locking throughout the system. They usually have power oscillator transmitters and may or may not have the receiver local oscillator locked to the transmitter by AFC. They have no COHO. Figure 2-7 is a block diagram of non-coherent frequency generation.

Important variation: In the previous discussions of frequency generation, it was assumed that the transmit frequency was the sum of the STALO and COHO frequencies, and that the receiver local oscillator (STALO) frequency was below that of the received echoes. There is no fundamental reason that the local oscillator frequency cannot be greater than that of the echo, as long as the difference between the illumination and STALO frequencies is the COHO frequency. The illumination frequency would now be the *difference* between those of the STALO and COHO. The one processing difference is that the Doppler shift is inverted; negative frequencies out of the I/Q demodulator are from closing targets (as an exercise, work

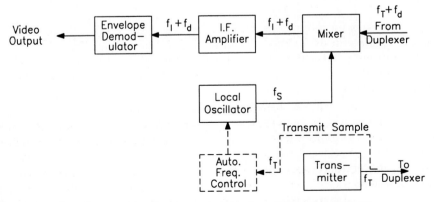

Figure 2-7. Frequency Generation in a Non-Coherent Radar.

76   Chap. 2. The Radar System

this out from Table 2-1). As long as this difference is accounted for, either scheme works equally well.

2-3. Transmitter Functions and Parameters

The radar's transmitter generates the RF (radio frequency) power which illuminates the targets. Transmitters are either coherent, quasi-coherent, or non-coherent. Figure 2-8a is a generic block diagram of a coherent transmitter. Figure 2-8b shows generic coherent-on-receive and non-coherent transmitters. The elements of the transmitter are described below.

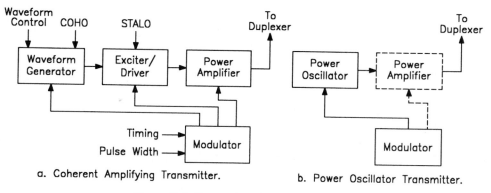

Figure 2-8. Transmitter Block Diagrams.

— *Waveform generator:* The waveform to be transmitted is generated in this block, often at the COHO frequency. Pulse width, pulse position, and any RF modulation are set in this block. The waveform generator is not present in this form in most quasi-coherent and non-coherent transmitters.

— *Exciter/Driver:* In this block, the waveform produced by the waveform generator is translated to the radar's illumination frequency and amplified to a level usable by the final power amplifier. If the illumination frequency is varied from pulse to pulse, look to look, or in some other fashion (frequency agility), the variation is applied in this block. This block also is not present in most coherent-on-receive and non-coherent transmitters.

— *Final power amplifier (FPA):* The illumination is amplified to its final level in this block. This block may not exist in oscillator transmitters, although some have amplifiers to boost the magnetron's power.

— *Modulator:* The modulator controls the transmitter in pulse radars, switching it on during transmission and off during the listening period. Modulation can be either low-level, furnishing a switching voltage to control elements in the transmitting devices, or high-level, furnishing the DC operating voltage and current for the transmitting device. Since transmitting devices often require tens of thousands of volts at tens of amperes current over a period of microseconds, high level modulators can be large and complex. The trend is toward using transmit tubes which have control electrodes, allowing low-level modulation.

Transmit parameters: Waveform and transmit parameters determine the energy with which targets are illuminated and the bandwidth of the illuminating waves. A typical transmitted waveform is very nearly rectangular in envelope shape, as shown in Fig. 2-9. Most views of signals in this book show an envelope, which is the shape of the wave with the sinusoidal cycles removed. Transmit parameters are given below and waveform parameters are given in Sec. 2-4.

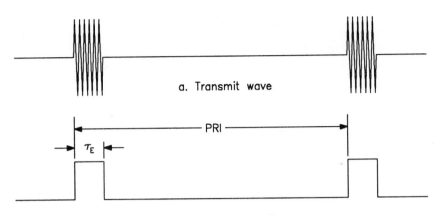

a. Transmit wave

b. Envelope of transmit wave

Figure 2-9. Transmit Waveform Envelope, Typical.

— *Pulse width*: Pulse width is the length of time the illuminating power is ON for each transmission. Radar transmitting amplifiers usually operate in saturation and most systems emit pulses which can be approximated as rectangular. In systems without pulse compression, the width of the echo pulses out of the signal processor is only slightly different from the transmitted pulse width. With pulse compression, the two pulse widths are markedly different, with the processed pulse width being less than that transmitted. Several pulse widths can occur in the system. They are:

τ = the pulse width, in seconds, for both the transmitted waveform and the processed echo. It is used in systems without pulse compression.

τ_E = the illumination (transmitted) waveform pulse width, in seconds, for systems having pulse compression. The subscript E is for *expanded*.

τ_C = the processed echo pulse width, in seconds, for systems with pulse compression. The subscript C indicates *compressed*.

The amount of pulse compression, if present, is described by the parameter compression ratio (CR), which is the ratio of the widths of the expanded and compressed pulses.

$$CR = \tau_E / \tau_C \qquad (2\text{-}1)$$

— *Pulse repetition frequency (PRF)*: The PRF is the number of illumination pulses transmitted per second. In some systems and modes, the PRF is a constant; in others it varies (Fig. 2-10). This PRF agility can be pulse-to-pulse, look-to-look, or scan-to-scan. Some systems use burst waveforms, where groups of pulses are transmitted. The rate within and between the groups may change.

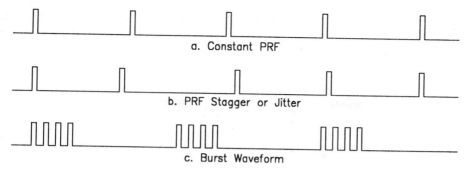

Figure 2-10. PRF Types.

PRF agility has several names and purposes. In moving target indicator (MTI) systems, the PRF is varied to prevent sampling at integer multiples of the Doppler shift. If this occurs, the radar senses a Doppler shift of zero and is blind to targets moving at these velocities (called blind velocities). PRF variation to prevent blind velocities is called PRF stagger.

PRF agility is also used to prevent jammers from locking to the radar's pulse rate, which allows some types of deceptive jamming not otherwise possible. Agility for this purpose is PRF jitter.

Range and Doppler ambiguities are often resolved (deghosted) with a variable PRF. This is also called jitter.

— *Pulse repetition interval (PRI)*: The PRI is the time between the start of consecutive transmissions and is the reciprocal of PRF. It is also called pulse period (T), pulse repetition time (PRT), and ranging interval.

$$PRI = 1 / PRF \text{ (seconds)} \tag{2-2}$$

— *Peak power (P_T)*: Peak power is the RMS signal power from the transmitter during the time the pulse is on.

— *Duty cycle (dc)*: The fraction of the total time that the transmitter is on is the duty cycle (dc). For fixed PRFs and rectangular envelope pulses, duty cycle is

$$dc = \tau_E / PRI \text{ (dimensionless)} \tag{2-3}$$

$$dc = \tau_E \, PRF \tag{2-4}$$

$$dc = \tau_E \, PRF_{AVG} \tag{2-5}$$

$$PRF_{AVG} = \frac{N_{PRF}}{(1/PRF_1) + (1/PRF_2) + \ldots + (1/PRF_N)} \tag{2-6}$$

For variable PRFs, the duty cycle relationship is:

$$dc = \frac{\text{Total transmitter ON time per PRF cycle}}{\text{Total time in one PRF cycle}} \tag{2-7}$$

$$dc = \frac{\tau_{E1} + \tau_{E2} + \ldots + \tau_{EN}}{(1/PRF_1) + (1/PRF_2) + \ldots + (1/PRF_N)} \tag{2-8}$$

2-3. Transmitter Functions and Parameters

The duty cycle correction factor (dccf) is the duty cycle expressed in negative decibels. It is the quantity which is added to an average power measurement (as most power meters measure average power) in dBm to find the peak power (in dBm).

$$dccf = -10 \log (dc) \qquad (2\text{-}9)$$

For pulse envelopes other than approximately rectangular, rarely used in radars, the duty cycle is calculated from the integral of the envelope over one pulse cycle.

— *Average power*: The power which a radar transmitter emits averaged over a long time is called average power. This power is, in most cases, more important for the detection of targets than peak power. Average power is, for pulses with rectangular envelopes, the peak power times the duty cycle.

$$P_{AVG} = P_T \, dc \text{ (Watts)} \qquad (2\text{-}10)$$
$$P_{AVG} = P_T \tau_E / PRI \qquad (2\text{-}11)$$
$$P_{AVG} = P_T \tau_E \, PRF \qquad (2\text{-}12)$$

If the pulse envelope is other than rectangular, which seldom occurs in radars, an integral must be used to calculate average power.

$$P_{AVG} = P_T / T_C \int_{T_0}^{T_0 + T_C} p(t) \, dt \qquad (2\text{-}13)$$

$p(t)$ = the power as a function of time
T_0 = the start of one PRF cycle
T_C = the length of one PRF cycle

— *Pulse energy*: Pulse energy is the energy in each transmitted waveform. It is peak power times expanded pulse width, or average power times PRI.

$$W_P = P_T \tau_E \text{ (Watt-seconds = Joules)} \qquad (2\text{-}14)$$
$$W_P = P_{AVG} \, PRI \qquad (2\text{-}15)$$
$$W_P = P_{AVG} / PRF \qquad (2\text{-}16)$$

— *Look energy*: The total energy emitted by the transmitter during any one signal processing data-gathering period (a look or dwell) is look (or dwell) energy. It is the pulse energy times the number of pulses in a look.

$$W_L = W_P N_L \text{ (Joules)} \qquad (2\text{-}17)$$
$$W_L = P_T \tau_E N_L \qquad (2\text{-}18)$$
$$W_L = P_{AVG} N_L / PRF \qquad (2\text{-}19)$$
$$W_L = P_{AVG} T_L \qquad (2\text{-}20)$$

N_L = the number of pulses in a look
T_L = the total time in one look (the dwell time T_D)

Of all the transmit power and energy parameters, look energy is in most cases the most important for determining whether or not targets will be detected.

— *Efficiency*: The efficiency of a transmitter is the ratio of the average RF power out of the transmitter to the total average input power (DC, AC, and RF drive).

$$\eta_T = \frac{P_{out(RF)}}{P_{in(TOTAL)}} \quad (2\text{-}21)$$

η_T = the transmitter efficiency
$P_{out(RF)}$ = the average RF power out of the transmitter
$P_{in(TOTAL)}$ = the total power into the transmitter (DC power supplies, AC power supplies, and RF input)

— *Transmitter stability*: For good Doppler performance, the transmitter must generate spectrally pure waves, having only those modulations intended. Spurious modulations result in unaccounted-for sidebands which may cause stationary targets to be interpreted as moving. Section 2-8 has the details.

Example 2-2: A transmitter emits a signal of 500 kW peak power. The pulse width is 1.5 μs, and the PRF is 1500 pps. Signal processing in this radar sums 16 consecutive target hits. The total DC and AC power supplies deliver is 3,200 W to the transmitter. The RF drive power is 1 W. Find:

Pulse repetition interval (pulse period)
Duty cycle
Duty cycle correction factor
Average power
Pulse energy
Look energy
Efficiency

Solution: The PRI is the reciprocal of the PRF (Eq. 2-2), and is 0.000667 s (667 μs). The duty cycle (Eq. 2-4) is 0.00225. The duty cycle correction factor (Eq. 2-9) is 26.5 dB. From the duty cycle and peak power, the average power is found (from Eq. 2-10) to be 1,125 W. The energy in each transmitted pulse is, from Eq. 2-14, 0.75 J. The look energy is the pulse energy times the number of pulses in a look (in this case 16), or 12 J (Eq. 2-17). The efficiency (Eq. 2-21) is 1125 W RF out divided by 3199 W total in, or 0.352 (35.2%).

2-4. Waveform Spectra and Bandwidths

Many different illumination waveform types are used in radar systems, depending on the radar's mission and on whether or not pulse compression is present. Pulse compression waveform design is predicated on simultaneously achieving wide pulse width for detection and wide bandwidth for range resolution. The spectrum of a waveform is thus a critical parameter.

The power spectrum of any waveform is the Fourier transform of its autocorrelation function. The waveform's autocorrelation function determines its ability to resolve in range (Ch. 13). Narrow autocorrelations, corresponding to wide bandwidths, are necessary for good range resolution. Bandwidth and autocorrelation functions are determined by the modulation present on the sinusoidal wave within the pulse.

The remainder of this section introduces commonly used radar waveforms, the modulations used to form them, and their spectra. A waveform evaluation method called the ambiguity function will be introduced in Ch. 13.

A generic time-domain description of a pulsed illumination waveform is given in Eq. 2-22. Using this relationship, the various types of waves differ only in the form of the phase function $\phi_T(t)$. In almost all cases, the absolute transmit phase is by definition zero and is locked to the system phase reference (the COHO). The only exception occurs in monopulse tracking (Ch. 7).

$$v_{TX}(t) = [u(0) - u(\tau)] A_T \cos[\phi_T(t) + \phi_{T0}] \text{ (volts)} \quad (2\text{-}22)$$

$u(0)$ = a unit step occurring at time zero. $u(t)$ is a unit step occurring at time τ_E. Their difference is a rectangular pulse starting at zero time and having the width of the transmitted pulse τ_E.

$v_{TX}(t)$ = the time domain description of the transmit waveform

A_T = the peak amplitude of the wave (volts)

$\phi_T(t)$ = the phase of the transmit waveform as a function of time

ϕ_{T0} = the absolute phase of the wave (usually defined as zero)

Gated CW pulse waveform: A commonly used radar waveform is gated CW, which is a sinusoid gated to a rectangular time envelope in the time domain. Its frequency is constant across the pulse (Fig. 2-11). The time domain wave is shown in Fig. 2-12a, and its calculated spectrum is given in Fig 2-12b. The wave is described by Eq. 2-22, with the phase term set to create a constant frequency f_0.

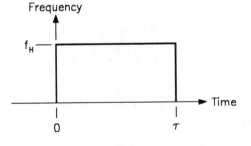

Figure 2-11. Gated CW Time versus Frequency.

$$f(t) = f_0 \text{ (cycles per second)} \quad (2\text{-}23)$$

$f(t)$ = the frequency of the wave as a function of time

f_0 = the frequency of the sinusoid in the pulse

Phase is the time integral of radian frequency.

$$\phi(t) = \int \omega(t)\, dt \text{ (radians)} \quad (2\text{-}24)$$

$$\omega(t) = 2\pi f(t) \text{ (radians per second)} \quad (2\text{-}25)$$

$\omega(t)$ = the radian frequency as a function of time

a. Time domain

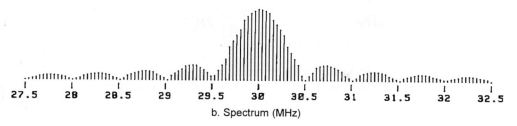

b. Spectrum (MHz)

Figure 2-12. Gated CW Wave and Spectrum.
Center frequency = 30 MHz, pulse width = 2.0 μs

Integrating Eq. 2-24 gives the phase of the gated CW wave as a function of time, which is applied to Eq. 2-22 to generate the wave.

$$\phi_{GCW}(t) = 2\pi f_0 t \text{ (radians)} \tag{2-26}$$

$\phi_{GCW}(t)$ = the phase term in Eq. 2-22 for gated CW
t = the time since the pulse began (seconds)

With gated CW, bandwidth is developed strictly from pulse width, requiring a compromise between detection and range resolution. The spectrum of gated CW is related to the time domain waveform parameters as follows:

− The envelope of the spectrum is a sinc function related to pulse width. A sinc function is of the form

$$\text{Sinc}(X) = \text{Sin}(\pi X) / (\pi X) \tag{2-27}$$

$$E(f) = \text{Sin}[\pi(f-f_0)\tau] / [\pi(f-f_0)\tau]. \tag{2-28}$$

$E(f)$ = the envelope of the spectrum as a function of frequency
f = the frequency (independent spectrum variable) (Hertz)
f_0 = the center frequency (Hertz or seconds^{-1})
τ = the pulse width (seconds)

− The center of the spectrum (f_0) is the frequency of the sinusoid being gated.
− The spectral nulls are located at the center frequency plus and minus integer multiples of the pulse width's reciprocal. The main spectral lobe is 2/τ wide.
− The bandwidth of the matched filter (Sec. 2-6) is approximately the reciprocal of the pulse width (about half the width of the main spectral lobe).

$$B \approx 1/\tau \text{ (Hertz)} \tag{2-29}$$

$$\tau B \approx 1 \text{ (dimensionless)} \tag{2-30}$$

B = the "matched bandwidth" of the signal (approximately the 3-dB bandwidth)

2-4. Waveform Spectra and Bandwidths

- The periodicity of the pulse train causes the spectrum to be composed of lines, each line representing an individual sinusoid (called a line spectrum). The spectral lines are the PRF apart.
- The peak amplitude of the spectrum is proportional to the area under the time domain wave's envelope, and the peak amplitude of the time domain wave is proportional to the area under the spectrum. This phenomenon of spectrum analysis makes possible the later (Ch. 5) explanation of an interesting signal detection technique, called Dicke-fix, for discriminating between target echoes and certain types of interference.

Linear frequency modulation (LFM): Another common radar waveform is linear FM, where the frequency is swept across the pulse. It is so named because the frequency versus time characteristic of the transmitted pulse is a line (Fig. 2-13). The LFM wave is also known as a chirp waveform, and the particular one shown is down-chirp, since its frequency changes in a downward direction. If the frequency sweep were from low to high, it would be known as up-chirp. This and other modulated waveforms achieve their bandwidths through the modulation, not pulse width. They can have wide bandwidths and wide pulse widths simultaneously.

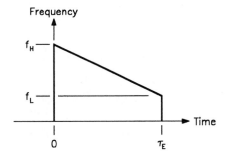

Figure 2-13. Linear FM Waveform Time / Frequency Characteristic.

The LFM wave is described in time by Eq. 2-22 with the phase term derived as follows. The frequency across the pulse is a linear function of time; thus its time derivative is a constant.

$$df_{LFM}(t)/dt = (f_B - f_F)/\tau_E \qquad (2\text{-}31)$$

$f_{LMF}(t)$ = the instantaneous frequency across the pulse as a function of time

d/dt = the time derivative

f_B = the frequency at the beginning of the pulse

f_F = the frequency at the end (fall) of the pulse

τ_E = the expanded (transmitted) pulse width

If the pulse is down-chirp, as shown in Figs. 2-13 and 2-14, the beginning frequency is the high frequency (f_H) and the ending frequency is the low (f_L). The instantaneous frequency across the pulse is the integral of Eq. 2-31, with a constant of integration equal to the beginning frequency.

$$f_{LFM}(t) = (f_F - f_B)(t/\tau_E) + f_B \tag{2-32}$$

By Eqs. 2-24 and 2-25, the phase if the LFM wave is found to be

$$\phi_{LFM}(t) = \pi f_{LFM}(t)\, t^2 + 2\pi f_B\, t + \phi_{T0} \tag{2-33}$$

The LFM wave and its calculated spectrum are shown in Fig. 2-14. This waveform's spectrum has the following characteristics:

a. Time domain

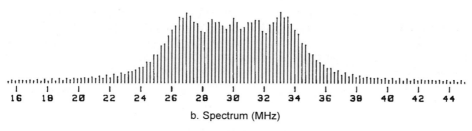

b. Spectrum (MHz)

Figure 2-14. Linear FM Wave and Spectrum.
Start frequency = 35.0 MHz, stop frequency = 25.0 MHz, pulse width = 2.0 µs

- The envelope of the spectrum is more or less rectangular. The wider the pulse width / bandwidth product, the more it approaches a rectangle.
- The center of the spectrum (f_0) is the mean of the high and low frequencies.

$$f_0 = (f_H + f_L)/2 \tag{2-34}$$

f_H = the high frequency in the sweep across the pulse
f_L = the low pulse sweep frequency

- The bandwidth of the matched filter is independent of the expanded pulse width and is related only to the sweep bandwidth. The compressed pulse width is determined by the bandwidth.

$$\tau_E B \gg 1 \tag{2-35}$$

$$\tau_C B \approx 1 \tag{2-36}$$

$$B \approx f_H - f_L \tag{2-37}$$

$$\tau_C \approx 1/B \tag{2-38}$$

- The spectral lines are the PRF apart.

Non-linear frequency modulation (NLFM): The non-linear FM waveform consists of a frequency sweep across the transmitted pulse wherein frequency is not a linear function of time. There are many different NLFM waveforms, each with its own frequency versus time relationship and spectrum. A typical frequency/time characteristic is shown in

2-4. Waveform Spectra and Bandwidths

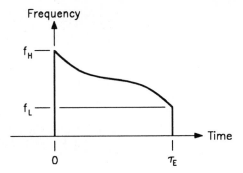

Figure 2-15. Non-Linear FM Waveform (Quadratic).

This shape is a non-symmetrical quadratic — a parabola split at its vertex and one-half inverted. The wave shown, described in Eq. 2-39, down-chirps. The first term of the equation is the first half of the pulse and the second term is the remainder of the pulse.

$$f_{NLQ}(t) = + \{[u(-\tau_E/2) - u(0)] \frac{4(f_B - f_0)}{\tau_E^2} t^2 + f_0\}$$

$$+ \{[u(0) - u(+\tau_E/2)] \frac{4(f_F - f_0)}{\tau_E^2} t^2 + f_0\} \quad (2\text{-}39)$$

$f_{NLQ}(t)$ = the frequency of a non-linear quadratic wave as a function of time (Hertz)

$u(t)$ = a unit step whose value is zero from negative infinite time to time t, and unity thereafter (dimensionless)

τ_E = the expanded pulse width (seconds)

f_B = the starting frequency (Hertz)

f_0 = the frequency at the center of the pulse (Hertz)

f_F = the frequency at the finish of the pulse (Hertz)

t = time (seconds)

The relationship between the beginning, finish, and center frequencies for the above quadratic function is given in Eq. 2-40.

$$f_0 = (f_B - f_F) / 2 \quad (2\text{-}40)$$

An advantage of non-linear FM is that windowing is often not necessary in processing for pulse compression. See Ch. 11 for information on windowing and its effect on signal processing. Other functions besides quadratics are used in NLFM. The 40-dB Taylor function is popular (Ch. 13).

The spectrum of a non-linear FM signal is dependent on the frequency/time characteristic. The quadratic waveform described above and its calculated spectrum areis shown in Fig. 2-16. Its properties are listed below.

— The envelope of the spectrum is determined by the frequency-versus-time function of the waveform.

— The center of the spectrum (f_0) lies between the high and low frequencies, not necessarily in the center, determined by the frequency-versus-time function.

a. Time domain

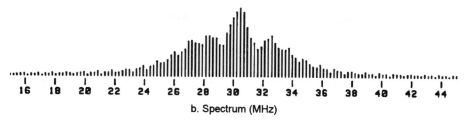

b. Spectrum (MHz)

Figure 2-16. Non-Linear Quadratic FM Wave and Spectrum.
Beginning frequency = 35 MHz, finish frequency = 25 MHz, pulse width = 2.0 μs

— The bandwidth of the matched filter is independent of the pulse width and is related to the sweep bandwidth and the frequency-versus-time function.

$$\tau_E B \gg 1 \tag{2-41}$$

$$B < f_H - f_L \tag{2-42}$$

$$\tau_C \approx 1/B \tag{2-43}$$

— The spectral lines are the PRF apart.

V-FM: The V-FM waveform consists of a frequency sweep in one direction across one-half of the transmitted pulse and a sweep in the other direction across the other half. The sweep is usually, but not necessarily, a linear function of time. The frequency-versus-time characteristic of a simple V-FM waveform is shown in Fig. 2-17a and a more complex function in Fig. 2-17b.

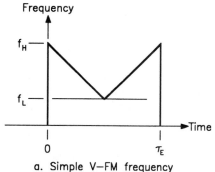
a. Simple V–FM frequency versus time function

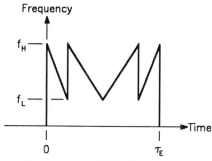
b. Complex V–FM frequency versus time function

Figure 2-17. VFM Frequency versus Time Functions.

2-4. Waveform Spectra and Bandwidths

The spectrum of the simple V-FM waveform of Fig. 2-17a is similar to the LFM wave of Fig. 2-14. The signal shown has the following spectral properties:

- The envelope of the spectrum is approximately rectangular if the sweeps are linear functions of time; otherwise the spectral envelope is sweep-dependent.
- The center of the spectrum (f_0) is the mean of the high and low frequencies if the sweeps are linear; otherwise, it is determined by the frequency *vs* time function.

$$f_0 = (f_H + f_L) / 2 \text{ (linear sweep)} \qquad (2\text{-}44)$$

- The bandwidth of the matched filter is independent of the pulse width and is related to the sweep bandwidth and the frequency-versus-time function.

$$\tau_E B \gg 1 \qquad (2\text{-}45)$$

$$B \approx f_H - f_L \text{ (linear sweep)} \qquad (2\text{-}46)$$

$$\tau_C \approx 1/B \qquad (2\text{-}47)$$

- The spectral lines are the PRF apart.

Phase-coded waveforms: The previously discussed waveforms are sometimes called the analog waveforms, since they are, in the main, processed by analog means. The digital waveforms usually consist of a pulse of a monotonic sinusoid which is divided into subpulses, with the phase of the sinusoid varied between subpulses. The phase variations are usually applied using biphase coding, where the subpulse is either 0° or 180° with respect to the reference (COHO). Biphase coding is preferred in many applications because for a given sequence, it produces the greatest bandwidth, desirable for range resolution.

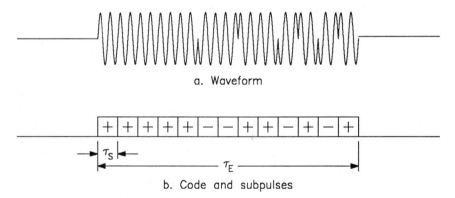

Figure 2-18. Phase-Code Waveform and Subpulses — 13-Bit Barker Sequence.

Figure 2-18a shows a wave biphase coded with a 13-bit biphase sequence. This particular one is a Barker sequence — see discussion following and Ch. 13.

The illumination pulse (τ_E) consists on a group of N_S subpulses, each having a width of τ_S. Figure 2-18b shows the relationship between the transmit pulse and its subpulses. In this figure, a "+" indicates a transmitted phase of 0°, while a "−" indicates 180°. The compressed pulse width is approximately the same as that of the subpulses (actually slightly wider).

$$\tau_C \approx \tau_S \text{ (seconds)} \qquad (2\text{-}48)$$

τ_C = the compressed pulse width

τ_S = the subpulse width (the width of one code segment)

Several classes of code sequences are commonly used in phase-coded waveforms. The Barker sequences form one group. They are unique in having very well-behaved autocorrelation functions (Ch. 13), allowing pulse compression to take place without windowing the signal. Windowing, discussed in Ch. 11, introduces an inherent information loss. There are two disadvantages to using Barker sequences. First, the longest known Barker sequence is thirteen bits, severely limiting the amount of pulse compression possible. The second is that there are only a limited number of Barker sequences and they are well known. This virtually precludes their use if signal security is a consideration.

Chapter 13 lists all the known Barker sequences. A wave coded with the 13-bit sequence and its spectrum are shown in Fig. 2-19.

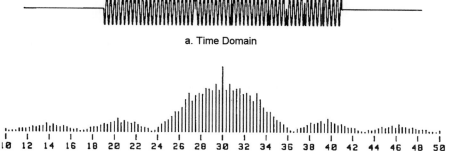

Figure 2-19. Barker-Coded Wave and Its Spectrum.
Center frequency = 30.0 MHz, pulse width = 2.0 μs

Another group of coded waveforms uses pseudorandom sequences, in which ones and zeros appear to be in random order, but the sequence can be reproduced. The advantage is that the codes can be as long as desired, limited only by transmit energy, signal bandwidth, and minimum range considerations. This type can be made secure, transmitting a different code on each pulse or group of pulses. The disadvantage is that processing them requires windowing, with the attendant reduction of the recoverable information in the signal.

An infinite number of pseudorandom sequences exist. Those with minimal responses outside the compressed pulse width (minimal range leakage — Ch. 13) are known as optimal binary codes. Several are listed in Ch. 13. Figure 2-20 shows a wave coded with one of the 222 optimal 35-bit pseudorandom sequences, with its spectrum. The spectrum of phase-coded waves is highly code-dependent.

Other code types are discussed in Ch. 13. The properties of phase-coded waveforms are:

— The envelope of the spectrum is code-dependent.
— The center frequency (f_0) is the frequency of the coded sinusoid.
— The bandwidth of the matched filter is approximately the reciprocal of the subpulse width and is independent of the expanded pulse width.

2-4. Waveform Spectra and Bandwidths

$$\tau_E B \gg 1 \qquad (2\text{-}49)$$
$$\tau_C B \approx 1 \qquad (2\text{-}50)$$
$$B \approx N_S / \tau_E \qquad (2\text{-}51)$$

N_S = the number of code bits in the sequence

– The spectral lines are the PRF apart.

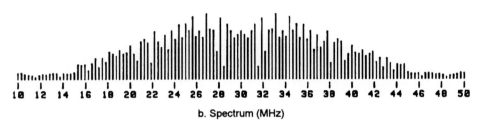

Figure 2-20. 35-Bit Optimal Pseudorandom Sequence and Its Spectrum. Center frequency = 30.0 MHz, pulse width = 2.0 μs

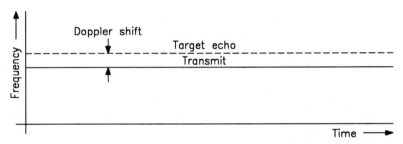

Figure 2-21. Unmodulated CW Wave.

Continuous wave (CW): The most basic continuous wave (CW) is monotonic, unmodulated CW. It functions well in recovering Doppler shifts, but since it has no time-dependent wave variation, range cannot be found directly. Figure 2-21 shows the frequency/time characteristic of an unmodulated CW wave.

To find range, the CW wave must be modulated by some time-variant signal. The wave is usually frequency modulated (FM). Some systems use sinusoidally frequency modulated CW (FMCW), shown in Fig. 2-22. Others use triangularly modulated FMCW (Fig. 2-23), and yet others phase code the CW. Recovery of Doppler and range from CW echoes is covered in Ch. 6.

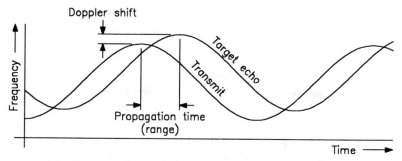

Figure 2-22. Sinusoidally Frequency Modulated CW Wave.

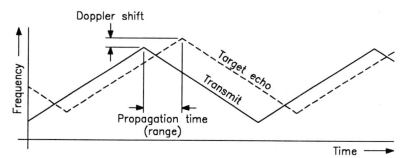

Figure 2-23. Triangularly Frequency Modulated CW Waveform.

The spectrum of CW waves depends on the type and amount of modulation. An unmodulated (non-ranging) sinusoidal CW signal's spectrum ideally is a single frequency, shown in Fig. 2-24. In practice, there are inadvertent small modulations on the wave, which degrade the Doppler performance of the radar. These instabilities are discussed in Sec. 2-8.

If the transmit signal is frequency modulated for ranging, the spectrum is dependent on the type of modulation (sinusoidal, triangular, or other), the modulation frequency, and the amount of carrier deviation caused by the modulation. Sinusoidal modulation is described in the following equations.

$$v(t) = A\ \text{Sin}[2\pi f_C t + m_F \text{Sin}(2\pi f_m t)] \tag{2-52}$$

$$m_F = \delta f_C / f_M \tag{2-53}$$

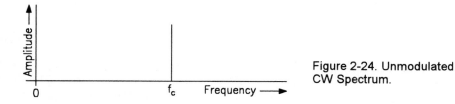

Figure 2-24. Unmodulated CW Spectrum.

2-4. Waveform Spectra and Bandwidths

$v(t)$ = the waveform voltage as a function of time
A = the peak amplitude of the modulated wave
f_C = the carrier frequency
m_F = the frequency modulation index
f_M = the frequency of the sinusoidal modulation
δf_C = the carrier deviation, the peak frequency change from center caused by the modulation

The spectrum consists of a carrier (the center frequency) and sideband pairs. The sidebands are separated from the carrier by multiples of the modulation frequency. The factors $J_0, J_1, \ldots, J_n$ are Bessel functions. J_0 is the zero-order Bessel function of the first kind, J_1 is the first order, J_2 is the second order, and so on. The spectrum of a sinusoidally modulated carrier is given as

$$\begin{aligned} v(f) = \ & A\, J_0(m_F)\, \mathrm{Sin}(2\pi f_C t)\ + \\ & A\, J_1(m_F)\, \{\mathrm{Sin}[2\pi(f_C + f_M)t] + \mathrm{Sin}[2\pi(f_C - f_M)t]\}\ + \\ & A\, J_2(m_F)\, \{\mathrm{Sin}[2\pi(f_C + 2f_M)t] + \mathrm{Sin}[2\pi(f_C - 2f_M)t]\}\ + \\ & A\, J_3(m_F)\, \{\mathrm{Sin}[2\pi(f_C + 3f_M)t] + \mathrm{Sin}[2\pi(f_C - 3f_M)t]\}\ + \\ & \ldots \end{aligned} \qquad (2\text{-}54)$$

$J_n(m_F)$ = the Bessel function of the first kind, nth order, with argument m_F

AJ_0 is the amplitude of the carrier. AJ_1 is the value of the first pair of sidebands, separated from the carrier by the modulation frequency. AJ_2 is the amplitude of the second sideband pair, separated from the carrier by twice the modulation frequency, and so on. The appropriate Bessel functions are charted in Fig. 2-25. Their definitions can be found in any good mathematical handbook. See Ex. 2-8 and Fig. 2-26 for a sample spectrum.

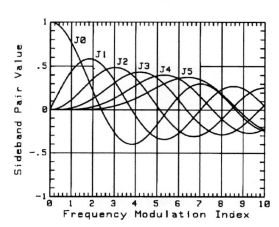

Figure 2-25. Bessel Functions of the First Kind (Order 0→5).

The approximate bandwidth of a sinusoidally modulated CW wave is

$$B \approx 2(m_F + 1) f_M \tag{2-55}$$

The spectrum of a triangular FM CW wave is found in the same way, except the modulation is more complex. A Fourier series expansion of a triangle wave shows that it is made up of an infinite number of harmonically related sinusoids. A periodic triangle wave with no DC term is given in the Eq. 2-56, in terms of its sinusoidal components. As shown, it contains sinusoids of the modulation frequency and odd harmonics (3 times the modulation frequency, 5 times the modulation frequency, and so forth).

$$\begin{aligned} v_M(t) = &\ (8 f_M / \pi^2) [\cos(2\pi f_M t) + 3^{-2} \cos(6\pi f_M t) \\ &+ 5^{-2} \cos(10\pi f_M t) + 7^{-2} \cos(14\pi f_M t) + \ldots] \end{aligned} \tag{2-56}$$

$v_M(t)$ = the modulation (triangular) voltage as a function of time

Each of the components of the modulation voltage produces it own sideband set according to its amplitude and frequency, found from the first-kind Bessel functions. The details of finding a triangular FM spectrum are left as an exercise.

Example 2-3: A gated CW waveform has a center frequency of exactly 1.250 GHz, a pulse width of 5.0 μs, and a PRF of 400 pps. Find the matched bandwidth, the spacing of the spectral lines, and describe the spectral envelope.

Solution: The bandwidth of this wave is related to the pulse width. The main spectral lobe extends from the center frequency minus the pulse width reciprocal (1,250,000 kHz − 200 kHz = 1,249,800 kHz) to the center frequency plus the reciprocal of the pulse width (1,250,200 kHz). The matched bandwidth is approximately half the width of the main spectral lobe, or 200 kHz (Eq. 2-29).

The spectral lines are separated by the PRF, or 400 Hz. The five spectral lines closest to the center of the spectrum have frequencies of, respectively, 1,249,999.2 kHz (center frequency − 2 PRF), 1,249,999.6 kHz (center − PRF), 1,250,000 kHz (center frequency), 1,250,000.4 kHz (center plus PRF), and 1,250,000.8 kHz (center plus 2 PRF).

The envelope of the spectrum is a sinc function (Eq. 2-28):

Sin[$\pi(f - 1{,}250{,}000{,}000)$ 0.000,005] / [$\pi(f - 1{,}250{,}000{,}000)$ 0.000,005]

Example 2-4: The waveform of the Ex. 2-3 is modified so that there is a linear frequency sweep across the pulse. At the start of the pulse, the frequency is 1,247,500,000 Hz and at the end it is 1,251,500,000 Hz. Find the new center frequency, matched bandwidth, spectral line spacing, and spectral envelope.

Solution: In the LFM waveform, the center frequency is the arithmetic mean of the start and stop frequencies (Eq. 2-34), or 1,249,500,000. This is the frequency which is used to calculate target Doppler shifts.

The matched bandwidth is now independent of pulse width and is approximately the difference between the high and low frequencies (Eq. 2-37), or 4.0 MHz.

The spectral line spacing is the PRF, as it is with any periodic waveform, and the spectral envelope is approximately rectangular.

2-4. Waveform Spectra and Bandwidths

Example 2-5: A phase-coded waveform has a center frequency of 10.324 GHz, a pulse width of 9.3 μs, and has 31 subpulses in the code sequence. The PRF is 2,400 pps. Assume that the phases are 0° and 180° and that the code is a proper one (some sequences are unsuitable - 10101010101 is an example). Find the subpulse width, the matched bandwidth, and the spectral line spacing.

Solution: The subpulses must total the transmitted pulse width, and are therefore 0.30 μs wide (9.3 μs divided by 31).

The matched bandwidth is the reciprocal of the *subpulse* width (Eq. 2-50), or 3.33 MHz.

The spectral line spacing, again is the PRF, or 2,400 Hz.

Example 2-6: A CW wave is modulated with a 100-Hz sinusoid such that the 10.0 GHz carrier is deviated ± 400 Hz from its center frequency. Assuming unity amplitude, find the spectrum of the wave, its bandwidth, and its range-resolving capability.

Solution: From Eq. 2-53, the modulation index is 4.0. Figure 2-25 or standard compilations of Bessel functions give $J_0(4.0) = -0.40$, $J_1(4.0) = -0.07$, $J_2(4.0) = +0.36$, $J_3(4.0) = +0.43$, $J_4(4.0) = +0.28$, $J_5(4.0) = +0.13$, $J_6(4.0) = +0.05$, $J_7(4.0) = +0.02$. These data indicate that the voltage of the carrier is -0.40 units. The negative indicates that its phase is 180° from unmodulated carrier. The first sideband pair, at ± 100 Hz from the carrier, has an amplitude of -0.07; the second sideband pair's (± 200 Hz) amplitude is +0.36; the third (± 300 Hz) is +0.43, and so forth. The magnitude of the spectrum is plotted as Fig. 2-26.

The bandwidth, from Eq. 2-55, is approximately 1000 Hz, giving a range resolving capability (Eq. 1-24) of about 150,000 meters.

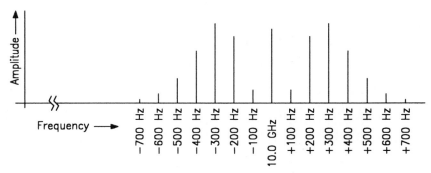

Figure 2-26. Spectrum of Example 2-6.

2-5. Antenna Principles, Functions, and Parameters

Antennas have several functions, depicted in Fig. 2-27.
- Act as a *transducer* and *impedance match* between the transmitter and the propagation medium, and between the medium and the receiver
- Provide *gain* to concentrate the transmitted signal in a preferred direction (Fig. 2-27ab)
- *Steer* the transmitted power to the desired angular position
- Provide *area* (*aperture*) to intercept and capture received echoes (Fig. 2-27c)
- Provide for effective reception over a small angular direction only and to move the response to the desired direction (*steer* the receive beam)

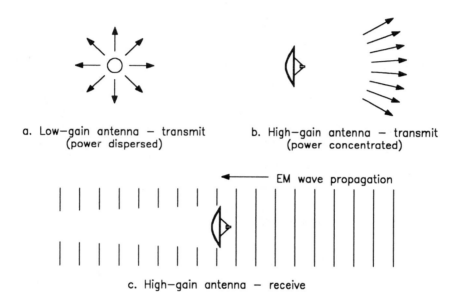

a. Low–gain antenna – transmit (power dispersed)

b. High–gain antenna – transmit (power concentrated)

c. High–gain antenna – receive

Figure 2-27. Antenna Functions.

This section discusses the antenna parameters from the standpoint of the radar system. Further details about antennas and arrays and their performance are found in Ch. 9. The parameters of importance covered in this section are:
- Radiation pattern
- Directivity and radiation efficiency
- Beamwidth and length efficiency
- Aperture: effective area and aperture efficiency
- Gain
- Efficiency
- Sidelobe definition and effects
- Field zones
- Polarization

Antenna radiation pattern: An antenna's radiation pattern is its response as a function of the angles from its principal axes. Antennas are normally treated in spherical coordinates. It will be demonstrated later that most practical radar operations are at sufficient distance from the antenna that the pattern is independent of range. This is one definition of the far field, discussed later in this section.

Two sample patterns are shown in Figs. 2-28 and 2-29. In each case, the radial axis from the antenna location expresses the relative response of the antenna, not distance. Figure 2-28 depicts the pattern of a small antenna whose size is related to the wavelength of the electromagnetic energy interacting with it. In this case the antenna is a dipole whose total length is 1/2 wavelength. On this type of antenna, the electromagnetic wave is resonant. It is called a resonant antenna.

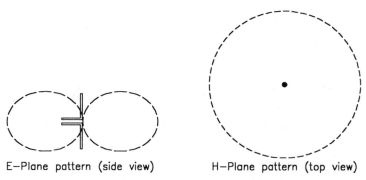

Figure 2-28. Vertical Dipole Antenna Pattern.

Figure 2-29 is a hypothetical pattern for an antenna which is much larger than the wavelength in all directions (an aperture antenna). It is composed of a main lobe, which is the principal lobe with the highest response, and sidelobes, which are minor responses in directions other than that of the main lobe. Sidelobes are one of the major sources of radar interference and, with very few exceptions, are undesirable.

Figures 2-28 and 2-29 are *one-way* antenna patterns, useful in evaluating communications, secondary radar, and ECM. Radar targets and clutter use the antenna's *two-way* pattern, which is the square of the one-way pattern. To radar targets and clutter, the 3-dB beamwidth becomes the 6-dB beamwidth, and sidelobes are lowered to double their dB value (Fig. 2-30).

Antennas are reciprocal. All parameters apply equally to transmit and receive; beamwidths, gains, sidelobe levels, and so on.

Antennas can be divided into several groups. Electrically small antennas are those whose operating dimensions are less than resonant, usually less than 1/4 wavelength. Resonant antenna dimensions are related to the wavelength, and they function as resonant circuits to the wavelengths at which they operate. Electrically large antennas are those whose principal dimension is large with respect to a wavelength. Resonant and large antennas are used in radar.

Line antennas are much larger in one dimension than the other. Wire antennas are of this type. Aperture antennas are large compared to the wavelength in all dimensions, and their areas are many square wavelengths.

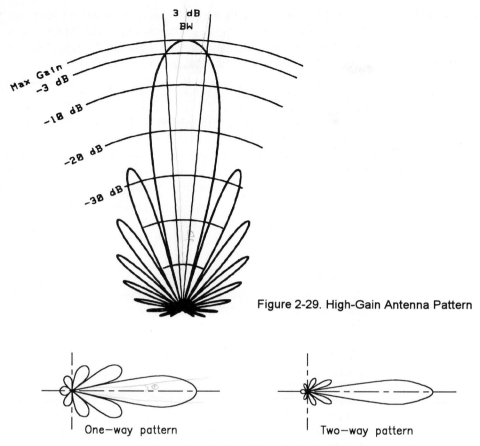

Figure 2-29. High-Gain Antenna Pattern

Figure 2-30. One- and Two-Way Patterns.

Parameters: Discussions in this chapter are at the parameter level, and antennas are treated by the way they affect the system's overall operation. The factors determining the various antenna parameters are summarized below and described later in this section. Discussions of beam development and antenna design are left to Ch. 9.

— *Beam pattern* is the transmit and receive response of the antenna as a function of azimuth and elevation angles from its axis. It is determined by the antenna's size, shape, illumination (see below), and frequency.

— *Beamwidth* is the angular width of the main antenna response (main lobe) and is a function of length, frequency, and illumination function.

— *Effective aperture* is the area with which the antenna, acting as a receiver, captures signal. It varies as a function of direction from the antenna's axis. An effective aperture specification, when given, is normally in the direction of the antenna's maximum response. It is determined by the antenna's projected area and illumination function.

— *Gain* is the factor by which the antenna concentrates transmitted signal as a function of direction, compared to an isotropic antenna (one with equal response in all directions — a spherical beam pattern). It is also the antenna's receive response as a function of direction, also compared to an isotropic antenna. When a single gain figure is published, it is the peak gain and in the direction of the antenna's maximum response. Peak gain is determined by the antenna's area, frequency, and illumination function.

— *Radiation efficiency* is the fraction of total power absorbed by the antenna which is radiated (transmit) or delivered to the antenna terminals (receive).

— *Sidelobes* are antenna responses in directions other than those of the main beam and are primarily a function of the illumination.

— *Field zones* describe the different responses of an antenna when observed close to the antenna (near field) and far from the antenna (far field). Most uses of antennas involve the far field only. The field zones are determined by the length of the antenna and the frequency.

— *Polarization* describes the preferential response of the antenna to electromagnetic waves whose fields are oriented in different directions. For example, horizontally polarized antennas respond best to EM waves whose electric fields are horizontal. Polarization is determined by the antenna type and orientation.

Beamwidth and beam shape: If an antenna responds primarily in a single direction, that major response is its main beam. The 3-dB beamwidth is the angular width of the main beam where the response is with 3 dB (1/2 power) of the peak response. The 3-dB beamwidth is an arbitrary measure of the width of the response. Antennas also have 6-dB beamwidths and null-to-null beamwidths (width of response between the minima surrounding the main beam). For most antenna responses, the null-to-null beamwidth is approximately double the 3 dB beamwidth. The beamwidth of an electrically large antenna is related to its length and the wavelength.

$$\theta_3 = \lambda / D_{\text{eff}} \text{ (radians)} \qquad (2\text{-}57a)$$

$$\theta_3 = (180 / \pi)(\lambda / D_{\text{eff}}) \text{ (degrees)} \qquad (2\text{-}57b)$$

D_{eff} = effective length of antenna in the plane of interest

λ = the wavelength (same units as D_{eff})

θ_3 = the 3-dB beamwidth in the same plane as the length

Large antenna surfaces are not used uniformly. This illumination taper is to suppress sidelobes (Ch. 9). Because the edges of an illumination-tapered antenna are seldom used as effectively as the center, its effective size is almost always smaller than the actual size. The effective length of large antennas is related to their physical length by a factor called length efficiency. Figure 2-31 shows the relationship between actual and effective dimensions. As a rule of thumb, *the length efficiency of most electrically large antennas is about 0.7.*

$$D_{\text{eff}} = \eta_L D \text{ (meters)} \qquad (2\text{-}58a)$$

D_{eff} = the effective length (meters)

η_L = the length efficiency (dimensionless)

D = the actual physical length (meters)

The beam shape of an aperture antenna is a function of the shape of the antenna (Fig. 2-32). Many radar and microwave communication antennas are circular in cross-

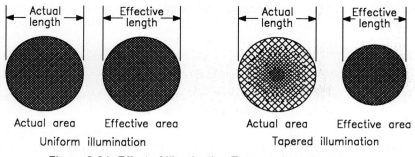

Figure 2-31. Effect of Illumination Taper on Length and Area.

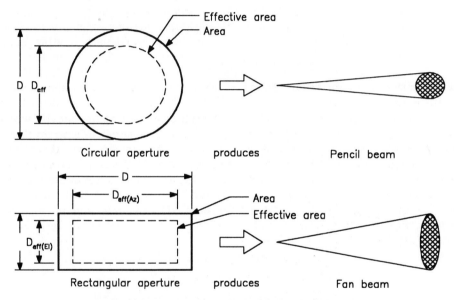

Figure 2-32. Antenna Shapes, Effective Dimensions, and Beams.

section, and have the same length in the azimuth and elevation directions. These antennas produce beams having circular cross-sections, called pencil beams. Other antennas are rectangular or ovoid in shape and produce different azimuth and elevation beamwidths. These beam shapes, common in search radars, are called fan beams.

$$D_{\text{eff}(Az)} = \eta_{L(Az)} D_{(Az)} \qquad (2\text{-}58b)$$

$D_{\text{eff}(Az)}$ = the horizontal effective length
$\eta_{L(Az)}$ = the azimuth length efficiency
$D_{(Az)}$ = the horizontal physical length

$$D_{\text{eff}(El)} = \eta_{L(El)} D_{(El)} \qquad (2\text{-}58c)$$

$D_{\text{eff}(El)}$ = the vertical effective length

2-5. Antenna Principles, Functions, and Parameters

$\eta_{L(El)}$ = the elevation length efficiency

$D_{(El)}$ = the vertical physical length

$\theta_{3(Az)}$ (radians) = $\lambda / D_{eff(Az)}$ (2-59)

$\theta_{3(Az)}$ (degrees) = $\lambda / D_{eff(Az)}$ $(180/\pi)$ (2-60)

The elevation beamwidth equations are the same as for azimuth, substituting the proper values.

Two-dimensional air search radar antennas require responses at higher elevation angles than are normally practical with fan beams. Targets near the horizon can be at much longer ranges than targets a high elevation angles, since the upper limit of the atmosphere is not very far away at high elevation angles. For example, discounting bending of the microwave beam (refraction — Ch. 6), a target at 50,000 ft (15,240 m) altitude at 0° elevation from a radar at sea level is at a range of about 240 nmi (440 km). That same target at 50,000 ft and 30° elevation is at a range of only 16.5 nmi (30.5 km). If the RCS of the target and the antenna gain remain the same, the signal from the horizon target is only 1/45,000 (−46.4 dB) from that of the 30° target because of the R^{-4} relationship. For this situation, about 23.2 dB more antenna gain is needed on the horizon than at 30° elevation, recalling that the antenna is used twice for a total of 46.4 dB.

Antennas producing fan beams cannot have sufficient gain on the horizon without having far more gain than needed at high elevation angles. The answer is to generate a shaped elevation beam, having high gain toward the horizon and a smaller (but not too small) response at high elevation angles. Such a beam is shown in Fig. 2-33. If the elevation pattern is shaped such that the echo signal power is constant with range for a constant altitude target, the beam is called a cosecant-squared beam. The effective area and gain lost by shaping the beam this way is usually about 1.5 dB. Most modern 2-D air search radar antennas have responses which approach cosecant-squared, and some direct even more power skyward than cosecant-squared.

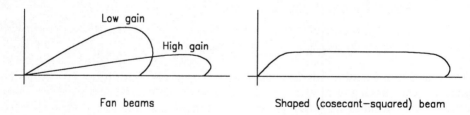

Figure 2-33. Two Ways of Achieving High Elevation Angle Coverage.

Effective area (effective aperture): The effective aperture of an antenna is that portion of its total projected area which effectively "captures" electromagnetic waves. It is related to the actual area by the aperture efficiency (Fig. 2-31).

$A_E = \eta_A A$ (2-61)

A_E = the effective aperture (square meters)

η_A = the aperture efficiency (dimensionless)

A = the antenna's actual projected area (square meters)

Aperture efficiency is related to length efficiency. For a circular (pencil beam) antenna, the aperture efficiency is the square of length efficiency.

$$\eta_A = \eta_L^2 \qquad (2\text{-}62)$$

The aperture efficiency of shaped-beam antennas is the product of the horizontal and vertical length efficiencies.

$$\eta_A = \eta_{L(Az)} \, \eta_{L(El)} \qquad (2\text{-}63)$$

The aperture efficiency of large antennas having equal horizontal and vertical illuminations is typically about 0.5, which is about 0.7 (their length efficiency) squared. For cosecant-squared antennas, the elevation length efficiency is nearer 0.5 than 0.7, giving an approximate aperture efficiency of 0.35.

Gain: An antenna's gain describes its ability to concentrate transmitted energy. Antenna gain is very nearly equal to the reciprocal of the fraction of the total sphere surrounding the antenna that its main beam occupies. An antenna whose main beam occupies 1/100 of a sphere has a gain very nearly equal to 100.

$$G \approx \frac{\text{Solid angle in a sphere (steradians)}}{\text{Solid angle in main beam (steradians)}} \qquad (2\text{-}64)$$

$$G \approx \frac{4\pi \text{ (steradians)}}{\text{Solid angle in main beam (steradians)}} \qquad (2\text{-}65)$$

$$G \approx \frac{32{,}400^*}{\theta_{3(Az)} \, \theta_{3(El)} \text{ (in degrees)}} \qquad (2\text{-}66)$$

* Authors disagree on the value to be used here. Johnson [7] uses 30,000. Stutzman and Thiele [8] suggest 26,000. Stegen [9] uses 35,000. All these values are equally correct; the actual number depends on the antenna's shape and illumination. The value in Eq. 2-66 is for a circular aperture uniformly illuminated.

The gains given in the previous equations are peak gains, describing the response of the antenna in the maximum direction. Gain is actually a function of direction (θ_{AZ} and θ_{EL}). The equations also actually define peak directivity, not gain. Gain includes the losses in the antenna itself, whereas directivity describes only the beam shape. They are related by a factor called radiation efficiency, defined below. The power lost heats the antenna. For large microwave antennas, the radiation efficiency is nearly unity and gain and directivity are virtually equal. For electrically small antennas the radiation efficiency can be very small.

$$G = \eta_R \, DIR \qquad (2\text{-}67)$$

G = gain (dimensionless)
η_R = radiation efficiency (dimensionless)
DIR = directivity (dimensionless)

Gain and effective aperture are related in all antennas, large, small, and resonant, by the following universal antenna relationship. It gives the correct response in all directions, not just for peak values.

$$G = \frac{4\pi}{\lambda^2} A_E \tag{2-68}$$

Example 2-7: A circular reflector antenna is 12.14 ft (3.7 m) in diameter and operates at 8.4 GHz. Its length efficiency is 0.7 and radiation efficiency is 1. Find the beamwidth, effective aperture, and gain.

Solution: The effective diameter of the antenna is 2.59 m (Eq. 2-58a). The wavelength is 3.57 cm. The beamwidth is 0.79° (Eq. 2-57b)

The aperture efficiency is found from Eq. 2-63 and is 0.49 (the azimuth and elevation length efficiencies of circular antennas are usually the same). The actual area of a 3.7-m diameter circle is 10.75 m^2. The effective area (Eq. 2-61) is therefore 5.27 m^2.

Gain is found from Eq. 2-66 or Eq. 2-68. Since Eq. 2-68 is exact and fundamental, its use is preferable. The gain is 51,900 (47.15 dB).

Example 2-8: A rectangular search radar antenna is 5.5 m wide by 2.4 m high. It has an azimuth length efficiency of 0.7, and elevation length efficiency of 0.45, and operates at 3.08 GHz. Find its azimuth and elevation beamwidths, effective aperture, and gain.

Solution: The effective size of the antenna (Eqs. 2-57 and 2-58) is 3.85 m by 1.08 m, for an effective area of 4.16 m^2 (the aperture efficiency — Eq. 2-63 — is 0.315). The wavelength (Eq. 1-30) is 9.74 cm. The beamwidths (Eq. 2-60) are 1.45° azimuth and 5.17° elevation. The peak gain, (Eq. 2-68) is 5,510 (37.4 dB).

Sidelobes and sidelobe effects: Antenna sidelobes are undesired responses in directions other than the main beam. They are formed in any antenna of finite length, and their severity is a function of antenna shape, illumination function, and aperture blockage (Ch. 9). Sidelobes are specified by their gain, in decibels, below the peak gain of the main lobe. The specification describes the largest sidelobe. For example, if the antenna of Fig. 2-29 has a peak gain of 43 dB and a sidelobe specification of 27 dB, it has at least one sidelobe whose gain is 16 dB (43 dB − 27 dB).

Sidelobes are major entry points for ECM, EMI, and clutter. With jamming and EMI, the interfering source is external to the radar and the one-way sidelobe levels determine how much interfering power enters the system. With clutter and targets, the two-way sidelobes prevail. Sidelobe levels in two-way patterns are double (in decibels) those of a one-way pattern. For example, an antenna with 27 dB sidelobes attenuates sidelobe ECM 27 dB compared to the main lobe, but attenuates sidelobe clutter 54 dB.

Airborne radars have special problems with sidelobe clutter (Sec. 4-13). These radars compensate for the velocity of main lobe clutter, which makes clutter from sidelobes appear as moving targets (Fig. 2-34). The altitude return, from a sidelobe pointed at the ground, makes airborne modes with little range resolution (high PRF) unsuited to tail-aspect targets (Secs. 4-13 and 6-6).

Many systems where sidelobe signals (clutter and ECM) are a problem must have methods of coping with them, including ultra-low sidelobe antennas, and sidelobe signal suppression. Both low sidelobe designs and suppression techniques, sidelobe blanking (SLB) and coherent sidelobe cancellation (CSLC) are covered in Ch. 9.

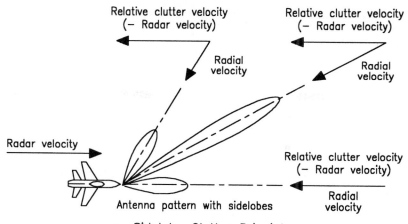

a. Sidelobe Clutter Principle.

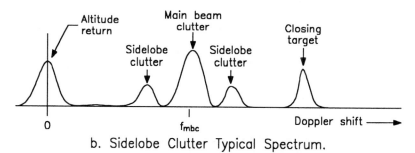

b. Sidelobe Clutter Typical Spectrum.

Figure 2-34. Antenna Sidelobes and Radar Clutter.

Field zones: The previous discussions about antenna parameters apply only in the antenna's far field, defined as that distance from the antenna at which, for all practical purposes, the electromagnetic wave fronts are planar (see Fig. 2-35).

Figure 2-35. Near- and Far-Field Waves.

Since the waves will always be spherical at less than infinite distances from the antenna, we must define flatness "for all practical purposes." This is conventionally defined as less than 1/16 of a wavelength of curvature across the extent of the antenna, as Fig. 2-36 illustrates. The far field is defined as that distance where the edge of a plane the size of the antenna and perpendicular to the axis at the antenna is no farther from the center than 1/16 of a wavelength.

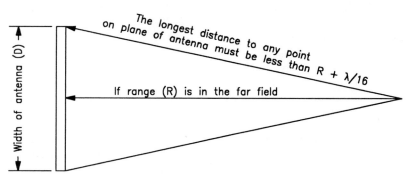

Figure 2-36. Geometry for Solving Far-Field Distance.

Solving the triangle in Fig. 2-36 for range gives the far-field distance. Ranges less than this value are said to be in the near field of the antenna. Note that the boundary between near and far fields is not absolute, and some applications can tolerate more or less phase curvature than the 1/16 wavelength criterion.

$$R_{FF} = 2 D^2 / \lambda \qquad (2\text{-}69)$$

R_{FF} = the range to the far field of the antenna

D = the antenna's length

Example 2-9: In Ch. 14, it will be shown that the most dangerous place to be from an RF radiation hazard standpoint is about 1/10 of the distance from the antenna surface to its far field. Find this distance for the antenna of Ex. 2-7.

Solution: The diameter of the antenna is 3.7 m and the wavelength is 3.57 cm. From Eq. 2-69, the distance from the antenna to its far field is 770 m (since the boundary of the far field is arbitrarily defined by Eq. 2-69, the distance found should not carry many significant figures). The most dangerous distance from the antenna is about 1/10 of the result of Eq. 2-69, which is 77 m, or about 250 ft.

Polarization: An antenna's polarization describes the orientation of the electric field component of the wave it radiates and the orientation of the EM wave that it responds best to on receive. A reflector antenna's polarization is set by it feed antenna which is usually a horn or dipole. An array's polarization is the polarization of its elements (Ch 9).

Possible antenna polarizations include linear, circular, and occasionally elliptical. In linear polarization, the electric field lies in a plane. If the plane is horizontal, the wave is said to be horizontally polarized. If vertical, the wave is vertically polarized. If at an angle, the wave is slant polarized. If the electric field is not planar, but describes an elliptical path through space, the wave is said to be elliptically polarized. Elliptical polarization contains two polarization components which are orthogonal (at right angles) to one another. If the components are equal in amplitude, the wave is circularly polarized.

Elliptical and circular waves are further divided by the direction of electric field rotation. If the rotation, viewed in the direction of propagation, is clockwise, the wave has a right-hand sense; if counterclockwise, left-hand.

Antennas respond best to waves of the same polarization as the antennas' design. If the received wave is of a different polarization, a polarization loss results. For linear waves, the polarization loss is given below.

$$L_{AP} = 20 \, \text{Log} \, [1 / \text{Cos}(\varphi)] \qquad (2\text{-}70)$$

L_{AP} = the antenna polarization loss

φ = the angle between the received wave's polarization and that of the antenna

The polarization loss between circularly polarized antennas and linear waves is 3 dB, since half of a circular wave energy is one linear polarization and the other half is at right angles. The loss between orthogonal linear polarizations approaches infinity, as does the loss between right-hand and left-hand circular. These last two conditions are known as cross-polarization. Table 2-2 gives the polarization loss for the principal wave polarizations.

Table 2-2. Polarization Losses (in dB).

Antenna Polarization:	Horizontal	Vertical	+45° Slant	−45° Slant	RHC	LHC
Wave Polarization:						
Horizontal	0	∞	3	3	3	3
Vertical	∞	0	3	3	3	3
+45° Slant	3	3	0	∞	3	3
−45° Slant	3	3	∞	0	3	3
RHC	3	3	3	3	0	∞
LHC	3	3	3	3	∞	0

Some signal conditions are best handled with linear polarization and some with circular. For example, horizontally polarized jamming is best suppressed with a vertically polarized antenna, giving a large polarization loss for the ECM. Rain clutter is suppressed by using circular polarization, because rain reflects the opposite circular polarization sense of the illumination (Ch. 4). For these reasons, many radar, communication, and EW systems have polarization diversity, or the availability of several different polarizations.

2-6. Receiver Description and Functions

The receiver amplifies and filters incoming echoes and interference and prepares them for signal processing. A radar's receiver has four functions.

— *Amplification function:* Echo signals from the antenna typically have very low amplitudes, ranging from perhaps −20 dBm (0.01 mW) for a very close target to −100 dBm (0.000,000,000,1 mW) for a minimally detectable target (these values vary from system to system — see Ch. 3). Before any use can be made of the signal, its amplitude must be increased by the receiver amplifiers.

— *Channel selection and out-of-channel filtering function:* Target echoes share their frequency band with signals from other radars and services. The receiver contains the filters which reject interfering signals occurring outside the bandpass of the received target echoes.

— *Matched filtering function:* The signal competes with thermal noise generated in the receiver itself. The matched filter's function is to admit the maximum signal with minimum noise to maximize the signal-to-noise ratio. Matched filters can be derived for many kinds of interference. In general, however, when the term "matched filter" is used, it refers to signals in white noise.

— *Demodulation function:* Information in communication and radar is "carried" on an electromagnetic wave whose frequency is higher than and independent of the information rate. The removal of this "carrier wave" and the recovery of the signal's information is called demodulation. Several kinds of demodulation are used, each recovering a different fraction of the information available. One kind, I/Q demodulation, recovers all the information in the signal.

Superheterodyne receiver introduction: Receivers must perform the above functions while adding the least practical amount of noise and distortion. Radar receivers are almost invariably of the superheterodyne design (Fig. 2-37), where the signals are offset downward by the local oscillator's (STALO) frequency and the bulk of the amplification and filtering takes place at a lower fixed frequency (the intermediate frequency — IF — equal to the COHO frequency). See Fig. 2-38. Superheterodyne receivers are easily tuned by changing the local oscillator frequency.

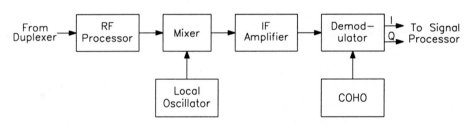

Figure 2-37. Superheterodyne Receiver.

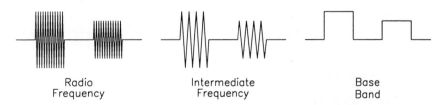

Figure 2-38. Receiver Frequency Relationships.

The major blocks of superheterodyne receivers have the following functions, explained in detail in Ch. 10.

— The *radio frequency (RF) processor* is where signal and interference are treated at the echo frequency. Its primary functions are to filter unwanted signals, to attenuate very strong signals which could saturate downstream circuits, and to amplify the signal plus interference. The RF processor is designed to generate minimal noise, but to amplify its generated noise enough so that noise from later stages does not materially affect the final signal-to-noise ratio.

— The *mixer*, in conjunction with the *local oscillator* (STALO), translates the signal and interference to the intermediate frequency (IF). The IF is the difference between the signal frequency and that of the local oscillator. At this lower frequency, amplifiers and filters can be implemented to match more precisely their requirements than at the higher radio frequency.

$$f_{IF} = f_{SIG} - f_{LO} \tag{2-71a}$$

or

$$f_{IF} = f_{LO} - f_{SIG} \tag{2-71b}$$

f_{IF} = the intermediate frequency

f_{SIG} = the signal frequency (including any Doppler shift)

f_{LO} = the local oscillator frequency

Example 2-10: A single conversion superheterodyne receiver has a nominal intermediate frequency of 30.0 MHz and is to receive target echoes caused by illumination at 5.700 GHz exactly. Find (a) the local oscillator frequency if the LO is below the echo frequency; (b) the local oscillator frequency if the LO is above the echo frequency; (c) the actual signal IF if the echo has a Doppler shift of +10,000 Hz and the LO is below the signal frequency; (d) the actual signal IF for +10,000 Hz Doppler and the LO above the signal.

Solution: (a) Using Eq. 2-71, the local oscillator is 5.670 GHz.

(b) Again using Eq. 2-71, the local oscillator is at 5.730 GHz.

(c) With a Doppler shift of 10 kHz, the echo frequency is 5.700010 GHz, or 10 kHz above nominal. The local oscillator of 5.670000 GHz gives an IF of 30.010 MHz, which is also 10 kHz above nominal.

(d) If the local oscillator is 5.730000 GHz, the Doppler shifted echo (10 kHz above nominal) has an intermediate frequency of 29.990 MHz, which is 10 kHz *below* nominal. *Note:* If the local oscillator is below the echo frequency, Doppler shift direction is preserved in the IF. If it is above, the *Doppler shift is inverted*, and positive Doppler from inbound targets becomes negative in the IF and demodulator. This is not a problem, as long as the position of the LO with respect to the echoes is known and accounted for.

One disadvantage of the superheterodyne design is that *two* incoming frequencies mix with the LO to produce the same IF. Using Ex. 2-10, a signal at 5.640 GHz, called the *image frequency*, also mixes with the 5.670-GHz LO to produce the 30-MHz IF. The image is a serious source of interference in some radars. It is twice the IF removed from the normal signal frequency.

$$f_{IMG} = f_{SIG} - 2f_{IF} \text{ (LO below signal frequency)} \tag{2-72a}$$

or

$$f_{IMG} = f_{SIG} + 2f_{IF} \text{ (LO above signal frequency)} \tag{2-72b}$$

2-6. Receiver Description and Functions

f_{IMG} = the image frequency

Images, unless suppressed, can interfere with normal signals. A second bandwidth's worth of noise, ECM, and EMI can enter the system at this frequency. The defense against images is to make the receiver insensitive to them. The principal method is to make the intermediate frequency high so that images are far removed from the signal frequency and can be filtered in the RF processor. A second suppression method is to use a cancelling mixer called an image-reject mixer (see Ch. 10).

Example 2-11: In Ex. 2-10, let the IF be 950 MHz, instead of 30 MHz. The signal frequency is above that of the LO. For the same signal frequency (5.700 GHz), find the new LO and image frequencies.

Solution: Equation 2-71 gives the LO frequency as 4.750 GHz. From Eq. 2-72a, the image response of the receiver is 3.800 GHz, which is easily filtered in the RF processor.

High intermediate frequencies are desirable from an image standpoint. Their disadvantage is that amplification and filtering are more difficult at the higher frequencies. The solution is to use multiple-conversion superheterodyne receivers. See Ch. 10 for details.

— The *intermediate frequency (IF) amplifier* is where the bulk of the amplification and filtering takes place and where gain controls are usually applied. Two types of filtering occur here. The channel-select filter accepts the signal and the interference within the signal bandwidth and rejects out-of-band interference. The matched filter treats the signal and residual interference in such a way as to pass the maximum possible signal and reject the maximum possible interference, optimizing the signal-to-interference ratio.

— The *demodulator* translates the signal and interference from the intermediate frequency to its information (or base-band) frequencies. Three types are commonly used in radars: envelope, synchronous, and I/Q. An envelope demodulator recovers the amplitude of the signal plus interference. A synchronous demodulator recovers a composite of the signal-plus-interference's amplitude and phase angle. An I/Q demodulator recovers two values which totally describe the amplitude and phase of the signal plus interference (Sec. 6-2).

$$v_S(t) = V_S \cos(2\pi f_C + \phi_S(t) - \phi_C(t)] \qquad (2\text{-}73)$$

$$V_{SE} = V_S \qquad (2\text{-}74)$$

$$V_{SS} = V_S \cos(\phi_S - \phi_C) \qquad (2\text{-}75)$$

$$V_{SI} = V_S \cos(\phi_S - \phi_C) \qquad (2\text{-}76a)$$

$$V_{SQ} = V_S \sin(\phi_S - \phi_C) \qquad (2\text{-}76b)$$

$v_S(t)$ = the instantaneous voltage of the signal plus interference, at the IF

V_S = the signal plus interference peak voltage

V_{SE} = the peak value of the envelope demodulated signal plus interference

V_{SS} = the peak value of the synchronously demodulated signal plus interference

V_{SI} = the peak value of the demodulated I component of signal plus interference

V_{SQ} = the peak value of the demodulated Q component of signal plus interference

f_C = the COHO (or IF) frequency

ϕ_S = the phase of the signal

ϕ_C = the phase of the COHO (usually zero by definition)

Envelope demodulation discards signal phase and is unsuitable for applications where the Doppler shift is to be sensed (MTI) or recovered.

Synchronous demodulation does not recover enough information about the signal to tell whether the Doppler shift is positive or negative. Only signal components which are in phase with the COHO are recovered; thus signals which are $\pm 90°$ to the COHO, called blind phases, are lost. Because of the loss of the quadrature signal components, the signal-to-noise ratio is lowered by 3 dB.

Only I/Q demodulation recovers all information available about the signal.

Key receiver parameters and specifications: Following are the key parameters and specifications for radar receivers, with discussions of the importance of each in target information recovery. This preliminary treatment of receiver parameters is to provide background needed in the following chapters.

Gain: Gain is by definition the ratio of the receiver's output power to its input.

$$G \equiv P_o / P_i \qquad (2\text{-}77)$$

G = the power gain

P_o = the receiver output power, whether target echo signal, interference, or both (Watts)

P_i = receiver input power (Watts)

Gain control: Several types of gain controls are present in radar receivers, and are described below.

— *Manual gain control (MGC):* MGC is an operator or system control which sets the receiver for a constant gain. With MGC, the receiver output power is directly proportional to the input power. See Fig. 2-39.

$$P_o = G P_i \qquad (2\text{-}78)$$

G = the gain, which is a parameter, or changeable constant

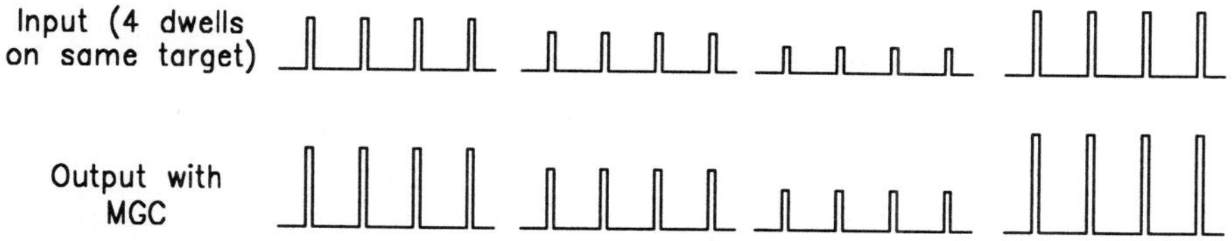

Figure 2-39. Manual Gain Control (MGC).

— *Automatic gain control (AGC):* AGC attempts to hold the amplitude at the output of the receiver constant. The gain is set by the signal amplitude itself, the result being that the receiver output power is constant. This method of gain control is prevalent in radars and modes which view only one target, such as single-target-track. See Fig. 2-40.

AGC usually sets receiver gain after averaging many pulses, and thus does not normally follow rapid changes in signal amplitude. Therefore Fig. 2-40 is valid only if the looks are separated by enough time for the AGC servo to settle. The AGC bandwidth sets the limits of how fast a target's amplitude can fluctuate with the AGC producing a constant output. It is typically on the order of 10 Hz.

$$P_o = K \qquad (2\text{-}79)$$

K = an adjustable constant, independent of RCS and range

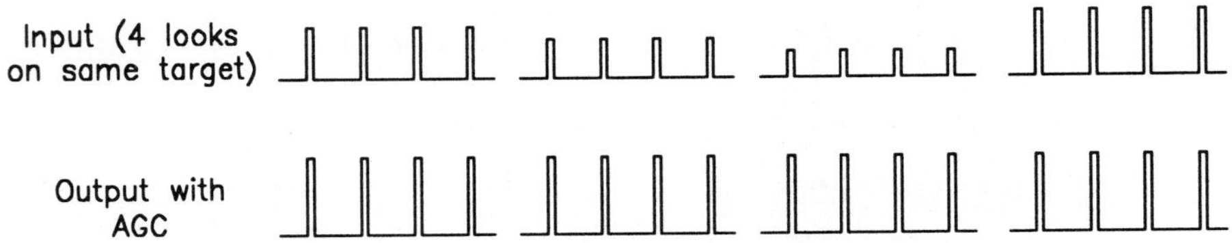

Figure 2-40. Automatic Gain Control (AGC).

— *Sensitivity time control (STC):* Point targets (those completely contained within the radar's resolution cell) at short ranges produce far more echo power at the receiver than at long ranges (proportional to R^{-4}). For example, a target at 1 mile range produces 1.6×10^9 (92 dB) more input power than the same target at 200-miles range. Thus large targets at close ranges may saturate receivers with gains set for weak targets at long ranges. STC varies the gain of the receiver with time, setting it low when the transmitter fires and gradually increasing it so that full gain is available for echoes from long ranges. With STC, receiver output power is proportional to input power, and gain is a function of the time elapsed since the last transmitted pulse. STC is primarily a search radar technique.

Figure 2-41 shows a typical STC gain curve. Note that if the STC increases the power gain by a factor of 16 (12 dB) for each doubling of time after the transmitted pulse, the range dependence of signal power output from the receiver vanishes. This particular STC curve, known as R^4 STC, causes a constant radar cross-section target to produce a constant receiver output power regardless of its range (Fig. 2-42). Some systems use R^3 STC where the gain

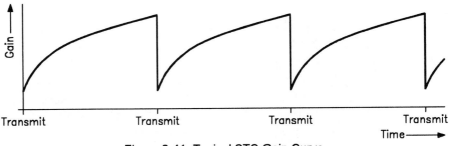

Figure 2-41. Typical STC Gain Curve.

increases by a factor of 8 (9 dB) for each range doubling, or R^2 STC where the factor is 4 (6 dB) per range doubling. R^3 STC is appropriate in area clutter and R^2 for volume clutter and chaff (see Chs. 3 and 4).

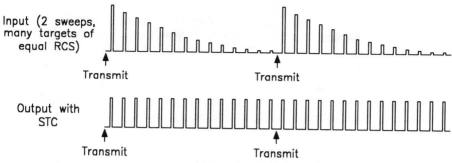

Figure 2-42. Effect of R^4 STC on Equal-RCS Point Targets.

Because the gain is reduced each time the transmitter fires, receiver STC is inappropriate for CW, medium PRF, and high PRF receivers, where echo arrival times are independent of the last time the transmitter fired. It can be and sometimes is implemented in the data processor in medium PRF systems, but only after range ambiguities are resolved. This is a technique used in some airborne radars to identify clutter entering the radar through antenna sidelobes [10].

$$P_o = K\sigma \qquad (2\text{-}80)$$

K = an adjustable constant, independent of range

σ = the target's RCS

— *Instantaneous AGC (IAGC) and fast time constant (FTC):* In some cases, signals are accompanied by interference which is large in extent (in the case of clutter, this is called extended clutter). IAGC and FTC cause the receiver output power to be proportional to the magnitude of the time rate of change of the input power, reducing the effect of extended interference. In Fig. 2-43, before FTC a threshold would have a difficult time detecting both targets without detecting the interference. Whereas AGC controls receiver gain by averaging signal over many PRIs (many milliseconds), IAGC and FTC control gain by averaging over a few range resolution intervals (a few microseconds).

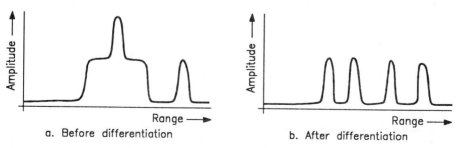

a. Before differentiation   b. After differentiation

Figure 2-43. Fast Time Constant (FTC).

2-6. Receiver Description and Functions

$$P_o \approx G |d(P_i)/dt| \qquad (3\text{-}81)$$

$d(P_i)/dt$ = the time derivative of input power

Sensitivity, noise, and temperature: Sensitivity is a measure of how small an echo signal can be successfully received. In communication systems, it is usually expressed in absolute terms, such as dBm or microvolts into 50 ohms. In radars, it is sometimes expressed absolutely and sometimes in terms of the interference type and level (such as signal-to-noise ratio, signal-to-clutter ratio, and so forth). Several sensitivity measures are used and are described below.

— *Noise:* Noise is, in most systems, the ultimate interference; it is lower in amplitude than other interfering sources and ultimately sets the maximum range at which targets can be detected. More detailed information than supplied here can be found in radio astronomy texts and in reference works such as Blake [11] and Barton [12].

All resistances at temperatures above absolute zero (0 °K) produce noise voltages across their terminals. If the resistors are terminated, power is generated. If the terminations are matched to the resistors, the noise power is maximum. Note that this noise is a real signal, capable of interfering with target echo signals, and is the result of a conversion of the resistor's thermal energy to electric energy. The mechanism is basically that at temperatures above absolute zero the atoms in the resistor are in motion, and statistically there is more charge at one end than the other at one instant, reversing the next. If electrical energy is extracted from the resistor, it reduces the resistor's thermal energy, cooling it (unless, of course, the thermal energy removed is replaced from the environment). The rms noise voltage across the resistor (open circuit) is

$$V_N = \sqrt{4 K T B_N R} \qquad (2\text{-}82)$$

V_N = the rms noise voltage across the open-circuited resistor terminals (volts)

K = Boltzmann's constant (1.38×10^{-23} Joule / ° Kelvin)

T = the resistor's temperature (° Kelvin)

B_N = the equivalent noise bandwidth over which the measurement is made (Hertz)

R = the resistance (Ohms)

If the resistor is terminated in the value R, the rms noise voltage becomes half its open circuit value.

$$V_{NT} = \sqrt{K T B_N R} \qquad (2\text{-}83)$$

V_{NT} = the rms noise voltage across the terminated resistor

The power delivered by the resistor into its termination (again, assuming the termination is R) is, from Ohm's law

$$P_{NT} = K T B_N \qquad (2\text{-}84)$$

P_{NT} = the noise power delivered to the termination (Watts)

It is expected, therefore, that any resistances in the receiver will generate noise. Since only resistances can generate or absorb power, at least one must be present; the antenna termination. If it were not there, all power from the antenna would reflect back out the antenna. The absolute minimum noise at the receiver's input is given by Eq. 2-84 where T is the temperature of the antenna termination.

The noise power generated increases linearly with the equivalent bandwidth over which the measurement is taken. In receivers, this relates to the frequency response. The noise equivalent bandwidth is defined in Eq. 2-85 and Fig. 2-44. It is the width of a rectangular filter whose peak response equals that of the actual filter, and whose area equals that under the actual filter.

$$B_N = \frac{\int_{-\infty}^{\infty} |H(f)|^2 \, df}{|H(f_0)|^2} \qquad (2\text{-}85)$$

$H(f)$ = the frequency response of the filter

$H(f_0)$ = the peak response of the filter

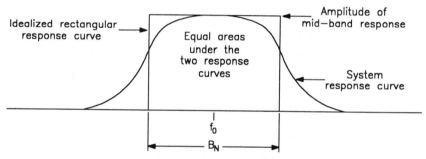

Figure 2-44. Equivalent Noise Bandwidth Definition.

In all cases, practical receivers generate more noise than predicted by Eq. 2-84. This is because there are other noise sources in addition to the termination. This results in additional noise competing with the signal, and is quantified as *minimum discernible signal*, *noise figure*, or *system temperature*.

— *Minimum discernible signal (MDS):* MDS is the minimum signal power which can be discerned from noise, usually by an operator. MDS is almost always the result of a simulated target injected into the system, either through a directional coupler or from a boresight tower. (A boresight tower is a tower with a signal source and antenna mounted on it. It is located in the far field of the antenna and provides signals for calibration of the radar — see Ch. 10.) MDS can be a good measure of the receiver's overall absolute sensitivity if the proper corrections are applied. With a trained operator, MDS occurs at a signal-to-noise ratio of about unity. MDS almost always is taken in the presence of noise interference only. The process of measuring MDS is described in Ch. 10.

A method similar to MDS for obtaining receiver sensitivity is the tangential sensitivity (TSS) measurement. It is also discussed in Ch. 10.

— *Noise factor / noise figure:* Noise factor / noise figure measures the thermal noise generated in the receiver compared to the noise produced (Eq. 2-84) by a perfect receiver at

290°K. This ideal receiver has a noise factor of 1. Outside of some exceptional conditions (e.g. cryogenic receivers), noise factor must be greater than unity. Noise factor (F) is a power ratio; noise figure (NF) is its decibel equivalent.

Noise factor/figure has several components, including the noise figure of the receiver itself, noise from the outside world (in the microwave bands, small but not insignificant), noise from resistances (losses) in the antenna, and noise from resistances (losses) in transmission lines. Breakout of the noise components is deferred to the discussion of system temperature later in this section.

Noise factor/figure as used in this book is often called *operating noise factor/figure*. It takes into account all noise sources and specifies the performance of the total system. The noise power at the output of a system is

$$P_{NO} \approx K T_0 B_N F G_0 \text{ (Watts)} \qquad (2\text{-}86)$$

K = Boltzmann's constant (1.38×10^{-23} J/°K)

T_0 = 290°K

B_N = the equivalent noise bandwidth

F = the operating noise factor, which may differ from the noise factor of the receiver alone

G_0 = the available gain from the system's input to its output.

The relationship given in Eq. 2-86 is an approximation because it assumes the noise is white (unity bandwidth produces the same noise power at all frequencies). It is effectively white for values of f/T (ratio of frequency to temperature) greater than about 108. For radars, Blake [11] indicates that the assumption is valid for frequencies less than 30 GHz and temperatures near or greater than 300°K.

Noise factor/figure can be defined in terms of its effect on signals as they pass through the system. Realizing that noise figure is a measure of how much noise is added by the system (KTB is always and unavoidably present), the following definition is appropriate.

$$F = \frac{(S/N)_I}{(S/N)_O} \qquad (2\text{-}87)$$

$(S/N)_I$ = the signal-to-noise ratio at the input to the system (before the added noise of the system influences it)

$(S/N)_O$ = the signal-to-noise ratio at the output of the system (after the system-generated noise is added)

If it is assumed that all the noise is at the system's input, the noise can be compared to the received signals calculated with the radar equation (Ch. 3). The equivalent input noise power is

$$P_N = K T_0 B_N F \qquad (2\text{-}88)$$

P_N = the system noise power reflected to the input of the system. This noise may be generated throughout the system, but is assumed to be concentrated at the input.

Operating (total) noise factor can be related to the noise produced as

$$F = \frac{P_N}{K T B_N} \qquad (2\text{-}89)$$

— *System temperature:* System temperature is another way of specifying the noise produced. It is the temperature at which an ideal system would have to operate to produce thermal (KTB_N) noise equal to the amount of noise the actual system is producing.

$$P_N = K T_S B \text{ (Watts)} \tag{2-90}$$

$$T_S = T_0 F \text{ (degrees Kelvin)} \tag{2-91}$$

$\qquad T_S$ = the system temperature (degrees Kelvin)
$\qquad T_0$ = 290°K

Example 2-12: The input to a receiver with a gain of 104.0 dB and bandwidth of 1.5 MHz is terminated. The noise out of the receiver is −2.5 dBm. Find the receiver's noise figure and system temperature.

Solution: The receiver's equivalent input noise power is −106.5 dBm (104.0 dB below −2.5 dBm), or 2.24 x 10^{-14} W. Equation 2-86 or 2-88 gives the noise factor as 3.73, and the noise figure as 5.7 dB. Equation 2-91 gives the system temperature as 1082°K. Note that the receiver is not at a temperature of 1082°K. It does, however, produce the same amount of KTB noise as an ideal receiver at a temperature of 1082°K.

— *Antenna, transmission line, and receiver equivalent temperatures:* System temperature has several components: external noise, antenna resistive noise, transmission line and microwave component resistive noise, and receiver noise. In terms of its components, the system temperature (from Blake [11]) is

$$T_S = T_A + T_R + L_R T_E \tag{2-92}$$

$\qquad T_A$ = the antenna temperature (degrees Kelvin)
$\qquad T_R$ = the equivalent temperature of the transmission line loss
$\qquad L_R$ = the transmission line loss
$\qquad T_E$ = the receiver equivalent temperature

The antenna temperature is itself made up of several components, all of which are given in degrees Kelvin. The first, T_a', is the temperature equivalent ($KT_a'B$) of the external sky noise power (solar and galactic) absorbed by the antenna. It is found from Fig. 2-45. The solid lines are for average galactic and solar noise, for an antenna with "average" sidelobes. The upper dashed line is for the same antenna with maximum galactic and solar noise. The lower dashed line is for the same antenna with minimum galactic and zero solar noise.

The second component is the contribution of the warm earth (or sea) under the antenna, viewed through ground-directed sidelobes. Third is the contribution of the losses in the antenna itself. The three components combine to form antenna temperature.

$$T_A = \frac{T_a'(1 - T_{ag}/T_{tg}) + T_{ag}}{L_a} + T_{ta}(1 - 1/L_a) \tag{2-93}$$

$\qquad T_a'$ = the noise contribution of solar and galactic sources (from Fig. 2-48)
$\qquad T_{ag}$ = the ground noise temperature component of antenna temperature

2-6. Receiver Description and Functions

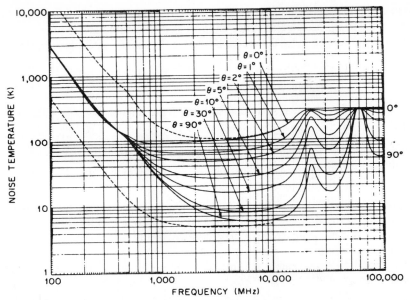

Figure 2-45. Noise Temperature of a Lossless Antenna.
(Reprinted from Blake [11]. © McGraw-Hill Book Company; used with permission)

T_{tg} = the effective noise temperature of the ground (approximately the ambient temperature)

$1 - T_{ag}/T_{tg}$, called the reduction factor, = the result of the antenna's entire pattern not being pointed at the sky.

L_a = the loss from antenna's aperture to its terminals

Note that this antenna temperature is for the antenna pointed away from the sun. If solar energy enters the antenna through its main lobe, the equivalent temperature increases to approximately that of the sun's surface, about 6000°K. The sun's noise is sometimes used to calibrate radar systems.

For typical values of $T_{ag} = 36\,°K$, $T_{tg} = 290\,°K$, and an approximately lossless antenna, Eq. 2-93 simplifies to

$$T_A \approx 0.876\, T_a' + 36 \quad \text{(degrees Kelvin)} \qquad (2\text{-}94)$$

The transmission line temperature is given by

$$T_R = T_{tr}(L_R - 1) \qquad (2\text{-}95)$$

T_R = the equivalent temperature of the transmission line loss (actually, it is the equivalent temperature of any loss)

T_{tr} = the thermal temperature of the transmission line (approximately ambient)

L_R = the loss of the transmission line (≥ 1.0)

The receiver equivalent temperature is the receiver's contribution to system temperature. It is given by

$$T_E = T_0 (F - 1) \qquad (2\text{-}96)$$

T_E = the receiver equivalent temperature, which is not a physically measurable temperature, but is merely a device to account for the noise power above K T B added by the receiver.

T_0 = 290 °K

F = the receiver's (not system's) noise factor

— *Comments on noise and sensitivity:* In the absence of other more powerful interference forms, the limit on how small an echo signal can be successfully received and detected is set by the radar's noise. Intuitively, it seems that the lower limit on noise power should be that of a resistor at the ambient temperature. A careful examination of Eq. 2-93 and its components (Eqs. 2-94, 2-95, and 2-96), however, shows that for realistic values of losses and receiver noise factor (<2), the system temperature when the antenna is pointed into space can be less than the reference temperature of 290 °K (17 °C or 62.6 °F) for systems at thermal temperatures equal to or greater than reference.

Matched filter: As stated earlier, one of the receiver's functions is matched filtering the signals and interference. The purpose is to filter out as much noise as possible, while at the same time retaining as much signal as possible. The definition of a matched filter is one which *optimizes the signal-to-noise ratio*. True "matched filters" are not realizable and those found in radar receivers are approximations. The approximation introduces a mismatch loss.

In many radar systems there are two functions known as matched filters, one in the receiver and one in the signal processor. The receiver matched filter's purpose is to optimize the signal-to-noise ratio. The signal processor matched filter performs pulse compression, where present. A difference between these filters (approximations, actually) is that the receive filter usually implements only the amplitude of the filter response, while the signal processor filter addresses the phase response. Signal processor matched filtering is covered in Chs. 11 and 13. In most receivers, the channel select filter and matched filter functions are performed by the same filter.

The fraction of the signal power passed through the receiver filters is a function of the filter bandwidth, the bandpass characteristic, and the spectrum of the received echo. The amount of noise passed through the same filter is a different function of the same parameters. If the spectral characteristics of the signal and noise lend themselves to the process, it may be possible to construct a filter which delivers a maximum signal-to-noise ratio. Figure 2-46 shows the concept. In it, the effect of the bandwidth of a rectangular filter on an echo with a rectangular envelope and on receiver noise is shown. In this case, the noise power is a linear function of bandwidth (Eq. 2-86), and the ratio of signal to noise peaks at a certain bandpass.

Actual matched filters have more complex responses than shown in the figure. Each transmitted waveform, echo Doppler shift, echo range, and interference type has its own matched filter characteristic. The so-called "matched filter" implemented in most radar receivers is an averaged approximation of the true matched filter and implements only the amplitude response. In the example of Fig. 2-46, the filter at its optimum bandwidth produces a signal-to-noise ratio which is about 1 dB worse than the true matched filter (Table 2-3). The matched filter's characteristic is given below. For a derivation, see [13].

$$H(f) = G_0 S^*(f) \exp(-j2\pi f t_1) \qquad (2\text{-}97)$$

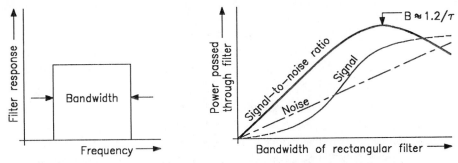

Figure 2-46. Matched Filter Principle using Rectangular Envelope
Signal and Rectangular Filter (non-matched approximation).

$H(f)$ = the transfer function of the matched filter
G_0 = the midband gain of the filter (a constant)
$S^*(f)$ = the complex conjugate of the spectrum of the received echo
t_1 = the time of maximum response of the filter (related to the time delay between transmit and echo reception)

The matched filter for most radar waveforms must be defined over all time and frequency. This is not practical (or even possible) in physically realizable systems, nor is it usually desirable in radar receivers. The matched filter given in Eq. 2-97 optimizes the *signal-to-white noise* ratio. Matched filtering in the presence of interference other than noise may not give the best signal-to-interference ratio. A true matched filter to rectangular pulses has an amplitude response in the form of a sinc function (Fig. 2-12), and thus responds to frequencies far removed from the bulk of the signal energy. This excessive response to out-of-band interference such as jamming and EMI precludes the use of true matched filters even if they were available.

The characteristics of several *approximately* matched filters are given in Table 2-3. All the waves are gated CW and the shapes are in amplitude only. Note that the case which results in a true match involves transmitted waves which are shaped in amplitude. Radar transmitters are usually operated in amplitude saturation, forcing the output pulse to be roughly rectangular.

A common misconception about the matched filter is that the echo signal out of the filter (in the time domain) "looks like" the transmit waveform. This is not true. To amplify a signal such that its form is unchanged in time would require a filter of very wide bandwidth and would introduce much noise. The matched filter limits noise by restricting bandwidth at the expense of the pulse shape. If the echo and transmit wave are rectangular, the pulse out of a matched filter is somewhat Gaussian in shape. See Fig. 2-47.

Matched filters for other than white noise can be found by the same technique used to derive the white-noise filter. It is easier, however, to recognize that the matched filter for any noise spectrum is

$$H(f) = \frac{G_0 \, S^*(f) \, \exp(-j2\pi f t_1)}{[N_i(f)]^2} \qquad (2\text{-}98)$$

$N_i(f)$ = the input noise spectrum

Table 2-3. Matched Filter Approximations
(After Skolnik [13])

Transmitted wave shape (gated CW)	Filter amplitude shape	Optimum $B_N \tau$	S/N loss compared to matched filter (dB)
Rectangular	Rectangular	1.37	0.85
Rectangular	Gaussian	0.72	0.49
Gaussian	Rectangular	0.72	0.49
Gaussian	Gaussian	0.44	0 (matched)
Rectangular	Single-tuned circuit, one stage	0.4	0.88
Rectangular	Single-tuned circuit, two cascaded stages	0.613	0.56
Rectangular	Single-tuned circuit, five cascaded stages	0.672	0.5

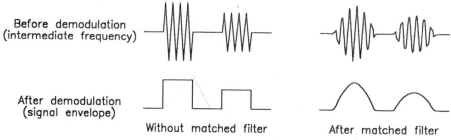

Figure 2-47. Signal Changes through Matched Filter.

For white noise, the power spectrum $[Ni(f)]^2$ is unity and Eq. 2-97 results. For noise spectra other than white, Eq. 2-98 can be modified as follows.

$$H(f) = \frac{1}{N_i(f)} \frac{G_0 \, S^*(f) \exp(-j2\pi f t_1)}{N_i^*(f)} \qquad (2\text{-}99)$$

In Eq. 2-99, the first fraction acts to make the noise white and is called a whitening filter. The second fraction is the matched filter for white noise.

Linearity and dynamic range: The receiver amplifies and filters a composite of target echoes and interfering signals (noise, clutter, jamming, EMI, and spillover). These composite signals enter the receiver as the superposition (sum) of their components, and signal processing can separate the components only if they remain a superposition. This requires that they be amplified and filtered linearly.

Non-linearity causes three problems. First, intermodulation products are created, forming new signals which may interfere with the desired target echoes. Second, it enhances the "leakage" of strong interfering signals into Doppler and range bins which they do not

2-6. Receiver Description and Functions

in themselves occupy, where they can interfere with legitimate targets (see the limited Kaiser-Bessel window in Ch. 11). Third, MTI processors may process non-linearly amplified clutter in such a way that it appears as a moving target (see discussion of MTI limiting in Ch. 12). In all cases, non-linearities may preclude the separation of target echo signal from interference in the signal processor.

In a linear system, the output is directly proportional to the input (gain is not a function of signal amplitude) and no new signal components are introduced. If the input is the sum of multiple signals (or signal-plus-interference), as below

$$v_{in}(t) = v_s(t) + v_i(t) \tag{2-100}$$

$v_{in}(t)$ = the input signal-plus-interference voltage as a function of time
$v_s(t)$ = the signal voltage as a function of time
$v_i(t)$ = the interference voltage as a function of time

then the output is the same signal sum amplified by the gain, or

$$v_{out}(t) = G_v [v_s(t) + v_i(t)] \tag{2-101}$$

$v_{out}(t)$ = the output voltage of time
G_v = the amplifier's voltage gain (in impedance matched systems, the square root of the power gain)

In a non-linear system, for the same input, the output also contains the gain times the sum of the inputs, but in addition has components of the product of the inputs (which is the intermodulation distortion).

$$v_{out}(t) = G_v [v_s(t) + v_s(t) + K_2 v_s(t) v_i(t) + \text{others}] \tag{2-102}$$

K_2 = the second order intermodulation product constant
"Others" include the higher order intermodulation products

Dynamic range is defined as the ratio of the largest signal a process or system can handle to the smallest signal that same process or system can handle (Fig. 2-48). A measure of the dynamic range of a receiver is its 1 dB compression point, which is the output power at which the gain of the receiver is reduced one decibel by saturation. In Fig. 2-48, the slope of the power-out / power-in curve is the receiver's gain, and the 1 dB compression point is shown. In many radar receivers, operation at this point results in too much distortion for good performance, but it gives a means of comparing receiver performance.

The result of passing a signal through a non-linear receiver is shown in Fig. 2-49. This signal is from a search radar scanning by a constant RCS target.

Non-linear amplification results in intermodulation distortion, which creates intermodulation products — new signal components at new frequencies. They result when signals are multiplied together (a non-linear process), rather than summed as in linear amplification. The frequencies of the intermodulation products are

$$f_{out} = m f_1 + n f_2 \tag{2-103}$$

f_{out} = an output frequency
f_1 = one of the input frequencies

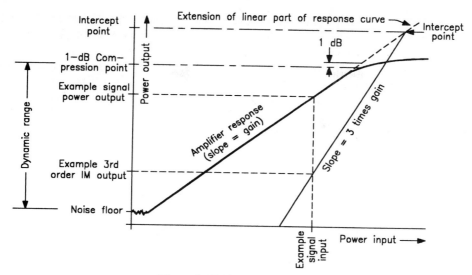

Figure 2-48. Dynamic Range.

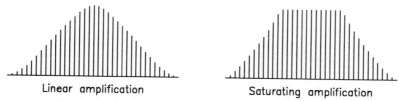

Figure 2-49. Effect of Non-Linearity on Received Signal.

f_2 = the other input frequency

m and n = integers (positive or negative)

When narrow bandwidth bandpass filters are involved, as they usually are in radar receivers, only third-order and higher-order intermodulation products can fall within the bandpass and appear as interfering signals.

$$\text{IM Order} = |m| + |n| \qquad (2\text{-}104)$$

Example 2-13: An airborne radar receiver's IF amplifier has a center frequency of 120 MHz and a bandwidth of 3.0 MHz. Two echoes are present: main beam clutter with 20,500 Hz of up-Doppler and a tail-aspect target with 1,400 Hz of up-Doppler. If the receiver is non-linear, find all intermodulation products through third order which fall within the receiver's bandpass.

Solution: The echo frequency is 120,001,400 Hz and the clutter is at 120,020,500 Hz. The bandpass of the receiver is 118,500,000 Hz to 121,500,000 Hz. The two first-order IM products are the two signals, at 120,001,400 and 120,020,500 Hz.

Eight second-order IM products exist: 240,051,000 Hz ($m = 0, n = 2$), 240,002,800 Hz ($m = 2, n = 0$), 240,021,900 Hz ($m = 1, n = 1$), 19,100 Hz, and their additive inverses. None lie within the receiver's bandpass.

There are twelve third-order IM products: 360,061,500 Hz ($m = 0, n = 3$), 360,004,200 Hz ($m = 3, n = 0$), 360,042,400 Hz ($m = 2, n = 1$), 360,023,300 ($m = 1, n = 2$), 120,039,600 Hz ($m = -1, n = 2$), 119,982,300 Hz ($m = 2, n = -1$), and their additive inverses. Of these, four (120,039,600, 119,982,300, −120,039,600, and −119,982,300 Hz) fall within the receiver's bandpass and will interfere with the legitimate signals.

A large number of higher order intermodulation products also lie within the receiver's bandpass. Since the amplitude of IM products falls off rapidly with higher order, their effect is usually negligible. In most cases, only third-order IM product must be considered in radar receivers.

The amplitude of intermodulation products can be found from the intercept point specification of the system (Fig. 2-48). The intercept point is plotted on the receiver's gain curve. From the intercept point, a line is drawn having a slope equal to the intermodulation order times the linear gain of the receiver. For example, to find the amplitude of the third-order IM products, a line having a slope of three times the linear gain is drawn from the intercept point; for the fifth-order products, the slope is five times the linear gain, and so forth. For the given input power, the linear gain line gives the output power at the input frequency; the three-times slope line gives the power out in third-order IM products.

Log receivers: Signals and interference are most difficult to separate when they occur simultaneously in time. An example is where a moving target echo arrives at the receiver simultaneously with clutter. This situation requires a linear receiver, also know as a normal receiver. If target echoes and interference occur at different times, intermodulation is not possible and non-linear amplification can occur without penalty. An example of the usefulness of this non-linear amplification in radars involves log receivers.

In a log receiver, the amplitude of the output is proportional to the *logarithm* of the amplitude of the input. If small echo signals and large interfering signals are applied to a log receiver, but in different range bins, the small targets may be detectable in the presence of the large interference because the log receiver compresses the output dynamic range. An example would be the two signals shown in Fig. 2-50. The −80-dBm target echo in range bin 17 would be very difficult to see at the output of a normal receiver (Fig.2-50a) in the presence of the −30-dBm interference in range bin 19 (the difference of 50 dB is a voltage ratio of 1:316. With a log receiver, the outputs might be 2 volts for the target and 7 volts for the interference, and they would both be easily seen (Fig. 2-50b).

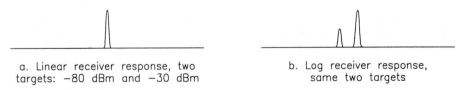

a. Linear receiver response, two targets: −80 dBm and −30 dBm

b. Log receiver response, same two targets

Figure 2-50. Linear and Log Receiver Response Example.

The primary use of log receivers is in situations where non-linear effects are less important that being able to simultaneously view very large and very small signals. Examples include weather displays, displays where certain types of ECM are present, and systems where small targets (such as missiles and bullets) may separate from large targets (such as aircraft and ships).

Linearity and Doppler processing for MTI: Doppler analysis is the method used to separate clutter from moving targets when the two may occur simultaneously in time. It requires very linear amplification to do well, and the radar's normal receiver is used for Doppler processing. However, receivers in some MTI systems, particularly those in older radars, purposely limit (non-linearly amplify) the signal prior to MTI processing. This is to reduce the dynamic range required of the processor and accommodate phase MTI processing (Ch. 12).

Limiting introduces the two MTI errors associated with non-linearity. First, intermodulation products from two strong signals with differing Doppler shifts can cause false targets which interfere with desired targets. This is particularly true with airborne radars which can, because of antenna sidelobe responses, simultaneously have clutter returns moving at different velocities (Ch. 4).

The second effect is a consequence of MTI processing. Moving targets are identified by the fact that one or more of their signal parameters changes from hit to hit. Modern MTIs look for both phase changes, caused by motion toward or away from the radar, and amplitude changes, caused by scintillation (Ch. 4). Fixed targets are cancelled because they lack these changes. The primary limitation on how well MTI performs is the signal variation introduced by the radar itself, with the search system's antenna scan (see Sec. 3-4). Non-linearities in the receiver, particularly those which cause limiting, magnify these antenna-induced changes and cause stationary targets to appear to be moving. This will be discussed further in Chs. 11 and 12.

2-7. Signal Processing Functions and Parameters

The signal processor's function is to treat target echo energy preferentially with respect to interfering signals. Signal processing involves dividing the signal space into segments (bins) in one or more dimension (for example, range and Doppler). Two conditions then allow improvements in signal detectability. First, the target echo signal concentrates into a single bin whereas some types of interference (any with random characteristics) are spread equally among all bins. The second condition is that the target echo signal concentrates into one bin and the interference concentrates into another.

In most signal processes, signal is gathered and stored from several target hits, and then summed in various ways. These signal integrations can be either coherent or non-coherent, depending on whether or not the signals' phase angles are considered in the summations. Many summation processes recover the Doppler shift of targets and interference.

The signal processor: Most modern signal processors are implemented in digital computers. A typical digital signal processor is shown in Fig. 2-51. Not all of the blocks are present in all signal processors nor are they always in the order shown.

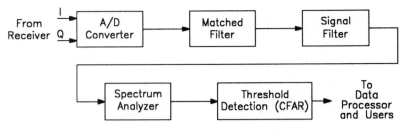

Figure 2-51. Signal Processor.

Virtually all modern signal processors work with the I/Q demodulated receiver output (Sec. 6-2), which is first digitized in the A/D converter. The digitized I and Q voltages are then applied to a matched filter, where pulse compression (if present) takes place.

The signal is then applied to the signal filter where a narrow band of frequencies is removed from it. This band reject filter, the response of which is shown in Fig. 2-52, has the task of removing clutter. Since clutter aliases to all multiples of the PRF (copies of the clutter spectrum appear at all multiples of the PRF — Sec 6-1), the signal filter must have a periodic frequency response. It is usually implemented as a delay-line digital filter (Ch. 11). Again, this block is not present in all signal processors.

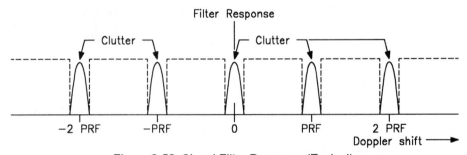

Figure 2-52. Signal Filter Response (Typical).

Next, the signal is applied to a spectrum analyzer, where several things happen (Fig. 2-53). First, target echoes are segregated by Doppler shift. This allows residual clutter (that which passed through the signal filter) to be further suppressed, since its Doppler shift should be different from that of targets. Second, noise or any other random interference is spread equally through all Doppler frequencies, reducing the interfering energy competing directly with a target (whose energy concentrates near a single Doppler frequency). Third, the segregation by Doppler frequency allows the velocity of moving targets to be determined.

Finally, the signal processor establishes the threshold by which detections are made (Sec. 1-6 and Ch. 5). The criterion for setting the threshold is usually to allow a fixed number of false detections to pass in a given time, or to have a constant false alarm rate (CFAR). This task is thus accomplished in the CFAR circuit or algorithm. Some authors place the detection function in the data processor. It really makes little difference which block houses it and it shall be treated here as pretty much a separate subject.

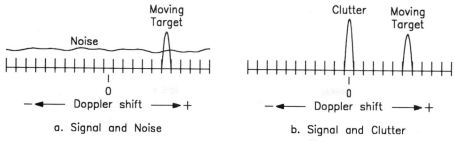

Figure 2-53. Spectrum Analyzer Action.

Signal processing parameters: Several key signal and signal processing parameters determine the effectiveness of the signal processor.

— *Information and signal bandwidths:* Two bandwidths are involved with signals carrying information. They are the signal bandwidth, which is that required to transfer the signal from one point to another, and the information bandwidth, which is the rate of the information carried by the signal. An example would be the transmission of the monaural component of music over a commercial broadcast FM radio station. The information bandwidth of the monaural music is 15 kHz; the highest music frequency allowed is 15,000 Hz. Fully modulated, this music is carried on a radio signal with a bandwidth of approximately 200 kHz, or 13 times the information bandwidth. This apparent waste of communication space allows signal processing to recover the information (music) in the presence of large amounts of interference (noise).

— *Information capacity of a communication channel:* In 1948, Claude E. Shannon published a paper (which later became a book [14]) in which he showed that the information capacity of a communication channel is determined by both its bandwidth and by its output signal-to-noise ratio. The classic relationship developed by him is

$$C = B \log_2(1 + S/N) \qquad (2\text{-}105)$$

C = the information capacity
B = the signal bandwidth
S/N = the power signal-to-noise ratio
$\log_2$ = the logarithm to the base 2

An examination of Eq. 2-105 shows that, given sufficient signal-to-noise ratio, a communication channel can carry information at rates far exceeding its own bandwidth! This apparent contradiction can be explained by the fact that with large S/N, very small signal changes are recoverable. If the information is encoded so that the rate of transmission of events does not exceed the bandwidth, but that each event contains more than one bit of information, the ability to use these small changes can be taken advantage of. For example, 19,200 bits of information can be communicated by transmitting 1,200 events, each containing 16 bits of information. Of course, each event can have 65,536 possible states, demanding a high S/N to decode all states without error. Since this side of the Shannon relationship is seldom applicable to radar, it will not be pursued further. An excellent reference is J.R. Pierce's *Symbols, Signals, and Noise* [15].

On the other side, Shannon's work also indicates that if the information rate is deliberately kept smaller than the signal bandwidth, this low-rate information can be recovered despite signal-to-noise ratios of much less than unity. For very low signal-to-noise ratios, the relationship (from Dixon [16]) becomes

$$C \approx 1.44 \, B \, S/N \tag{2-106}$$

This is the basic relationship which allows signal processing to be useful. If the information rate is held much lower than the signal bandwidth, the information can be recovered from signal-to-noise ratios of much less than unity. This situation is typical in radars.

— *Process gain:* The effectiveness of a signal process is quantified by two measures: the process gain and the jamming margin. Process gain is defined as the ratio of the signal processor's output signal-to-interference ratio to its input signal-to-interference ratio.

$$G_P \equiv \frac{S/I_o}{S/I_i} \quad \text{(dimensionless)} \tag{2-107}$$

G_P = process gain

S/I_o = the output signal power to interfering power ratio

S/I_i = the input signal-to-interference (power) ratio

In bandwidth terms, the process gain is the ratio of signal bandwidth to information bandwidth.

$$G_P = B / R_I \tag{2-108}$$

B = the signal bandwidth (Hertz)

R_I = the information bandwidth (bits per second)

— *Jamming margin:* Jamming margin is the ratio of interference-to-signal at the input of the signal processor which results in a minimally detectable signal at the output.

$$M_J = I/S_i \text{ for a given } S/I_o \tag{2-109}$$

M_J = the jamming margin

I/S_i = the input interference-to-signal (power) ratio

Jamming margin is related to process gain as

$$M_J = G_P / (S/I_{min}) \tag{2-110}$$

S/I_{min} = the minimum signal-to-interference ratio at the processor's output for detection

— *Process noise:* As with all physically realizable processes, the process itself introduces interference, called process noise. It is in general the result of the quantization of signals (quantization noise) and the fact that the process is carried out only at intervals (as it must be in digital systems). Further discussion of process noise is beyond the scope of this book.

Doppler processing parameters: The following definitions indicate the effectiveness of signal processing for suppressing clutter.

- *Moving target indicator improvement factor (MTI-I):* The MTI-I is the measure of how much the signal-to-clutter ratio is improved in the signal processor. Note that its definition is identical to that of process gain if the interference is clutter. MTI-I is thus the signal processor's process gain to signal versus clutter. It is also sometimes called the *improvement factor (I)*.

$$MTI\text{-}I = I \equiv \frac{S/C_o}{S/C_i} \qquad (2\text{-}111)$$

$MTI\text{-}I$ and I = designations for the moving target indicator improvement factor

S/C_o = the signal-to-clutter ratio out of the process

S/C_i = the signal-to-clutter ratio into the process

- *Sub-clutter visibility (SCV):* Sub-clutter visibility is the maximum ratio of clutter to signal which results in a "visible" target. Its definition is the same as that of jamming margin, with clutter as the interference. Sub-clutter visibility is thus the processor's jamming margin to signal versus clutter.

$$SCV \equiv C/S_i \text{ for a "visible" signal at the processor's output} \qquad (2\text{-}112)$$

C/S_i = the clutter-to-signal ratio into the processor

$$SCV = \frac{MTI\text{-}I}{S/C_{min}} \qquad (2\text{-}113)$$

S/C_{min} = the minimum signal-to-clutter ratio for a "visible" signal

Example 2-14: A CW radar receives echoes with a bandwidth of 200 kHz. The echoes consist of targets with Doppler shifts ranging from −100 kHz to +100 kHz. The targets are expected to have Doppler bandwidths of no greater than 1000 Hz. If the Doppler spectrum could be separated into 200 bins, each 1000 Hz wide, would the signal-to-noise ratio in each bin be improved over that from the 200,000-Hz bandwidth? If so, by how much?

Solution: Each target in this radar has an information bandwidth no greater than 1000 Hz (the implication is that range and range resolution, requiring large bandwidths, are not factors here). If the signals will always be found within a 1000-Hz bandwidth, it makes no sense to accept 200,000 Hz of noise (remember that noise power is directly proportional to bandwidth – Eq. 2-86). The signal-to-noise ratio to a target within one 1000-Hz bin is 200 times that of the same target within a 200,000-Hz bin. This is because the noise energy is spread equally among all 200 bins, whereas the target is concentrated into a single bin. The process gain of the process which develops the 1000-Hz bins from 200,000 Hz of bandwidth is 200, from Eq. 2-107. A Fourier transform with 1000-Hz bins would implement this very nicely.

Note that Eq. 2-108 is not operative here. In order for it to be valid, the 1000 Hz of Doppler information would have to be spread over the entire 200,000-Hz spectrum. Pulse compression waveforms and processing do this, but Doppler processing normally does not.

Example 2-15: Two targets are received by a radar, one at short range and one at long. The long-range target has a received power level of -64.0 dBm, and the short-range target's received power is -21.3 dBm. At the same range as the nearer range target, there is clutter which produces a received echo power of -4.8 dBm. Noise jamming is present at a received power level of -49.0 dBm at all ranges. The radar has a noise figure of 4.0 dB and a bandwidth of 250 kHz.

If after signal processing, the near target has a signal-to-clutter ratio of 20.0 dB and the far target has a signal-to-noise ratio of 31.0 dB, find the MTI improvement factor and the process gain of the system to noise.

Solution: Starting with the long-range target, two sources of interference are present: thermal noise at -116.0 dBm (Eq. 2-88) and noise jamming at -49.0 dBm. The noise jamming is identical to the thermal noise except for its amplitude, which is a factor of 67 dB greater (about 5 million). The thermal noise is very thoroughly swamped by the noise jamming and thus plays no significant role.

The signal-to-noise (jamming) ratio before signal processing is -15 dB (the noise jamming is greater than the signal by a power factor of 32). If the signal-to-noise ratio after signal processing is 31.0 dB, Eq. 2-107 shows that the process gain to noise (jamming) is 46.0 dB (40,000).

Three sources of interference affect the short range target: receiver thermal noise (-116.0 dBm), noise jamming (-45.0 dBm), and clutter (-4.8 dBm). The signal-to-interference ratios before signal processing are 94.7 dB to thermal noise, 27.7 dB to noise jamming, and -16.5 dB to clutter. Even without signal processing, the noise and noise jamming will have negligible effect. From Eq. 2-111, the MTI-I, which is the process gain to clutter, is 46.5 dB, a factor of 44,700.

Comment: In theory, an MTI-I of 46.5 dB is easy to implement. In practice, this improvement is easy to realize if the radar is monopulse tracking and always points the maximum gain of the antenna beam at the target. If the radar is searching, the gain pointed at the target varies as the antenna scans past it, and the signal received varies in amplitude even if the target has a constant RCS. This makes stationary targets appear to move to the MTI or MTD and dramatically reduces the available improvement factor (Ch. 12). Since phased array antennas can move beams to any position instantly, they do not have to continuously scan, but can dwell in one location for a look, then "step" the beam to the next scan location (known as step scanning). This would seem to make phased array radars inherently capable of higher improvement factors than those with mechanically scanned antennas.

Example 2-16: In the radar of Ex. 2-15, if a signal-to-clutter ratio of 16.0 dB is necessary for "visibility," find the sub-clutter visibility.

Solution: The system requires a signal-to-clutter ratio of 16.0 dB (40) for seeing targets in clutter. The MTI improves the signal-to-clutter ratio by a factor of 44,700. It follows (Eqs. 2-112 and 2-113) that the clutter can then be 1120 times the signal at the input to signal processing and the signal will be 40 times the clutter at the output. The sub-clutter visibility is therefore 1120, or 30.5 dB.

2-8. System Stability Requirements

The information about targets is contained in the amplitude and phase of their echo signals. The radar's ability to extract this information depends on having a precise knowledge of the amplitude and phase of the transmitted signal and of the internal oscillators. In the final analysis, the amplitude and phase stability of the transmitted and internal waveforms determine the radar's ability to extract Doppler information.

The high stability required in Doppler radars manifests itself in the production of spectrally pure waves in the frequency generation circuitry, transmitter, receiver, and I/Q demodulator. These waves must have only those modulations intended, such as pulse modulation for pulse radars and frequency modulation for ranging in CW radars. Spurious modulations in amplitude, frequency, and phase will result in unaccounted-for sidebands which may cause stationary targets to be interpreted as moving. Figure 2-54 illustrates the spectrum of clutter and a moving target for a stationary radar.

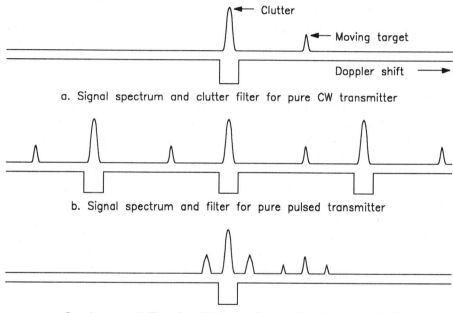

Figure 2-54. Effect of Transmit Spurious Modulations on Echo Spectrum.

In Fig. 2-54a, a spectrally pure CW transmitter (no spurious modulation sidebands) produces a single pure clutter spectrum, which is readily filtered. Figure 2-54b shows the result from a spectrally pure pulsed transmitter. The only difference is that the pulsing (sampling) of the target volume produces multiple copies of the clutter and target spectra (aliasing). The clutter is still filterable, albeit requiring a more sophisticated periodic filter.

Figure 2-54c shows a CW transmitter having spurious modulation sidebands. The clutter spectrum now also has sidebands which are outside the clutter filter and are interpreted as moving targets. Since in typical Doppler radars the clutter amplitude is normally tens of decibels greater than target amplitudes, it does not take large spurious modulation sidebands to markedly degrade the operation of the system. The same problem occurs in pulsed systems.

Amplitude stability requirements: Amplitude modulation (AM) of the transmitted waveform causes one pair of modulation sidebands. The spurious frequencies introduced are at the carrier plus the modulation frequency and the carrier minus the modulation frequency. See Fig. 2-55. Each modulation frequency can thus cause a stationary target to produce two spurious targets with apparent "Doppler" shifts of plus and minus the frequency of the modulation. For example, modulation at 120 Hz (the primary ripple frequency of 60-Hz single-phase power supplies) will cause clutter to produce moving targets at ± 120 Hz Doppler, corresponding to velocities of ± 30 Kt to an L-Band radar. The power in each AM sideband, as a fraction of the carrier power, is

$$G_{SB} = 0.25 \, m_A^2 \qquad (2\text{-}114)$$

G_{SB} = the power in one AM sideband as a fraction of the carrier power

m_A = the amplitude modulation index

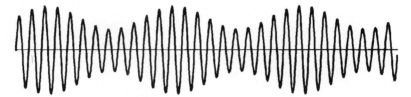

Figure 2-55. Amplitude Modulation.

For example, if a transmitter has amplitude modulation with an m_A of 0.5% (0.005), each clutter target will produce two "moving" targets which are 52 dB below the clutter (each sideband has 6.25×10^{-6} times the carrier power). Since many radars can detect moving targets in clutter 50+ dB stronger than the target, this spurious modulation would present a significant probability of false targets being declared.

Phase (angle modulation) stability requirements: To get an idea of the phase stability required, please refer to Fig. 2-56. The phasors shown represent a moving target's echo signal and a large stationary clutter return occurring simultaneously in time. The composite (sum) of these two signals is what the radar receives and processes. Assuming the clutter signal has a constant phase and the target echo can have any phase, the phase variation of the composite caused by the moving target is

$$\Delta\phi_T = 2 \sin^{-1}(V_{TGT}/V_{CL}) \qquad (2\text{-}115)$$
$$\Delta\phi_T = 2 \sin^{-1}\sqrt{S/C} \qquad (2\text{-}116)$$

$\Delta\phi_T$ = the change in phase of the composite caused by the interaction of the fixed clutter and the moving target

V_{TGT} = the rms voltage of the target echo

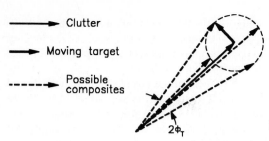

Figure 2-56. Phase Stability Concept.

Table 2-4
Composite Phase Change versus
Signal-to-Clutter Ratio

S/C	V_{TGT}/V_{CL}	$\Delta\phi_T$
−10 dB	0.316	35.1°
−20 dB	0.1	11.4°
−30 dB	0.032	3.6°
−40 dB	0.01	1.15°
−50 dB	0.0032	0.36°
−60 dB	0.001	0.12°
−70 dB	0.00032	0.036°
−80 dB	0.0001	0.012°

V_{CL} = the rms voltage of the clutter echo

S/C = the power signal-to-clutter ratio

Table 2-4 gives the phase variations expected of the composite signal-plus-clutter for various signal-to-clutter ratios.

If phase instabilities exist in the system, their main manifestation is to make fixed targets (clutter) appear to be moving. If Eq. 2-116 is solved for signal-to-clutter ratio and the system phase instability is substituted for composite phase variation, the result is an equation which gives the maximum amount of improvement in signal-to-clutter ratio which can be expected of a moving target indicator. In the presence of phase instability, fixed clutter appears to be a composite containing a moving target whose S/C is expressed by Eq. 2-117, which represents the lower limit of signal-to-clutter ratio which can be sensed for a given level of system phase instability.

$$S/C_{MIN} = [\sin(\Delta\phi_S/2)]^2 \qquad (2\text{-}117)$$

S/C_{MIN} = the minimum signal-to-clutter ratio which can be sensed with the given system phase instability

$\Delta\phi_S$ = the system phase instability

2-8. System Stability Requirements

Signal processing and system instability: Some types of signal processing, particularly spectrum analysis, will improve the detection of targets in the presence of the level of system instability discussed above. This is done by segregating the targets and instabilities into Doppler bins which are independent of one another. In this case, only the instability components which fall into the specific bandwidth (bin) of the target compete with it, and the remainder of the instability energy is rejected. If the instability is random, this type of signal processing improves the signal-to-instability energy ratio by the number of bins processed. For example, if 32 Doppler bins were used, the moving target energy would fall into one of them (or at worst be split between two), while the random system instability energy would be divided equally, with only 1/32 of it landing in the target's bin. The signal-to-clutter ration would therefore be improved by a factor of approximately 32 (15 dB), and now a -88.2-dBm target could be detected. If, on the other hand the instability were not random, it would be concentrated in one or more Doppler bins. If it were in a single bin, the target might be relatively clear of clutter energy, unless the instability energy were in the same bin as the target, but the clutter would then appear as a false moving target.

MTI processing usually has only two Doppler bins, one for stationary targets and one for moving targets. It is as though all the Doppler bins from a filter bank except the one containing stationary targets were summed to produce a single "moving target" bin (Fig. 2-57). If there is one moving target plus clutter, this latter bin contains the target echo from only one filter-bank bin, but the instabilities from all of them. This process of grouping bins together is called collapsing and results in a degraded signal-to-interference ratio. If 32 bins were used and one of them (the constant phase bin) were discarded, 31/32 of the instability energy would remain to compete with the moving target.

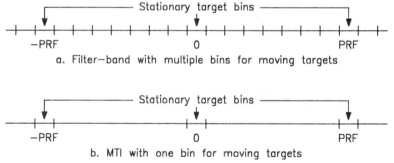

Figure 2-57. Effect of Doppler Collapsing in MTI.

Instability effects on system design: It will be shown that the spurious modulation phenomenon alone in many cases dictates the type of transmitter a radar must have. Some transmitter types (klystrons, travelling wave tubes, and solid state) are much more stable in phase than others (cross-field amplifiers and magnetrons). The problem shown here is most severe in look-down airborne radars, where good Doppler performance is essential. Virtually all look-down radars use travelling wave tube (TWT) transmitters for this reason.

The MTI Improvement factor (MTI-I) is limited by system stability. Table 2-5 gives the effect on MTI-I of several sources of instability in radar systems.

Table 2-5. Limits on MTI-I from Instabilities.
(After Shrader and Gregers-Hansen [17])

Pulse-to-pulse instability	Limit on MTI-I
Transmit frequency	$20 \log[1/(\pi \Delta f \tau)]$
STALO or COHO frequency	$20 \log[1/(2\pi \Delta f T_P)]$
Transmitter phase shift	$20 \log(1/\Delta\phi)$
COHO locking	$20 \log(1/\Delta\phi)$
Pulse timing	$20 \log\{\tau/[2^{1/2} \Delta t (B\tau)^{1/2}]\}$
Pulse width	$20 \log\{\tau/[\Delta\tau (B\tau)^{1/2}]\}$
Pulse amplitude	$20 \log(A/\Delta A)$
A/D jitter	$20 \log\{\tau[J (B\tau)^{1/2}]\}$
A/D jitter with pulse compression following A/D conversion	$20 \log[\tau/(JB\tau)]$

Δf = the interpulse frequency change
τ = the transmitted pulse width (same as τ_E)
T_P = the propagation time to target and back
$\Delta\phi$ = the interpulse phase change
Δt = time jitter
J = the A/D sample time jitter
$B\tau$ = the time-bandwidth product of the pulse
$\Delta\tau$ = the pulse-width jitter
A = the pulse amplitude
ΔA = the interpulse amplitude change

2-9. Further Information

The material presented in this chapter is basic to radar systems. As seen in Ch. 1, there are many general radar references. There are surprisingly few modern books dedicated to the radar's subsystems. The following books are recommended for further information about the subsystems.

General:

[1] M. I. Skolnik, *Radar Handbook* 2nd ed., New York: McGraw-Hill, 1990.

Waveforms: The following reference is a classic treatment of radar waveforms and their capabilities. It is highly detailed and analytical and is referred to more extensively in Ch. 13.

[2] A. W. Rihaczek, *Principles of High Resolution Radar*, Palo Alto, CA: Peninsula Press, 1977.

Transmitters: This reference is a thorough treatment of transmitters. It is unfortunately out of print, but many libraries have copies.

[3] G. W. Ewell, *Radar Transmitters*, New York: McGraw-Hill, 1981.

Antennas: A thorough engineering treatment of antennas is given in the following reference. Chapters 9, 16, 17, 19, 20, 32, 33, and 34 contain most of the radar information.

[4]   R. C. Johnson and H. Jasik, *Antenna Engineering Handbook*, New York: McGraw-Hill, 1984.

Receivers: Few books specifically addressing receivers, particularly radar receivers, are available. Most communication texts include receivers. One book which may be of help is

[5]   J. B. Tsui, *Microwave Receivers with Electronic Warfare Applications*, New York: John Wiley & Sons, 1986.

Signal processors: Many signal processing texts exist, but few target radars specifically. See Ch. 11 references for examples of useful texts.

Instabilities and spurious modulations: A book exists which treats this subject in detail.

[6]   S. J. Goldman, *Phase Noise Analysis in Radar Systems*, New York: John Wiley & Sons, 1989.

Cited references:

[7]   R. C. Johnson, Ch. 1 in R.C. Johnson and H. Jasik, *Antenna Engineering Handbook*, New York: McGraw-Hill, 1984, pp. 1-15.

[8]   W. L. Stutzman and G. A. Thiele, *Antenna Theory and Design*, New York: John Wiley & Sons, 1981, p. 397.

[9]   R. J. Stegen, "The Gain-Bandwidth Product of an Antenna," *IEEE Transactions on Antennas and Propagation*, vol. AP-12, July 1964, pp. 505-506.

[10] M.B. Ringel, D.H. Mooney, and W.H. Long III, "F-16 Pulse Doppler Radar (AN/APG-66) Performance," *IEEE Transactions on Aerospace and Electronic Systems*, vol. AES-19, no. 1, January 1983.

[11] L. V. Blake, Ch. 2 in M.I. Skolnik (ed.), *Radar Handbook*, 2nd ed., New York: McGraw-Hill, 1990.

[12] D. K. Barton, Ch. 25 in D.G. Fink (ed.), *Electronic Engineers' Handbook*, New York: McGraw-Hill, 1975.

[13] M. I. Skolnik, *Introduction to Radar Systems* 2nd ed., New York: McGraw-Hill, 1980, pp. 369-375.

[14] C. E. Shannon and W. Weaver, *A Mathematical Theory of Communication*, Urbana: University of Illinois Press, 1949.

[15] J. R. Pierce, *Symbols, Signals, and Noise*, New York: Harper and Row, 1961.

[16] R. C. Dixon, *Spread Spectrum Systems*, New York: John Wiley & Sons, 1976, p. 4.

[17] W. W. Shrader and V. Gregers-Hansen, Ch. 15 in M.I. Skolnik (ed.), *Radar Handbook* 2nd ed., New York: McGraw-Hill, 1990.

3

THE RADAR EQUATION

3-1. Radar Equation Introduction

Before target information can be extracted from an echo signal, that signal must be of sufficient magnitude to overcome the effects of interference. The radar equation is used to predict echo power and interfering power to assist in making the determination of whether or not this condition is met. Use of the radar equation accomplishes the following:

— Assists in the design of radar systems to meet the detection specifications set by the users
— Establishes the relationship between the signal power received and the radar and target parameters
— Describes the power received from interfering sources, including thermal noise, clutter, jamming, and EMI
— Provides a means for predicting signal-to-interference ratios, and for predicting the maximum range at which targets of a given RCS will produce a specified signal-to-interference ratio

Several parameters affect the signal and interfering power received by the radar system.

— The radar's operating parameters, including transmitted power, transmitted energy, transmitted waveform, antenna gain and effective aperture, receiver noise performance, radar system losses, and the minimum signal-to-interference ratio for detection
— Target parameters, including radar cross-section (RCS), RCS fluctuations, and the target's range
— The propagation medium parameters, including RF energy absorption by gasses and the scattering of RF energy by particles in the medium

In Ch. 1 it was shown how signals behave during transmission, forward propagation, reflection, backscatter propagation, and reception. That discussion led to Eq. 1-22, a form of the radar equation. That equation, however, was a special case. This chapter expands and generalizes, fills in missing detail, and ends in a set of radar equations that apply to differing radars and conditions. There are many more forms of the radar equation than are covered here, although this chapter includes the most common and useful. After studying this material, one should be able to derive appropriate relationships for other situations.

Note that although it is possible to calculate the predicted signal-to-interference ratios and detection ranges to great precision, few of the resulting digits are significant. Many uncertainties exist in the system, target, and propagation parameters which, if not worked out, can cause answers to be uncertain by 10%, 20%, or more.

The radar equation *is* good, though, for showing how changing parameters or scenarios affect system performance. For example, if a particular target is detectable in noise at a range of 50 mi, reducing the target's radar cross-section by a factor of 10 (no easy task) will only reduce the detection range by a factor of $10^{1/4}$ to 28 mi. If, however, the interference is sea clutter and the original target is detectable to 20 mi, reducing the RCS to 1/10 of its value reduces detection range in the same clutter by 10^1, to 2 mi.

3-2. Summary of the Radar Equation

Predicting the performance of radars using the radar equation involves first selecting the appropriate equation. Some of the multitude of equation forms result from the many different types of target/interference scenarios; other forms result because different authors use different parameters and notations. This is the reason for placing a summary at the beginning of the study.

The most familiar form of the radar equation is for point targets interfered with by the thermal noise generated in the radar's receiver. This form was discussed in Ch. 1. Equations 1-22, 1-26, and 1-27 are combined here as Eq. 3-1.

$$S/N = \frac{P_T G^2 \lambda^2 \sigma}{(4\pi)^3 R^4 L_S L_A K T_0 B F} \quad \text{(dimensionless, single echo)} \quad (3\text{-}1)$$

P_T = the transmit peak power (Watts)
G = the antenna gain (dimensionless)
σ = the target radar cross-section (square meters)
λ = the wavelength (meters)
R = the one-way range from radar to target (meters)
L_S = the system loss
L_A = the propagation path loss
K = Boltzmann's constant (1.38×10^{-23} J/°K)
T_0 = 290°K
B = the bandwidth (Hertz)
F = the system noise factor (dimensionless)

The signal-to-noise ratio is directly proportional to transmit power, directly proportional to target RCS, and inversely proportional to the fourth power of range, as has been shown. However, Eq. 3-1 is valid only if the target is a point target (see below) and if the interference is independent of range, such as noise generated internally within the radar. The factors for non-point targets and other forms of interference are different.

Point targets are those which are totally contained within the radar's resolution cell (Fig. 3-1). Most microwave radars see targets as points. Exceptions include high resolution ground mapping radars, diagnostic radars (which map targets to determine where they scatter), and laser radars. These radars usually see their targets as area targets, which are larger than the resolution cell and on which only a limited area is illuminated.

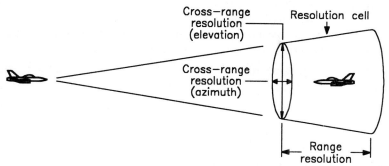

Figure 3-1. Point Target Definition.

Additionally, the detection range of some radars, including military radars used in hostile environments, are seldom limited by internally generated noise. In these, clutter and ECM are usually limiting. Others, such as air traffic control radars, may or may not be ultimately limited by noise, depending on target range; at close ranges, clutter is limiting — at long ranges, it is noise.

Table 3-1 summarizes the effect of various system and target parameters on signal-to-interference ratios and detection range. It shows how much (or how little) these parameters affect system performance and serves as the basis for the discussions to follow. It is by no means all-inclusive — there are many more situations, some of which will be examined later in this book.

Table 3-1. Radar Equation Summary — Effect on Detection

Effect on:	Range Rule	Transmit Power*	Antenna Gain†	Target RCS	See Section
Point target signal in internal noise[A]	R^{-4}	$P_T^{1/4}$	$G^{1/2}$	$\sigma^{1/4}$	3-3
Low-angle area target signal in noise[B]	R^{-3}	$P_T^{1/3}$	$G^{2/3}$	$\sigma^{1/3}$	3-10
High-angle area target signal in noise[C]	R^{-2}	$P_T^{1/2}$	G^1	$\sigma^{1/2}$	3-10
Volume target signal in noise[D]	R^{-2}	$P_T^{1/2}$	G^1	$\sigma^{1/2}$	3-11
Point target signal in low-angle area clutter[E]	R^{-1}	P_T^0	G^0	σ^1	3-10
Point target signal in high-angle area clutter[F]	R^{-2}	P_T^0	G^2	σ^1	3-10
Point target signal in volume clutter[G]	R^{-2}	P_T^0	G^2	$\sigma^{1/2}$	3-11
Point target signal in SP jamming[H]	R^{-2}	$P_T^{1/2}$	$G^{1/2}$	$\sigma^{1/2}$	3-12
Point target signal in SO jamming[I]	R^{-4}	$P_T^{1/4}$	$G^{1/2}$	$\sigma^{1/4}$	3-13
Point target signal in augmentation[J]	R^0	P_T^0	G^0	σ^0	3-14

3-2. Summary of the Radar Equation

Notes and definitions for Table 3-1: Each situation in Table 3-1 is described fully in the section given in the last column of the table.

* The range rule is the relationship between target range and signal-to-interference ratio. For example, for a point target in noise (R^{-4}), a doubling of range causes S/N to drop to 1/16 (2^{-4}) of its original value.

† The table assumes a single antenna. If two antennas are used, replace G with $G_T G_R$ (G_T is the transmit antenna gain, G_R the receive antenna gain).

A Internal noise is generated within the radar's receiver. Its value is independent of target range.

B A low-angle area target is one which is larger than the radar's resolution cell and which is viewed from a low grazing angle (Fig. 3-2). Its radar cross-section is proportional to the area within the resolution cell. An example is sea clutter in front of a low-flying aircraft.

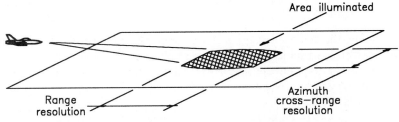

Figure 3-2. Low-Angle Area Target Definition.

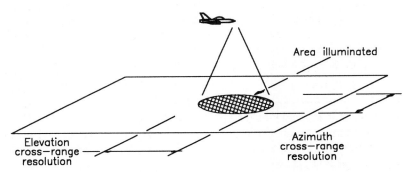

Figure 3-3. High-Angle Area Target Definition.

C A high-angle area target (Fig. 3-3) is one which is larger than the resolution cell (RCS is a function of area illuminated), viewed essentially perpendicular to the plane of the target. Sea clutter viewed by a radar pointed perpendicular to it (as with a radar altimeter) is a high-angle area target.

D A volume target is larger than the radar's resolution cell. It differs from an area target in that it has three dimensions and its radar cross-section is proportional to the volume of the resolution cell containing the target (Fig. 3-4). Common volume targets are weather and chaff (a form of ECM).

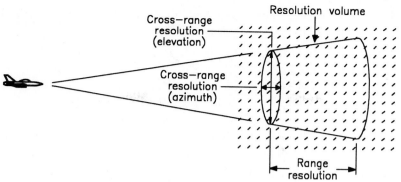

Figure 3-4. Volume Target Definition.

E Area clutter is unwanted interfering signals from an area target. The most common area clutter comes from land and sea (land clutter and sea clutter). If viewed from a low angle, it is low-angle area clutter (Fig. 3-2).

F High-angle area clutter is from the same sources as low-angle area clutter (land and sea), but viewed from a high angle (Fig. 3-3).

G Volume clutter is an unwanted interfering signal from a volume target such as rain, snow, and chaff (Fig. 3-4).

H Self-protection jamming (SPJ) is an interfering signal deliberately emitted by the vehicle the radar is trying to detect (hence the vehicle is protecting itself). See Fig. 3-5. It is also known as self-screening jamming.

I Stand-off jamming (SOJ — Fig. 3-6) is an interfering signal deliberately emitted by a vehicle other than the one the radar is trying to detect. These jamming vehicles normally stand away (stand off) from the battle beyond the range of enemy anti-radiation missiles (ARMs).

J Augmentation is a method of generating returns to the radar from the radar's own illumination signal. It has several functions, including making targets appear larger than they really are (for example, a small target drone can simulate an enemy bomber), and self protection jamming. Figure 3-7 shows a simple augmentor. If it is used to simply increase the RCS of a target, the modify block is absent. For jamming, the modify block may contain delay circuitry (for range deception) or frequency shift circuits (Doppler deception).

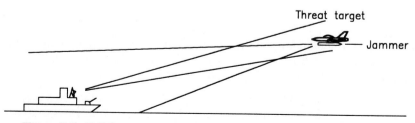

Figure 3-5. Self-Protection (also known as Self-Screening) Jamming.

3-2. Summary of the Radar Equation

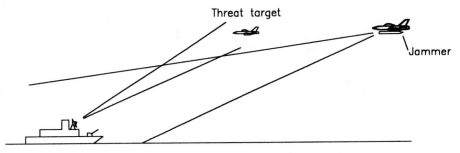

Figure 3-6. Stand-Off Jamming.

Figure 3-7. Augmentation.

3-3. Point Targets in Noise

The development of the elementary radar equation for point targets in noise was started in Sec. 1-4, resulting in Eq. 1-22, repeated here as Eq. 3-2.

$$P_R = \frac{P_T G \sigma A_E}{(4\pi)^2 R^4} \quad \text{(Watts)} \tag{3-2}$$

Equation 3-2 gives the signal power received from a target. If the received power from interfering sources is known, the signal-to-interference ratio (which is always a power ratio) is found by dividing signal power by interfering power.

$$S/I = \frac{P_R}{P_I} = \frac{P_T G \sigma A_E}{(4\pi)^2 R^4 P_I} \quad \text{(dimensionless)} \tag{3-3}$$

S/I = the signal-to-interference ratio

P_I = the interfering power level at the same place in the system receive power is taken

Equation 3-3 is the simplest look at predicting received signal-to-interference ratio from radar parameters, target parameters, and range. It is for a single hit on the target. It is sometimes useful in our thinking to represent this version of the radar equation in a slightly different form, emphasizing its development.

$$S/I = P_T G \underbrace{\frac{1}{4\pi R^2}}_{\substack{\text{Radar ERP} \\ \text{P/A of radar at target}}} \sigma \underbrace{\frac{1}{4\pi R^2}}_{\substack{\text{ERP of target} \\ \text{P/A of target power at radar}}} A_E \underbrace{\frac{1}{P_I}}_{\substack{\text{Power from target captured by radar} \\ \text{Signal-to-interference ratio}}} \quad (3\text{-}4)$$

In Eq. 3-4, the radar's single antenna is represented by two different but related parameters: gain and effective aperture. They are direct functions of one another, as shown in Eq. 2-68. Solved for effective aperture, this relationship is

$$A_E = \frac{\lambda^2}{4\pi} G \quad \text{(square meters)} \quad (3\text{-}5)$$

A_E = the antenna's effective aperture in a certain direction (m²)
λ = the wavelength (meters)
G = the antenna's gain in the same direction as A_E

Substituting Eq. 3-5 into Eq. 3-4 gives a form of the radar equation which uses only one antenna parameter.

$$S/I = \frac{P_T G^2 \lambda^2 \sigma}{(4\pi)^3 R^4 P_I} \quad \text{(dimensionless)} \quad (3\text{-}6)$$

Equation 3-6 is the lossless radar equation for one hit (pulse).

Losses: Three kinds of losses exist in radars: losses within the system itself (system loss), losses in the propagation medium (propagation path loss), and losses caused by multiple signal paths (ground plane loss). They are covered in Sec. 3-4. If losses are considered, the radar equation becomes

$$P_R = \frac{P_T G^2 \lambda^2 \sigma}{(4\pi)^3 R^4 L_S L_A L_{GP}} \quad \text{(Watts)} \quad (3\text{-}7)$$

$$S/I = \frac{P_T G^2 \lambda^2 \sigma}{(4\pi)^3 R^4 P_I L_S L_A L_{GP}} \quad \text{(dimensionless)} \quad (3\text{-}8)$$

L_S = the system loss (> 1)
L_A = the propagation medium loss (> 1)
L_{GP} = the ground plane loss (This factor can be less than unity, increasing signal power.)

Multiple hits: If more than one hit is processed (the usual case), the signal-to-interference ratio can be increased. This is the function of signal processing. In this case, a process gain is applied to the radar equation (Sec. 2-7). The signal-to-interference form of the radar equation then becomes

3-3. Point Targets in Noise

$$S/I = \frac{P_T G^2 \lambda^2 \sigma G_P}{(4\pi)^3 R^4 P_I L_S L_A L_{GP}} \quad \text{(dimensionless)} \quad (3\text{-}9)$$

G_P = the process gain (usually >> 1)

In the special case where the interfering power is internally-generated noise, the process gain is a function of the number of hits processed together. Noise was discussed briefly in Chs. 1 and 2, and will be considered further in Ch. 4. The noise power generated within the radar, taken as though it were a source at the input to the antenna, is (from Eq. 2-88)

$$P_{Ni} = K T_0 B F \quad (3\text{-}10)$$

P_{Ni} = the noise power, taken as though it were at the input to the antenna

K = Boltzmann's constant (1.38×10^{-23} J/°K)

T_0 = 290 °K

F = the noise factor of the radar (noise figure, if in dB)

The signal-to-noise ratio for multiple hits processed together is

$$S/N = \frac{P_T G^2 \lambda^2 \sigma G_P}{(4\pi)^3 R^4 K T_0 B F L_S L_A L_{GP}} \quad (3\text{-}11)$$

The process gain for noise interference is given below, and will be discussed in Sec. 3-5.

$$G_{P(N)} = \frac{N_L}{L_i} \quad \text{(dimensionless)} \quad (3\text{-}12)$$

$G_{P(N)}$ = the process gain to noise

N_L = the number of hits processed together as a look

L_i = the integration loss (Sec. 3-5)

Look energy and average power: It is instructive and useful to consider the relationship between pulse width and bandwidth for the gated CW waveform. Equation 2-29, repeated here, shows that pulse width and bandwidth for this wave are reciprocals of one another.

$$B \approx 1/\tau \quad (3\text{-}13)$$

τ = the gated CW pulse width (seconds)

B = the matched bandwidth (Hertz)

Substituting Eqs. 3-12 and 3-13 into Eq. 3-11 and rearranging produces a most useful form.

$$S/N = \frac{P_T \tau N_L G^2 \lambda^2 \sigma}{(4\pi)^3 R^4 K T_0 F L_S L_A L_{GP} L_i} \quad (3\text{-}14)$$

The first three factors of the numerator are the look energy. It will be shown in Sec. 3-6 that signal-to-noise or signal-to-noise jamming ratio is directly proportional to look energy, *regardless of the waveform* used. This means that the S/N is independent of peak power, as long as sufficient energy is produced — in other words, pulse width can be traded for peak power.

$$S/N = \frac{W_L G^2 \lambda^2 \sigma}{(4\pi)^3 R^4 K T_0 F L_S L_A L_{GP} L_i} \qquad (3\text{-}15)$$

Equation 3-15 can be modified using the relationships from Sec. 2-3 so that the role of average power is evident.

$$S/N = \frac{P_{AVG} T_D G^2 \lambda^2 \sigma}{(4\pi)^3 R^4 K T_0 F L_S L_A L_{GP} L_i} \qquad (3\text{-}16)$$

P_{AVG} = the average (CW equivalent) power
T_D = the dwell or look time

Equation 3-16 is useful with pulsed, CW, and pulse Doppler waveforms. However, as shall be shown in Sec. 3-9, often CW radars are not limited by internal noise but by spillover from the transmitter to receiver. Those which are noise limited often have the transmitter and receiver widely separated.

Using an energy view of signal-to-noise ratio is one of the three keys to low probability of intercept (LPI — see Ch. 14) radar, the concept of which is to allow a radar to be used without alerting the enemy of the radar's presence (it does little good to have "stealthy" targets if their radar transmitters announce their presence for hundreds of miles). LPI is based on the fact that target echo detection is done primarily with illumination energy and interception is done with peak power. The three primary LPI factors are:

— Low peak power, high-energy waveforms
— Encyphered illumination waveforms, with correlation detection based on the illumination code. If the other side doesn't know the code, the radar has a large process gain advantage over the interceptor
— Keeping the radar transmitter off the air except when absolutely necessary, using so-called EMCON - *EM*ission *CON*trol

Example 3-1: A multimode airborne radar has the following specifications in one of its modes (high PRF single target track):

Transmit power	10 kW
Pulse width	1.2 μs
PRF	250,000 pps
Antenna gain	35 dB
Frequency	10.5 GHz (wavelength = 0.0286 m)
Receiver noise figure	3.5 dB
System loss	1.4 dB
Propagation path loss	1.6 dB (assume for all ranges)
Ground plane loss	0 dB

Find the single-hit signal-to-noise ratio for a 2.0 m² target at 55 nmi.

Solution: Several equations developed so far will work for this problem, in particular Eq. 3-11 with process gain equal to one — correct for a single hit without pulse compression — or Eq. 3-14 with $N_L = 1$. All factors must be in ratio form, rather than dB, and all units must be uniform. The antenna gain is 3,160, the noise factor is 2.24, the system loss is 1.38, the propagation loss is 1.45, the range is

3-3. Point Targets in Noise

101,860 m, and the ground plane loss is 1.00. The answer, from either equation, is 0.051 (-12.9 dB). This, as will be shown in Ch. 5, is insufficient for detection. Generally, an S/N of 12 to 16 dB is required for reliable detection.

Example 3-2: If the radar in Ex. 3-1 can coherently integrate (process together using phase) 2048 pulses with an L_i of 1.6 dB, find the new S/N and the integration time (the time of one look).

Solution: Using Eq. 3-12 or 3-14, the single-hit S/N is multiplied by 1,420, which is 2048 less 1.6 dB. The integrated S/N is 72, or 18.6 dB. Except for the most extreme conditions, this is sufficient for reliable detection of the target.

Monostatic radars with separate antennas: To find the radar equation for a monostatic radar which has separate transmit and receive antennas, simply substitute the product of the transmit and receive antenna gains for the square of the single antenna's gain.

$$\text{Replace } G^2 \text{ with } G_T G_R \tag{3-17}$$

G_T = the transmit antenna gain

G_R = the receive antenna gain

The signal-to-noise ratio for multiple hits then becomes (from Eq. 3-11)

$$S/N = \frac{P_T G_T G_R \lambda^2 \sigma G_P}{(4\pi)^3 R^4 K T_0 B F L_S L_A L_{GP}} \tag{3-18}$$

Detection range prediction: One of the primary uses of the radar equation is the prediction of the maximum range at which a particular target can be detected. To accomplish this, the appropriate form of the radar equation is simply solved for range. Once this is done, the maximum range is found where the signal-to-interference ratio is the minimum for detection. Using Eq. 3-34, the maximum range at which a target can be detected becomes

$$R_{max}^4 = \frac{P_T G^2 \lambda^2 \sigma G_P}{(4\pi)^3 P_I L_S L_A L_{GP} S/I_{min}} \tag{3-19}$$

R_{max} = the maximum detection range

S/I_{min} = the minimum S/I for detection

Example 3-3: Find the maximum range at which the radar of Exs. 3-1 and 3-2 can detect a 10 m² target if the minimum signal-to-noise ratio for detection is 16 dB. Find this range for (a) a single hit, and (b) a 2048-hit look.

Solution: Using Eq. 3-19 for the single-hit conditions (G_p =1) and Eq. 1-26 for the interfering power, the maximum detection range for this target from a single hit is 28,800 m (15.6 nmi). For 2048 hits, the process gain becomes 1420 (Ex. 3-2) and the detection range is 176,800 m (95.5 nmi).

Signal-to-noise examples: The detection process will be covered in Ch. 5, but it is useful at this point to observe what the radar sees for different signal-to-interference ratios. Figure 3-8 shows signal plus noise for different signal-to-noise ratios. In each case, there is a target echo in the center of the display. The signal-to-noise ratios are given below the figure; the number in parentheses is the ratio of the rms signal voltage to the rms noise voltage and is the square root of S/N. For low S/N (up to about 16 dB), it would be difficult to reliably place a detection threshold (a horizontal line, in these figures) which would intercept the signal without also intercepting noise peaks, thus producing false alarms.

The signals and noise in Fig. 3-8 have been passed through an approximation of a matched filter (Sec. 2-6). Note that the signal and noise peaks look exactly alike except for amplitude. The amplitude of a signal, after signal processing, is the only mechanism by which it can be separated from interference.

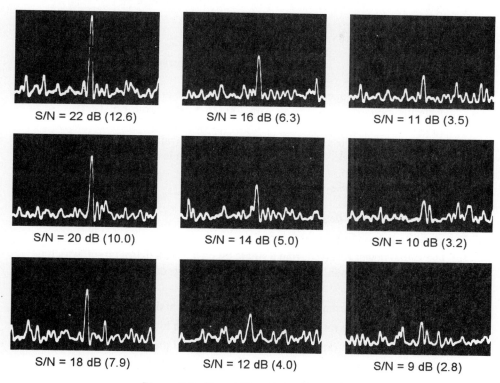

Figure 3-8. Signal Plus Noise Examples.
(Diagrams courtesy of D.L. Mensa. U.S. Government work
not covered by U.S. copyright)

Space gain/loss forms: A factor of the radar equation can be isolated which represents all system parameters which are inconvenient to express in decibels, called the space gain, or space loss, depending on whether it is in the numerator or denominator. Taking Eq. 3-9 as the base point, the radar equation can be reorganized as

$$S/I = \frac{P_T G^2 \sigma G_P}{P_I L_S L_A L_{GP}} \frac{\lambda^2}{(4\pi)^3} \frac{1}{R^4} \tag{3-20}$$

The *space gain* is therefore

$$G_{R\text{-}SPACE} = \frac{\lambda^2}{(4\pi)^3 R^4} \quad (1/m^2) \tag{3-21}$$

3-3. Point Targets in Noise

The *space loss* is

$$L_{R\text{-}SPACE} = \frac{(4\pi)^3 R^4}{\lambda^2} \quad \text{(square meters)} \tag{3-22}$$

When expressed in decibels, the space gain is dB/sm (decibels in terms of the reciprocal of square meters). The space loss is in dBsm (decibels reference one square meter).

Example 3-4: A hypothetical high power early warning radar has the following specifications:

Transmit power	15 MW
Type	Monostatic
Antenna	94-ft diameter array
Frequency	1.00 GHz
Antenna effective aperture	320 m² (half the area)
Antenna gain	44,700 (46.5 dB — from Eq. 3-62)
System Loss	1.0 dB (ratio = 1.26)
Propagation path loss	approx. 1.3 dB (ratio = 1.35)
Ground plane loss	0 dB (1.0) — no loss
Receive bandwidth	1.0 MHz
Process gain to noise	29 dB (800)
Noise figure	1.1 dB (1.29)
Minimum S/N for detection	14 dB (25.1)

Estimate the following for this radar:

(a) The signal power received from a single hit on a -11.4 dBsm target, which is a metal sphere about the size of a basketball, at a range of 2,000 km (about 1,080 nmi)

(b) The noise power present in the radar (assumed to be at the input to the antenna)

(c) The signal-to-noise ratio for this target for a single hit

(d) The signal-to-noise ratio for this target for a signal processing look

(e) The maximum range at which this radar can detect the specified target (a metallic basketball)

Solution: Various equations developed in this section will be used to generate the answers to this exercise, as follows.

(a) Equation 3-18 is used to find the received power from a single hit. The result is 3.62×10^{-15} W (-114.4 dBm).

(b) Equation 1-26 gives the noise power at the input to the antenna as 5.16×10^{-15} W (-112.9 dBm).

(c) The signal-to-noise ratio is the ratio of answer a to answer b (Eq. 1-16), or from Eq. 3-8. It is 0.702, or -1.5 dB (-114.4 dBm $- (-112.9$ dBm$)$). This is not sufficient for detection, as will be shown in Ch. 5.

(d) A signal processing look results in a process gain of 800. This is probably the result of about 1200 hits processed together. The S/N is now 800 times that of part c, or about 560 (27.5 dB, which is -1.5 dB $+ 29$ dB). This is more than sufficient for detection.

(e) Equation 3-19 allows estimation of the maximum range at which a particular target can be detected. The result is 4,350 Km, or about 2,350 nmi. This value can be found in another way by realizing that at 2,000 km, the S/N is 560 and at maximum range, it is 25.1 (from the specifications). The received power can thus be reduced by a factor of 560/25.1=22.3 and still detect the target. Since signal is inversely proportional to the fourth power of range, range can be increased by the fourth root of 22.3, or by a factor of 2.17. The maximum range is about 4,350 km, as found from Eq. 3-18.

3-4. Losses in the Radar Equation

As noted in the previous section, there are three classes of losses which must be considered in predicting received power and estimating the maximum range at which a target will be detected. They are

— *System losses*, losses within the radar system itself
— *Propagation medium losses*, losses in the propagation medium (usually the atmosphere) from radar to target and from target to radar
— *Ground plane losses*, caused by multiple signal paths from radar to target and target to radar, usually from interaction of the transmit and echo electromagnetic waves with the ground or sea

System loss: Losses within the system itself are from many sources. Several are described below, but the reader is cautioned that this list is by no means exhaustive. An expression for the total system loss, possibly incomplete, is given below. Losses in Eq. 3-45 are in dB; if expressed as ratios, they are multiplied.

$$L_S \text{ (dB)} = L_{PL} + L_{PO} + L_{AP} + L_{PW} + L_{SQ} + L_{LIM} + L_C + L_{OP} + L_{NE} \quad (3\text{-}23)$$

L_{PL} = the plumbing loss
L_{PO} = the polarization loss
L_{AP} = the antenna pattern loss, also known as the scan loss
L_{PW} = the pulse width loss
L_{SQ} = the squint loss
L_{LIM} = the limiting loss
L_C = the collapsing loss
L_{OP} = the operator loss
L_{NE} = the non-ideal equipment loss

— *Plumbing loss:* From the transmitter to the duplexer, duplexer to receiver, and duplexer to the antenna, there are various transmission lines and microwave components. Some, such as isolators (to protect the transmitter from reflections) and directional couplers (for testing the radar) are absorbing. They take small (hopefully) amounts of power in the transmission lines and either convert them to heat or divert them for other purposes. Others are reflecting, in that they inadvertently reflect a small (again hopefully) amount of power in the opposite direction from that desired. All transmission line systems exhibit both types of signal diversions. The losses caused by absorption and reflection in the transmission line system of the radar are lumped together as the plumbing loss.

— *Polarization loss:* In some systems, the receive polarization may differ from that transmitted (either slightly or grossly). This results in a loss caused by the transmitted wave polarization not being absorbed completely by the receive antenna. This is more a problem of systems which use separate feeds or antennas for transmit and receive. The polarization loss does not include the lower received signal power caused by twisting of polarization by the target. Polarization twists by the target should show up as a lower measured RCS.

— *Antenna pattern loss (scan loss):* As a search radar antenna scans across a target, it does not view the target with its peak gain for all echo pulses. If the system integrates several echo pulses, they will be of varying magnitudes caused by the two-way antenna beam pattern. Since the radar equation usually contains peak antenna gain, a loss results. See Fig. 3-9 for an example of a lossless scan and a real scan past a constant RCS target. Some authors, particularly L.V. Blake [3], include a propagation pattern factor F_a in the radar equation which describes the antenna's one-way electric field pattern factor (or F_t and F_r if separate antennas are used). This causes the radar equation to be modified. The multiple-hit radar equation forms used to now give the signal-to-noise ratio proportional to

$$S/N \propto G^2 / L_{AP} \qquad (3\text{-}24)$$

Equation 3-24 using the propagation pattern factor would be of the form

$$S/N \propto G^2 F_a^4 \qquad (3\text{-}25)$$

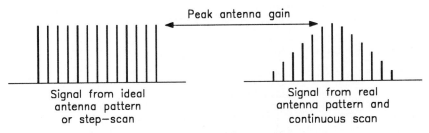

Figure 3-9. Antenna Pattern Loss Principle.
(Received pulse amplitude as search antenna scans across target)

Here, the pattern factor to the fourth power (the factor is in voltage terms, must be squared to get power, is one-way, and must be used twice) replaces the antenna pattern portion of the system loss. Since this form is most useful in search radars, it will be used later in this chapter when a specialized radar equation for searching is written (Sec. 3-6).

Again, signal anomalies resulting from target amplitude fluctuations are not contained in an antenna pattern loss and must be accounted for in the detection characteristics of the radar.

— *Pulse width loss:* There are certain radar antennas in which the transmission line structure distributing power over the surface of the antenna is much longer than the extent of the antenna itself. This is necessary to, for example, electronically position beams by changing the transmit frequency (AN/SPS-48 in Fig. 1-1). Because of this long transmission line structure, it takes a significant time to cause the entire antenna surface to commence (and

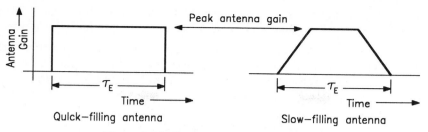

Figure 3-10. Pulse Width Loss Principle.

cease) radiating, sometimes a significant fraction of the pulse. This is called filling the antenna (Fig. 3-10). During the fill time, the antenna does not radiate efficiently and a loss called the pulse width or antenna fill loss occurs. Note that this same phenomenon occurs on receive; a fraction of the receive pulse is lost while the signals from all parts of the antenna make their way through the transmission line system, some more quickly than others.

— *Squint loss (Crossover loss):* In Ch. 7, it will be shown that some types of tracking radars, when following a target on the axis of the antenna, view it with an antenna beam which is squinted (displaced slightly) from the axis. The result is that the antenna gain in the direction of the target is lower than the peak gain (Fig. 3-11). These are the conical scan and lobing radars. In typical systems of this type, the losses are somewhere between 3 dB and 9 dB.

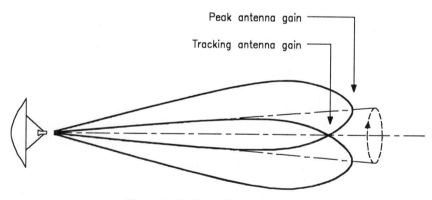

Figure 3-11. Squint Loss Principle.

— *Limiting loss:* If limiting occurs in the receiver, that portion of the signal energy above the saturation point is lost. Limiting is usually undesirable, although a few radars and signal processors introduce it deliberately. See Fig. 3-12 for an example of limiting and its effect.

— *Collapsing loss:* Radars resolve targets in several dimensions, such as range, azimuth angle, elevation angle, and Doppler shift. Within a resolution cell a target produces a certain signal-to-interference ratio. If one or more dimensions are omitted in a display or process, the system is said to be collapsed in that dimension. For example, consider the radar of Fig. 3-13 connected to a plan position indicator (PPI) display (Ch. 10). Even if the radar resolves in elevation, a PPI cannot display this resolution. Hence, an azimuth/range cell on

3-4. Losses in the Radar Equation

Figure 3-12. Limiting Loss Concept.

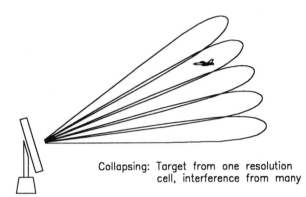

Collapsing: Target from one resolution cell, interference from many

Figure 3-13. Collapsing Principle.

a PPI may contain a target echo from one elevation resolution cell, *but it contains the interference from all elevation cells*, hence the S/I is worse than in data where all resolved dimensions are present.

— *Operator loss:* In many systems, especially those without automatic target detection, the operator plays an integral role in the detection and location of targets. The operator can influence several of the radar parameters, such as gains and signal processing. If poorly trained, distracted, or fatigued, the human operator may not be able to use the full potential of the radar. As can probably be guessed, this is an extremely difficult loss to quantify.

— *Non-ideal equipment loss:* As radar systems age, they may undergo equipment degradations, such as loss of transmit power, increased receiver noise figure, and the like. The aging of system components, particularly if maintenance is neglected, causes the radar to perform differently than it did when new.

Propagation medium loss: Microwave energy does not transit the earth's atmosphere unimpeded. Like light, it is subject to losses, although because the radar's wavelength is much longer than light's, the losses are smaller. In other words, as to light, the atmosphere appears *hazy* to the radar.

Two phenomena cause the propagation path loss: absorption by atmospheric gases and scattering by particles in the atmosphere. The gases in the atmosphere absorb microwave power, especially at frequencies where resonances occur. It is difficult to show completely, in one or two figures, the effect of atmospheric absorption. Figure 3-14 shows the trend of losses caused by the microwave absorption by atmospheric gasses. Detailed data can be found in the writings of Blake [3].

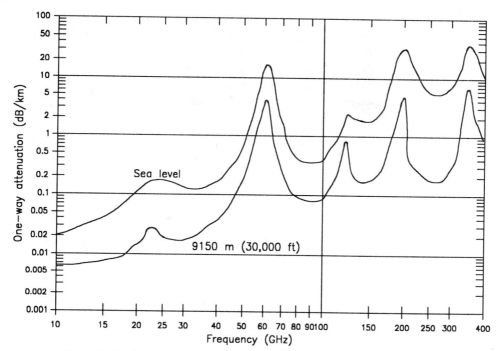

Figure 3-14. Atmospheric Absorption of Microwave Signals (One-Way).
(Data from [10])

The second cause of signal attenuation in the atmosphere is scattering of the signal by precipitation, primarily rainfall. Figure 3-15 shows the principle. Precipitation (which for radar purposes is many scatterers, each smaller than a wavelength) scatters ("deflects") electromagnetic energy traveling through it. The signal level in the organized beam is

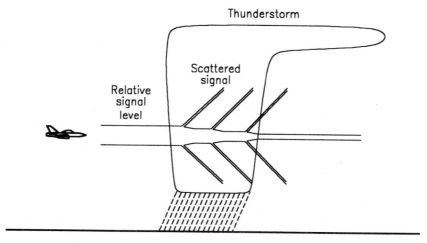

Figure 3-15. Scattering of Signal by Precipitation.

3-4. Losses in the Radar Equation

reduced by the amount of this scattered power. The scattering by particles smaller than a wavelength is strongly frequency sensitive, with the scattering effect of a single particle directly proportional to the fourth power of frequency. Section 4-4 explores this effect.

Figure 3-16 gives an estimate of average attenuations for various conditions. The loss caused by rainfall can be severe at higher frequencies and is one reason long range radars (or communication systems) are seldom found above S-Band, and why the X-Band radars found in aircraft can sometimes be effectively blinded by intense rain. This, incidentally, is one reason the National Weather Service lowered the frequency of the new NEXRAD weather radars from C-Band to S-Band. Even though weather RCS is much lower at S-Band, this signal loss can be made up with larger antennas and more powerful transmitters. The weather penetration of S-Band signals is much better than C-Band.

$$L_{rain} \approx K_{rain}\, r\, R \text{ (dB)} \tag{3-26}$$

L_{rain} = the loss due to rainfall

K_{rain} = the rainfall attenuation factor (from Fig. 3-20 and Eq. 3-27) in dB/nmi/mm/hr

r = the rainfall rate (mm/hr)

R = range in nmi

$$K_{rain} \approx 0.0013\, f_{GHz}^{2} \tag{3-27}$$

f_{GHz} = the frequency in GigaHertz

The attenuation measurements summarized in the Fig. 3-16 give wide variations for a single frequency, reflecting the difficulty in making and applying this loss. The problem arises because rain is seldom of uniform intensity over much distance. This makes it hard to estimate accurately the loss for any particular set of conditions. Another problem is that, at higher frequencies, attenuation is not a linear function of rainfall rate. Nathanson [4] states that the X-Band attenuation is proportional to $r^{1.3}$, where r is the rainfall rate (mm/hr).

Example 3-5: Estimate the total two-way attenuation of a 10-GHz radar viewing a target at 20 nautical miles in uniform rainfall of 15 mm per hour. How much does this amount of attenuation affect the radar's detection capability, in the sense of reducing the range at which this target can be detected?

Solution: Figure 3-16 shows that, on average, rain attenuates signals at 10 GHz 0.13 dB per nautical mile per millimeter per hour of rain. The total attenuation is therefore 0.13 dB per nmi per mm/hr of rain times 20 nmi times 15 mm/hr, for a total attenuation of approximately 39 dB.

Since maximum detection range halves for every 12 dB of signal loss (power factor of 16 or 2^4), the 39-dB loss decreases detection range by a factor of $2^{3.25}$, or a factor of 9.5. If the radar could barely detect the given target at 20 nmi in 155 mm/hr rain, it could detect the same target to approximately 190 nmi without the rain.

Comment: Rain attenuation is a major problem with higher frequency radars, but much less of a problem at lower frequencies. A rule of thumb is that the constant K_{rain} increases as approximately the square of frequency. This means that an L-Band radar would have a K_{rain} of about 0.002, which is 1/64 that of X-Band (X-Band is about 8 times the frequency of L-Band). The same 20 nmi target in 15 mm/hr rain would attenuate less than 1 dB to an L-Band radar.

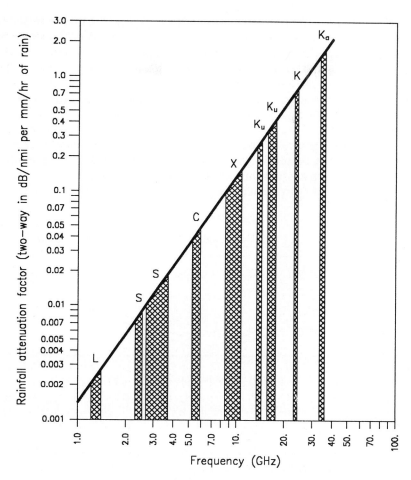

Figure 3-16. Rainfall Attenuation Coefficient K_{rain}
(Adapted and extrapolated from Nathanson [4])

The implication in the above discussion is that rain attenuation makes detection much more difficult. Detection in rain is actually impaired by more than attenuation; rainfall produces *clutter*. If the target is clear of rain and the attenuation is calculated from the distance the beam moves through the rain, the predictions above can be fairly accurate. When the target is in the rainfall, weather clutter may be the limiting interference and may cause a much greater reduction in detection range than is caused by attenuation alone (Ch. 4).

Snow also causes attenuation. Equation 3-28, from Gunn and East [5], can be used to estimate the attenuation from snow in dB / nmi.

$$\alpha_{snow} \approx \frac{0.00188\, r^{1.6}}{\lambda^4} + \frac{0.00119\, r}{\lambda} \quad \text{dB / nmi} \qquad (3\text{-}28)$$

α_{snow} = the attenuation caused by snow in dB per nmi

3-4. Losses in the Radar Equation

    r = the snowfall rate in equivalent rainfall water content (mm/hr). (As a general rule, there is about a 10:1 ratio between snowfall and its water equivalent in rain. 10 mm/hr of snow has the same amount of water as 1 mm/hr of rain and r would be 1 mm/hr)

    λ = the wavelength in cm (non-standard for wavelength — be careful of the units)

Ground plane loss: Figure 3-17 shows the principle of the ground plane gain or loss. If a ground plane is present, two distinct signal paths exist: direct and reflected. With radar, the signal can follow four paths: transmit direct, receive direct (called the d-d path); transmit direct, receive reflected (d-r); transmit reflected, receive direct (r-d); and transmit reflected, receive reflected (r-r).

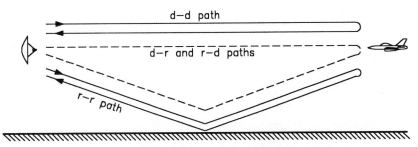

Figure 3-17. Signal Paths for the Ground Plane Loss.

The magnitude of the reflection can range from zero to unity. A reflection from a horizontal conducting surface will be phase-shifted 180° to horizontal polarization and 0° for vertical (Fig. 3-18). Since a circularly polarized wave contains equal horizontally and vertically polarized waves and only one is shifted, its polarization should be sense-reversed; right hand becoming left hand and vice versa. It is found, however, that for land and water, the reflection is 180° for small grazing angles (< about 5°), regardless of polarization. The implication is that the so-called "reflection" is actually refraction [6].

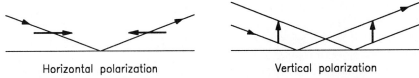

    Horizontal polarization   Vertical polarization

Figure 3-18. Reflection of Linear Polarization from a Conducting Surface.

The amount of signal returned to the radar with the ground plane reflection can range from zero (if the four signal paths are of such magnitude and phase that they cancel) to 16 (the four paths add voltage-wise, with a power gain of the voltage gain squared). This produces a ground plane gain of from 0 to 16, and a ground plane loss from 1/16 to infinity. It will be shown in Ch. 7 that this loss may vary slowly enough to markedly interfere with the radar's ability to see and track low-flying targets.

The defenses against the ground plane loss are (1) an antenna having a narrow enough elevation beamwidth to prevent illuminating the reflection point (Fig. 3-19), and (2) frequency agility to force the loss to change rapidly, allowing the radar to average it. Rapid frequency agility causes the target amplitude to vary much more rapidly than without agility, and it will be shown in Ch. 5 that rapidly fluctuating targets are easier to detect than those which vary slowly.

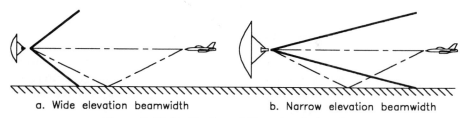

a. Wide elevation beamwidth   b. Narrow elevation beamwidth

Figure 3-19. Multipath and Elevation Beamwidth.

Another problem with ground reflections is that when they are present, the radar cannot locate the target in elevation angle as well as without them. This is a particular problem with tracking systems, in that it magnifies the elevation error of low-flying targets by orders of magnitude. Again, very narrow antenna beams help because they exclude the reflection point from being sensed. Also, frequency agility causes the apparent elevation position to vary so rapidly that the antenna servos cannot follow, with the servos averaging the target's position. A system like Phalanx (a radar-aimed gun for protection against anti-ship missiles) requires an accurate elevation track, since bullets must strike the target to destroy it. It uses an antenna which is vertically large, to produce a small elevation beamwidth. See Ch. 7 for more information on the elevation accuracy effects of multipath.

3-5. Signal Processing and the Radar Equation

Signal processing has the effect of increasing the signal-to-interference ratio. Its effectiveness is measured by process gain (G_P), defined as

$$G_P \equiv \frac{(S/I)_O}{(S/I)_I} \qquad (3\text{-}29)$$

$(S/I)_O$ = the signal-to-interference ratio out of the process

$(S/I)_I$ = the signal-to-interference ratio into the process

Signal integration is one method of achieving process gain. It is the summation of the signal contents of several samples of the same range bin in order to increase the signal-to-interference ratio. The integration number (N_L) is the number of samples summed in the process. If an integrator has an integration number of eight, eight consecutive samples of range bin 1 will be summed to form the integrated signal for range bin 1, eight samples of range bin 2 will be summed, and so on. The concept is shown in Fig. 3-20.

As will be shown in Ch. 11, summation of several target samples (hits) of signal plus interference increases the signal-to-interference ratio for certain kinds of interference. In the

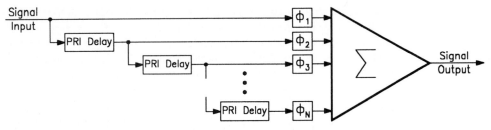

Figure 3-20. Signal Integration Principle.

case of noise or noise jamming, the signal-to-interference ratio is increased by the integration number adjusted for an integration loss. The result is a signal-to-interference ratio increase by the effective integration number. See Eq. 3-30, which gives the effective integration number; it is the system's process gain when noise is the interference.

$$N_{eff} = N_L / L_i \qquad (3\text{-}30)$$

N_{eff} = the effective integration number (the number by which the single-hit signal-to-interference ratio is multiplied to find the integrated (multi-hit) signal-to-interference ratio

N_L = the number of pulses in the look (integration number)

L_i = the integration loss.

Signal integration can be either coherent, or non-coherent. Coherent signal integration occurs when both the amplitude and phase of the signal are accounted for in the sum. Coherent integration requires some knowledge of the target's motion to compensate for the phase changes caused by that motion. This process is usually implemented as a complex correlation, for pulse compression, or a Fourier transform for Doppler analysis. With the Fourier transform, *all possible target motions* are tried and a summation made for each. See Ch. 11.

Non-coherent integration is the summation of signal amplitudes only and does not require knowledge of target motion. The phase-shifters shown in Fig. 3-20 are not present with non-coherent integration.

Coherent summation is the more effective of the two, having an integration loss which is determined primarily by the window used in the process (see below and Ch. 12). The coherent integration loss seldom exceeds 2, and the effective integration number is usually more than half the actual number. Most signal integration in radar systems having digital signal processors is coherent.

Non-coherent integration is less effective than coherent. Its effective integration number is determined by the number of pulses integrated, the detection and false alarm probabilities, and target fluctuations. For large values of integration number, the effective integration number approaches the square root of the actual number of pulses integrated.

$$N_{eff(NC)} \to (N_L)^{1/2} \qquad \text{(for large } N_L, \text{ non-coherent integration)} \qquad (3\text{-}31)$$

Figure 3-21 shows an example of non-coherent signal integration. Each of the eight top traces is a single amplitude versus range sweep. Each has a target echo with a signal-to-noise ratio of 10 dB. It is readily apparent that it would be difficult without prior knowledge to reliably identify it as a target. The bottom trace was generated by allowing multiple sweeps

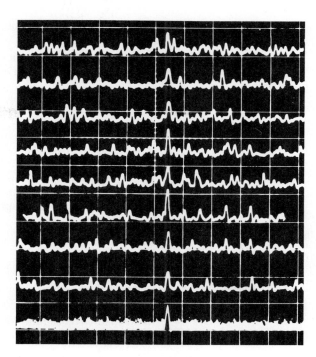

Figure 3-21. Non-Coherent Integration Example.

(Diagram courtesy of D.L. Mensa; U. S. government work not covered by U. S. copyright)

to overlay one another. It can easily be seen that the effective signal-to-noise ratio of the integrated bottom trace is much higher than those of the individual sweeps.

Integration loss: The integration loss for non-coherent integration has been explored in great detail by authors in the past. Two good sources of this analysis are the first edition of the *Radar Handbook* [7], and Meyer and Mayer [8]. The latter gives an extensive treatment and can be used for specific design problems. The integration loss is discussed further in Chs. 5 and 11.

The coherent integration loss, discussed in Ch. 11, is a function of the signal processing window. Most coherent signal processes involve windowing, whereby the signal being processed is modified to minimize artifacts introduced by the process itself. These artifacts interfere with legitimate target echoes (see spectral leakage and range leakage in Chs. 11, 12, and 13). Each type of window has a degradation factor associated with it which will, in effect, become the integration loss. As with non-coherent integration, the loss thus found is applied to the integration number to produce an effective integration number.

3-6. Radar Equation with Pulse Compression

Pulse compression is the process of transmitting a wide pulse (pulse width τ_E) and processing it into a narrow pulse (τ_C). The goal is to use the energy in the wide transmitted pulse for detection and the bandwidth of the narrow pulse for range resolution. The process of compression correlates the echo pulse with a delayed copy of the transmitted waveform. The echo signal can be considered as though it were a group of subpulses, each having the width of the compressed pulse. The subpulses are delayed in time so that they align and are then summed. See Fig. 1-27 and Ch. 13.

The signal-to-noise ratio without pulse compression is Eq. 3-32, a combination of Eqs. 3-11 and 3-12. In look energy form, it is Eq. 3-14, repeated here for convenience as Eq. 3-33.

$$S/N = \frac{P_T G^2 \lambda^2 \sigma N_L}{(4\pi)^3 R^4 K T_0 B F L_S L_A L_{GP} L_i} \qquad (3\text{-}32)$$

$$S/N = \frac{P_T \tau N_L G^2 \lambda^2 \sigma}{(4\pi)^3 R^4 K T_0 F L_S L_A L_{GP} L_i} \qquad (3\text{-}33)$$

τ = the pulse width without compression (both transmitted and processed echo)

Using Eq. 3-33, the signal-to-noise ratio for each compressed pulse segment width (subpulse) in a pulse compression echo can be defined.

$$S/N'_{PC} = \frac{P_T \tau_C G^2 \lambda^2 \sigma}{(4\pi)^3 R^4 K T_0 F L_S L_A L_{GP}} \qquad (3\text{-}34)$$

S/N'_{PC} = the signal-to-noise ratio for each subpulse, or compressed pulse width, of a pulse compression echo

τ_C = the compressed pulse width

From Fig. 1-27, it can be seen that the signal-to-noise ratio for the entire pulse results from the summation of CR (the compression ratio — number of compressed pulse widths in the echo wave) wave segments. The summation is coherent and if an integration loss occurs, it is accounted for in the compression ratio. The signal-to-noise ratio for the entire echo wave is therefore

$$S/N_{PC} = \frac{P_T CR \tau_C N_L G^2 \lambda^2 \sigma}{(4\pi)^3 R^4 K T_0 F L_S L_A L_{GP} L_i} \qquad (3\text{-}35)$$

S/N_{PC} = the signal-to-noise ratio for N_L hits from the entire echo wave

CR = the compression ratio

Substituting the expanded pulse width for the compression ratio times the compressed pulse width (Eq. 2-1) gives

$$S/N_{PC} = \frac{P_T \tau_E N_L G^2 \lambda^2 \sigma}{(4\pi)^3 R^4 K T_0 F L_S L_A L_{GP} L_i} \qquad (3\text{-}36)$$

τ_E = the expanded (transmitted) pulse width

Recognizing the reciprocal relationship between bandwidth and compressed pulse width, Eq. 3-35 becomes

$$S/N_{PC} = \frac{P_T N_L G^2 \lambda^2 \sigma}{(4\pi)^3 R^4 K T_0 (B/CR) F L_S L_A L_{GP} L_i} \qquad (3\text{-}37)$$

A comparison of Eqs. 3-33 and 3-36 shows the value of pulse compression. For a given set of radar parameters, as long as the transmitted pulse width remains the same, the signal-to-noise ratio remains the same, *regardless of the signal bandwidth*. This means that, with pulse compression, detection can be maintained and range resolution improved by keeping the same transmitted pulse width but increasing the bandwidth. Or, range resolution can remain the same and detection can be improved by increasing the transmitted pulse width and maintaining the same bandwidth. With pulse compression, the classic detection/range resolution paradox is no more. The system is, of course, more complex with pulse compression, but modern technology makes this a small price to pay.

The major differences between compression and non-compression radars, other than the signal generation filter (the expansion filter) and the compression filters themselves, is that the receiver filter shapes are different and the transmit and receive bandwidths may be marginally larger. See Chs. 2 and 13 for the filter and bandwidth requirements.

Example 3-6: Using the radar of Ex. 3-4 and assuming it had no compression, find the required transmit peak power necessary to detect the same targets at the same ranges if the transmit pulse width were increased to 2000 μs.

Solution: If no compression is used in Ex. 3-4, the transmit pulse width is 1.0 μs (Sec. 3-3). This means that the effective look energy is 3,000 J (15 MW x 1.0 μs x 200 effective pulses per look). If detection is to remain the same, signal-to-noise ratio and thus look energy must remain the same. If the pulse width is increased to 2000 μs and nothing else other than transmit power is changed, the transmit power can be reduced by a factor of 2000 to 7.5 kW peak. Equation 3-36 or 3-37 gives the same result.

3-7. The Radar Equation for Search Radars

Search radars present a special case of the radar equation for two reasons, both related to their search pattern. They have a limited time-on-target (T_{OT} — the time for each scan that a particular target is within the beam of the antenna — see below), and they do not point the peak gain of the antenna at targets for all hits in a look.

Search radars must scan rapidly to cover their assigned scan volume in as short a time as possible. Also, short scan times reduce the target maneuvering possible between scans and help maintain more accurate target position reporting. However, they must scan slowly enough so that the required number of hits are received from within the antenna's beamwidth.

The radar equation for search radars is unique primarily because of the need to use antenna scan rate, beamwidth, and PRF to account for the antenna scan loss and the required integration number. In systems where the signal processor determines integration number (all digital signal processors), the scan rate can be set so that the number of pulses transmitted as the antenna beam scans past a given point matches the integration number in the signal processor.

$$N_{SC} = (\theta_{3(Az)} / \omega) \, PRF \qquad (3\text{-}38)$$

N_{SC} = the number of pulses transmitted per antenna beamwidth as the beam scans

$\theta_{3(Az)}$ = the 3 dB azimuth beamwidth of the antenna (degrees)

ω = the antenna scan rate in degrees per second

If the number of pulses in the beam matches the integration number (the normal case), the signal-to-noise ratio for the search radar becomes

$$S/N = \frac{P_T G^2 \lambda^2 \sigma}{(4\pi)^3 R^4 K T_0 B F L_S L_A L_{GP}} \frac{\theta_{3(Az)} PRF}{\omega L_i} \quad (3\text{-}39)$$

It is interesting that in search radars, antenna gain is often set by scan rate considerations and not for detection. Few search radars have antenna gains of much greater than 5,000, unless the radar searches in three dimensions. If gain is fixed by scan rate, signal-to-noise ratio is proportional to wavelength squared. Hence, long range search radars tend to operate at lower frequencies. Other factors, such as rainfall attenuation and target characteristics, also favor lower frequencies.

As mentioned above, there is a parameter in traditional radar lore called time-on-target (T_{OT}). It is the time the antenna beam requires to scan one antenna beamwidth. Another parameter with which we are familiar is data gathering time (T_d), also known as dwell time (T_d) integration time (T_I), and look time (T_L). If the beamwidth, scan rate, PRF, and signal processing match one another, T_{OT} and T_d are equal. In any case, T_d is the parameter which determines the integration number and contributes to detection, not T_{OT}. If the scan and integration parameters match,

$$T_d = T_{OT} = \theta_{3(Az)} / \omega \quad (3\text{-}40)$$

Example 3-7: A search radar has the following characteristics:

Type	2-D search
Azimuth beamwidth	2.2°
PRF	630 pps
S/N_{min} for detection	17 dB

The radar receives a 7-dB signal-to-noise ratio from a 2-m² target at 92 nmi range. What is the maximum rotation rate of the antenna if the scan is matched to signal integration? Assume an integration loss of 2 dB.

Solution: Integration must improve the signal-to-noise ratio by 10 dB (17 dB required and 7 dB present). The integration loss of 2 dB requires that the integration process gain without the loss be 12 dB. This is a power ratio of 16, so the integration number must be 16.

Equation 3-38 says that with an integration number of 16, a beamwidth of 2.2°, and a PRF of 630 pps, the scan rate cannot exceed 86.6°/s. This is the scan rate. The antenna requires 4.2 s to scan a circle, or scans at the rate of 14.3 scans per minute.

To this point, the antenna gain variation during the scan across a target has been handled using an antenna pattern (or scan) loss as part of the system loss, as in Eq. 3-39. Another method is to introduce a pattern propagation factor (F_A), defined in the *IEEE Standard Radar Definitions* [9] as the "Ratio of the strength that is actually present at a point in space to that which would have been present if free-space propagation had occurred with the antenna beam directed toward the point in question. This factor is used in the radar equation to modify the strength of the transmitted or received signal to account for the effect of multipath propagation, diffraction, refraction, and pattern of an antenna." Since the pattern propagation factor is in field strength (voltage) terms, it must be squared for each use of the antenna. Using the pattern factor, Eq 3-39 becomes

$$S/N = \frac{P_T G^2 F_A^4 \lambda^2 \sigma}{(4\pi)^3 R^4 K T_0 B F L'_S L_A L_{GP}} \frac{\theta_{3(Az)} PRF}{\omega L_i} \quad (3\text{-}41)$$

L'_S = the system loss less its antenna pattern component

If separate transmit and receive antennas are used, Eq. 3-41 becomes

$$S/N = \frac{P_T G_T G_R F_T^2 F_R^2 \lambda^2 \sigma}{(4\pi)^3 R^4 K T_0 B F L'_S L_A L_{GP}} \frac{\theta_{3(Az)} PRF}{\omega L_i} \quad (3\text{-}42)$$

G_T and G_R = the transmit and receive antenna peak gains, respectively

F_T and F_R = the transmit and receive antenna propagation pattern factors

3-8. The Radar Equation for Tracking Radars

In tracking radars, time-on-target is essentially infinite and irrelevant, since the antenna is always pointed at the target. The integration time is determined by either the signal processor or by the servo bandwidths, depending on how integration is achieved. If the signal processor determines the integration number, it is fixed by the process. If integration is done in the track filters or algorithms, the integration number (at least for the tracking channels) is by

$$N_E = PRF \, T_d \approx PRF / B_S \quad (3\text{-}43)$$

N_E = the integration number (in the error channels)
PRF = the pulse repetition frequency (pulses per second)
T_d = the data gathering or integration time
B_S = the tracking servo bandwidth (Hertz)

The signal-to-noise ratio then becomes

$$S/N = \frac{P_T G^2 \lambda^2 \sigma}{(4\pi)^3 R^4 K T_0 B F L_S L_P L_{GP}} \frac{PRF}{B_S L_i} \quad (3\text{-}44)$$

Physical size is in, many cases, the antenna limiting factor in tracking radars, as opposed to gain in search systems. It is thus useful to express tracking S/N as a function of antenna effective aperture, instead of gain. Note that in tracking radars, antenna beamwidths can be as narrow as desired, as long as an acquisition source — some pointing source to assist the radar in finding its targets — of sufficient accuracy is available.

$$S/N = \frac{P_T A_E^2 \sigma}{4\pi \lambda^2 R^4 K T_0 B F L_S L_P L_{GP}} \frac{PRF}{B_S L_i} \quad (3\text{-}45)$$

This form of the radar equation shows that short wavelengths are favored for high signal-to-noise ratios. Weather and targets still favor long wavelengths, however, leaving the choice of frequency not as clear-cut as it is for search radars.

Example 3-8: Find the maximum possible integration number for the angle tracking servo of a radar whose PRF is 320 pps and whose angle servo bandwidths are 5.0 Hz.

Solution: The radar can integrate signal in its angle error channels for up to 0.2 s (the reciprocal of the servo bandwidth). During this time, 64 pulses are emitted by the transmitter. Equation 3-43 gives the same result.

Comment: Although 64 is the maximum possible integration number for this example, the S/N improvements experienced may not be nearly this great. The reason is that many track radars do not coherently integrate error-channel signals, although some do. Non-coherent integrations in analog track filters are subject large integration losses.

3-9. The Radar Equation for CW and Pulse Doppler Radars

CW and FMCW radars present an entirely different problem regarding the prediction of signal-to-interference and maximum detection/tracking ranges than do pulsed radars. The primary difference is that, while in pulse systems, receiver noise is the ultimate limiting interference, in CW and FMCW radars this is not so. The ultimate limiting interference in these radars is usually contamination of the receiver input by the transmitter. This form of interference is called spillover and can have several causes.

One path is internal coupling between the transmitter and receiver. This spillover path, shown at the left of Fig. 3-22, is of a fixed length and thus can be cancelled using a variable attenuator and a variable delay to inject a fraction of the transmitter wave back into the receiver at a magnitude equal to that of the spillover and a phase of 180° from the spillover. If the spillover cancellation is to be broadband, the cancellation path length must match that of the spillover, or the delay and attenuator must be matched to each frequency used.

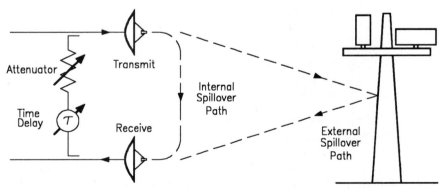

Figure 3-22. Spillover Paths.

This arrangement can theoretically eliminate both spillover which is purely sinusoidal and that which contains transmitter noise, provided the cancellation and spillover paths have the same length. If the lengths do not match, cancellation occurs for only the sinusoidal portion of the wave.

Other spillover paths are reflections of transmitted signals back into the antenna from structures near the radar, such as masts, flagpoles, buildings, and the like (right side of Fig. 3-22). These spillover paths are of unpredictable length, perhaps even varying with antenna pointing angles, and cannot be cancelled. The defense against this form of spillover is prevention. In some extreme cases radars must be relocated to minimize this phenomenon. In others, the offending structures can be treated with RAM (radar absorbing material) to reduce the reflection, or screened to assure that reflections do not return to the radar. This source of spillover is also made worse by transmit instabilities, which may cause the interfering signal to appear to be moving, thus foiling signal processing efforts to minimize its effect. See Sec. 2-8.

Close-in clutter can behave in much the same way as structural reflections, with no opportunity to treat the source. The only defense is Doppler clutter suppression, which is again contaminated by transmit instabilities.

If a CW radar has sufficient isolation between transmitter and receiver and sufficient transmit stability, noise again becomes limiting, and the systems can be treated much the same as if they were pulsed. In this case, the total energy in the signal and interference is contained within the processed Doppler spectrum. Doppler processing, as shown in Ch. 11, concentrates the signals, both moving and clutter, into a few Doppler bins, and spreads random interference equally over all bins. Thus the signal-to-interference ratio in the bin containing the desired signal is larger after Doppler processing than before. The signal is spread slightly because of the processing window (Ch.11), and this spreading is accounted for by making the effective noise bandwidth somewhat greater than the bin width. The width of each Doppler bin is simply the digital sample rate divided by the number of bins. Thus the signal-to-noise ratio for a CW radar limited by receiver noise is

$$S/N = \frac{P_{CW} T_L G^2 \lambda^2 \sigma}{(4\pi)^3 R^4 K T_0 F L_S L_A L_{GP} L_W} \quad (3\text{-}46)$$

P_{CW} = the CW power
T_L = the look (dwell) time
L_W = the loss caused by the window used in the Doppler process (found in Table 11-1)

If a fast Fourier transform algorithm is used to implement the Doppler filters, the effective dwell time (from Ch. 11) is

$$T_L = N_L / f_S \quad (3\text{-}47)$$

N_L = the number of samples in a dwell
f_S = the sample frequency

Pulse Doppler: Pulse Doppler systems generally do not have the spillover problems of CW, but do suffer from system instabilities converting clutter to apparently moving targets. If transmit noise is limiting, the analysis is as shown in Sec. 2-8. If receiver noise is limiting, Eq. 3-46 is usable, with P_{CW} replaced with P_{avg}. The pulse Doppler dwell time is simply Eq. 3-47 with f_S replaced by PRF.

Example 3-9: A pulse Doppler radar transmits 3,000 W of average power at a PRF of 300,000 pulses per second. Its frequency is 10.0 GHz and its antenna gain is

35.4 dB. It processes the echoes in a 2048 point FFT using a 60-dB Dolph-Chebyshev window. Losses, except for the window loss, are assumed negligible. Noise is limiting, and the receiver noise figure is 4.0 dB. Find the maximum range at which a 2.0 m² target produces a 15-dB signal-to-noise ratio.

Solution: The dwell time, from Eq. 3-47, is 6.83 ms. The window loss is found is the column marked Equivalent Noise Bandwidth in Table 11-1. The 60 dB D-C window is the one with an alpha of 3.0. The window loss is 1.51. Solving Eq. 3-46 for range and applying the stated parameters gives a maximum range for this target of 147 km, or 74 nmi.

3-10. Radar Equation for Area Targets and Clutter

Area targets are those which are larger in two dimensions than the radar's resolution cell. Their radar cross-section is proportional to the area illuminated and processed by the radar (Figs. 3-2, 3-3, and 3-23). To microwave radars with relatively wide antenna beamwidths, area targets are usually land or sea clutter. To laser radars, with very small beamwidths, many objects are area targets.

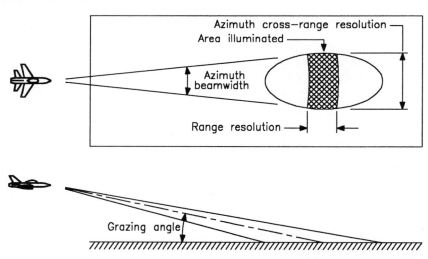

Figure 3-23. Area Target Illumination Geometry.

The signal echo power at the radar from one hit on any target is given by Eq. 3-7, repeated here as Eq. 3-48.

$$P_R = \frac{P_T G^2 \lambda^2 \sigma}{(4\pi)^3 R^4 L_S L_A L_{GP}} \quad \text{(Watts)} \tag{3-48}$$

The radar cross-section of area targets is proportional to the area of the target (usually clutter) illuminated. The radar cross-section of Eq. 3-48 thus becomes, for area targets

$$\sigma_A = \sigma^0 A_I \quad \text{(square meters)} \tag{3-49}$$

σ^0 = the area target radar cross-section per unit area illuminated. [The units are usually square meters per square meter, often in dB form. *Be careful* — some authors use units of square centimeters (RCS) per square meter (area illuminated) and their numbers are 40 dB bigger than normal.] The unit is pronounced "sigma zero."

A_I = the area illuminated within the resolution cell. [As shown in Figs. 3-2 and 3-23, one dimension of A_I is usually azimuth cross-range. The other is either related to the range resolution (small grazing angles) or elevation beamwidth (large — nearing 90° — grazing angles).]

The area illuminated is roughly rectangular, and one dimension is the azimuth cross-range resolution, given by Eq. 3-50. The other is the lesser of either the range resolution modified for the grazing angle (Fig. 3-24a and Eq. 3-51), or the elevation cross-range resolution modified for the grazing angle (Fig. 3-24b and Eq. 3-52).

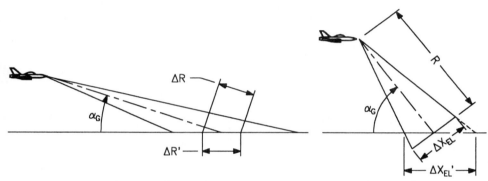

Figure 3-24. Area Target Correction for Low and High Grazing Angles.

$$\Delta X_{AZ} = \frac{\lambda R}{\eta_{AZ} D_{AZ}} \quad \text{(meters)} \quad (3\text{-}50)$$

ΔX_{AZ} = the azimuth cross-range resolution (meters)
λ = the wavelength (meters)
R = the range from radar to intersection of antenna axis with the target (meters)
η_{AZ} = the azimuth antenna length efficiency
D_{AZ} = the length of the antenna in azimuth (meters)

$$\Delta R' = \frac{c}{2 B \cos \alpha_G} \quad (3\text{-}51)$$

$$\Delta X'_{EL} = \frac{\lambda R}{\eta_{EL} D_{EL}} \quad \text{(meters)} \quad (3\text{-}52)$$

3-10. Radar Equation for Area Targets and Clutter

$\Delta R'$ = the range resolution adjusted for the grazing angle — Fig. 3-24 (meters)

c = the propagation velocity

B = the bandwidth matched to the echo waveform

α_G = the grazing angle (Fig. 3-24)

$\Delta X_{EL}'$ = the elevation cross-range resolution adjusted for grazing angle (meters)

η_{EL} = the azimuth antenna length efficiency

D_{EL} = the length of the antenna in the azimuth direction (meters)

Two equations emerge for area target radar cross-section. One is range-limited and applied to modern radars at low grazing angles. The other is elevation-beamwidth-limited and applies to modern radars at high grazing angles and older radars (with little range discrimination) at moderate grazing angles. The correct relationship is the one producing the smaller RCS. The factor $\pi/4$ in Eq. 3-56 is because the area is elliptical in shape.

$$\sigma_A \approx \sigma^0 \, \Delta X_{AZ} \, \Delta R' \quad \text{(low angles)} \tag{3-53}$$

$$\sigma_A \approx \sigma^0 \, \Delta X_{AZ} \, \Delta X_{EL}' \quad \text{(high angles)} \tag{3-54}$$

$$\sigma_A \approx \sigma^0 \, \frac{\lambda R}{\eta_{AZ} D_{AZ}} \, \frac{c}{2 B \cos \alpha_G} \quad \text{(low angles)} \tag{3-55}$$

$$\sigma_A \approx \sigma^0 \, \frac{\pi \lambda^2 R^2}{4 \, \eta_{AZ} D_{AZ} \, \eta_{EL} D_{EL} \cos \alpha_G} \quad \text{(high angles)} \tag{3-56}$$

The echo power from one hit on an area target is therefore given by Eq. 3-57 or 3-58, whichever is smaller.

$$P_{RA(Low)} = \frac{P_T G^2 \lambda^3 c}{(4\pi)^3 R^3 L_S L_A L_{GP} \eta_{AZ} D_{AZ} \, 2 B \cos \alpha_G} \tag{3-57}$$

$$P_{RA(High)} = \frac{\pi P_T G^2 \lambda^4}{4(4\pi)^3 R^2 L_S L_A L_{GP} \eta_{AZ} D_{AZ} \eta_{EL} D_{EL} \cos \alpha_G} \tag{3-58}$$

$P_{RA(Low)}$ = the power received from one hit on an area target at low grazing angle

$P_{RA(High)}$ = the power received from one hit on an area target at high grazing angle

In a form where antenna beamwidths are used instead of antenna dimensions, Eqs. 3-57 and 3-58 become

$$P_{RA(Low)} = \frac{P_T G^2 \lambda^2 \theta_{3(AZ)} c}{(4\pi)^3 R^3 L_S L_A L_{GP} \, 2 B \cos \alpha_G} \tag{3-59}$$

$$P_{RA(High)} = \frac{\pi P_T G^2 \lambda^2 \theta_{3(AZ)} \theta_{3(EL)}}{4 (4\pi)^3 R^2 L_S L_A L_{GP} \cos \alpha_G} \tag{3-60}$$

$\theta_{3(AZ)}$ = the *2-way* azimuth 3-dB beamwidth in radians

$\theta_{3(EL)}$ = the *2-way* elevation 3-dB beamwidth in radians

Point targets in area target interference: In most microwave radar cases, area targets are clutter and interfere with point targets. The signal-to-clutter ratio of a point target in area clutter is calculated by dividing Eq. 3-48 (point target echo power) by the appropriate expression for area clutter echo power. Using Eq. 3-59 and 3-60, two possible expressions for signal-to-clutter ratio are formed.

$$S/C_{(Low)} = \frac{2\sigma B \cos\alpha_G}{\sigma^0 c\, \theta_{3(AZ)} R} \qquad (3\text{-}61)$$

$$S/C_{(High)} = \frac{4\sigma \cos\alpha_G}{\pi \sigma^0 \theta_{3(AZ)} \theta_{3(EL)} R^2} \qquad (3\text{-}62)$$

σ = the point target's radar cross-section

The above equations give the signal-to-clutter ratios for a single hit. For multiple hits, this is multiplied by the process gain. It is known from Sec. 2-7 that the process gain for point targets in clutter is the Moving Target Indicator Improvement factor, MTI-I. The multi-hit S/C expressions are therefore

$$S/C_{(Low)} (\text{Look}) = \frac{2\sigma B \cos\alpha_G}{\sigma^0 c\, \theta_{3(AZ)} R} \text{MTI-I} \qquad (3\text{-}63)$$

$$S/C_{(High)} (\text{Look}) = \frac{4\sigma \cos\alpha_G}{\pi \sigma^0 \theta_{3(AZ)} \theta_{3(EL)} R^2} \text{MTI-I} \qquad (3\text{-}64)$$

Beamwidths in these equations are in radians — multiply beamwidths in degrees by $\pi/180$ to convert to radians before calculating S/C values.

Note that the signal-to-clutter ratio is independent of several factors which are present in other forms of the radar equation, notably transmit power, antenna gain (except as related to beamwidth), frequency, and most losses. This is because, of course, the clutter echo power is affected by these factors exactly the same as point target echo power. Note also that at low grazing angles (the usual case), the range dependence of S/C is R^{-1} rather than R^{-4}. If the radar cross-section of a point target is reduced by a factor of 5, the detection range in noise is reduced by a factor of 1.50 (the fourth root of 5). If that same target is flying low in clutter, its detection range is reduced by a factor of 5. Military tactics usually require that aircraft not fly at high altitudes, where radars see them in noise. To avoid detection, they should fly low in ground and sea clutter.

Example 3-10: An airborne radar with two-way antenna beamwidths of 2.9° in both axes and range resolution of 75 m (2.0 MHz bandwidth) looks down at a target of radar cross-section 1.0 m². The grazing angle (also the depression angle if the aircraft containing the radar is flying level and the ground is level) is 2.0°. The target and clutter cannot be discriminated in range or angles. The minimum signal-to-clutter ratio for reliable detection is 20 dB. Find (a) the maximum detection range for a single hit on this target in sea clutter with σ^0 of -27 dB, and (b) the maximum detection range in the same clutter if signal processing applies an MTI-I of 35 dB.

3-10. Radar Equation for Area Targets and Clutter

Solution: (a) The area of clutter illuminated is the range resolution adjusted for the grazing angle times the azimuth cross-range resolution. The adjusted range resolution is 75 m (see Eq. 3-50). The azimuth cross-range resolution is $0.0506R$ (0.0506 rad $= 2.9°$). The radar cross-section of the clutter cannot exceed 0.010 m^2 (20 dB S/C). The area illuminated to give 0.010 m^2 of -27 dB (1/500) sigma-zero clutter is 5.0 m^2 (500 down 27 dB). If the range resolution is 75 m, the cross-range resolution must be 0.067 m. A $2.9°$ (0.0506 rad) beam spreads to 0.067 m at a range of 1.3 m. This minimum detection range won't do much for us in a tactical situation and besides, it's well into the near field of the antenna where the above calculations are invalid.

(b) The MTI-I of 35 dB increases the S/C by a factor of 3160. Now the clutter RCS can be 31.6 m^2 (35 dB above 0.010 m^2). The illumination area for this clutter RCS is 27 dB above 31.6 m^2, or $15{,}800$ m^2. With a range resolution of 75 m, the beam can be 210 m wide. A $2.9°$ beam spreads to 210 m at a range of 4,200 m. This is the new detection range.

Note: Equations 3-61 and 3-63 solved for R give the same results, but solving problems without resorting to recipes is more fun.

3-11. Radar Equation for Volume Targets and Clutter

In this section, the signal echoes from volume targets, such as weather and chaff (Ch. 4), and point targets in volume interference are described. Volume targets are those whose extent is greater than the radar's resolution cell in range, azimuth, and elevation (Fig. 3-4). Equation 3-65, like previous equations, gives the echo power received from one hit from a point target.

$$P_{SIG} = \frac{P_T G^2 \lambda^2 \sigma}{(4\pi)^3 R^4 L_S L_A} \qquad (3\text{-}65)$$

Equation 3-66 gives the echo power from one hit from a volume target. The two equations are identical except for the symbology for radar cross-section.

$$P_{VC} = \frac{P_T G^2 \lambda^2 \sigma_{VC}}{(4\pi)^3 R^4 L_S L_A} \qquad (3\text{-}66)$$

σ_{VC} = the radar cross-section of the volume clutter

The radar cross-section of a volume target is proportional to the volume illuminated. The radar cross-section per unit volume is given by the symbol $\Sigma\sigma$, pronounced "summation sigma." Its units are m^2 per m^3. *Caution:* Some authors give $\Sigma\sigma$ in units of cm^2 per m^3. If data in cm^2/m^3 are used, 40 dB must be subtracted to use the relationships stated here.

$$\sigma_{VC} = \Sigma\sigma \times \text{Volume Illuminated} = \Sigma\sigma\, V_I \qquad (3\text{-}67)$$

$\Sigma\sigma$ = the RCS of volume targets per unit volume (m^2 / m^3)

V_I = the volume of clutter illuminated by the radar

The volume illuminated can be approximated as an elliptical disk, whose axes are azimuth cross-range resolution and elevation cross-range resolution, and whose thickness is the range resolution. The volume of an elliptical disk with these dimensions is given in Eq. 3-68. Using the relationships of Eqs. 3-69 through 3-71 for the disk's dimensions, the volume is found from Eq. 3-72, and the volume's radar cross-section is found from Eq. 3-73.

$$V_I = \Delta R \, \Delta X_{AZ} \, \Delta X_{EL} \, (\pi/4) \tag{3-68}$$

ΔR = the range resolution (meters)
ΔX_{AZ} = the azimuth cross-range resolution (meters)
ΔX_{EL} = the elevation cross-range resolution (meters)
$\pi/4$ = a factor for the volume of an elliptical disk

$$\Delta R = \frac{c}{2B} \tag{3-69}$$

c = the velocity of propagation
B = the waveform matched bandwidth

$$\Delta X_{AZ} = \frac{R\lambda}{D_{eff(AZ)}} \tag{3-70}$$

$$\Delta X_{EL} = \frac{R\lambda}{D_{eff(EL)}} \tag{3-71}$$

$D_{eff(AZ)}$ and $D_{eff(EL)}$ = the antenna effective lengths in azimuth and elevation, respectively

$$V_I = \frac{c \lambda^2 R^2 \pi}{8 B \, D_{eff(AZ)} D_{eff(EL)}} \tag{3-72}$$

$$\sigma_{VC} = \frac{\Sigma\sigma \, c \lambda^2 R^2 \pi}{8 B \, D_{eff(AZ)} \, D_{eff(EL)}} \tag{3-73}$$

The power received from one hit from a volume target is therefore

$$P_{VC} = \frac{P_T G^2 \lambda^4 \Sigma\sigma \, c}{32 \, (4\pi)^2 \, R^2 \, L_S L_A \, B \, D_{eff(AZ)} \, D_{eff(EL)}} \tag{3-74}$$

P_{VC} = the echo power received from the volume clutter

If a point target is present in volume interference, such as weather, the signal-to-clutter ratio from one hit is given in Eq. 3-75. Since volume clutter is usually suppressed using Doppler, the multi-hit signal-to-clutter ratio is described in Eq. 3-76. Note that S/C is independent of many of the radar's parameters, such as transmit power and antenna gain. This is because volume interference is a real target producing a real echo. The interfering received power is affected by these parameters exactly the same as the target echoes. *Note that the relationships in this section and the previous one (on area clutter) are valid only if the signal-to-noise ratio is high.*

3-11. Radar Equation for Volume Targets and Clutter

$$S/C_V = \frac{P_{SIG}}{P_{VC}} = \frac{2\pi\, B\, \sigma\, D_{eff(AZ)}\, D_{eff(EL)}}{4\, \Sigma\sigma\, c\, \lambda^2\, R^2} \quad \text{(one hit)} \tag{3-75}$$

S/C_V = the signal-to-volume-clutter ratio

$$S/C_V = \frac{2\pi\, B\, \sigma\, D_{eff(AZ)}\, D_{eff(EL)}\, MTI\text{-}I_V}{4\, \Sigma\sigma\, c\, \lambda^2\, R^2} \quad \text{(multiple hits)} \tag{3-76}$$

$MTI\text{-}I_V$ = the MTI improvement factor for volume clutter, which may be different from the factor for land or sea clutter because of the special spectral characteristics of wind-borne clutter — see Ch. 4

Example 3-11: A radar has an antenna of width 3.2 m and height 2.0 m. The length efficiency in both directions is 0.65. The transmit wave is a gated CW pulse of width 2.4 μs. The transmit frequency is 3.05 GHz. Weather with $\Sigma\sigma$ of -83 dBmet^{-1} (units are meters^{-1}) is present at range 22 nmi. Find the illumination volume at 22 nmi. Find the radar cross-section of the weather. Find the RCS of a target at 22 nmi which would produce a 26-dB S/C ratio.

Solution: The illumination volume is found from Eq. 3-73 or from the resolutions. The bandwidth is 417 kHz (reciprocal of the pulse width). The range resolution is 360 m. The azimuth cross-range resolution at 22 nmi (40,744 m) is 1930 m. The elevation resolution at the same range is 3080 m. The volume of an elliptical disk with these dimensions (Eq. 3-68) is 1.68×10^9 m^3 (92.3 dBmet^3). The weather RCS is 83 dB below the illumination volume, or 9.3 dBsm. The RCS of a target for a 26 dB S/C is 35.3 dBsm (9.3 dBsm plus 26 dB).

3-12. Radar Equation for Self-Protection Jamming

Self-protection jamming is ECM emitted from the vehicle the jammer is protecting. It is also known as self-screening jamming — Fig. 3-5. The relationships derived here also hold approximately for escort jamming, where the vehicle carrying the jammer remains near the target it is protecting. The following discussions assume that the range from the target to the radar and from the jammer to the radar are equal, and that the jammer's power out is constant.

The power from a single hit on a point target is repeated here as Eq. 3-77.

$$P_R = \frac{P_T\, G^2\, \lambda^2\, \sigma}{(4\pi)^3\, R^4\, L_S\, L_A\, L_{GP}} \quad \text{(Watts)} \tag{3-77}$$

The power received by the radar from the jammer is the result of a one-way communication and is expressed below.

$$P_{SPJ} = \frac{P_J\, G_J}{4\pi\, R^2}\, \frac{\lambda^2\, G}{4\pi}\, \frac{B_R}{B_J}\, \frac{1}{L_J\, L'_A\, L'_{GP}} \quad \text{(Watts)} \tag{3-78}$$

P_{SPJ} = the power received by the radar from the self-protection jammer (Watts)

P_J = the jammer transmit power (Watts)

G_J = the jammer antenna's gain in the direction of the radar
G = the radar receive antenna's peak gain
R = the range from the target to the radar and from the jammer to the radar (meters)
L_J = the jammer system loss
L'_A = the propagation path loss from the jammer to the radar
L'_{GP} = the ground plane loss from the jammer to the radar
B_R = the bandwidth of the radar's receive filters (Hertz)
B_J = the effective bandwidth of the jammer's emission (the power per unit bandwidth is assumed constant across the jammer's spectrum) (Hertz)

The first group of factors in Eq. 3-78 is the power density of the jammer at the radar antenna. The second is the effective aperture of the radar's receive antenna. The third is the ratio of radar bandwidth to jammer bandwidth. The fourth is losses associated with the jammer.

The result is exactly what one would expect from communications except for the B_R/B_J factor. This factor compensates for the fact that in many cases the jammer must emit a broader bandwidth than that of the radar signal. The reason is either that the jammer must interfere with more than one radar, or that the radar is frequency-agile and the jammer cannot predict the frequency of the next radar emission. This works to the radar's advantage; if the jammer can be forced to emit a wide-band signal, the radar's receive filters will reject most of it.

The bandwidth ratio is never greater than unity; if the jammer bandwidth is less than that of the radar, all the jammer's power is within the radar's bandpass and the ratio is set to one.

Dividing Eq. 3-77 by Eq. 3-78 gives the self-protection signal-to-jamming ratio.

$$S/SPJ = \frac{P_T G \sigma}{4\pi P_J G_J R^2} \frac{B_J}{B_R} \frac{L_J L'_A L'_{GP}}{L_S L_A L_{GP}} G_P \qquad (3\text{-}79)$$

S/SPJ = the signal-to-self-protection-jamming power ratio
G_P = the process gain to the type of jamming used (if noise, the process gain is N_L / L_i)

When the target/self-protection jammer is at a long distance from the radar, the jammer predominates and the S/SSJ ratio is less than one. At short ranges, the target echo is often strong enough to make the signal-to-jamming ratio greater than one. A range exists where the signal and jamming have the same power; it is called the crossover range. It is also known, occasionally, as the burnthrough range, although this terminology is confusing in that some people associate it with detection of the target in jamming, as "the radar burned through the jamming." Because of the ambiguity in terminology, it is better to use *crossover range* as the range at which the signal-to-jamming ratio is unity. The range at which the S/SSJ ratio is sufficient for target detection in the interference will be called the *detection range*. The crossover range is found by setting the left side of Eq. 3-79 equal to one and solving for range.

$$R_C^2 = \frac{P_T G \sigma}{4\pi P_J G_J} \frac{B_J}{B_R} \frac{L_J L'_A L'_{GP}}{L_S L_A L_{GP}} G_P \qquad \text{(square meters)} \qquad (3\text{-}80)$$

3-12. Radar Equation for Self-Protection Jamming

R_C = the crossover range, where the S/SPJ ratio is unity

Detection occurs when the signal-to-jamming ratio is sufficient for detection of the target in the interference, given by

$$R_D = R_C / \sqrt{S/SSJ_{MIN}} \qquad (3\text{-}81)$$

R_D = the detection range in self-protection jamming

S/SSJ_{MIN} = the minimum S/SSJ for detection

The relationship between target echo power, jamming power, and range is shown in Fig. 3-25.

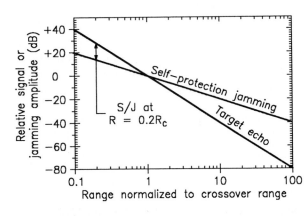

Figure 3-25. Target Echo Signal in Self-Protection Jamming.

Example 3-12: A radar is subject to interference by a self-protection jammer emitting noise. The radar and jammer characteristics are:

Radar transmit power	50 kW
Radar antenna gain	35 dB
Radar pulse width (no compression)	1.5 μs
Radar process gain to noise	16 dB
Jammer power	200 W
Jammer antenna gain in direction of radar	10 dB
Jammer bandwidth	50 MHz
All losses (radar and jammer)	0 dB

Find the crossover range for a 10-m² target. If the required signal-to-noise ratio for detection is 17 dB, find the detection range.

Solution: Convert all dB to ratios. The radar bandwidth is 667 kHz. Eq. 3-80 gives the crossover range as 13,700 m. Eq. 3-81 gives the detection range as a factor of 7.08 less (−8.50 dB), or 1,940 m.

3-13. Radar Equation for Standoff Jamming

Standoff jamming is ECM emitted from a vehicle at a range longer than that of the target being protected. It is particularly useful in situations where the side with the radar also has weapons with home-on-jam modes, since the jamming vehicle can stand at a distance beyond the range of these weapons. Standoff jamming from dedicated ECM vehicles can also be more sophisticated and powerful than self protection. The relationships discussed here also apply to stand-forward jamming, where the ECM vehicle is closer to the radar than the target it is trying to protect. As in the previous section, the relationships developed here assume that the power out of the jammer is constant.

The power from a single hit on a point target is given as Eq. 3-82.

$$P_R = \frac{P_T G^2 \lambda^2 \sigma}{(4\pi)^3 R_T^4 L_S L_A L_{GP}} \quad \text{(Watts)} \quad (3\text{-}82)$$

R_T = the range from the radar to the target

The power received by the radar from the jammer is expressed as

$$P_{SOJ} = \frac{P_J G_J}{4\pi R_J^2} \frac{\lambda^2 G_{RJ}}{4\pi} \frac{B_R}{B_J} \frac{1}{L_J L'_A L'_{GP}} \quad \text{(Watts)} \quad (3\text{-}83)$$

P_{SOJ} = the power received by the radar from the standoff jammer (Watts)
P_J = the jammer transmit power (Watts)
G_J = the jammer antenna's gain in the direction of the radar
G_{RJ} = the radar receive antenna's gain in the direction of the jammer — possibly sidelobe
R_J = the range from the jammer to the radar (meters)
L_J = the jammer system loss
L'_A = the propagation path loss from the jammer to the radar
L'_{GP} = the ground plane loss from the jammer to the radar
B_R = the effective bandwidth of the receive filters (Hertz)
B_J = the effective bandwidth of the jammer's emission (the power per unit bandwidth is assumed constant across the jammer's spectrum — Hertz)

As with self protection, the first group of factors in the equation is the power density of the jammer at the radar antenna, the second is the effective aperture of the radar's receive antenna, the third is the ratio of radar bandwidth to jammer bandwidth, and the fourth is losses associated with the jammer. Unlike self-protection, standoff jamming enters the radar primarily through its antenna's sidelobes. Hence the modified radar antenna gain factor.

Again, it is to the jammer's advantage for the factor B_R/B_J to be unity (spot jamming), while the radar's advantage is best if the ratio is as small as possible (broadband jamming). As with self-protection, radar designs in the form of frequency agility can in some circumstances force a large ratio.

Dividing Eq. 3-82 by Eq. 3-83 gives the signal-to-standoff jamming ratio.

$$S/SOJ = \frac{P_T G^2 R_J^2 \sigma}{4\pi P_J G_J G_{RJ} R_T^4} \frac{B_J}{B_R} \frac{L_J L'_A L'_{GP}}{L_S L_A L_{GP}} G_P \qquad (3\text{-}84)$$

S/SOJ = the signal-to-self-protection-jamming power ratio

G_P = the process gain to the type of jamming used (if noise, the process gain is N_L / L_i)

Unlike self protection where the jamming signal in the radar is a function of target range, the ECM power at the radar is independent of target range with standoff jamming. The signal-to-standoff-jamming ratio thus decreases as the fourth power of the target's range. At the range where the signal and jamming at the radar are equal, crossover occurs.

$$R_{C(SOJ)}^4 = \frac{P_T G^2 R_J^2 \sigma}{4\pi P_J G_J G_{RJ}} \frac{B_J}{B_R} \frac{L_J L'_A L'_{GP}}{L_S L_A L_{GP}} G_P \quad (\text{meters}^4) \quad (3\text{-}85)$$

$R_{C(SOJ)}$ = the crossover range, where the S/SOJ ratio is unity

Detection occurs when the signal-to-jamming ratio is sufficient for detection of the target in the interference. For standoff jamming, it is given by

$$R_{D(SOJ)} = R_C / \sqrt[4]{S/SOJ_{MIN}} \qquad (3\text{-}86)$$

$R_{D(SOJ)}$ = detection range in standoff jamming

S/SOJ_{MIN} = the minimum S/SOJ for detection

The relationship between target echo power, standoff jamming power, and range is shown in Fig. 3-26.

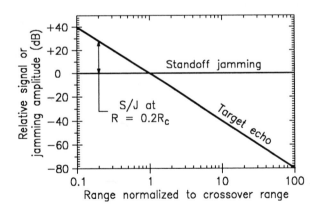

Figure 3-26. Target Echo Signal in Standoff Jamming.

Example 3-13: The radar of Example 3-12 is subject to interference by a standoff jammer transmitting 5,000 W, with a jamming antenna of 30-dB gain, from a range of 12 nmi. All other jammer specifications remain the same. The radar views the jammer through 25-dB antenna sidelobes. Find the crossover range to a 10-m^2 target. If the required detection S/J is 17 dB, find the detection range.

Solution: The radar antenna's gain in the direction of the jammer is 10 dB (35-dB peak less 25-dB attenuation in the sidelobe). From Eq. 3-85, the crossover range is 10,400 m, or 5.6 nmi. Detection, from Eq. 3-86, occurs at 2.1 nmi.

3-14. Radar Equation in Active Augmentation

Active augmentors capture a portion of the radar's illumination signal, amplify and possibly modify it, and transmit it back to the radar. Passive augmentors use enhanced reflectors, such as corner reflectors and Luneburg lenses, and are covered in Ch. 4. Active augmentors are used to effectively increase a small target's RCS and are commonly used to make small pilotless drones simulate larger aircraft, both for weapon testing and as decoys against radars and radar-guided weapons. An active augmentor is shown in Fig. 3-7. When the modification block is present, augmentors can be used as deceptive jammers, generating false targets with incorrect ranges and/or Doppler shifts. Augmentation is usually used in self protection, which will be assumed here.

The active augmentor intercepts (receives) signal from the radar's illumination according to one-way communication relationships.

$$P_{AR} = \frac{P_T G A_{EA}}{4\pi R^2} \quad (3\text{-}87)$$

P_{AR} = the power received by the augmentor from the radar's illumination

P_T = the radar's transmit power

G = the gain of the radar's transmit antenna

A_{EA} = the effective aperture of the augmentor's receive antenna

R = the distance from the radar to the augmentor and target, assuming self protection

If the augmentor's receive antenna gain is used, Eq. 3-87 becomes

$$P_{AR} = \frac{P_T G G_{AR} \lambda^2}{(4\pi)^2 R^2} \quad (3\text{-}88)$$

G_{AR} = the augmentor's receive antenna gain

The effective radiated power transmitted by the augmentor (ERP_A) is

$$ERP_A = \frac{P_T G G_{AR} \lambda^2 G_{AA} G_{AT}}{(4\pi)^2 R^2} \quad (3\text{-}89)$$

G_{AA} = the overall gain of the augmentor, from its receive antenna terminals to its transmit antenna terminals

G_{AT} = the gain of the augmentor's transmit antenna

The power received by the radar from the augmentor (P_{RA}) is

$$P_{RA} = \frac{P_T G^2 G_{AR} \lambda^4 G_{AA} G_{AT}}{(4\pi)^4 R^4} \quad (3\text{-}90)$$

G = the gain of the radar antenna (assumes one antenna)

The received echo power from one hit from a point target is

$$P_{SIG} = \frac{P_T G^2 \lambda^2 \sigma}{(4\pi)^3 R^4 L_S L_A L_{GP}} \quad \text{(Watts)} \quad (3\text{-}91)$$

If a signal-to-augmentation ratio is formed by dividing Eq. 3-90 by Eq. 3-91, an interesting relationship emerges.

$$S/AUG = \frac{\sigma}{G_{AR}(\lambda^2/4\pi) G_{AA} G_{AT}} \qquad (3\text{-}92)$$

S/AUG = the signal-to-augmentation ratio

The radar parameters are missing! They have exactly the same effect on the signal received from the augmentor as they have on the signal received from the target echo. The augmentor thus has an equivalent radar cross-section, given by the denominator of Eq. 3-92.

$$\sigma_{A(EQ)} = [G_{AR}(\lambda^2/4\pi)] [G_{AA}] [G_{AT}] \qquad (3\text{-}93)$$

$\sigma_{A(EQ)}$ = the equivalent RCS of the augmentor

The first group of factors (in brackets) of Eq. 3-93 is the effective aperture of the augmentor receive antenna. The second is the augmentor gain, and the third is the gain of the augmentor transmit antenna. Equation 3-93 is valid if the augmentor is not saturated. If it is, its transmit power becomes fixed and Eq. 3-79 describes the signal-to-augmentation ratio, assuming self protection.

Example 3-14: An augmentor has a receive antenna with 15-dB of gain, a transmit antenna of 12-dB gain, and an overall internal gain of 35 dB. It operates from a target of 8-m² radar cross-section against an X-Band radar with transmit frequency of 10.25 GHz. If the augmentor is not saturated, find the signal-to-augmentation ratio at the radar.

Solution: Equation 3-93 gives the equivalent RCS of the augmentor, with a wavelength of 0.0293 m, as 108 m² (+20.3 dBsm). The signal-to-augmentation ratio is thus 0.074 (the target RCS divided by the augmentor equivalent RCS), or −11.3 dB.

3-15. Bistatic Radar Equation, including Missile Illumination

Bistatic radar places the transmitter and receiver in different locations (Ch. 1 and Fig. 1-6). Hence, the range from transmitter to target is different from the range from target to receiver. Also differing from monostatic is the radar cross-section of the target. For small bistatic angles, the bistatic RCS is similar to monostatic. However, as the bistatic angle approaches 180°, the radar cross-section of most target shapes increases. At 180°, the target radar cross-section approaches that of the peak of a flat plate whose area is the projected area seen by the transmitting antenna. This phenomenon will be discussed further in Ch. 4. The development of the bistatic radar equation below applies also to semi-active radar homing (SARH).

The power density of the radar's transmission at the target is given in Eq. 3-94, and is the same as for monostatic radar.

$$P/A_{tgt} = \frac{P_T G_T}{4\pi R_T^2} \quad \text{(Watts per square meter)} \qquad (3\text{-}94)$$

P/A_{tgt} = the power density at the target

P_T = the transmitted power (peak)
G_T = the gain of the transmitting antenna
R_T = the range from transmitter to target

The effective isotropically radiated power reflected by the target in the direction of the receiver is

$$P_{tgt} = P/A_{tgt}\, \sigma_B \quad \text{(Watts)} \tag{3-95}$$

P_{tgt} = the effective power reflected by the target in the direction of the radar

σ_B = the bistatic radar cross-section of the target

The power density at the receive antenna is

$$P/A_R = \frac{P_{tgt}}{4\pi R_R^2} \quad \text{(Watts per square meter)} \tag{3-96}$$

P/A_R = the power density at the receive antenna
R_R = the range from target to receive antenna

The power delivered by the receive antenna to the receiver (lossless) is

$$P_R = P/A_R\, A_{ER} = \frac{P_T G_T G_R \lambda^2 \sigma_B}{(4\pi)^3 R_T^2 R_R^2} \quad \text{(Watts)} \tag{3-97}$$

P_R = the received power
G_R = the gain of the receive antenna

If losses are considered, the power received is

$$P_R = \frac{P_T G_T G_R \lambda^2 \sigma_B}{(4\pi)^3 R_T^2 R_R^2 L_{ST} L_{SR} L_{AT} L_{AR} L_{GPT} L_{GPR}} \tag{3-98}$$

L_{ST} = the transmit system loss
L_{SR} = the receive system loss
L_{AT} = the propagation medium loss from transmitter to target
L_{AR} = the propagation medium loss from target to receiver
L_{GPT} = the ground plane loss from transmitter to target
L_{GPR} = the ground plane loss from target to receiver

Example 3-15: An aircraft radar illuminates a target for a semi-active radar homing missile. The radar has the following parameters:

Transmit power	4,000 W CW
Antenna gain	33 dB
Frequency	10.0 GHz

The missile has the following specifications:

Receive antenna	Circular array, 6" in diameter, η_{AZ} is 0.5
Receiver bandwidth	500 Hz effective bandwidth
Receiver noise figure	9.0 dB

3-15. Bistatic Radar Equation

Losses   Negligible
Signal integration   None (it is in the effective bandwidth)
S/N minimum   26.0 dB

If a 4-m² target (bistatic RCS) is 25 nmi from the illuminator (transmitter), and the missile is 15 nmi from the target, find the signal-to-noise ratio of the illumination reflection at the missile. If this value is insufficient for missile operation, find the range to which the missile must close before achieving a satisfactory S/N.

Solution: From Ch. 2, the effective aperture and gain of the missile receiving antenna is 0.0091 m² and 127 (21.0 dB), respectively — don't forget to convert inches to meters. Range from transmitter to target is 46,300 m and from target to receiver is 27,780 m. Equation 3-94 shows the power density at the target to be 2.97×10^{-4} W/m². Effective radiated power from the target is 1.19×10^{-3} W (Eq. 3-95).

Equation 3-96 gives the power density at the receive antenna to be 1.23×10^{-13} W/m². Multiplying by the effective aperture of the receiving antenna gives the power received, 1.12×10^{-15} W.

The noise power in the missile receiver is 1.60×10^{-17} W (KT_0BF), giving a signal-to-noise ratio of 70 (18.4 dB). This is insufficient for proper operation of the missile.

Assuming the illumination range does not change, the power effectively radiated from the target remains at 1.19×10^{-3} W. The required receive power is 26 dB (400) above the receiver noise, or 6.40×10^{-15} W. Solving the first part of Eq. 3-97 for power density gives the required power density at the missile antenna to be 7.03×10^{-13} W/m². Solving any of the power density equations in this chapter (Eq. 3-83 works nicely) using the existing parameters gives the range at which the S/N is 26 dB to be 11,600 m, or 6.3 nmi.

3-16. Beacon Equation

Secondary radar, where the illuminator (now the interrogator) interrogates a transponder and receives its reply was introduced in Ch. 1. Figure 3-27 is a block diagram of a generic transponder. The types and uses of secondary radar are discussed in Chs. 1 and 14.

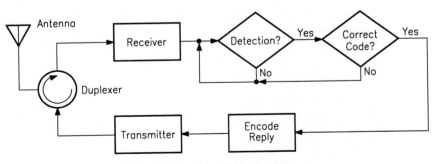

Figure 3-27. Generic Radar Beacon.

The beacon in Fig. 3-27 is representative of the types addressed here, in that whenever an interrogation is detected, the beacon always transmits at the same power level. This is the only type of beacon studied here, but readers should be aware that beacons exist where the transmit power is not constant, and where the relationships developed herein do not hold.

With these constant-power beacons, the beacon equation becomes two one-way communications, one of which will undoubtedly function over a longer range than the other. The beacon detection range is simply the shorter of the two links.

The interrogation link (the up-link) is examined first. The interrogation power density at the beacon is

$$P/A_{bcn} = \frac{P_I G_{TI}}{4\pi R^2} \quad \text{(Watts per square meter)} \quad (3\text{-}99)$$

P/A_{bcn} = the interrogation power density at the beacon
P_I = the interrogator transmitted power (peak)
G_{TI} = the gain of the interrogator transmit antenna
R = the range from radar to beacon

The power received by the beacon is

$$P_{R(bcn)} = \frac{P_I G_{TI}}{4\pi R^2} \frac{\lambda_I^2 G_{BR}}{4\pi} \quad \text{(Watts)} \quad (3\text{-}100)$$

$P_{R(bcn)}$ = the power from the interrogator transmitter received by the beacon
G_{BR} = the gain of the beacon receive antenna
λ_I = the interrogation wavelength
R = the range from radar to beacon

If losses are considered, Eq. 3-100 becomes

$$P_{R(bcn)} = \frac{P_I G_{TI} \lambda_I^2 G_{BR}}{(4\pi)^2 R^2 L_{SI} L_P L_{GPI}} \quad \text{(Watts)} \quad (3\text{-}101)$$

L_{SI} = the interrogator system loss
L_P = the one-way propagation path loss
L_{GPI} = the interrogation ground plane loss

Noise is usually the limiting interference in secondary radar. The signal-to-noise ratio of the interrogation in the beacon is

$$S/N_{bcn} = \frac{P_I G_{TI} \lambda_I^2 G_{BR} G_{PB}}{(4\pi)^2 R^2 K T_0 B_I F_B L_{SI} L_P L_{GPI}} \quad (3\text{-}102)$$

S/N_{bcn} = the signal-to-noise ratio from the interrogator transmitter at the beacon
K = Boltzmann's constant
T_0 = 290° K
B_I = the matched bandwidth of the interrogation signal

3-16. Beacon Equation

F_B = the noise factor of the beacon receiver

G_{PB} = the beacon's process gain to noise (usually unity)

Solving the previous equation for range, the maximum range at which the beacon will be successfully interrogated is

$$R_{I(max)}^2 = \frac{P_I G_{TI} \lambda_I^2 G_{BR} G_{PB}}{(4\pi)^2 K T_0 B_I F_B L_{SI} L_P L_{GPI} S/N_{bcn(min)}} \quad (3\text{-}103)$$

$R_{I(max)}$ = the maximum range at which the beacon can be interrogated

$S/N_{bcn(min)}$ = the minimum signal-to-noise ratio required for interrogation

Examine now the one-way communication which is the beacon's reply (the down-link). Using the same reasoning as with the up-link, the power received by the interrogator receiver from the beacon is

$$P_{R(int)} = \frac{P_B G_{TB} \lambda_B^2 G_{RI}}{(4\pi)^2 R^2 L_{SB} L_P L_{GPB}} \quad \text{(Watts)} \quad (3\text{-}104)$$

P_B = the beacon transmitter power (peak)

G_{TB} = the beacon transmit antenna gain

λ_B = the beacon reply wavelength

G_{RI} = the interrogator receive antenna gain

L_{SB} = the beacon system loss

L_P = the one-way propagation path loss

L_{GPB} = the interrogation ground plane loss

The signal-to-noise ratio of the beacon reply at the interrogator is

$$S/N_{int} = \frac{P_B G_{TB} \lambda_B^2 G_{RI} G_{PB}}{(4\pi)^2 R^2 K T_0 B_B F_I L_{SB} L_P L_{GPB}} \quad (3\text{-}105)$$

S/N_{int} = the signal-to-noise ratio from the beacon transmitter at the interrogator

K = Boltzmann's constant

T_0 = 290° K

B_B = the matched bandwidth of the reply signal

F_I = the noise factor of the interrogator receiver

G_{PI} = the interrogator's process gain to noise

The maximum range at which the beacon reply will be successfully detected by the interrogator is

$$R_{B(max)}^2 = \frac{P_B G_{TB} \lambda_B^2 G_{RI} G_{PB}}{(4\pi)^2 K T_0 B_B F_I L_{SB} L_P L_{GPB} S/N_{int(min)}} \quad (3\text{-}106)$$

$R_{B(max)}$ = the maximum range at which the interrogator receives a usable signal from the beacon

$S/N_{int(min)}$ = the minimum signal-to-noise ratio required for the interrogator to function properly

The maximum range at which the interrogator produces a useful reply is, therefore, the shorter of the two ranges found from Eqs. 3-103 and 3-106. A comparison of these equations shows that, if both the interrogator and beacon are monostatic, the antennas do not play a role in determining which is the weaker link. The antennas in bistatic interrogators and beacons do help determine which link fails first. Frequency and receiver design also help determine the shorter maximum range, but in most systems the transmit powers are usually the determining factors.

When the interrogator is separate from and not used in a primary radar, such as with ATC and IFF transponders, the up-link and down-link designs are usually balanced so that up-link and down-link fail at about the same range. This is done by making the transmit powers comparable to one another. For example, in ATC interrogators the transmit power is about 2000 W peak and the beacon power is in the 500-W area. The extra interrogation power makes up for the less sensitive receiver in the transponder. The antennas are both shared, with the result that the maximum range for both links is about the same.

Instrumentation and tactical beacons, on the other hand, are interrogated by transmitters which are also used for primary radar, meaning they are considerably more powerful that the beacon transmitter. Both antennas are usually monostatic. For these reasons, the interrogation link is normally much stronger than the reply, and these beacons can be interrogated at longer ranges than their replies can be detected.

Cross-band interrogators use different antennas for transmit and receive. The relative strength of each link depends on the individual design. In the ACLS (automated carrier landing system), the interrogator is also a primary radar and the interrogation transmit antenna usually has more gain than the receive antenna. The result is that the down-link usually fails first. ACLS is, however, a short-range system and performs with high signal-to-noise ratios in both links at the ranges at which it is normally used.

Example 3-16: A C-Band instrumentation radar system has the following parameters:

Transmit power	1.0 MW peak
Transmit frequency	5.70 GHz
Antenna gain	43 dB at interrogation frequency (monostatic)
Transmit pulse width	0.25 μs
System loss	1.0 dB
Receiver bandwidth	1.2 MHz (matched to beacon reply)
Receiver noise figure	1.8 dB
Process gain to noise	Unity
Minimum S/N for operation	20 dB

The beacon with which the above radar works has the following specifications:

Beacon power	1000-W peak
Beacon transmit freq	5.60 GHz
Beacon pulse width	1.0 μs

Beacon antenna gain	4.8 dB at interrogation frequency (monostatic)
Beacon system loss	2.5 dB
Beacon receiver bandwidth	4.8 MHz (matched to interrogation)
Beacon noise figure	7.0 dB
Beacon process gain	Unity
Minimum S/N for interrogation	23 dB

Assume that the propagation path characteristics are:

Propagation path loss	1.5 dB each way
Ground plane loss	0 dB (Unity)

Find the maximum range for which this radar/beacon pair can be successfully operated.

Solution: Both the radar and beacon are monostatic, so the antenna effective areas are the same for transmit and receive. The interrogation (up-link) wavelength is 0.0526 m. The reply (down-link) wavelength is 0.0536 m. The antenna gains for the interrogation frequency are 20,000 radar and 3.0 beacon. The radar antenna's reply gain is 19,300 and the beacon's reply gain is 2.9.

Equation 3-103 gives a maximum interrogation range of 294,200 km. Equation 3-106 gives the maximum reply range, which is 23,600 km. The maximum range over which this system can be used is therefore 23,600 km, or 12,700 nmi.

3-17. Further Information

General references: All radar books treat the radar equation, as do many communication texts (communication specialists call signal prediction "link analysis"). Two useful treatments are:

[1] L.V. Blake, Ch. 2 in M.I. Skolnik (ed.), *Radar Handbook* 2nd ed., New York: McGraw-Hill, 1990.

[2] G.W. Stimson, *Introduction to Airborne Radar*, El Segundo, CA: Hughes Aircraft Company, Radar Systems Group, 1983, Ch. 11.

Cited references:

[3] L.V. Blake, Ch. 2 in M.I. Skolnik (ed.), *Radar Handbook*, 2nd ed., New York: McGraw-Hill, 1990.

[4] F.E. Nathanson, *Radar Design Principles*, New York: McGraw-Hill, 1969, pp. 195-199.

[5] K.L.S. Gunn and T.W.R. East, "The Microwave Properties of Precipitation Particles," *Quarterly Journal of the Royal Meteorological Society*, vol. 80, October 1954, pp. 522-545.

[6] D.J. Povejsil, R.S. Raven, and P. Waterman, *Airborne Radar*, Princeton, NJ: Van Nostrand, 1961.

[7] L.V. Blake, Ch. 2, and C.M. Johnson, Ch. 37 in M.I. Skolnik (ed.), *Radar Handbook*, 1st ed., New York: McGraw-Hill, 1970.

[8] D.P. Meyer and H.A. Mayer, *Radar Target Detection: Handbook of Theory and Practice*, New York: Academic Press, 1973.

[9] *IEEE Standard Radar Definitions*, Institute of Electrical and Electronic Engineers Standard 686-1982, New York: 1982.

[10] The Microwave System Designer's Handbook issue of *Microwave System News*, Palo Alto CA: EW Communications Inc.; vol. 15 no. 6, May 1985.

4

TARGETS AND INTERFERING SIGNALS

Introduction

In the previous chapter, it was shown how radar and target parameters affect the echo power received by the radar. In this chapter, we will explore the mechanisms of target scattering, examine radar cross-section fluctuations, and look at the RCS characteristics of some simple and complex shapes. Additionally, we will look at signals from radar interferers — noise, clutter, and ECM — and will see their overall RCS parameters and their statistical descriptions. Once this knowledge is gained, we can discuss target detection in interference (Ch. 5).

4-1. Radar Cross-Section (RCS) Definition and Fundamentals

Radar cross-section (RCS) measures the target's reflection of signals in the direction of the radar receiving antenna. A target's RCS is defined as the ratio of its effective isotropically reflected power to the incident power density.

$$\sigma = 4\pi \frac{\text{Reflected power / Unit solid angle}}{\text{Incident power / Unit area}} \qquad (4\text{-}1)$$

σ = radar cross-section

The concept shown in Eq. 4-1 is illustrated in Fig. 1-38. Note that only in the model does the target radiate isotropically, and the model is valid only in the direction of the radar. *Only the power reflected toward the radar is part of RCS.*

Example 4-1: A target at range 5000 m reflects power such that -58 dBm appears at the output of an antenna with an effective area of 10 m^2. The illumination power density at the target is 20 mW/m^2. Find the radar cross-section of the target.

Solution: The surface area of a sphere of radius 5000 meters is 3.142×10^8 m^2. The echo power density at the radar antenna is calculated as 1.58×10^{-10} W/m^2. Therefore the total effective isotropically radiated power from the target is 0.0498 W (the power density times the area of the sphere). The effective isotropically radiated power per unit solid angle is 0.00396 W/steradian, which is the total power for the sphere divided by 4π steradians per sphere. The illumination power per unit area at the target is given as 0.020 W/m². Therefore the radar cross-section, from Eq. 4-1, is 2.49 m². This number can be confirmed by applying the given illumination power density at the target and an RCS of 2.49 m² to the radar equation (Ch. 3), the result of which is the same received power as that given.

RCS components: Radar cross-section is made up of three components: the area of the target, the reflectivity of the target at the polarization of the radar's receive antenna, and the antenna-like "gain" of the target.

$$\sigma = |A_{tgt} \Gamma_{tgt} G_{tgt}| \qquad (4\text{-}2)$$

$$\Gamma_{tgt} = \frac{P_{Refl(tgt)}}{P_{Impg(tgt)}} \qquad (4\text{-}3)$$

A_{tgt} = the projected area of the target as viewed from the radar (simple for monostatic, more difficult to define for bistatic)

Γ_{tgt} = the reflectivity of the target at the polarization of the radar

G_{tgt} = the antenna-like "gain" of the target in the direction of the radar

$P_{Refl(tgt)}$ = the power reflected by the target (in all directions)

$P_{Impg(tgt)}$ = the illuminating power impinging on the target (within its projected area)

RCS units: Radar cross-section is, as shown in Eq. 4-1, in area units (angles are dimensionless). The area is usually given in square meters, although some authors persist in using square centimeters and square feet. The decibel equivalent of square meters is dBsm, of square centimeters dBscm, and of square feet dBsf. To convert from dBscm to dBsm, subtract 40.0 dB. To convert from dBsf to dBsm, subtract 10.32 dB.

Target field zones: Targets, like antennas, have near and far fields. In the near field, the waves reflected from the targets are spherical. In the far field, the waves are planar to within 1/8 of a wavelength. This is different from antennas, where the maximum error is 1/16 wavelength. Radar targets can have phase errors of 1/16 wavelength each way, summing to 1/8 wavelength round trip.

$$R_{FF} > \frac{2D^2}{\lambda} \qquad (4\text{-}4)$$

R_{FF} = the distance one must be from the target to be in its far field

D = the target's linear extent normal to the axis of the radar antenna

λ = the wavelength

4-2. RCS Fluctuations

Target RCS fluctuates in both amplitude and phase. Amplitude fluctuation, called scintillation, results in variations in received target echo power. Phase fluctuation, called glint, affects the wave fronts echoing from the target. Non-glinting far-field targets reflect approximately planar waves. Scintillation and glint are related to one another and are discussed later in this chapter. Both can occur slowly or rapidly, as defined below.

RCS fluctuation speeds: Slow target fluctuations (Fig 4-1) are those where the time constant of the variation is greater than the radar's look time. Their RCS does not vary significantly over the time of a look, but does vary from scan to scan in search radars. In tracking systems and modes, slow fluctuations are insignificant within a look, but are significant over longer times. For slow fluctuations, pulse to pulse RCS variations are insignificant. Rapid fluctuations are those where the target RCS varies significantly from pulse to pulse.

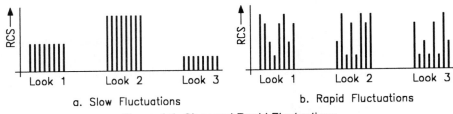

Figure 4-1. Slow and Rapid Fluctuations.

Fluctuation mechanisms: All discussions in this section assume that the radar is low resolution, meaning that the radar's resolution cell size is larger than the target (point targets). Later in this chapter, we will look at area targets (land and sea clutter) and volume targets (weather clutter and chaff). Definitions of point, area, and volume targets are in Sec. 3-2. In this discussion, it helps to remember that the echo fields at the radar are the vector sums of the fields from all scatterers within the radar's resolution cell. The published RCS of a particular target is normally the result of low resolution measurements, meaning that the measuring radar is low resolution to the target being measured.

— *Fluctuation of extended targets:* The first cause of target fluctuation is shown in Fig. 4-2. A single extended far-field target, such as a thin cylinder, can be considered to be made up of a number of small elemental scatterers. Its total RCS is the vector sum of the contribution of all elemental scatterers and depends on the angle at which the radar views them. The viewing angle in the horizontal plane is known as the target's aspect angle. In the vertical plane it is the tilt angle. With aircraft targets the nose is usually considered zero aspect and the roll-pitch plane is zero tilt.

At zero aspect (normal to the cylinder in this case), all elemental scatterers are the same distance from the radar, and the signal echoes from the elemental scatterers arrive in-phase with one another (middle summation in Fig. 4-2). This produces the maximum RCS for this particular target.

As the aspect angle changes, the relative distances from the elemental scatterers change, causing different vector summations at the radar. Note that in the second summation from

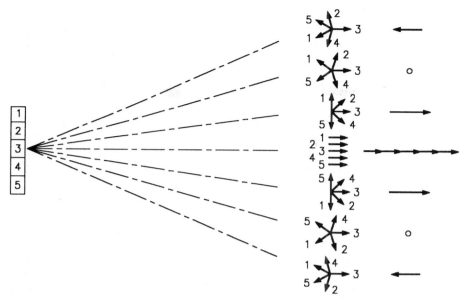

Figure 4-2. Fluctuations of a Single Extended Scatterer.

the top, the elemental signals sum to a null and the RCS of the target becomes zero. Different angles cause different summations, in this case all less than that of zero aspect.

— *Fluctuation of arrays of scatterers versus aspect:* A second cause of RCS fluctuation is the aspect angle from which a complex target of many scatterers is viewed. Most aircraft targets, for example, have an RCS which is the summation of numerous individual scattering centers on the target (see Fig. 4-11 for an example). The summation at the radar of these scattering centers is highly aspect dependent, particularly if the scatterers are many wavelengths apart.

A simple target is shown in Fig. 4-3. Two equal isotropic scatterers (scatter equally in all directions) are separated from one another by 1 m. Isotropic scatterers have the same RCS, regardless of aspect — only spheres and targets much smaller than the wavelength are isotropic. In this case, zero aspect is defined as in Fig. 4-3a, with the scatterers separated only in range. The radar does not resolve the scatterers, and the forward scatterer does not shade the rear one. The frequency is 3.00 GHz (wavelength is 0.10 m) and the scatterers are separated by 20.0 wavelengths round trip.

At zero aspect, the signals from each scatterer arrive at the radar in-phase and the total low resolution RCS is a maximum. If the RCS of the individual scatterers is one unit, the low resolution RCS of the pair is four units, since fields (corresponding to voltages and currents) sum. This gives twice the "voltage," but RCS is in power terms, and power is proportional to voltage squared. RCS components sum as follows:

$$RCS_{TOTAL} = \left| [\sqrt{\mathbf{RCS_1}} + \sqrt{\mathbf{RCS_2}} + \sqrt{\mathbf{RCS_3}} + ...] \right|^2 \qquad (4\text{-}5)$$

RCS_{TOTAL} = the scalar composite RCS of all the scatterers

$\mathbf{RCS}_i$ = the vector (with phase) radar cross-section of each individual scatterer

4-2. RCS Fluctuations

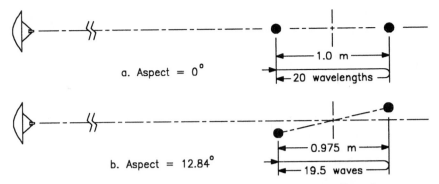

Figure 4-3. Aspect Dependence of RCS for Complex Targets.

Assume now that the scatterer pair is rotated so the radar views them at an aspect of 12.84° (Fig. 4-3b). They are now only 0.975 m apart in range, and the electrical distance between them is 19.5 wavelengths (round trip). The signal vectors in the direction of the radar are out of phase, short circuiting one another. The resulting RCS at the radar is zero.

At 12.84° aspect, the total signal echo from the two scatterers is not zero, but only sums to a null in the direction of the radar, which gives zero RCS *to the radar*. In other directions, the signal sums to a non-zero value. If viewed by a bistatic radar where the receiver is at an angle of 28.32° from the transmitter (15.48° from zero aspect), the RCS of this target is a maximum; four units if each scatterer has one unit of RCS (Fig. 4-4). The calculation is left to the reader.

— *Fluctuation of complex targets as a function of frequency:* The third cause of RCS fluctuation results from viewing a target at different frequencies. Figure 4-5a shows the same two equal isotropic scatterers of Fig. 4-3a. At a frequency of 3.00 GHz, the scatterers are separated by 20 wavelengths round trip and the RCS is maximum. At a frequency of 2.925 GHz ($\lambda = 0.1026$ m), the scatterers are 19.5 round trip wavelengths apart and the RCS is zero.

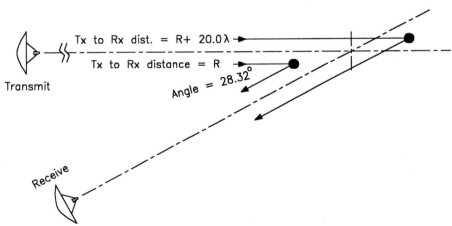

Figure 4-4. Bistatic RCS of Target of Previous Figure.

Note again that this is RCS only in the direction of the radar. To a bistatic angle of 15.94°, the RCS at this frequency is maximum.

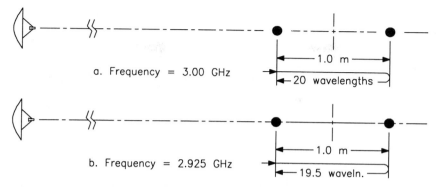

Figure 4-5. Frequency Dependence of RCS.

The spacing of the scatterers and the wavelength determines the amount of fluctuation for a given frequency change. Figures 4-6 and 4-7 show two isotropic 1.0-m² scatterers which are separated by 0.2 m and 1.0 m, respectively, and illuminated in the X-Band. Note that the closely spaced scatterers (Fig 4-6) require a large frequency change for significant RCS fluctuation. The wider the spacing, the greater the RCS fluctuation with small frequency changes (Fig. 4-7). The nulls in both functions actually go to zero ($-\infty$ dB). The apparent lack of complete nulls in the figures is the result of not taking samples at the actual frequencies of the nulls.

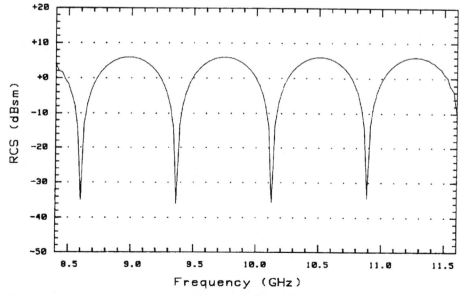

Figure 4-6. Frequency Response of Two 1.0 m² Isotropic Scatterers Separated 0.2 m.

4-2. RCS Fluctuations

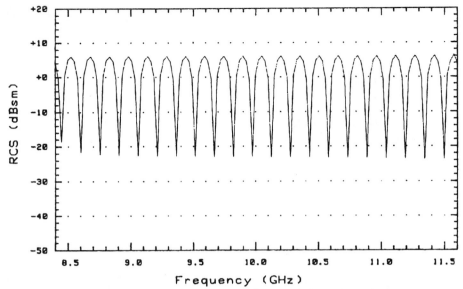

Figure 4-7. Frequency Response of Two 1.0 m² Isotropic Scatterers Separated 1.0 m.

— *Fluctuations from multipath:* The fourth cause of fluctuation is multipath, shown in Fig. 4-8. Four signal paths connect radar and target; d-d, d-r, r-d, and r-r (Sec. 3-4). As the range changes, the sum of the four signals changes. The radar cross-section of the target can thus vary from 0 (paths short-circuit one another) to 16 times that of the target without reflection (4 times the voltage, 16 times the power). This 12 dB gain is used to advantage in making RCS measurements with what are called ground-plane ranges.

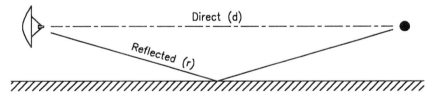

Figure 4-8. Target Fluctuations Caused by Multipath.

In the illustration, assume the range from radar to target is 3,000.0 m and the target and radar are both 15.0 m above a perfectly reflecting surface. The reflected path length is 3000.15 m. If the frequency is 3.00 GHz (λ = 0.100 m), the d-d path (round trip) is 60,000 wavelengths (21,600,000°) long. The other paths are d-r = r-d = 60,001.5 wavelengths (21,600,720°, including a reflection of 180°) and r-r of 60,003 wavelengths (21,601,440°, including two reflections of 180° each). All four are in phase and as long as they are not resolved, the low resolution RCS is 16 times that of the target.

If the range falls to 2,250 m, the new path lengths are: d-d of 45,000 wavelengths (16,200,000°), d-r and r-d of 45,002 wavelengths (16,200,900°, including the 180° reflection) and r-r of 45,004 wavelengths (16,201,800° with two reflections). By removing the 360° multiples, it is shown that the d-d and r-r signals arrive in-phase and the signals from the r-d and d-r paths arrive at 180° phase from the other two, resulting in a monostatic RCS of zero.

4-3. Target Fluctuation Models

There are many models of RCS fluctuation. The most commonly used are those of Marcum [7] and Swerling [8]. Five Marcum/Swerling models describe the behavior of a wide variety of targets. Their concept is illustrated in Fig. 4-9.

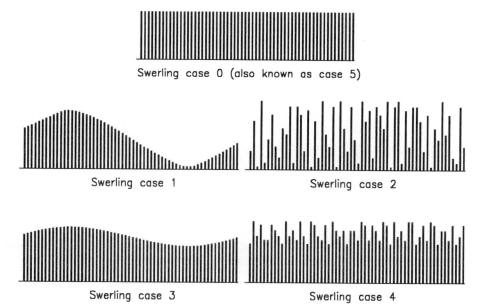

Figure 4-9. Marcum and Swerling Cases.

— *Marcum*, also known as *Swerling case 0* and *Swerling case 5* targets, are non-fluctuating. They are modeled as single isotropic scatterers. As mentioned earlier, only spheres and targets much smaller than the wavelength of the illumination signal are Marcum targets.

— *Swerling case 1* targets exhibit large fluctuations, typically several orders of magnitude, with the fluctuations occurring slowly. They are modeled as several scatterers of approximately equal radar cross-section. Case 1 is generally a good fit for complex targets such as aircraft to radars without pulse-to-pulse frequency agility, at either long ranges or flying toward or away from the radar. In either case, there is little aspect change over a look.

— *Swerling case 2* targets exhibit large fluctuations, occurring rapidly. This case is modeled the same as case 1. It fits complex targets where aspect is changing rapidly, as at close range. It also fits complex targets for radars with pulse-to-pulse frequency agility.

Case 1 and case 2 targets are described statistically as shown in the equations below, and their probability density is given in Fig. 4-10.

$$P(\sigma) = (1/\mu)\ e^{-\sigma/\mu} \qquad (4\text{-}6)$$

$P(\sigma)=$ the probability density of a certain RCS

$\sigma\ \ =$ the RCS (independent variable)

$\mu\ \ =$ the median RCS

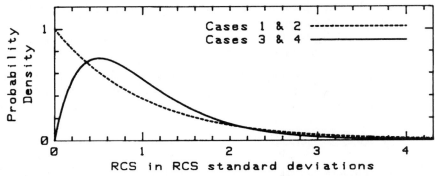

Figure 4-10. Probability Density for the Swerling Cases.

— *Swerling case 3* targets exhibit smaller fluctuations, typically an order of magnitude or less, and fluctuate slowly. They are modeled as one prominent scatterer with several smaller scatterers. This model fits simpler targets, such as some missiles, at ranges where aspect changes are small over a look. It also fits complex targets which are augmented (have their RCS made artificially larger) with a single augmentor. An example is a small target drone aircraft augmented with a single Luneburg lens (see Sec. 4-4).

— *Swerling case 4* is the same as case 3 except that the fluctuations occur rapidly, either from rapid aspect changes or from radar frequency agility. The statistical description of cases 3 and 4 is given below and in Fig. 4-10.

$$P(\sigma) = (4\sigma/\mu^2)\ e^{-2\sigma/\mu} \qquad (4\text{-}7)$$

Example 4-2: Using the geometry of Fig. 4-3, find the maximum and minimum RCS of two scatterers separated by 2.20 m and illuminated at 3.05 GHz. One of the scatterers has an RCS of 1.0 m² and the other is 2.0 m². If zero aspect is as shown in Fig. 4-3a, find one of the angles which produces maximum RCS and one of the angles which produces minimum RCS. Assume the scatterers do not shade one another.

Solution: The wavelength is 9.836 cm (0.09836 m). At zero aspect, the scatterers are 22.367 wavelengths apart (one way). The smallest aspect which produces maximum RCS would have the scatterers separated by 22.000 wavelengths in range. The smallest aspect for maximum RCS is therefore $\text{Cos}^{-1}(22.000/22.367)$, or 10.39°. RCS is in power terms but summations are of fields, which are treated as voltages. The contribution to the summation of each scatterer is the square root of its RCS. Therefore the sum is 1.000 + 1.414, or 2.414. To find the maximum RCS, this sum must revert to power terms by being squared, giving 5.83 m².

Case changes through target augmentation: It is interesting to see what happens to a complex target, such as a small pilotless drone, when it is augmented to give it the radar cross-section of a much larger aircraft (Fig. 4-11). For example, a BQM-34 target drone can be augmented to the radar cross-section of a Soviet Bear bomber. The BQM-34 is inherently case 1 or 2, depending on the look time and the aspect change. If augmented with a single large scatterer, it becomes case 3 or 4. In most circumstances, case 3 and 4 targets are easier to detect and track accurately than case 1 and 2 (Chs. 5 and 7). Thus, the entire character of the target is changed through the augmentation. Figure 4-12 shows the expected signal return before and after augmentation.

Question: If you were required to augment a small target to ten times its natural RCS and have it remain case 1 or 2, how would you do it? *Hint:* See the case definitions.

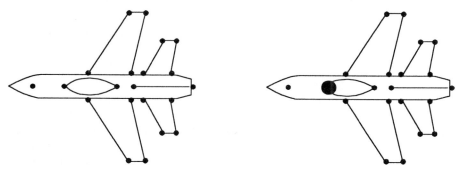

Figure 4-11. Augmentation of a Small Target with a Large Single Scatterer.

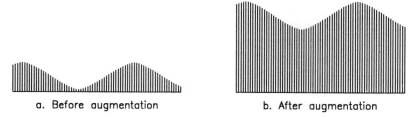

a. Before augmentation		b. After augmentation

Figure 4-12. Hypothetical RCS of Target (Fig. 4-11) Before and After Augmentation.

RCS fluctuations and Nyquist sampling: In order to fully describe information from samples, each sinusoidal cycle of the information must be sampled more than twice; the samples must be less than 180° apart (Nyquist sampling criterion — Ch. 6). Radar target RCS is information and its description must obey the same laws of information as any other signal. Therefore, to fully describe a target, RCS measurements must be made such that the composite target echo phase changes less than 180° between samples in aspect (horizontal rotation), tilt (vertical rotation), and frequency. The next example illustrates.

4-3. Target Fluctuation Models

Example 4-3: Assume an aircraft target is 60 ft long, has a wingspan of 40 ft, and is illuminated at 10.0 GHz. Find the number of samples and the sample spacing necessary to fully describe the target as it is rotated 360° in one plane only (assume aspect) and at one frequency. Find the number of samples necessary to fully describe the aircraft as it is rotated in both aspect and tilt at the same frequency. What is the maximum spacing between sample frequencies to fully describe the target to a frequency agile radar?

Solution: The longest dimension and the wavelength govern the sample frequency. Sixty feet is 18.288 m. The wavelength is 0.0300 m. Scatterers at the extremes must move in range (and thus phase at the radar) between samples less than 180°. The target extremities are 18.288 m apart, and changes between samples must be less than 1/4 wavelength, or 0.0075 m. (The physical motion of 1/4 wavelength results in a phase change of the echo signal of 180° round trip.) Basic trigonometry shows that the angular motion between samples must be less than 0.02359°, $\tan^{-1}(0.0075/18.288)$. Thus there must be more than 15,320 samples to Nyquist sample the target over 360° of rotation in one plane.

In two angular dimensions, the samples must each occupy less than 0.0005520 square degrees (0.02359 degree squared). There are 41,253.0 square degrees in a sphere. Thus, 74,731,592 samples are required to fully describe the target — at one frequency.

The greatest target span is 609.6 wavelengths (18.288 m divided by 0.03 m). The maximum frequency change must be such that this number changes by less than 1/4 wavelength (round trip of less than 1/2 wavelength). To the next higher sample frequency, the target can be no greater than 609.85 wavelengths, and at the next lower frequency, no less than 609.35 wavelengths. The higher frequency has a wavelength of no less than 0.0299877 m, or a frequency of 10.004101050 GHz, and the lower's wavelength must be no greater than 9.995898950 GHz. The frequency samples should be no more than about 4.0 MHz apart.

Comment: This example illustrates the problem encountered in trying to fully describe the RCS of a radar target. The difficulty is compounded by the fact that probably no two examples of an aircraft type are identical, and they may change enough in flight (control surfaces, flaps, aerodynamic heating, exhaust plumes, and so forth) to further complicate matters. Obviously, simplifications are needed to make the problem manageable.

4-4. Radar Cross-Section of Fundamental Shapes

Table 4-1 is a summary of some shapes which are useful in visualizing the RCS of more complex targets. A discussion of these and other fundamental targets follows the table.

Sphere radar cross-section: Spheres are isotropic radiators. As long as the circumference is much greater that the wavelength, the sphere has an antenna-like gain of unity. Thus its RCS, from Eq. 4-2, is

$$\sigma_{SPH} = \pi d^2 / 4 \tag{4-8}$$

Table 4-1. RCS of Fundamental Shapes
(Adapted from Barton [9])

Shape of Object	Maximum RCS	Minimum RCS	Number of Lobes	Width of Major Lobe
Sphere ($\pi d \gg \lambda$)	$\pi d^2/4$	$\pi d^2/4$	1	2π rad.
Ellipsoid ($a, b \gg \lambda$)	πa^2	$\pi b^4/a^2$	2	$\approx b/a$
Cylinder ($L, r \gg \lambda$)	$\pi dL^2/\lambda$	0	$8L/\lambda$	λ/L
Flat Plate (Area $\gg \lambda^2$)	$4\pi A^2/\lambda^2$	0	$8L/\lambda$	λ/L
Dipole ($L = \lambda/2$)	$0.88\lambda^2$	0	2	$\pi/2$

σ_{SPH} = the radar cross-section of an optical sphere (square meters)

d = the sphere's diameter (meters)

If the wavelength is much less than the circumference, the sphere's RCS is the same as its optical cross-section. If the wavelength approaches or is greater than the circumference, the RCS is as shown in Fig. 4-13. In the resonant region, where the circumference is from about 0.5 to 5 wavelengths, the RCS can vary from about 0.3 to 3.5 times the optical cross-section. The exact RCS of a sphere has been solved by several authors [10] and is given below. This is the only analytically exact RCS to be examined here.

$$\sigma / \pi r^2 = (j/kr) \sum_{n=1}^{\infty} (-1)^n (2n+1)(b_n - a_n) \qquad (4\text{-}9)$$

$$a_n = \frac{J_n(kr)}{H^{(1)}_n(kr)} \qquad (4\text{-}10)$$

$$b_n = \frac{kr\, J_n(kr) - nJ_n(kr)}{kr\, H_{n-1}(kr) - nH^{(1)}_n(kr)} \qquad (4\text{-}11)$$

$$H^{(1)}_n(kr) = J_n(kr) + jY_n(kr) \qquad (4\text{-}12)$$

$$kr = 2\pi r/\lambda \qquad (4\text{-}13)$$

$\sigma/\pi r^2$ = the ratio of the radar cross-section to optical cross-section

r = the sphere's radius

j = $\sqrt{-1}$

n = the summation counter

λ = the wavelength

$J_n(kr)$ = the spherical Bessel function of the first kind with order n and argument kr

$Y_n(kr)$ = the spherical Bessel function of the second kind with order n and argument kr

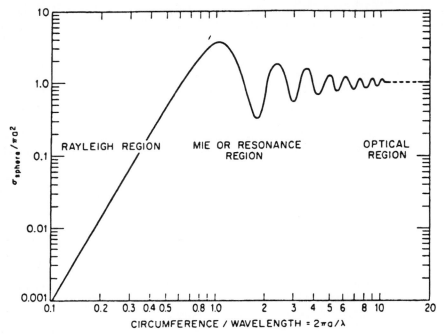

Figure 4-13. Sphere RCS versus Frequency. (Reprinted from Knott, [11]; © 1990 McGraw-Hill Book Company, used with permission)

For a sphere whose circumference is much greater than the wavelength, the right-hand side of Eq. 4-9 becomes unity. This is the sphere's optical region. For circumferences between about 0.5 wavelength and 10 wavelengths, the ratio of RCS to optical cross-section oscillates about unity, with its peak value occurring at $kr \approx 1$. This is the resonant, or MIE region. For circumferences much less than a wavelength, the ratio of RCS to optical cross-section becomes proportional to $(kr)^4$. This is the Rayleigh scattering region.

The Rayleigh scattering region, where RCS of each drop is proportional to the fourth power of its diameter, is the premise on which weather radars function. The largest possible raindrops are about 1/4 in. in diameter and are Rayleigh scatterers at frequencies up to about X-Band; in higher bands they become resonant. It is assumed that violent weather is associated with large drops, since the vertical currents must be high to hold the water in the cloud long enough for it to coalesce into the large drops. Thus violent weather has more radar cross-section than the same amount of water in a less violent environment. Rayleigh scattering, where the RCS of a drop is proportional to the fourth power of frequency, is also the reason weather clutter is much more of a problem at higher frequencies. Drops which have one unit of RCS at L-Band (new D-Band) have over 4,000 units in the X-Band, since X-Band has about one-eighth the wavelength of L-Band and the drops are electrically eight times larger at X-Band; $8^4 = 4,096$.

Optical-region spheres are case 0 targets and their radar cross-section has no aspect dependence, making them attractive calibration targets. Spheres are often flown attached to balloons to provide known RCS targets for testing tracking radars. To obtain Doppler shift, they are tossed out of airplanes (sphere drops) and towed behind airplanes (sphere tows).

Example 4-4: A hollow aluminum sphere calibration target is 6" in diameter. Find its radar cross-section (optical region) and the lowest frequency at which it can be used as an optical radar target.

Solution: The area of a 6" circle (0.152 m) is 0.0182 m². This is its RCS, −17.4 dBsm. A sphere is optical if its circumference is more than 5 to 10 wavelengths. Using the 5-wavelength criterion, the minimum frequency is 3.1 GHz. For a 10-wavelength circumference, frequency is 6.3 GHz.

Cylinder RCS: Cylinders (right circular cylinders are considered here) have a peak broadside antenna-like gain proportional to their length in wavelengths, assuming both the circumference and length is much larger than the wavelength.

$$G_{CYL} = \pi L / \lambda \tag{4-14}$$

G_{CYL} = the "gain" of an optical cylinder
L = the length of the cylinder
λ = the wavelength

$$A_{CYL} = d L \tag{4-15}$$

A_{CYL} = a cylinder's broadside projected area
d = its diameter

From Eq. 4-2 and assuming perfect conduction ($\Gamma = 1$), the peak broadside radar cross-section of the cylinder is

$$\sigma_{CYL} = \pi d L^2 / \lambda \tag{4-16}$$

σ_{CYL} = the cylinder's RCS

Cylinder RCS is highly aspect-dependent and is shown in Fig. 4-14. Lobe widths are given in Table 4-1. It is interesting that the broadside (0°) aspect sidelobes are 13.5 dB below the peak RCS. This is correct for a rectangularly illuminated object, which the cylinder at

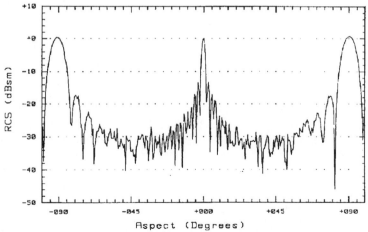

Figure 4-14. Cylinder RCS versus Aspect.

4-4. Radar Cross-Section of Fundamental Shapes

broadside aspect is. The ends, at $\pm 90°$, are circular flat plates with 17-dB sidelobes. The sidelobe level for circular illumination is 17 dB. See Chs. 9 and 11 for illumination functions and sidelobes.

Cylinder RCS has a first-order frequency dependence. Rewritten in frequency terms, the peak RCS of a conducting cylinder is

$$\sigma_{CYL} = \pi f d L^2 / c \tag{4-17}$$

f = the frequency
c = the propagation velocity

Flat plate RCS: The gain of a collimated antenna which is much larger that the wavelength is, from Ch. 3

$$G = 4\pi A_E / \lambda^2 \tag{4-18}$$

G = the gain of any antenna
A_E = the antenna's effective area
λ = the wavelength

The effective area of a flat plate which is uniformly illuminated is its actual area. From Eq. 4-2, therefore, the peak broadside radar cross-section of a flat plate is

$$\sigma_{FP} = 4\pi A^2 / \lambda^2 \tag{4-19}$$

σ_{FP} = the flat plate RCS
A = the area of the flat plate

Flat plate RCS is highly aspect dependent, as the end plates ($\pm 90°$) of the cylinder of Fig. 4-14 show. Main lobe width is given in Table 4-1.

Flat plate RCS has a second order frequency dependence, and in frequency terms the peak RCS is

$$\sigma_{FP} = 4\pi f^2 A^2 / c^2 \tag{4-20}$$

Example 4-5: An interesting calibration target is a hollow aluminum cylinder constructed such that its broadside (cylinder) and end (flat plate) radar cross-sections are the same. Design a cylinder for a broadside and end RCS of 1.0 m² at 10.0 GHz.

Solution: The diameter of the cylinder sets the RCS of the end plates. Equation 4-20 gives the area of a peak 1.0 m² RCS flat plate target at 10.0 GHz as 0.00846 m². The diameter of a circle with this area is 0.104 m (4.1"). The length of a cylinder with RCS of 1.0 m², a diameter of 0.104 m, and a frequency of 10.0 GHz is 0.303 m (11.94") (Eq. 4-17).

Dipole RCS: The radar cross-section of a dipole as a function of frequency is shown in Fig. 4-15. Its peak value of 0.88 λ^2 occurs when the dipole is one-half wavelength (on the dipole) long and is oriented in the direction of the E-Fields and perpendicular to the direction of propagation. The average value for all orientations and polarizations is about 0.15 λ^2. Dipoles make up chaff, a form of ECM. Chaff is discussed later in this chapter.

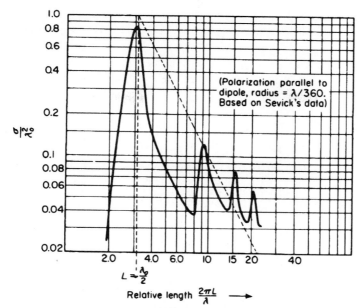

Figure 4-15. Dipole RCS versus Frequency.
(Reprinted from Barton [9]; © 1976 Artech House, used with permission)

Corner reflector RCS: Corner reflectors are plates arranged such that they reflect electromagnetic energy over a wide aspect. The principle is shown in Fig. 4-16. Figure 4-17 gives the RCS *versus* aspect for a triangular corner reflector. The peak RCS is

$$\sigma_{CR3} = 4\pi a^4 / (3 \lambda^2) \qquad (4\text{-}21)$$

σ_{CR3} = the RCS of a triangular corner reflector

a = the length of one of the "seams" between the plates

Corner reflectors behave very much as flat plates, with similar RCS and frequency response. The difference is that, for a given size, the peak RCS lobe of the corner reflector is much wider than the plate. When used deliberately, they are target augmentors. When they occur accidentally, as at the intersection of a wing and fuselage of an aircraft, they contribute to higher RCS.

Luneburg lens RCS: Luneburg lenses [12] are devices which reflect power in the direction from which it arrived. They are similar to corner reflectors in this respect, but have wider main lobes. The construction of a Luneburg lens is shown in Fig. 4-18. It is a dielectric

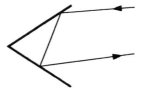

Figure 4-16. Corner Reflector Principle.

4-4. Radar Cross-Section of Fundamental Shapes

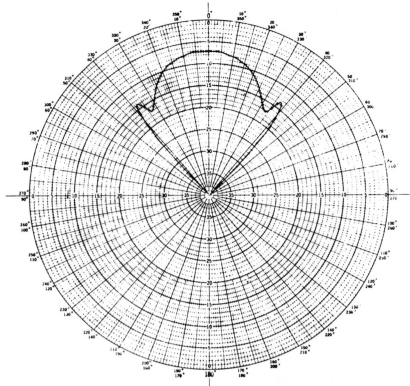

Figure 4-17. Triangular CornerReflector RCS. (Diagram courtesy of D. L. Mensa; U. S. Government work not protected by U.S.copyright)

body with varying index of refraction. Incident electromagnetic waves from a particular direction focus to a spot on the back surface of the lens, where they reflect off a conducting cap and exit the same direction from which they entered. Hence they behave as flat plates, but with much wider lobes. They are widely used as augmentors. The augmentation of Fig. 4-12 could, for example, be done with a Luneburg lens.

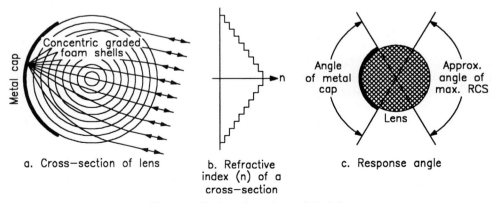

Figure 4-18. Luneburg Lens Principle.

To achieve the desired effect, the refractive index across the lens must obey the following relationship:

$$n = \sqrt{2 - (r/r_0)} \qquad (4\text{-}22)$$

n = the index of refraction at radius r from the center
r = the radius for which the refractive index is calculated
r_0 = the radius of the lens

The peak radar cross-section of a Luneburg lens is approximately that of a flat plate of the same size, ignoring losses in the dielectric, and is

$$\sigma_{LL} \approx 4\pi A^2 / \lambda^2 \qquad (4\text{-}23)$$

σ_{LL} = the RCS of the Luneburg lens
A = the optical cross-sectional area of the lens ($\pi d^2 / 4$ for a spherical lens)
d = the diameter of a spherical lens
λ = the wavelength

The width of the main response lobe is approximately the subtended angle of the metal cap, with a limit of about 120°, beyond which the cap shades the lens excessively (Fig. 4-18c). The radar cross-section *versus* aspect of a Luneburg lens is shown in Fig. 4-19.

Example 4-6: Luneburg lenses are commonly used as augmentors and are available commercially in several sizes. Estimate the peak RCS of a lens 8" in diameter to an X-Band (10.0 GHz) radar.

Solution: The area of an 8" (0.203 m) circle is 0.0324 m². The wavelength is 0.03 m. Equation 4-19 or Eq. 4-23 gives the RCS of a plate of this size, which is 14.7 m², or +11.7 dBsm.

4-5. Radar Cross-Section of Complex Objects

The interaction of multiple scatterers, shown in Sec. 4-2, accounts for most target fluctuations. Complex targets can be modeled as arrays of individual scattering centers (Fig. 4-11). Some target models treat the individual scatterers as isotropic (N-Point models) and others treat them as non-isotropic simple shapes, such as cylinders and flat plates (N-Shape models). If the individual scatterers making up a target are roughly equal in RCS, a Swerling case 1 or 2 target results. If one scatterer predominates, the composite is a case 3 or 4 target.

The spacing of lobes in a target's response depends on the scatterer spacing in wavelengths. For antenna arrays, the number of lobes produced by elements spaced S apart, and the width of each lobe are given as

$$N_{LA} = 4 S / \lambda \qquad (4\text{-}24)$$

N_{LA} = the number of lobes in a 360° pattern for a two-element antenna array S long, where S is in the same units as λ

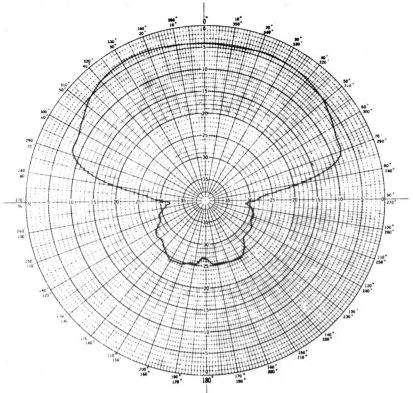

Figure 4-19. Luneburg Lens RCS. (Diagram courtesy of D. L. Mensa; U. S. Government work not protected by U.S. copyright)

$$\theta_{LA} \approx \lambda / S \, (180/\pi) \text{ (degrees)} \qquad (4\text{-}25)$$

θ_{LA} = the width of each lobe in an antenna's pattern, broadside aspect

For radar targets, the array is effectively twice as long as the antenna because of radar's two-way propagation.

$$N_{LT} = 8 L / \lambda \qquad (4\text{-}26)$$

N_{LT} = the number of lobes for a radar target composed of two scatterers spaced L apart, where L is in the same units as λ

$$\theta_{LT} \approx \lambda / (2L) \, (180/\pi) \text{ (degrees)} \qquad (4\text{-}27)$$

θ_{LT} = the width of each lobe in a target's pattern, broadside aspect

Figure 4-20 shows a T-33A jet training aircraft, which is about 30 ft long, has a wingspan of about 40 ft, and has a large cylindrical fuel tank on each wingtip. Figure 4-21 is the RCS versus aspect scintillation pattern of this aircraft at 425 MHz. The scintillations, which occur over about six orders of magnitude, are the result of many roughly equal scatterers spaced many wavelengths apart. Its wingspan is about 17 wavelengths at this frequency, predicting nose aspect lobe widths of about 1.7° (Eq. 4-27). To a constant-frequency radar this target is Swerling case 1, except at very short ranges where it may be case 2. To a pulse-to-pulse frequency agile radar it is case 2.

Figure 4-20. T-33A Aircraft.

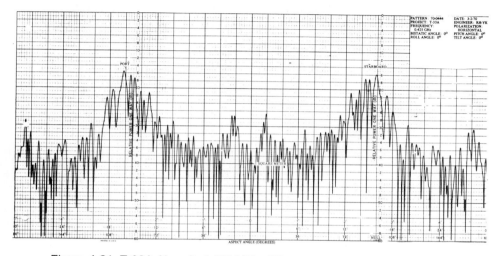

Figure 4-21. T-33A Aircraft at 425 MHz. (Diagram courtesy of D. L. Mensa; U.S. Government work not protected by U.S. copyright)

Figure 4-22 shows the same aircraft at 3 GHz. The wingspan is now about 120 wavelengths, for a lobe width at nose aspect of about 0.23°. The RCS pattern has the same form, but the period of the scintillation is much shorter.

The fine structure of the RCS *versus* aspect of a large target like an aircraft sometimes hides major features and trends in the information. For this reason, data for a target of this type are usually presented as its *median* RCS over a certain aspect span. This is known as medianized RCS data. Figure 4-23 is an example of medianized hypothetical data (not T-33A), showing 10° medians.

4-6. Glint

Radar cross-section of targets fluctuates in both amplitude (scintillation) and in phase (glint). Glint causes the far-field phase fronts of the reflected wave to be non-planar. Figures 4-24 and 4-25 show a target model which is useful for visualizing glint. It is two

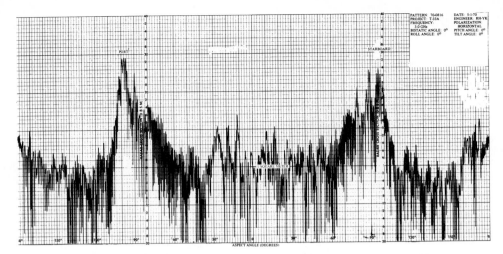

Figure 4-22. T-33A Aircraft at 3.0 GHz. (Diagram courtesy of D. L. Mensa; U.S. Government work not protected by U.S. copyright)

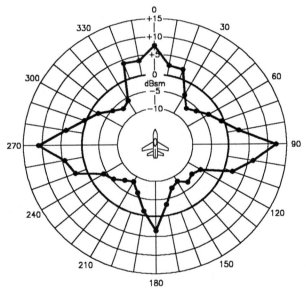

Figure 4-23. Hypothetical Medianized RCS Data.

isotropic scatterers separated by one wavelength, one scatterer having an RCS of 1.0 unit and the other an RCS of 0.81 unit. Electric field response is in the ratio of 1:0.9. Figure 4-24 is the amplitude response versus aspect angle. The sum of the two scatterers fluctuates from 0.01 to 3.61 RCS units (0.1 to 1.9 electric field units).

By Eq. 4-26 there are eight lobes, as seen in the figure. The phase of each lobe alternates between 0° (broadside and end lobes) and 180°. At each null, the phase changes and it is these changes that cause the target glint.

Figure 4-25 plots for one time instant the echo wave fronts propagating away from the target. In the transitions between 0° and 180° phase, the wave fronts are not perpendicular

204   Chap. 4. Targets and Interfering Signals

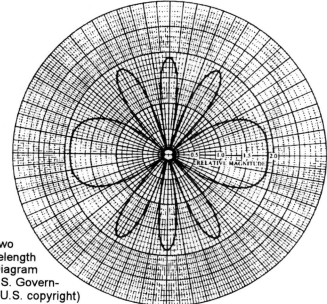

Figure 4-24. Amplitude of Two Isotropic Scatterers, 1 Wavelength Apart, 1:0.81 RCS Ratio. (Diagram courtesy of D. L. Mensa; U. S. Government work not protected by U.S. copyright)

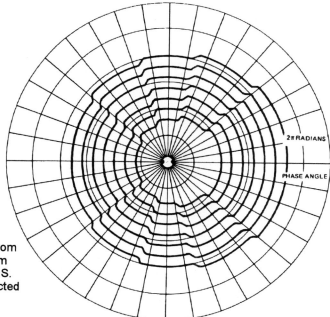

Figure 4-25. Wave Fronts from Target in Fig. 4-24. (Diagram courtesy of D. L. Mensa; U. S. Government work not protected by U.S. copyright)

4-6. Glint

205

to the direction of propagation. Since all antennas, including our eyes, identify a signal's angle-of-arrival (AOA) as coming from the direction perpendicular to the waves, an angle error results.

Figure 4-26 shows the apparent AOA of a glinting target at different ranges. Since the phase change for two scatterers is constant (180° for two scatterers – different for other numbers of scatterers), the wave tilt angle becomes shallower at longer ranges. Regardless of range, the apparent location of a glinting target remains constant, unless of course the target aspect or illumination frequency changes.

Figure 4-27 depicts the wave fronts from a hypothetical complex glinting target. In this case, the glint appears as "noise" on the waves which contributes to the uncertainty in the target location.

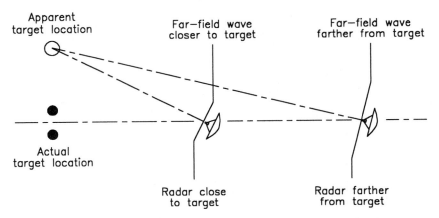

Figure 4-26. Apparent Target Location Shift Caused by Glint.

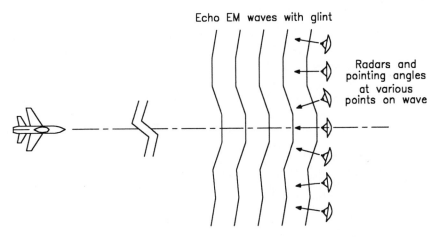

Figure 4-27. Wave Fronts from Complex Glinting Target.

Chap. 4. Targets and Interfering Signals

Figure 4-28 shows experimental data from a scale-model aircraft. The model, shown on the plot in its actual size relative to the plot scales, was placed on a positioner and a tracking radar locked to it. The target was then rotated slowly enough so that the tracking servos followed the target's apparent location. The plot is of this apparent location, where the radar "thought" the target was. Had the target rotation been stopped, the radar would have held its position. A cumulative distribution of apparent target positions taken from this data would show that the radar was pointed off the target in 5 percent to 15 percent of the samples, depending on the range of aspect angles analyzed.

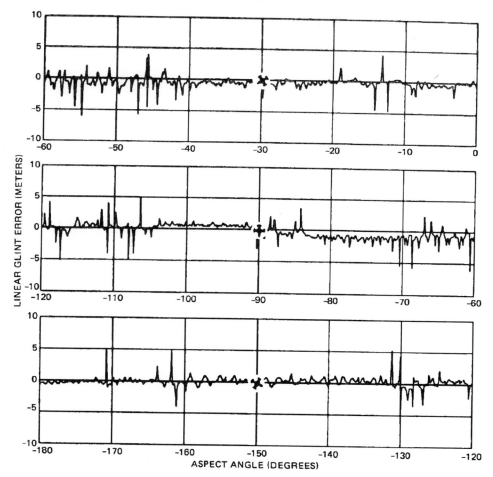

Figure 4-28. Apparent Location of the Aircraft Model. (Diagram courtesy of D. L. Mensa; U. S. Government work not protected by U.S. copyright)

Glint linear errors are small, a few target spans at most. At short ranges, however, angular errors can be large. The systems most affected are tracking radars where high accuracy is required, such as precision instrumentation trackers, missile seekers, radars directing guns, and automated aircraft landing systems.

4-6. Glint

The rapidity of fluctuations very much affects glint-induced position errors. If the target fluctuates slowly (cases 1 and 3), the tracking servos can follow the target's apparently shifting position and the errors are substantial. If the fluctuations are rapid (cases 2 and 4), the servos, with bandwidths usually less than 10 Hz, average the target location and produce a much more accurate track. One way of turning a case 1/3 target into case 2/4 is with pulse-to-pulse frequency agility. See Fig. 4-29. Many radar gun director modes are in fact pulse-to-pulse frequency agile. Glint as it affects track accuracy is examined further in Ch. 7.

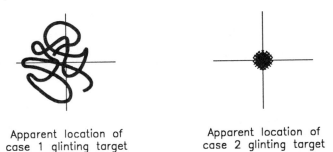

Apparent location of case 1 glinting target

Apparent location of case 2 glinting target

Figure 4-29. Effect of Glint Rapidity on Apparent Position to a Track Radar.

4-7. Bistatic Radar Cross-Section

Most targets behave differently to bistatic radars than to monostatic. Figures 4-3 and 4-4 gave an example. At the aspect where the target depicted had a minimum RCS to a monostatic radar, it had a maximum RCS to a bistatic system with a bistatic angle of 28.32°. Figure 4-30 shows a flat plate oriented so that it presents a low RCS to a monostatic radar but high RCS to bistatic.

Bistatic geometry and definitions are shown in Sec. 3-15. More information on bistatic RCS can be found in Morchin [5] and other references (see Further Information section).

It is well known that when the bistatic angle approaches 180°, the RCS of all targets, *including absorbing targets*, becomes large. This is because of diffraction from the "holes" punched in the electromagnetic waves by the target, shown in Fig. 4-31. In the limit (bistatic angle equal to 180°) all targets, including absorbers, have the RCS of a flat plate the size of the EM wave "holes" oriented for maximum response.

$$\sigma_{B(max)} = 4\pi A^2 / \lambda^2 \tag{4-28}$$

$\sigma_{B(max)}$ = the maximum bistatic RCS of a target (bistatic angle = 180°)

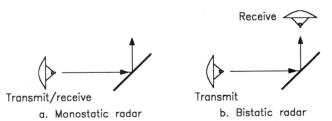

a. Monostatic radar

b. Bistatic radar

Figure 4-30. Flat Plate — Monostatic and Bistatic.

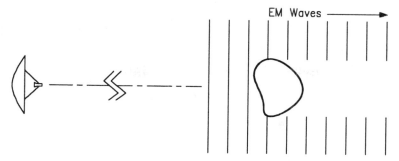

Figure 4-31. EM "Holes" at Large Bistatic Angles.

4-8. Target Doppler Spectra

Target echoes have a Doppler spectrum caused by different parts of the target moving at different speeds. This is usually the result of rotating engine and other propulsion parts. Doppler spectra may be useful for identifying targets.

The spectrum is composed of lines separated from the airframe Doppler by the blade rate of the rotating components. For example, a propeller-driven aircraft with a four-blade propeller turning 1800 rpm has a blade rate of 120 blades per second (30 revolutions per second times 4 blades per revolution). If the front fan of a jet engine has 25 blades and turns 9000 rpm, the frontal blade rate is 3,750 per second. The rear aspect blade rate is probably different, since the radar would be viewing a different engine "spool," with a different number of blades and a different rotational velocity.

The blades modulate the signal in both amplitude and frequency. Amplitude modulation is caused by, among other things, rotating blades alternately covering and uncovering stationary parts of the engine, as shown in Fig. 4-32a. Frequency modulation comes from several factors, one (Fig. 4-32b) being that unless the radar is looking down the engine's axis, blades move toward the radar for half their revolution and away the other half (relative to the airframe). Another is that blades are angled to the engine axis to produce thrust. From a frontal aspect, any intervening structure causes the radar to see advancing blades as it enters and leaves the obscuration (Fig. 4-32c).

The result of simultaneous amplitude and frequency modulation at the same rate is a spectrum which contains unbalanced sidelobes, as shown in Fig. 4-33. There may be more than one set of sidelobes if the frequency modulation index is high. Multi-engined aircraft produce more complex spectra.

The percentage of the total echo energy which appears in the sidebands depends on the target, ranging from a fraction of a percent to more than 50 percent. Table 4-2 gives some Doppler spectral characteristics of a few older aircraft. Modern jet aircraft will tend toward larger fractions of their RCS in the Doppler sidebands, particularly at frontal aspects, because their modern engines are often high-bypass turbofans and are a large fraction of the total aircraft frontal area. Turbofans, which are essentially a turbine engine attached to a large ducted propeller, have a much larger front fan structure than turbojets, where the front structure is simply a compressor for the jet engine. Most of the air goes around the engine core in a turbofan, all of it goes through a turbojet.

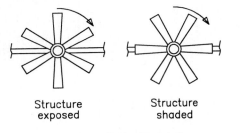
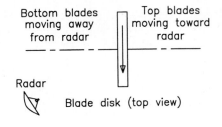

a. RCS Amplitude Modulation

b. RCS Frequency Modulation

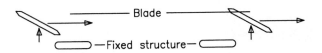

c. Frequency Modulation

Figure 4-32. Some Mechanisms of RCS Modulation (figure needs revision *****).

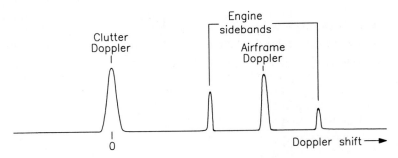

Figure 4-33. Target Doppler Spectrum.

Table 4-2. Echo energy characteristics of some older aircraft. (After Nathanson [4])

Aircraft	Aspect	% Energy in sidebands		
		Minimum	Average	Maximum
T-28 single engine prop.	Nose	12	29	75
Lear jet, turbojet	0-30°	0	22	59
Medium multi-engine jet	Nose	4	11	17
DC-8 / B-707 4-engine turbojet	Nose	17	28	38

4-9. Polarization Scattering Matrix

Echoes from simple targets such as spheres, raindrops, and the like are strongly co-polarized (have the same polarization as) with the incident wave. Complex targets such as aircraft typically reflect waves with polarization components cross-polarized to the illumination. The scattering characteristics of a target can be described as having several scattering coefficients, each representing the ratio of scattered fields to incident fields. These scattering coefficients are similar to Γ in Eq. 4-2, but each is for a particular combination of incident wave and reflected wave polarizations. The coefficients together make up the target's polarization scattering matrix.

$$\begin{bmatrix} E_{SH} \\ E_{SV} \end{bmatrix} = \begin{bmatrix} \alpha_{HH} & \alpha_{HV} \\ \alpha_{VH} & \alpha_{VV} \end{bmatrix} \begin{bmatrix} E_{IH} \\ E_{IV} \end{bmatrix} \qquad (4\text{-}29)$$

E_{SH} = the horizontal component of the electric field scattered by the target

E_{SV} = the vertical component of the electric field scattered by the target

E_{IH} = the horizontal component of the electric field impinging on the target

E_{IV} = the vertical component of the electric field impinging on the target

α_{HH} = the scattering coefficient (similar to Γ in Eq. 4-2) of the target for horizontally polarized transmitted signal and horizontally polarized received echo

α_{VH} = the scattering coefficient of the target for horizontal transmitted signal and vertical received echo

α_{HV} = the scattering coefficient of the target for vertical transmitted signal and horizontal received echo

α_{VV} = the scattering coefficient of the target for vertical transmitted signal and vertical received echo

The coefficients given above are oriented along the horizontal and vertical axes. The axis set can be in any directions, as long as the two components are orthogonal. A similar matrix exists for circular polarization, with coefficients for right- and left-hand incident and reflected waves.

The polarization matrix provides, among other things, an estimate of how effective cross-polarized systems would be in suppressing interference such as rainfall, while retaining the ability to detect aircraft and ship targets. Table 4-3 shows some data on the attenuation of RCS to cross-polarized systems for precipitation. Table 4-4 gives the same information for aircraft and ship targets.

In the precipitation table, note that rainfall has a higher RCS to circular cross-polarization than to co-polarization (using RHC transmit and LHC receive gives a higher RCS than RHC transmit and RHC receive). The precipitation's RCS to RHC/LHC is about

Table 4-3. RCS Cross-Polarization Properties of Precipitation.
(after Nathanson [4])

Freq. band	Precipitation type	Elevation angle (degrees)	σ_{RL}/σ_{RR} (dB)
L	Rain		22-30
L	Wet snow		15
S	Rain	<10	18
S	Rain over sea	0-10 avg	20*
S	Rain over marsh	0-10 avg	24*
S	Rain over land	0-10 avg	27*
S	Rain over desert	0-10 avg	34*
C	Rain	0-10	17
X	Thunderstorm		15
X	Rain	high	26-28
X	Bright band		13-20
X	Fine snow		26
K_u	Rain	high	26
K_a	Rain	30	17-18
K_a	Bright band	30	5-11
K_a	Dry snow	30	12-16

* Theoretical.

Table 4-4. RCS Cross-Polarization Properties of Targets.
(after Nathanson [4])

Freq. band	Target type	Target aspect angle (degrees)	Ratio of σ_{HH} or σ_{VV} / σ_{RR} or σ_{LL}	Ratio of σ_{HH} or σ_{VV} / σ_{HV} or σ_{VH}
L	Piston aircraft		6-8 dB	
L	Land	−8 El		7-8 dB
C	Land	−8 El		6-10 dB
C	Vehicles	−3 El		5-16 dB
C	Trees	−3 El		0-6 dB
C	Ships	−3 El		−(2-8) dB
C	Ships	−6 El		3-15 dB
X	Aircraft	Nose and tail	2.5 dB avg	2.5-12 dB
X	Jet aircraft	Nose	4.5-6 dB	10 dB
X	Ships	0 El	6 dB	
X	Ships	−(3-6) El		7-15 dB
Ka	Jet and piston acft.	Tail	3-5 dB	7-11 dB
Ka	Jet and piston acft.	Nose	2-4 dB	8-10 dB
Ka	Trees	−3 El		4
Ka	Vehicles	−3 El		8-14

the same as using linear co-polarization. This is the basis of polarization suppression of weather clutter. Aircraft and ships, on the other hand, tend to have a moderately higher RCS to co-polarization than cross-polarization, although it can be seen that their RCS is somewhat lower to circular co-polarization than to linear.

All in all, weather RCS is much more sensitive to co- and cross-polarization than are complex targets. Many radars intended to detect targets in precipitation, such as terrain avoidance (TA) and terrain following (TF) system are circularly polarized, because the precipitation is attenuated more by circular co-polarization than are the complex targets (in this case, the ground).

Example 4-7: A jet aircraft's X-Band nose aspect RCS is measured by a linearly co-polarized (HH) measurement system as +10.0 dBsm. Using the above tables, find the approximate improvement in signal-to-clutter ratio for high-angle rain weather clutter (rain) by using circular co-polarization (RHC/RHC) versus using linear co-polarization.

Solution: Table 4-3 shows that high angle rain is attenuated about 26 to 28 dB (use 27 dB) by using RHC/RHC *versus* using RHC/LHC. This is approximately the same attenuation as would be found by comparing RHC/RHC to linear co-polarized waves. Table 4-4 shows that the aircraft target is attenuated about 5 dB (4.5 to 6 dB) using circular co-polarization *versus* linear. Therefore the improvement in S/C brought about by using circular co-polarization is about 22 dB (27 dB loss in clutter RCS and 5 dB loss in the target) as opposed to using linear co-polarization.

4-10. Interfering Signals — Noise

Noise is a random interfering signal, generated by the random motion of electrical charge, usually in the receiver. Noise parameters are covered in detail in Sec. 2-6.

Statistical distribution of noise and signal plus noise: Noise, being random, must be described either by its overall effect, as in the above equations, or statistically. Pre-demodulation noise in the RF and IF parts of the receiver is Gaussian distributed. Noise is usually dealt with in radars, however, post-demodulation. The distribution of noise alone after I/Q demodulation is described by the Rayleigh function of Eq. 4-30 and Fig. 4-34.

$$P(n) = (R / \psi_0) \exp(-R^2 / 2\psi_0)] \qquad (4\text{-}30)$$

$P(n)$ = the probability of noise exceeding the threshold value R, where for our purposes R can be considered a voltage threshold.

R = the variable (usually voltage)

ψ_0 = the RMS value of the Gaussian noise

If a sinusoidal signal is added to the noise, the distribution is called "modified Rayleigh" or "Ricean" and is described in Eq. 4-31 and Fig. 4-35. Note that Eq. 4-30 (noise alone) is simply Eq. 4-31 with the signal (a) set to zero.

$$P(s+n) = (R / \psi_0) \{\exp[-(R^2 + a^2) / 2 \psi_0]\} I_0\{[aR / \psi_0]\} \qquad (4\text{-}31)$$

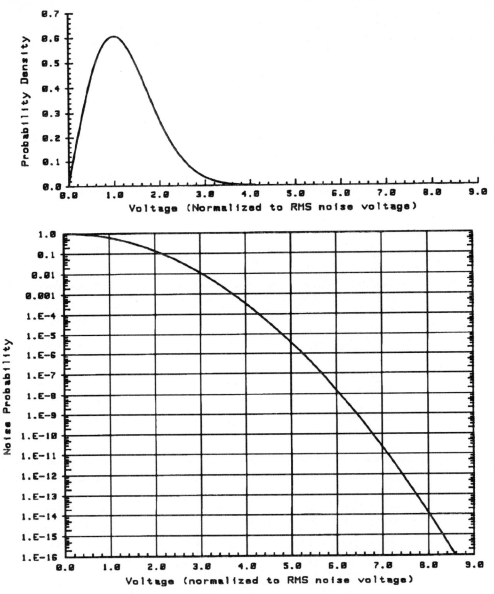

Figure 4-34. Rayleigh Probability Function and Distribution of Noise.

$P(s+n)$ = the probability of signal plus noise exceeding a threshold of value R

a = the peak amplitude of the sinusoidal signal

I_0 ( ) = the zero order Bessel function of the first kind with imaginary argument

The modified Rayleigh distributions for signal plus noise have as parameters the peak signal voltage (V_P in $V_P \text{Cos}(2\pi ft+\phi)$) and RMS noise voltage. The power signal-to-noise ratio relates to the distribution parameters by

$$S/N = (a/\psi_0)^2 / 2 \qquad (4\text{-}32)$$

Use of the Rayleigh and Ricean distributions in signal detection is discussed in Ch. 5.

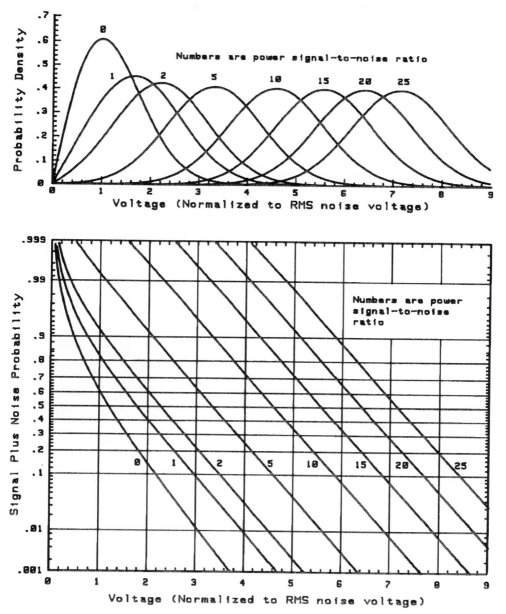

Figure 4-35. Modified Rayleigh (Ricean) Function and Distribution of Signal Plus Noise.

4-11. Interfering Signals — Surface Clutter

Surface, or area, clutter is signal echo from approximately planar surfaces whose area exceeds the radar's resolution cell on the clutter surface. Sea and land clutter are area clutter. The radar cross-section of area clutter depends on the clutter characteristics and the area of clutter within the radar's resolution cell. The RCS of area clutter, assuming the clutter is uniformly distributed over the resolution area, (from Ch. 3) is

$$\sigma_{AC} = \sigma^0 A_I \tag{4-33}$$

σ_{AC} = the RCS of the area clutter

σ^0 = the "figure of merit" of the clutter, which is it RCS per unit area illuminated (m² RCS per m² illuminated)

A_I = the area of clutter illuminated within the radar's resolution cell

The factor σ^0 depends on several factors, including:

— *Grazing angle:* Grazing angle (α_G) is the angle from horizontal at which the illumination energy strikes the clutter surface (Sec. 3-10). With smoother seas and land, σ^0 increases with increasing grazing angle. With rougher surfaces, very high grazing angles produce less σ^0 than at lower angles.

— *Vertical texture of the clutter:* Rough surfaces have larger σ^0 than smooth for low grazing angles; at very high grazing angles (near 90°), smooth surfaces have higher σ^0 than rough.

— *Wavelength of the radar's illumination:* σ^0 is a function of vertical texture in wavelengths. Thus shorter wavelengths generally cause larger is σ^0.

Clutter regions and models: Three distinct regions of clutter behavior are recognized, separated by the grazing angle: the low grazing angle region, the plateau region, and the high grazing angle region. See Morchin [5] and Fig. 4-36.

The low grazing angle region extends from zero to a critical angle determined by the RMS height, in wavelengths, of surface irregularities. This critical angle is the grazing angle below which a surface is "smooth" by Rayleigh's definition, and above which it is "rough." By definition, a surface is smooth if

$$\Delta h \sin\Theta < \lambda/8 \tag{4-34}$$

Δh = the RMS height if surface irregularities

Θ = the angle between incident rays and the average surface irregularities

λ = the wavelength

From the above expression, the critical angle can be derived as

$$\sin \theta_C = \lambda/4\, h_E \tag{4-35}$$

θ_C = the critical angle, below which the surface is smooth and above which it is rough

h_E = the RMS surface-height irregularity (meters)

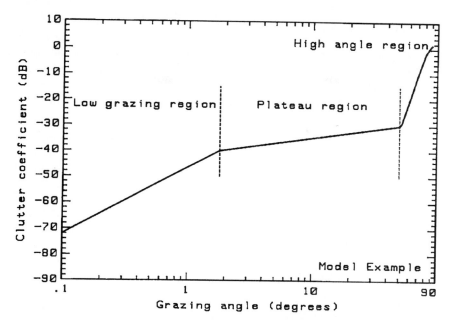

Figure 4-36. Example of Clutter Regions.

According to Morchin, the sea surface roughness is related to the sea state by

$$h_E \approx 0.025 + 0.046\, SS^{1.72} \text{ (meters)} \quad (4\text{-}36)$$

SS is the sea state (see Table 4-5)

Below the critical angle, the backscatter depends on a propagation factor expressing the constructive/destructive interference of the direct and reflected waves.

$$|F|^4 = |4\pi\,(h/\lambda)\,\sin\theta_g|^4 \quad (4\text{-}37)$$

F = the propagation factor
h = the surface irregularity height
θ_g = the grazing angle

In power terms, clutter RCS is proportional to F^4. Thus for grazing angles less than critical, the clutter σ^0 is proportional to the grazing angle raised to the fourth power.

In the plateau region, the scattering is by facets smaller than the wavelength and the grazing angle dependence of σ^0 is much less than in the low angle region. Katzin's [14] expression of the backscatter coefficient for sea clutter in this region is

$$\sigma^0 = A\,N_0\,\lambda^{-1}\,W_S^{3/4}\,(1 \pm 2\times10^{-3}\,W_S) \quad (4\text{-}38)$$

A = 0.009 when wind speed is in knots), 0.015 in meters per second
W_S = the wind speed in knots or meters per second
N_0 = a constant relationship between average facet areas and number of facets per unit area (Katzin uses 7.2×10^{-5})

4-11. Interfering Signals — Surface Clutter

The high grazing angle region is one of specular scattering, with diffuse clutter components disappearing. It is characterized by the fact that, at angles nearing 90°, the surface acts as a specular "mirror" and clutter coefficient is larger for smooth terrain and seas than for rough. The clutter coefficient for the high angle region is given, from Beckmann and Spizzichino [15], as

$$\sigma^0 = (u / \tan^2 \beta_0) \exp(-\tan^2 \beta / \tan^2 \beta_0) \qquad (4\text{-}39)$$

u = the reflectivity of the clutter at a 90° grazing angle. [It is essentially unity for sea clutter and ranges from about 0.06 to 0.6 for land clutter (Morchin [5]). It is a function of frequency and is averaged for many land clutter types in Eq. 4-40].

β = the bisector of the angle between the incident and reflected rays (for horizontal surfaces, it is $90° - \theta_g$)

β_0 = related to the mean slope of the surface irregularities, and is estimated in Eq. 4-41

$$u \approx -10 \log (4.7 / \sqrt{f}) \qquad (4\text{-}40)$$

u = the high-angle reflectivity of the clutter surface — unity for sea clutter

f = the frequency in GHz

$$\beta_0 \approx 2.44 \, (SS + 1)^{1.08} \, (\pi/180) \text{ for sea clutter} \qquad (4\text{-}41)$$

β_0 = clutter slope irregularity factor (radians) calculated for sea clutter, 0.14 radian for desert, 0.2 farmland, 0.4 woodlands, 0.5 mountains (Morchin)

SS = the sea state

Generalized sea clutter model: Using isolated models of the three area clutter regions, Morchin proposes a consolidated model for sea clutter, averaged over all polarizations, viewing angles, and wind direction (up-wind, down-wind, or cross-wind). The following model is adapted from Morchin.

$$\sigma^0 = 10 \log \{10 \wedge [-6.4 + 0.6(SS+1) - \log \lambda + \log(\sin \theta_g) + 0.1 \, \sigma^0_C]$$

$$+ \operatorname{ctn}^2 \beta_0 \exp\left[\frac{-\tan^2 (\pi/2 - \theta_g)}{\tan^2 \beta_0}\right] \} \quad \text{(dB)} \qquad (4\text{-}42)$$

$$\sigma^0_C = 10 \, K \log (\theta_C / \theta_g) \qquad (4\text{-}43)$$

σ^0_C = the clutter coefficient at the critical angle θ_C, taken as zero above that angle

K = a constant from 1 to 4; Morchin suggests 1.9

Sea states: Sea states are a method of describing the sea, including wave height. Standard sea states are given in Table 4-5. Wind speed and sea state are related by Nathanson [4] as

$$W_S = 10^{(B + 0.14SS)} \qquad (4\text{-}44)$$

$B = 0.65$ if wind speed is in knots, 0.36 if in meters per second

Table 4-5. Sea states (after [16]).

Sea state	Wave height (ft)	Wave period (s)	Wave length (ft)	Wave velocity (Kt)	Particle velocity (ft/s)	Wind velocity (Kt)	Required fetch (mi)
0. Not a recognized sea state, but commonly used to describe a flat sea							
1. Smooth	0-1	0-2	0-20	0-6	0-1.5	0-7	0-25
2. Slight	1-3	2-3.5	20-65	6-11	1.5-2.8	7-12	25-75
3. Moderate	3-5	3.5-4.5	65-110	11-14	2.8-3.5	12-16	75-120
4. Rough	5-8	4.5-6	110-180	14-17	3.5-4.2	16-19	120-190
5. Very rough	8-12	6-7	180-250	17-21	4.2-5.2	19-23	190-250
6. High	12-20	7-9	250-400	21-26	5.2-6.7	23-30	250-370
7. Very high	20-40	9-12	400-750	26-35	6.7-10.5	30-45	370-600
8. Precipitous	40+	12+	750+	35+	10.5+	45+	600+

Notes: Numbers are approximate.

Numbers shown are limits in the steady state. It is possible to have a particular sea state outside the numbers given. For example, SS-3 could occur with a higher wind than shown, but a shorter fetch. Assumes deep water.

Wave velocity determines clutter Doppler. Particle velocity is how fast a particular particle of water moves.

The data given applies only to waves; swells are generated at long distances by other wind systems and complicate the issue.

Period, wavelength, and wave velocity apply to swells and waves.

Sea clutter σ^0 data: Figures 4-37 through 4-43 give average values for σ^0 for many different conditions. The different figures are for different frequency bands. The left-hand column is for vertical polarization and the right-hand column is horizontal. The horizontal axis is grazing angle and the different lines are different sea states.

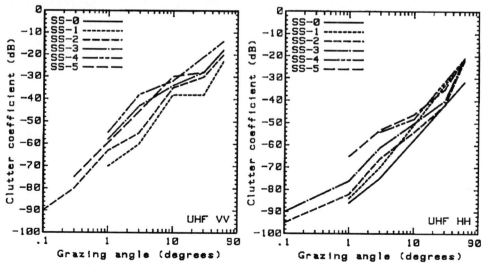

Figure 4-37. Sea Clutter Coefficient Data — 425 MHz.
(Data from Nathanson [4])

4-11. Interfering Signals — Surface Clutter

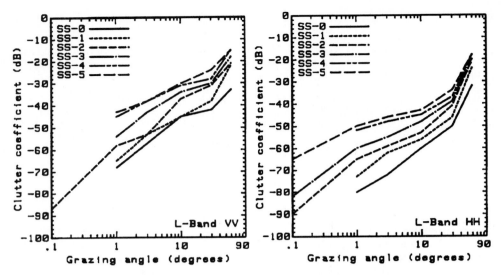

Figure 4-38. Sea Clutter Coefficient Data — 1.25 GHz.
(Data from Nathanson [4])

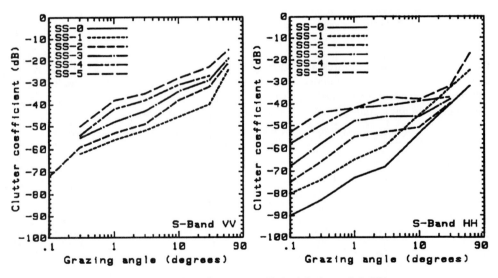

Figure 4-39. Sea Clutter Coefficient Data — 3.0 GHz.
(Data from Nathanson [4])

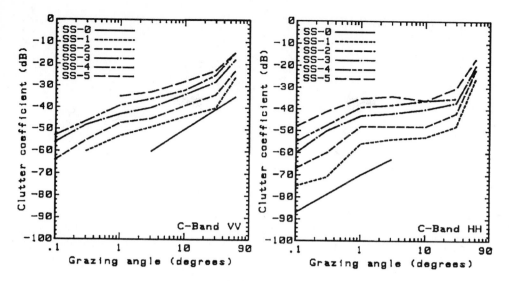

Figure 4-40. Sea Clutter Coefficient Data — 5.6 GHz.
(Data from Nathanson [4])

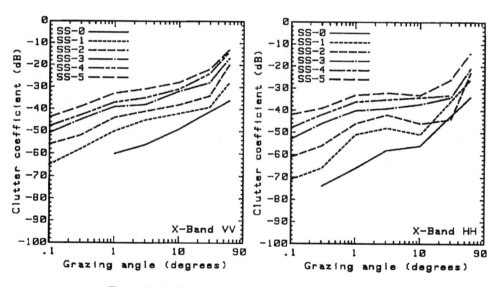

Figure 4-41. Sea Clutter Coefficient Data — 9.3 GHz.
(Data from Nathanson [4])

4-11. Interfering Signals — Surface Clutter

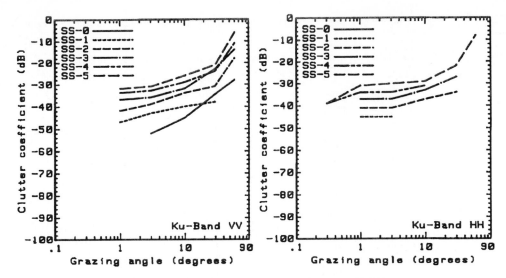

Figure 4-42. Sea Clutter Coefficient Data — 17 GHz.
(Data from Nathanson [4])

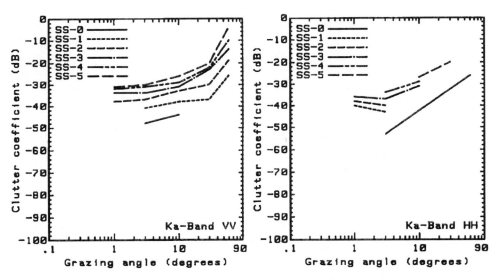

Figure 4-43. Sea Clutter Coefficient Data — 35 GHz.
(Data from Nathanson [4])

Sea clutter generalities: The following "rules of thumb" apply to sea clutter (adapted from Nathanson [4]):

— Sea clutter echoes using vertical polarization equal or exceed those from horizontal, and the differences increase with lower sea states, lower grazing angles, and lower frequencies.

— The clutter coefficient increases as θ^n for angles up to about 20°. The factor n decreases with increasing angle, frequency, and sea state.

— With horizontal polarization and angles below critical, the clutter coefficient increases with frequency as f^m, with m varying from 3 for low frequencies (below 2 GHz), very low angles (<1°) and low sea states (<3), and decreasing toward zero for higher frequencies, angles, and sea states.

— At low sea states and frequencies, σ^0 increases as much as 10 dB per sea state, dropping to very small changes with higher frequencies and sea states.

— With airborne radars at small antenna depression angles, a correction to grazing angle may be necessary because of the earth's curvature.

Sea clutter horizon: Sea clutter is primarily visible only to ranges shorter than the radar horizon (the distance beyond which the earth's curvature hides the clutter). Beyond that range, clutter still exerts a small effect through diffraction. At ranges shorter than the radar horizon, clutter has an R^{-3} characteristic, and signal-to-clutter ratios are affected as R^{-1}. Beyond the horizon, the clutter range characteristic is R^{-7} and S/C *increases* with increasing range as R^{+3}. The radar horizon can be found from figures in Sec. 6-8, setting the target altitude to zero.

Land clutter: Like sea clutter, land clutter exhibits the three regions; low angle, plateau, and high angle. The integrated model (adapted from Morchin [5]) is described below.

$$\sigma^0 = 10 \log \{10 \wedge [0.1 A - \log \lambda + \log (\sin \theta_g) + 0.1\sigma^0{}_C]$$

$$+ \operatorname{ctn}^2 \beta_0 \exp \left[\frac{-\tan^2 (B - \theta_g)}{\tan^2 \beta_0}\right] + 10 \wedge (u/10) \} \quad \text{(dB)} \quad (4\text{-}45)$$

$\sigma^0{}_C = -10 \log (\theta_C / \theta_g)$, desert only where $\theta_g < \theta_C$ (dB) (4-46)

$\theta_C = \sin^{-1}(\lambda / 4\pi h_E)$ (4-47)

$h_E = 9.3 \beta_0{}^{2.2}$ (4-48)

$A\ =\ -29$ for desert (4-49)
$\ =\ -24$ for farmland
$\ =\ -19$ for wooded hills
$\ =\ -14$ for mountains

u is defined in Eq. 4-40

$B\ =\ \pi/2$ for all terrain except mountains (4-50)
$\ =\ 1.24$ for mountains because of their net slope

$\beta_0\ =\ 0.14$ for desert (radians) (4-51)
$\ =\ 0.2$ for farmland

= 0.4 for wooded hills

= 0.5 for mountains

In general, land clutter behaves similarly to sea clutter in that the more vertical texture the land has, the higher is σ^0 for low grazing angles. Also, higher frequencies result in higher clutter coefficients. Land clutter is much more location-specific than sea clutter, and analyses may require clutter data from the location where the radar is to be used. Some of the information given on statistical distributions later in this section applies to land clutter.

Land clutter generalities: As with sea clutter, there are some general rules for estimating land clutter [4]. As stated earlier, land clutter is far more location-specific than sea clutter, and the generalities are less likely to fit a particular data set.

— The statistical distribution of land clutter of a particular type cannot be found as easily as sea clutter and is not as consistent (sea clutter is sea clutter, no matter where found — farmland in Iowa may be very different from that in Tibet).

— Land clutter amplitude distributions are not described by the Rayleigh function as clearly as sea clutter, because of the shadowing from mountains, trees, and cultural features.

— The land clutter σ^0 may be altered by soil moisture content and snow cover.

— Measurements for land clutter are different when made from an airplane versus those from a land-based radar, in that the land-based system performs a time average on the data and the airborne radar performs a spatial average (the same can be said of sea clutter).

Spiking clutter: All of the above discussions and models of sea and land clutter assume that the number of clutter facets per radar resolution area is large and that individual clutter scatterers do not predominate (in Swerling terms, the clutter is cases 1 or 2). In cases where individual scatters are prominent, the clutter is called spiking clutter, and the above data, generalities, and models may not hold.

Clutter statistical distributions: Just as noise is statistically distributed (Sec. 4-10), so also is clutter. The distributions describing clutter depend on the character of the clutter itself, the frequency, and the grazing angle.

Several distributions have been suggested as empirically fitting sea and land clutter, including the Rayleigh distribution, the log-normal distribution, the K distribution, the Weibull distribution, and others. In this section, we will define those distributions listed and give some of the conditions under which sea clutter fits them. A further discussion is beyond our purpose. Morchin [5] and Nathanson [4] give good overviews and references for further study.

— *Rayleigh distribution:* The Rayleigh distribution empirically fits sea and land clutter which is composed of many small scatterers within the resolution cell (not spiky). This function was examined earlier for noise and is given below in clutter terminology.

$$P(x) = (x / 2x_0^2) \exp(-x^2 / 2x_0^2) \quad \text{for } x \geq 0 \tag{4-52}$$
$$= 0 \quad \text{for } x < 0$$

x = the independent variable normalized to the RMS clutter voltage

$2x_0^2$ = the average value of x^2

- *Log-normal distribution:* The log-normal distribution empirically fits some low angle land clutter, and also fits high resolution samples of sea clutter in the plateau region [18].

$$P(x) = x\sigma(2\pi)^{0.5} \exp\left[\frac{(\ln x - \ln x_m)^2}{2\sigma^2}\right] \quad \text{for } x \geq 0 \quad (4\text{-}53)$$
$$= 0 \quad \text{for } x < 0$$

x = the independent variable
x_m = the median value of x
σ = the standard deviation of ln x

- *K distribution:* This distribution has been suggested (by Ward and Watts [19]) as fitting sea clutter.

$$P(x) = [2b/\Gamma(v)](bx/2)^v K_{v-1}(b\sigma) \quad (4\text{-}54)$$
$$\log v = 2/3 \log \theta_g + 5/8 \log(\Delta X) + d - 1 \quad (4\text{-}55)$$

x = the clutter power
K_v = the modified Bessel function of the third kind, vth order
b = a scale parameter
v = a shape parameter, describing the spiking nature of the clutter
θ_g = the grazing angle ($0.1° < \theta_g < 90°$)
ΔX = the cross-range resolution (100m $< \Delta X <$ 800m)
d = $-1/3$ for swells moving toward and away from the radar
  $+1/3$ for cross-swell aspects
  0 for no swells

- *Weibull distribution:* The Weibull distribution empirically fits low angle clutter, in the range 0.5° to 5.0°, for frequencies of 1 to 10 GHz [20].

$$P(\sigma_1) = (b\sigma_1^{b-1}/a)\exp(-\sigma_1^b/a) \quad \text{for } \sigma_1 > 0 \quad (4\text{-}56)$$
$$= 0 \quad \text{for } \sigma_1 < 0$$

a = $\sigma_m^b / \ln 2$ (4-57)
σ_m = $[x/(\ln 2)^b \Gamma(1+1/b)]$ (4-58)

σ_1 = the clutter backscatter normalized
b = a shape parameter (Table 4-6)
x = the mean clutter cross-section

Clutter Doppler: Doppler shift is a very important component of clutter. As will be seen in Sec. 6-6, radars, particularly low PRF radars, are severely limited in the bandwidth of the clutter Doppler they can accommodate. In most cases, the absolute velocity of the clutter can be compensated for and is not a problem (Ch. 12). It is the *clutter bandwidth*, or Doppler spectrum, that causes high residues to escape the clutter processor.

4-11. Interfering Signals – Surface Clutter

Table 4-6. Weibull distribution shape parameters (*b*) for a clutter patch of 9,290 square meters (from Morchin [5])

Sea state or terrain	Wind direction	UHF H	UHF V	L H	L V	S H	S V	C H	C V	X H	X V
SS-3	Up					1.52				1.75	
	Cross					1.59					1.99
	Down					1.48					
SS-4	Up										
	Cross									1.85	1.94
	Down										1.95
SS-5	Up	1.82	1.82	1.82	1.87			1,81	1.65	1.92	1.88
	Cross										
	Down	1.81	1.74	1.71	1.78			1.76	1.88	1.88	1.72
Farmland						1.85					
Woodlands						1.63					
Desert						1.83					

Table 4-7 shows some examples of RMS clutter bandwidths, with a discussion of the mechanisms for the different types of clutter to follow.

Table 4-7. Summary of Standard Deviations of the Clutter Spectrum (from Barton [21])

Clutter source	Wind speed (Kt)	Velocity Standard Deviation (m/s)
Sparse woods	Calm	0.017
Wooded hills	10	0.04
Wooded hills	20	0.22
Wooded hills	25	0.12
Wooded hills	40	0.32
Sea echo	not given	0.7
Sea echo	not given	0.75-1.0
Sea echo	8-20	0.46-1.1
Sea echo	"windy"	0.89
Chaff	not given	0.37-0.91
Chaff	25	1.2
Chaff	not given	1.1
Rain clouds	not given	1.8-4.0
Rain clouds	not given	2.0

Sea clutter Doppler: Sea clutter has two velocity parameters which must be considered: net velocity and spectral width. The net Doppler shift of sea clutter is determined by the wave height, water depth, and a few other factors, which in turn determine the wave velocity. In cases where strong local winds blow spray off the tops of waves (called feathering), the net clutter Doppler has a component at approximately the local wind velocity. Net sea clutter Doppler shift, being relatively constant for a particular clutter condition, is often removed in MTI radars. The process, called clutter locking, senses the net velocity of the clutter and offsets the local oscillator or COHO so that the clutter becomes the zero Doppler shift reference to which other signals are compared. See Ch. 12 for a discussion of clutter locking.

The spectral width of sea clutter determines whether or not it will be removed by Doppler processing, particularly in low PRF radars where ambiguities may cause its aliases to be spread over the entire Doppler spectrum (Sec. 6-4). Sea clutter spectrum is primarily dependent on the locally generated capillary waves and is thus dependent on local winds. Figure 4-44 shows some measured sea clutter spectral data.

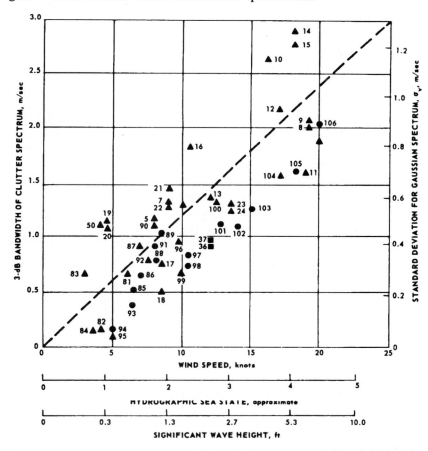

Figure 4-44. Sea Clutter Spectrum. Points 5-18 from [26], 19-24 from [27], 36-37 from [28], 81-106 from [29] and [30]. ▲ = horizontal polarization, ● = vertical, ■ = circular. (Reprinted from Nathanson [4]; © 1969 McGraw-Hill Book Company, used with permission)

4-11. Interfering Signals – Surface Clutter

Sea clutter's velocity statistical distribution is generally assumed to be Gaussian, with a standard deviation of [4]

$$\sigma_V = 0.101 \, W_S \qquad (4\text{-}59)$$

σ_V is the standard deviation of the clutter motion in some velocity unit

W_S is the wind velocity in the same unit

Combining Eq. 4-59 with Eq. 4-44 gives clutter bandwidth in terms of sea state.

$$\sigma_V = 0.101 \times 10^{B + 0.14 \, SS} \qquad (4\text{-}60)$$

B is 0.65 if σ_V and W_S are in knots, 0.36 if meters per second

Nathanson [4] and Long [3] (with others) have shown that in some cases the spectral distribution of sea and land clutter is expressed by

$$P(f) = 1 / [1 + (f/f_C)^n] \qquad (4\text{-}61)$$

f = the clutter spectral frequency of interest

f_C = the half-power frequency

n = a constant, generally taken to be 3

Some investigators have found a difference in Doppler spectrum for a given sea state among different polarizations. The validity of these data for all conditions and the mechanism for the polarization difference is in some question.

Sea clutter echoes are periodic, with values for the period given in Table 4-5. Some radars take advantage of this periodicity by using signal integration times of up to several wave periods. In these long times, the clutter becomes noise-like and sums less completely than do target echoes (Ch. 11). Figure 4-45 gives sea clutter autocorrelation data for various dwell times. Note that minimum correlation (and maximum clutter rejection in a signal integrator) corresponds to dwell times of one-half wave period plus an integer number of periods.

An example is the AN/APS-116, in the U.S. Navy's anti-submarine S-3A aircraft. In it, integration times of several seconds enhance coherent targets such as periscopes with respect to sea clutter. Figure 4-46 shows performance data for this radar. Note the constancy of required RCS with range. As range increases, the resolution cell becomes larger as the first power of range (Ch. 3), and signal-to-clutter ratio decreases as the negative first power of range. However, as the resolution cell size and the area of clutter illuminated increase, the randomness of the clutter increases, increasing the signal-to-clutter ratio. The two effects virtually cancel one another.

Land clutter Doppler: The Doppler spectrum of land clutter differs from that of sea clutter in several important aspect. It is caused by vegetation moving in the wind, blowing dust and sand, birds and insects, and cultural factors (vehicles, and so forth) Land clutter spectrum is often taken to be distributed as shown in Eq. 4-61, and several investigators have reported values for the 3 dB frequency. For example, Pollon (as reported in Morchin [5]) gives a value of f_C for wooded terrain of

$$f_C = 1.33 \exp(0.0272 \, W_S) \qquad (4\text{-}62)$$

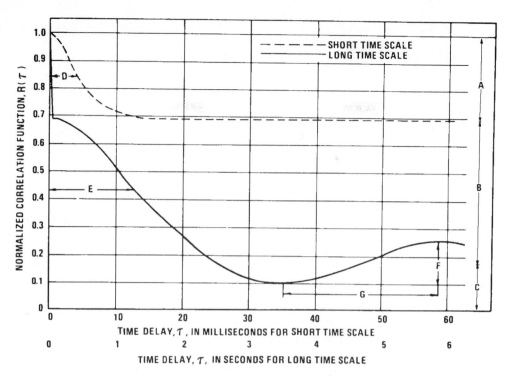

Figure 4-45. Sea Clutter Autocorrelation. (Reprinted from Long [31]; © 1974 IEEE, used with permission)

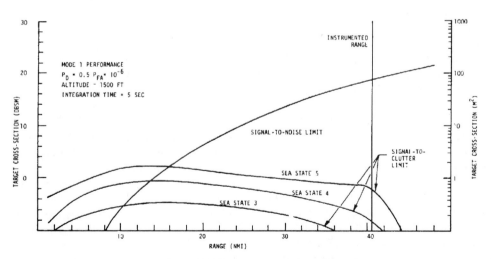

Figure 4-46. AN/APS-116 Performance. (Reprinted from J. M. Smith and R. H. Logan [22]; © 1980 IEEE, used with permission)

4-11. Interfering Signals — Surface Clutter

Some authors have considered land clutter spectra as being Gaussian distributed, with various standard deviations. Two reported values are given, one from Nathanson [4] for "vegetated terrain" and one from Barton [9] for woodlands.

$$\sigma_V = a\, W_S^b \qquad (4\text{-}63)$$

σ_V = the standard deviation of the velocity in some unit

W_S = the wind speed in the same unit

a = 0.007 for vegetated terrain [4]
 = 0.0045 for woodlands [9]

b = 1.28 for vegetated terrain [4]
 = 1.4 for woodlands [9]

Once σ_V is known, calculation of the RCS as a function of Doppler shift requires that the velocity probability distribution of the clutter be known. This study is beyond the scope of this book. Long [2] is a good place to start.

The power spectrum of clutter does not for some measurements fit the Gaussian approximation (Fishbein, et al. [31]). They suggest an empirical relationship which works well for their data.

$$P(f) = 1 / [\, 1 + (f/f_C)^3\,] \qquad (4\text{-}64)$$

$P(f)$ = the power spectrum of the clutter as a function of frequency

f = the frequency of interest

f_C = a cutoff frequency given by $f_C = 1.33 \exp(0.1356\, W_S)$, where W_S is in Knots

4-12. Airborne Clutter: Weather, Chaff, Insects, and Birds

This type of clutter, if of large extent, behaves as volume clutter (Ch. 3), and its radar cross-section is dependent on a clutter coefficient and the volume of the radar's resolution cell. The volume clutter coefficient ($\Sigma\sigma$) is usually in m^2 RCS per m^3 resolution volume. As was pointed out in Ch. 3, some authors use other units and one must be careful.

Weather clutter: Weather is unique in that it is composed of very simple elemental scatterers. Raindrops can be modeled closely as spheres, which scatter electromagnetic waves strongly co-polarized (linear polarization) with the illumination (Table 4-3). The primary suppression technique, then, is to cross-polarize the radar's transmit and receive antennas.

One way to accomplish this is to use circular polarization. Assume an antenna which is right-hand circularly (RHC) polarized. The illumination wave is strongly RHC polarized, and the wave scattered from the raindrops is strongly co-polarized. Because the backscattered energy retains the same wave rotation as the illumination but its direction of propagation is reversed, the backscatter becomes left-hand circularly (LHC) polarized. Since the LHC backscatter is cross-polarized to the antenna, the weather backscatter is attenuated. Complex targets do not exhibit this strong co-polarized scattering, and significant energy from, for example, an aircraft is received by the antenna (Table 4-4).

Figure 4-47 shows the results. In Fig. 4-47a, a storm is viewed by an air search radar (ARSR-1) with linear polarization. In Fig. 4-47b, the same radar views the storm with circular polarization. The attenuation of circular over linear polarization is about 20 dB, limited principally by the antenna's polarization purity and the ellipticity of the drops as they fall through the atmosphere.

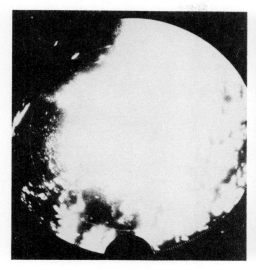
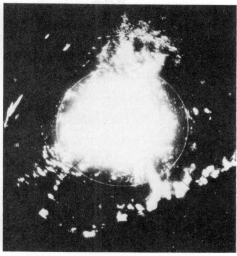

a. Linear polarization   b. Circular polarization

Figure 4-47. Storm Using Linear and Circular Polarization. (Reprinted from Shrader [23]; © 1970 McGraw-Hill Book Company, used with permission)

For the reasons given above, many ground- and ship-based air search radars have circular polarization capability, including the newer ASR and ARSR radars. Many airborne navigation and targeting radars meant to "see" through weather also have circular polarization.

The remaining weather clutter is suppressed with moving target indication (MTI). See Ch. 12 and the weather clutter spectra discussion later in this section.

The parameter $\Sigma\sigma$, the RCS per unit resolution volume. It is the summation of the RCS of the individual scatterers within the volume.

$$\Sigma\sigma = \Sigma\ \sigma_i \qquad (4\text{-}65)$$

$\Sigma\sigma$ = the reflectivity (m² / m³)

σ_i = the RCS of the i-th scatterer

The total RCS of one resolution volume, assuming uniform distribution of the drops throughout the volume, is

$$\sigma = \Sigma\sigma\ V_I \qquad (4\text{-}66)$$

V_I = the volume of the radar's resolution cell (Ch. 3)

Drop sizes in clouds and precipitation are a function of the precipitation rate, with higher rates associated with larger drops (Nathanson [4] gives some data on this). Clouds and rain

4-12. Airborne Clutter: Weather, Chaff, Insects, and Birds

at rates less than those found in severe thunderstorms (which is called excessive rainfall) are composed of drops much smaller than the wavelength, at least up through X-Band. Thus they behave as Rayleigh scatterers and their backscatter is proportional to the sixth power of drop diameter, or the third power of drop volume.

To describe the weather clutter coefficient, meteorological radar people use a factor called Z, related to drop diameter rather than drop RCS.

$$Z = \Sigma D_i^6 \tag{4-67}$$

Z = the reflectivity (mm^6 / mm^3)

D_i = the diameter of the ith drop

Z, for temperate latitudes, is usually estimated as

$$Z \approx 200 \, r^{1.6} \tag{4-68}$$

r = the rainfall rate in mm / hr

The factor Z relates to $\Sigma\sigma$ as [24]

$$\Sigma\sigma = (\pi^5 / \lambda^4) |K|^2 Z \times 10^{-18} \tag{4-69}$$

$$|K|^2 = [(m^2 - 1) / (m^2 + 2)]^2 \tag{4-70}$$

m = the complex index of refraction

$$|K|^2 \approx 0.93 \text{ for drops in the water phase} \tag{4-71}$$

$$|K|^2 \approx 0.20 \text{ for particles in the ice phase} \tag{4-72}$$

The 10^{-18} scales from drop size in mm to RCS in m^2.

Information on cloud, rain, and snow reflectivity can be found in meteorological radar texts. The measured clutter coefficients for limited weather types are summarized in Table 4-8.

Table 4-8. Clutter coefficient of uniform clouds and rain. (After Nathanson [4])

Type of cloud or precipitation	Radar band and frequency (GHz)						
	P (0.45)	L (1.25)	S (3.0)	C (5.6)	X (9.3)	K (24)	Ka (35)
Heavy stratocumulus clouds 4 gm water / m^3 cloud	−140	−140	−118	−108	−98		
Fog, 100 ft visibility			−120	−110	−100		
Drizzle, 0.25 mm/hr			−102	−91	−82	−64*	−57*
Light rain, 1 mm/hr		−107	−92	−81.5	−72	−54	−47
Moderate rain, 4 mm/hr		−97	−83	−72	−62	−46	−39
Heavy rain, 16 mm/hr			−73	−62	−53		−32*

Notes: All valued in dB(m^2 / m^3).
Tropospheric attenuation not included.
Add 98 dB to convert to m^2 / nmi^3.
*Up to 3 dB error.

Chaff: Chaff is artificially generated volume clutter, used in an attempt to hide legitimate targets from the radar, or to fool the radar into thinking the target is in some location where it isn't. It is a useful ECM tool against some seeker and fuze radars. It is extensively used to protect both ships (from anti-ship missiles) and aircraft (from radar-guided air-to-air and surface-to-air missiles).

Chaff is basically a large number of thin dipole reflectors ejected (dispensed) from an aircraft or rocket. These dipoles, being resonant, have very high radar cross-sections for their size. Thus a large RCS can be generated from a small package.

Figure 4-15 showed the RCS of a dipole oriented parallel with the wave polarization. Chaff has the same characteristic except that it peaks at a lower value because chaff dipoles (sometimes called needles) are randomly oriented. The maximum RCS occurs when the length of the dipole is one-half wavelength *on the dipole*, slightly shorter than 1/2 wave in free space. The RCS of resonant chaff is given by the following equations.

$$\sigma_{Ch\text{-}Res} \approx 0.7 L^2 \qquad (4\text{-}73)$$

$\sigma_{Ch\text{-}Res}$ = the RCS of one dipole, randomly oriented, at its resonant frequency (m^2)

L = the resonant length of the dipole (meters)

Since the dipole is resonant when it is slightly less than 1/2 free space wavelength long, the resonant RCS is about

$$\sigma_{Ch\text{-}Res} \approx 0.15 \lambda_0^2 \qquad (4\text{-}74)$$

λ_0 = the free space wavelength

The total RCS of a chaff cloud is usually taken as the RCS of one dipole multiplied by the number of dipoles in the radar's resolution volume. This assumes no electromagnetic interaction between dipoles and is a good assumption if the dipoles are more than about two wavelengths apart.

$$\sigma_{Ch\text{-}Tot} \approx 0.15 \lambda_0^2 N_{Dip} \qquad (4\text{-}75)$$

$\sigma_{Ch\text{-}Tot}$ = the total RCS of all resonant and randomly oriented dipoles within the radar's resolution volume (square meters)

N_{Dip} = the number of dipoles in the resolution volume

Chaff dispensed from aircraft is often fired from a cartridge very much resembling a shotgun shell. From ships, chaff is usually dispensed from a rocket or shell, to get it away from the ship quickly.

In the past, chaff was made of aluminum foil. Now it is usually a glass fiber with a conductive coating. The reason for the change is that the fibers are lighter per unit of RCS and they are more rigid, giving each dipole a larger RCS than if it were bent or crumpled. The rapid dispersal ("blooming") of chaff is a major consideration in its design and the design of the dispensers.

Since air-dispensed chaff quickly slows to the ambient air speed, it can often be resolved from the dispensing aircraft with Doppler.

Doppler spectra of weather and chaff: The motion of weather and chaff is similar to the motion of the air in which it is contained. The Doppler spectrum is the result of four phenomena (from Nathanson[4]): wind shear (velocity change with altitude), air turbulence, beam broadening (tangential velocities across a finite beam width have radial components — see Figs. 4-66 through 4-69), and distribution of fall velocities (caused by variations in precipitation drop sizes). Equation 4-76 gives the total spectral variance predicted by Nathanson's model.

$$\sigma_v^2 = \tau_{shear}^2 + \tau_{turb}^2 + \tau_{beam}^2 + \tau_{fall}^2 \qquad (4\text{-}76)$$

$\tau_{shear} = 0.00733\, k_{shear}\, R\, q_{2(El)}$ m/s, limited to 6 m/s (4-76a)

k_{shear} is the shear constant (Nathanson suggests 5.7 in the direction of primary high-altitude winds and 4.0 for an all-azimuth average — see Fig. 4-48 for τ_{shear} for these values), R is range (m), $\rho_{2(El)}$ is the antenna's two-way elevation 3-dB beamwidth (degrees).

$\tau_{turb} = 1.0$ m/s (suggested by Nathanson) (4-76b)

$\tau_{beam} = 0.0073\, V_0\, \rho_{2(Az)} \sin\beta$ (4-76c)

V_0 is the wind velocity at beam center (m/s), $\rho_{2(Az)}$ is the two-way azimuth beamwidth (degrees), $\rho_{(Az)}$ is the azimuth angle relative to the wind velocity.

$\tau_{fall} = 1.0 \sin\rho_{(El)}$ (m/s) (4-76d)

At long ranges, shear is limited in the presence of a strong *bright band*, which dominates the echo spectrum. Bright band is defined as a layer of small elevation extent which gives greater intensity echoes (by 10-15 dB) than the areas above and below it.

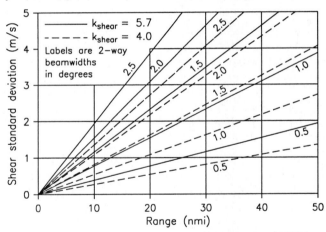

Figure 4-48. Weather and Chaff Doppler Spectral Width.

Angel clutter: Birds, insects, dust, and other flying particles make up a class of volume clutter called angel clutter. Figure 4-49 shows the clutter caused by a large cloud of insects, in this case the annual Chad Fly hatch at Lake Champlain.

Vaughn has summarized the RCS data for birds and insects [25]. Figure 4-50 relates the RCS versus weight of virtually all bird and insect data published to the date of the figure (1985), plus reference objects (see figure caption).

The best fit of these data, averaging frequency and polarization, gives a relationship between the weight of a bird or insect and its RCS [5].

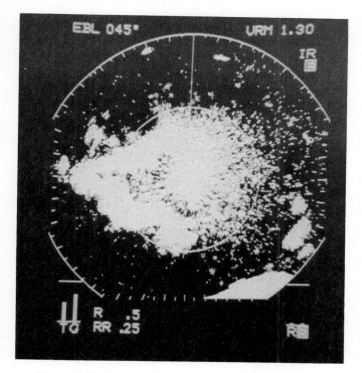

Figure 4-49. Insects: Chad Flies on Lake Champlain.

$$\sigma_{bi} \approx -46 + 5.8 \log W_{bi} \qquad (4\text{-}77)$$

σ_{bi} = the RCS of individual birds or insects (dBsm)

W_{bi} = the weight of the individual bird or insect (grams)

Bird and insect RCS is a function of illumination frequency. Table 4-9 gives some data on three bird species.

Table 4-9. Bird RCS (data in dBsm). (after Vaughn [25])

Radar Band	Pigeon	Starling	Western Sandpiper
UHF	−36	−48	−57
S	−26	−32	−36
C	−22	−28	−33
X	−27	−30	−33

4-12. Airborne Clutter: Weather, Chaff, Insects, and Birds

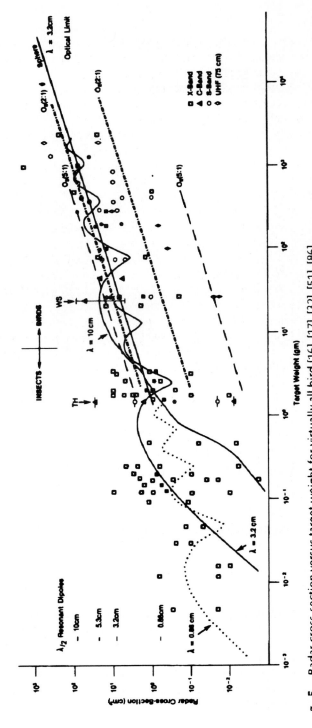

Figure 4-50. RCS versus Weight for Birds and Insects. Includes for reference RCS of perfectly conducting half-wave dipole (=0.86 λ^2), optical RCS of a water sphere, and 2:1 and 5:1 prolate spheroids at side and end views, as well as water spheres at λ=0.86, 3.2, and 10.0 cm. (Reprinted from Vaughn [25]; © 1985 IEEE, used with permission)

4-13. Moving clutter

Airborne radars encounter a special problem in dealing with surface clutter; the apparently moving clutter, and radar ambiguities can cause the clutter to have the same parameters as a moving target, and the two may not be resolvable in any dimension.

Consider a medium PRF radar, ambiguous in both range and Doppler, resolving in four dimensions (azimuth, elevation, range, and Doppler). Assume an airborne target at a particular range, Doppler, azimuth, and elevation. Some of the clutter from under the radar will be at the same *apparent* range as the target (the same propagation time from the *last transmit pulse* — see Chs. 1 and 6). In the presence of this ambiguous range clutter, the target and clutter cannot be resolved in range, whether or not they are at the same *actual* range.

Likewise, the target and clutter may, because of ambiguities, appear to have the same *apparent* Doppler shift and not be resolvable in frequency.

If the target is viewed through the antenna's main lobe and the clutter through sidelobes, they will appear to be at the same angles, and are not resolvable in these dimensions.

It is obvious that we have run out of dimensions with which to separate the target and clutter, and *the target may not be resolvable from the clutter in any dimension*. This phenomenon presents a problem only if there are simultaneously ambiguities in range and/or Doppler and the clutter is viewed through an antenna sidelobe. This is the normal situation in moving radars of all PRF classes, and is a particular problem in airborne radars.

Let us examine this problem from the standpoint of range first. Figure 4-51 is a profile of an airborne radar observing an airborne target without range ambiguities (low PRF). There is an area on the ground (or sea) that has the same range as the target, and its ground range is slightly less than that to the target.

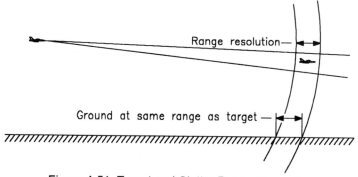

Figure 4-51. Target and Clutter Range Geometry.

Figure 4-52 shows the same geometry in three dimensions. The locus of points on the ground having the same range from the radar is on a cone perpendicular to the clutter surface. Where the cone intersects the surface, a ring forms, the width of which represents the projected range resolution of the radar.

A two-dimensional top view is given in Fig 4-53, showing the relative ground ranges to target and same-range clutter. This figure shows low PRF, with no range ambiguity and a single ring. The rings are known as *iso-ranges*.

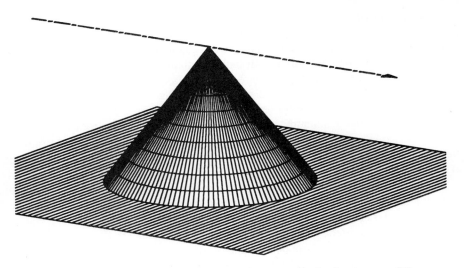

Figure 4-52. Locus of Points of Constant Range. Radar is at apex of Cone.

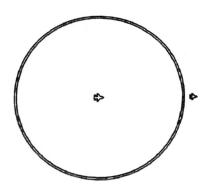

Figure 4-53. Iso-Range for Low PRF.

PRF = 2,000 pps,
target range = 22 km,
range resolution = 400 m,
radar altitude = 10,000 m,
target altitude = 8,000 m.

The width of the ring relates geometrically to the range resolution of the radar (Fig. 4-51) as developed below. *Note:* All views given in this section assume the clutter surface is planar and horizontal; the geometry is considerably more complicated if it is not.

$$R_{GT}^2 = R^2 - (h_R - h_T)^2 \qquad (4\text{-}78)$$

R_{GT} = the horizontal range from the radar to the target, seen by viewing the radar/target from the top

R = the slant range from radar to target

h_R = the height of the radar from the clutter surface

h_T = the height of the target from the clutter surface

$$R_{GG}^2 = R^2 - h_R^2 \qquad (4\text{-}79)$$

R_{GG} = the horizontal range from the radar to the point on the ground at the same slant range as the target

The width and area of the ring, the clutter grazing angle, and the radar cross-section of the clutter within it are given below, as is the signal-to-clutter ratio expected for a target competing with this sidelobe clutter. The equations given assume the ring width is limited by the radar's range resolution. If limited by the antenna elevation angle, the geometry shown must be adapted to that condition.

$$\Delta R_G = \Delta R / \cos \theta_{DG} \quad (4\text{-}80)$$

ΔR_G = the range resolution cell width on the clutter surface

ΔR = the radar's range resolution

θ_{DG} = the depression angle from the radar to the clutter surface (It is also the clutter grazing angle if the surface is horizontal)

$$A_{RG} = \pi [2 R_G \Delta R_G + (\Delta R_G^2 / 2)] \quad (4\text{-}81)$$

A_{RG} = the area of the ring on the ground

$$\sigma_{RG} = \sigma^0_{RG} A_{RG} \quad (4\text{-}82)$$

σ_{RG} = the radar cross-section of the ring

σ^0_{RG} = the average clutter coefficient of the clutter in the ring

$$S/C_{RG} = \frac{G^2 \sigma G_P}{G_{DG}^2 \sigma^0_{RG} A_{RG}} \quad (4\text{-}83)$$

The points on the ground within the ring are in the same range bin as the target. If not resolved in some other dimension, interference results.

At high PRF, range is highly ambiguous, and many iso-range rings exist. Each ring shown in Fig. 4-54 is at the same apparent range from the radar. The clutter area and RCS

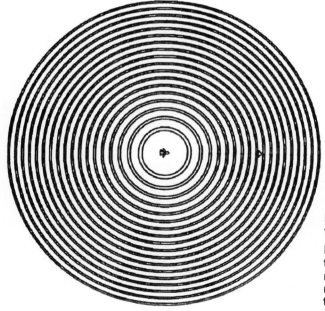

Figure 4-54. Iso-Range for High PRF.

PRF = 100,000 pps, target range = 22 km, range resolution = 400 m, radar altitude = 10,000 m, target altitude = 8,000 m.

4-13. Moving Clutter

are calculated in the same manner as above, but the total clutter echo is the vector sum of the echoes from the individual rings. Medium PRF, which is less ambiguous than high, is shown in Fig. 4-55.

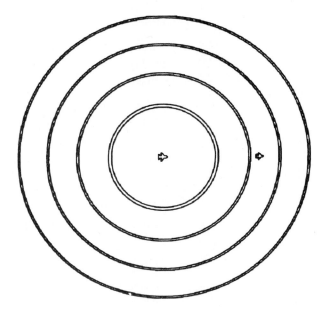

Figure 4-55. Iso-Range for Medium PRF.

PRF = 24,000 pps,
target range = 22 km,
range resolution = 400 m,
radar altitude = 10,000 m,
target altitude = 8,000 m.

Moving clutter Doppler shift can be represented in a similar fashion (Fig. 4-56). Points in space which have the same Doppler shift to the radar are on a figure whose surface is at

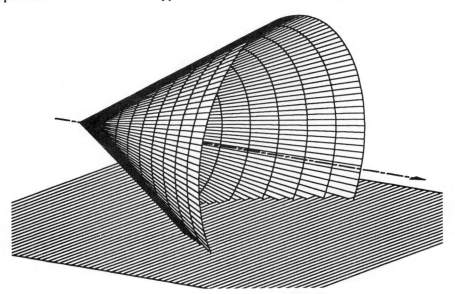

Figure 4-56. Isodop Formation at High PRF.

a constant angle from the velocity vector of the radar. This condition is met by a cone whose axis is the velocity vector of the radar and whose angle is determined by the Doppler shift of the target of interest. The intersection of the Doppler cone and the clutter surface is the locus of all points on the ground having the same Doppler shift as the target. These lines are known as *isodops*. Figure 4-56 shows the cone, clutter surface, and the intersection. The radar is at the apex on the cone.

The cone points in the direction of the radar's velocity vector for positive target Doppler shifts and in the opposite direction for negative shifts. The cone's axis is coincident with the radar's velocity vector, and is not a function of antenna azimuth angle. The angle of the cone's apex is a function of azimuth angle, radar velocity, and target position, velocity, and direction (the factors which determine the Doppler shift).

The cone shown in Fig. 4-56 has no Doppler ambiguities (high PRF). Figure 4-57 shows its intersection with the horizontal clutter surface and the isodop which forms. The intersection of a cone and a plane parallel to its axis is a hyperbola. The double hyperbolas are caused by the Doppler resolution cell width — all points on the ground between the two are within the specified Doppler bin. If the target is in the same bin, interference results.

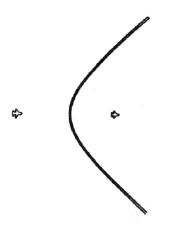

Figure 4-57. Isodop for High PRF.

PRF = 100,000 pps,
target range = 22 km,
range resolution = 400 m,
radar altitude = 10,000 m,
target altitude = 8,000 m,
radar velocity = 1458 Kt,
target velocity = 350 Kt.

If ambiguities exist, multiple cones and isodops form, as seen in Fig. 4-58. Figures 4-59 and 4-60 show the isodops for the ambiguous PRF classes, low and medium. As seen from the isodops, the cones can form in either direction along the radar's velocity vector.

Figures 4-61 through 4-63 show the location on the clutter surface of all points which have the same apparent range *and* Doppler as the target of interest. These are the areas within the intersections of the iso-ranges and the isodops. If these intersections are viewed through antenna sidelobes, they also appear at the same azimuth and elevation angles as the target. *They thus cannot be discriminated from the target in any dimension and are interfering signals which cannot be separated from the target.*

This is primarily a problem with airborne radars in the tail aspect. If the target is in the nose aspect (closing on the radar), there is no point on the clutter surface which can have a Doppler shift as high as that of the target, and thus there are no isodops. These interfering signals generally prevent high PRF from being useful with tail aspect targets in clutter (Ch. 6).

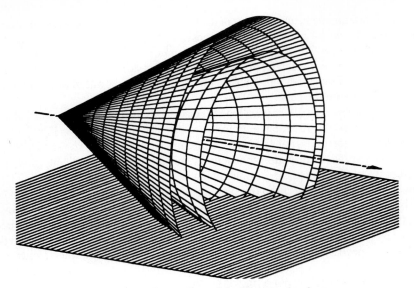

Figure 4-58. Formation of Ambiguous Isodops.

Figure 4-59. Isodop for Low PRF.

PRF = 2,000 pps,
target range = 22 km,
range resolution = 400 m,
radar altitude = 10,000 m,
target altitude = 8,000 m,
radar velocity = 1458 Kt,
target velocity = 350 Kt.

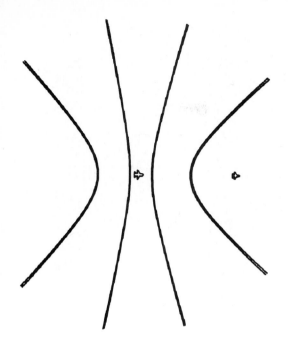

Figure 4-60. Isodop for Medium PRF.

PRF = 24,000 pps,
target range = 22 km,
range resolution = 400 m,
radar altitude = 10,000 m,
target altitude = 8,000 m,
radar velocity = 1458 Kt,
target velocity = 350 Kt.

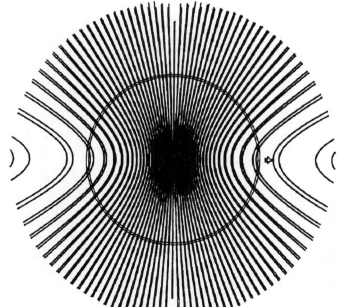

Figure 4-61. Clutter Range and Doppler at Low PRF.

PRF = 2,000 pps,
target range = 22 km,
range resolution = 400 m,
radar altitude = 10,000 m,
target altitude = 8,000 m,
radar velocity = 1458 Kt,
target velocity = 350 Kt.

4-13. Moving Clutter

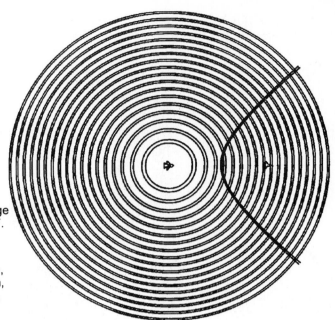

Figure 4-62. Clutter Range and Doppler at High PRF.

PRF = 100,000 pps,
target range = 22 km,
range resolution = 400 m,
radar altitude = 10,000 m,
target altitude = 8,000 m,
radar velocity = 1458 Kt,
target velocity = 350 Kt.

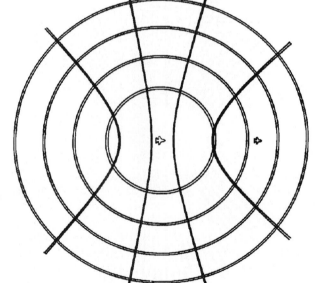

Figure 4-63. Clutter Range and Doppler at Medium PRF.

PRF = 24,000 pps,
target range = 22 km,
range resolution = 400 m,
radar altitude = 10,000 m,
target altitude = 8,000 m,
radar velocity = 1458 Kt,
target velocity = 350 Kt.

Medium PRF seems intuitively best of the three from the standpoint of minimizing sidelobe clutter, and this is true. We shall see in Ch. 6 that medium PRF allows ambiguities in both range and Doppler to be resolved, restricting the sidelobe clutter to two points: the two intersections of the correct target range and the correct target Doppler.

Figure 4-64 shows the situation which results in the altitude return (zero Doppler shift) problem. In this case, the closing rate between radar and target is negligible. The range resolution of the radar is coarse, if existent at all; pulse Doppler radars and modes usually have little range discrimination. In the figure, the areas on the ground which the radar "sees" in range are the wider rings. The large circle below the radar is within its range response.

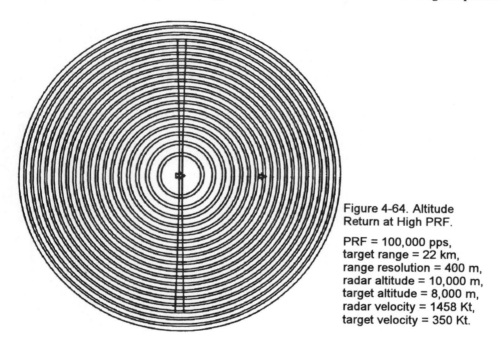

Figure 4-64. Altitude Return at High PRF.

PRF = 100,000 pps,
target range = 22 km,
range resolution = 400 m,
radar altitude = 10,000 m,
target altitude = 8,000 m,
radar velocity = 1458 Kt,
target velocity = 350 Kt.

Clutter spectrum from antenna beam shape: Finally, we must address the spectral spreading of clutter caused by the moving radar's antenna beam shape. If the beam is at any angle other than directly down the velocity vector over the ground, scatterers in different parts of the beam have different Doppler shifts. This causes even absolutely stationary clutter to have a Doppler spectrum.

In air-to-air modes, this spreading of the clutter spectrum forces the clutter filter rejection notches to be wide, interfering with the detection of moving targets (Chs. 1 and 6). It is interesting that this phenomenon is used to advantage in some airborne radars. If sufficient Doppler resolution is available, ground scatterers in the same azimuth/range bin can sometimes be resolved, producing higher resolution radar ground maps than would be available with the antenna's real beam alone. This is the principle of Doppler beam sharpening (DBS), a technique falling into the general category synthetic aperture radar

4-13. Moving Clutter

(SAR). It can also be used air-to-air to resolve formations of aircraft, in so-called raid assessment modes. See Ch. 13 for details.

Figure 4-65 shows the beam-shape geometry. Two scatterers are shown in the same angle/range bin. The two, because of the geometry, have a slightly different radial velocity to the radar, and thus a slightly different Doppler shift.

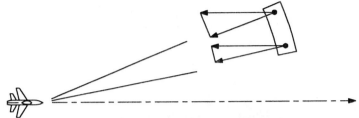

Figure 4-65. Doppler Spectrum Geometry.

Figures 4-66 through 4-69 show the spectrum from a particular antenna beam shape (sinc^2) for several different beam angles from the radar's velocity vector. All plots show 3,500 Hz of bandwidth and the angles are total, including both the azimuth and depression angles. Note that the further from the radar's flight path the beam is placed, the wider the clutter spectrum becomes.

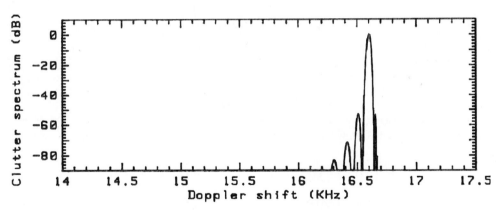

Figure 4-66. Spectrum of Clutter with a Finite Width Antenna Beam — Angle = 5°.
Azimuth pointing angle = 5°, radar velocity = 486 Kt, beamwidth = 2°, frequency = 10 GHz

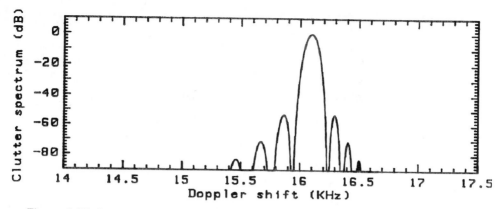

Figure 4-67. Spectrum of Clutter with a Finite Width Antenna Beam — Angle = 15°.
Azimuth pointing angle = 15°, radar velocity = 486 Kt, beamwidth = 2°, frequency = 10 GHz

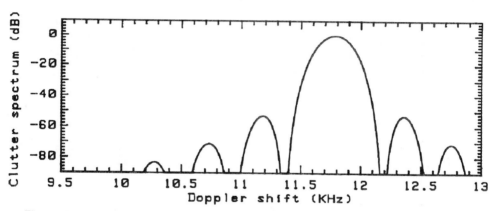

Figure 4-68. Spectrum of Clutter with a Finite Width Antenna Beam — Angle = 45°.
Azimuth pointing angle = 45°, radar velocity = 486 Kt, beamwidth = 2°, frequency = 10 GHz

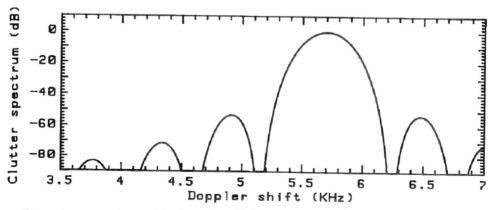

Figure 4-69. Spectrum of Clutter with a Finite Width Antenna Beam — Angle = 70°.
Azimuth pointing angle = 70°, radar velocity = 486 Kt, beamwidth = 2°, frequency = 10 GHz

4-13. Moving Clutter

4-14. Further Information

General references: Several books have been written on the analysis of radar cross-section, including Knott [1]. Information on clutter RCS is widely scattered. A relatively recent book is Long [2], which includes extensive clutter RCS information and has some information on clutter Doppler. A book by Ulaby and Dobson [3] gives much statistical information on statistical modeling of land clutter, but does not address the Doppler spectrum of clutter. Much good summary clutter and radar signal environment information is found in Nathanson [4], and Morchin [5] gives an excellent overview of clutter and clutter modeling. The *Radar Handbook* [6] has sections on targets (Ch. 11, by E.F. Knott), land clutter (Ch. 12, by R.K. Moore), sea clutter (Ch. 13, by L.B. Wetzel, and meteorological effects (Ch. 23, by R.J. Serafin).

[1] E. F. Knott, J.F. Shaeffer, and M.T. Tuley, *Radar Cross-Section*, Norwood, MA: Artech House, 1985.

[2] M. W. Long, *Radar Reflectivity of Land and Sea*, Norwood, MA: Artech House, 1983.

[3] F.T. Ulaby and M.C. Dobson, *Handbook of Radar Scattering Statistics for Terrain*, Norwood, MA: Artech House, 1989.

[4] F.E. Nathanson, *Radar Design Principles*, New York: McGraw-Hill, 1969.

[5] W.C. Morchin, *Airborne Early Warning Radar*, Norwood, MA: Artech House, 1990.

[6] M.I. Skolnik, *Radar Handbook* 2nd ed., New York: McGraw-Hill, 1990.

Cited references:

[7] J.I. Marcum, "A Statistical Theory of Detection by Pulsed Radar, and Mathematical Appendix," *IRE Transactions*, vol. IT-6, pp. 59-267, April 1960.

[8] R. Swerling, "Probability of Detection for Fluctuating Targets," *IRE Transactions*, vol. IT-6, pp. 269-308, April, 1960.

[9] D.K. Barton, *Radar System Analysis*, Norwood, MA: Artech House, 1976.

[10] G.T. Ruck, et al., *Radar Cross-Section Handbook*, New York: Plenum Press, 1970.

[11] E.F. Knott, Ch. 11 in M.I. Skolnik, *Radar Handbook* 2nd ed., New York: McGraw-Hill, 1990.

[12] R.K. Luneburg, *A Mathematical Theory of Optics*, Berkeley, CA: University of California Press, 1964.

[13] L.V. Blake, Ch. 2 in M.I. Skolnik, *Radar Handbook*, 2nd ed., New York: McGraw-Hill, 1990.

[14] M. Katzin, "On the Mechanisms of Radar Sea Clutter," *Proceedings of the IRE*, vol. 5, no. 1, January, 1957.

[15] P. Beckmann and A. Spizzichino, *The Scattering of Electromagnetic Waves from Rough Surfaces*, Norwood, MA: Artech House, 1987.

[16] The sea state number scale is also known as the Douglas scale.

[17] I. Katz in M.I. Skolnik, *Introduction to Radar Systems*, New York: McGraw-Hill, 1980, Ch. 13.

[18] W.C. Morchin, *Airborne Early Warning Radar*, Norwood, MA: Artech House; 1990, p. 161.

[19] K.D. Ward and S. Watts, "Radar Sea Clutter," *Microwave Journal*, vol. 29, no. 6, June 1985, pp. 109-121.

[20] R.R. Boothe, "The Weibull Distribution Applied to the Ground Clutter Backscatter Coefficient," USAMC Report RE-TR-69-15, June 1969.

[21] D.K. Barton, *Radar System Analysis*, Englewood Cliffs, NJ: Prentice Hall; 1964.

[22] J.M. Smith and R.H. Logan, "AN/APS-116 Periscope-Detecting Radar," *IEEE Transactions on Aerospace and Electronic Systems*, vol. AES-16, no. 1, January 1980, pp. 66-73.

[23] W.H. Shrader, Ch. 17 in M.I. Skolnik, *Radar Handbook* 1st ed., New York: McGraw-Hill; 1970.

[24] R.J. Serafin, Ch. 23 in M.I. Skolnik, *Radar Handbook* 2nd ed., New York: McGraw-Hill; 1990.

[25] C.R. Vaughn, "Birds and Insects as Radar Targets: A Review," *Proceedings of the IEEE*, vol. 73, no. 2, pp. 205-227, February 1985.

[26] B.L. Hicks, et al., "Sea Clutter Spectrum Studies Using Airborne Coherent Radar III," *Control Sys. Lab./Univ. Illinois Rept.* R-105, May 1958.

[27] J.J. Kovaly, et al., "Sea Clutter Studies Using Airborne Coherent Radar," *Univ. Illinois Control Sys. Lab. Rept.* 37, June 1953.

[28] G.R. Curry, "A Study of Radar Clutter in Tradex", *M.I.T./Lincoln Lab., Group Rept.* 1964-29, May 1964.

[29] V.W. Pidgeon, "Bastatic Cross Section of the Sea for a Beaufort 5 Sea," *Space Sys. Planetary Geol. Geophys.*, American Astronautical Society, May 1967.

[30] B.W. Pidgeon, "The Doppler Dependence of Radar Sea Return," *J. Geophys. Res.*, February 1968.

[31] M.W. Long, "On a Two-Scatterer Theory of Sea Echo," *IEEE Transactions on Antennas and Propagation*, vol. AP, September, 1974.

5

TARGET ECHO INFORMATION EXTRACTION:
Part 1 – Detection

5-1. Detection Introduction

Detection is the process of deciding whether or not a target is present. It is accomplished by comparing the signal-plus-interference, after all processing has occurred, to a threshold. If the signal-plus-interference, or interference alone, crosses the threshold, a detection is declared. If not, no detection occurs. As shown in Ch. 1, four detection states exist (Table 1-1 and Fig. 1-6). Two of these states are correct:

— No target is present, the composite signal does not cross the threshold, and detection does not occur, and

— A target is present, the composite signal crosses the threshold, and a detection occurs.

The two other states are detection errors:

— No target is present, but interference alone exceeds the threshold, and a false detection, called a false alarm, occurs, and

— A target is present but is either too weak or the threshold is too high, the composite does not cross the threshold, and no detection occurs.

Detection, because of the unpredictability of interference and target fluctuation behavior, must be described as a stochastic process. Thus the performance of detection in the various forms of interference will be treated as probabilities and all detection processes and parameters are thus described.

Note: Detection has in the past been the subject of much study and mathematical modeling. It is far beyond the scope of a general purpose text such as this to present the detail necessary for an in-depth understanding of the subject. In this treatment, principles are presented in a manner that should satisfy all but those who must actually analyze and design detection schemes. Fortunately, all the necessary information for a detailed look at detection in a noise environment has been compiled into a single volume by Meyer and Mayer, *Radar Target Detection: Handbook of Theory and Practice* [1]. This definitive work treats coherent and non-coherent detection for Swerling target fluctuation models in thermal noise interference. The theory is presented in such a way that it can be extended to other fluctuation models and other forms of interference. Much of the information presented in Secs. 5-2 and 5-3 is based on Meyer and Mayer's data.

In all radars and before any other information can be recovered, the detection decision must be made. In search radars, detection is accomplished either by an operator or automatically, and the action which causes the antenna position encoders and the range timer to be sampled gives the target's angular position and range. In tracking radars, detection is necessary before automatic track can be initiated, in a process called track acquisition. Detection of targets is always complicated by the presence of interfering signals.

Detection compares the amplitude of the total signal, a composite of target echoes plus interference, to a predetermined threshold value. This can be a comparison of voltages, a comparison of digital words, or an evaluation of cathode ray tube (CRT) spot intensity by an operator. If the composite signal exceeds the threshold, a detection occurs, whether or not a target was present. Detection takes place after signal processing has enhanced the signal-to-interference ratio.

The placement of the threshold sets the rate at which false alarms occur (the false alarm rate). The threshold's value depends on the statistical distribution of the interference and the desired false alarm rate. Many radar designs assume that the interference probability distribution is well enough known to accurately set the desired false alarm rate by setting the threshold. In these systems, thresholds are set automatically by sensing the average interference level and setting the threshold based on it, so that a small constant rate of false alarms occurs. This threshold-setting process, known as CFAR (constant false alarm rate), is discussed later in this chapter.

Detection goal of radar: Because the threshold setting is normally a function of the interference level, the signal quantity controlling detection is the signal-to-interference ratio. The goal of radar designers and operators is to obtain a high enough signal-to-interference ratio to give reliable detection without having the threshold so low as to result in an inordinately high false alarm rate.

Detection goal of radar interferers: The goal of those who design and operate systems which interfere with target detection is to create detection errors. This may be accomplished by creating so many false alarms as to overload operators and data processing systems with targets, causing them to miss legitimate targets. To counter this overloading, operators and automatic detection systems increase the level of the threshold. This causes the lower level target echoes to be missed. In either case, the goal of the interferer is met. Low observable targets exacerbate this problem, allowing interfering signals to be more effective.

Another method of hiding targets from detection is to use tactics to take advantage of naturally occurring interference, such as clutter, to force the threshold higher and hamper the detection process. Low-flying aircraft and missiles are more difficult to detect than those which fly high because the land and sea clutter residues cause detection errors.

Detection errors and radar design philosophy: In digital communications, thresholds are set so that the two detection errors ($0 \rightarrow 1$ and $1 \rightarrow 0$) are equally likely. In radar, detection thresholds are biased so that the false alarm error ($0 \rightarrow 1$) is much less likely than the missed-target error ($1 \rightarrow 0$). This is because false alarms harm the radar process more than missed targets. Many (sometimes millions) of detection trials are made each second.

If only a small fraction of them are false alarms, the number of false alarms per second can be large enough to badly confuse an operator or overload a computer.

On the other hand, a missed target is of little consequence in many radars, because there will be more opportunities to detect it. In fact, radar detection is normally specified so that the likelihood of detecting a target (the probability of detection P_D) at extreme range seems inordinately low. As, however, the target approaches the radar, the signal-to-interference ratio increases, increasing detection probability. For example, suppose that a 4 m^2 target at 100 nmi produces a S/N of 12 dB, and suppose further that this S/N gives a 50-50 chance of detection (P_D = 0.5). If it is critical that the target be detected reliably, this probability seems quite low. Remember, though, that this is only a specification used to evaluate the radar for a tightly controlled set of conditions. If this same target closes to within 84 nmi of the radar, its signal-to-noise ratio increases to 15 dB. This yields a P_D of 0.99 (see Fig 5-5 and its explanation). The target which was detected in only half the trials at 100 nmi is detected with virtual certainty at 84 nmi.

Definitions: The following quantities define detection:

— *Probability of Signal* (P_S): This is the probability that on any given *single test* of signal-plus-interference and threshold, the result will be a threshold crossing if a target was present.

— *Probability of Detection* (P_D): This is the probability that for any given evaluation of signal-plus-interference and a threshold, the result will be a detection if a target was present. The difference between P_S and P_D is that P_S is for the evaluation of a single signal echo or detection trial, and P_D is the result of many consecutive signal echoes processed together (a look or dwell) or is the final result of a compound detection trial (see M of N detection in Sec. 5-5).

— *Probability of Noise* (P_n): This is the probability that interference (noise in this case) alone will cross the threshold, for a single test.

— *Probability of False Alarm* (P_{FA}): False alarm probability is the probability that interference alone will cross the threshold for a look or compound test. P_n and P_{FA} are related in the same way as are P_S and P_D.

— *False Alarm Number* (FAN): The number of tests per false alarm is the *FAN*. FAN is the reciprocal of false alarm probability P_{FA}.

$$FAN = 1 / P_{FA} \tag{5-1}$$

— *False Alarm Time* (FAT): The mean time between noise threshold crossings (false alarms) is the *FAT*. It is found from [2] as

$$FAT = \lim_{N \to \infty} 1/N \sum_{i=1}^{N} T_i \tag{5-2}$$

T_i = the time between the ith and (i+1)th threshold crossing by noise

The average time the noise spends above the threshold is

$$\tau_M = \lim_{M \to \infty} 1/N \sum_{j=1}^{M} \tau_j \tag{5-3}$$

τ_M = the average noise-above threshold pulse width

τ_j = the time the jth noise crossing spends above the threshold

The average noise pulse width above the threshold is the reciprocal of the system bandwidth at the point of the test.

$$\tau_M \approx 1/B \tag{5-4}$$

B = the system bandwidth at the point of the test

The probability of false alarm (or probability of noise, for one test) is the ratio of Eqs. 5-2 and 5-3, modified by Eq. 5-4.

$$P_{fa} = \frac{1}{B\,FAT} \tag{5-5}$$

— *False Alarm Rate* (*FAR*): *FAR* is the average number of false alarms per second. It is the product of P_{FA} and the number of tests per second. With little error, the number of tests per second can be approximated as the signal bandwidth at the point of the test [1]. It is the reciprocal of *FAT*.

$$FAR = P_{FA} R_{DT} \approx P_{FA} B = 1/FAT \tag{5-6}$$

R_{DT} = the rate at which detection tests occur (1/s), and is equal to the bandwidth at the point of the test. (It is often the range bin rate)

Example 5-1: A radar's processed echo pulse width is 0.8 μs. At the output of the receiver is a threshold test which produces a false alarm probability of 3.0×10^{-7}. Find the false alarm number, the false alarm rate, and the false alarm time.

Solution: The false alarm number is the reciprocal of the false alarm probability (Eq. 5-1), or 3.33×10^6 tests per false alarm. The false alarm rate is the false alarm probability times the number of tests per second, equal to the range bin rate — Eq. 5-6. Each range bin is approximately the pulse width (0.8 μs) and there are 1.25×10^6 of them per second. The false alarm rate is therefore 0.375 false alarms per second. The false alarm time (Eq. 5-3) averages 2.67 seconds per false alarm.

5-2. Detection in Noise

The detection of radar signals in interference has been explored in depth by many authors since the 1940s. The Rayleigh probability distribution, discussed in Sec. 4-10, describes signal probabilities in noise interference. Figures 4-34 and 4-35, given in modified form as Figs. 5-1 and 5-2, show the Rayleigh distribution for noise and the modified Rayleigh (Ricean) distribution for signal-plus-noise. Figure 5-1 (top) is the probability density function for thermal noise, and Fig. 5-1 (bottom) is the noise probability distribution. The distribution yields the probability that noise will exceed the independent variable, and yields gives P_n. The independent variable is voltage, normalized to the root-mean-square (rms) voltage of the noise.

Figure 5-2 (top) is the probability density for signal plus noise for several different signal-to-noise ratios. Figure 5-2 (bottom) is the integral of these signal-plus-noise densities (the distribution) and yields P_S. For a signal-to-noise ratio of zero (no signal), the distribution is Rayleigh. With moderate signal-to-noise ratios, it is known as modified Rayleigh. For high signal-to-noise ratios, the distribution approaches Gaussian. As in Fig. 5-1, the independent

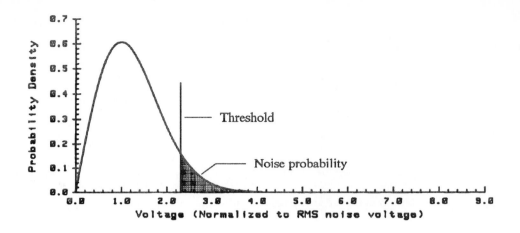
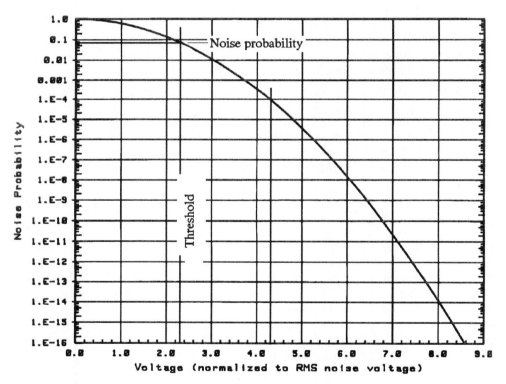

Figure 5-1. Rayleigh Probability Function and Distribution for Thermal Noise.

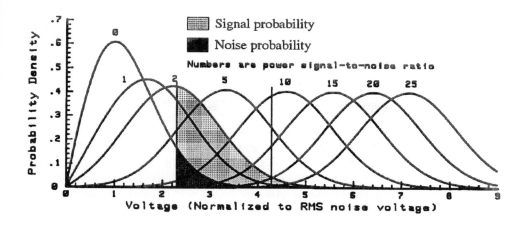
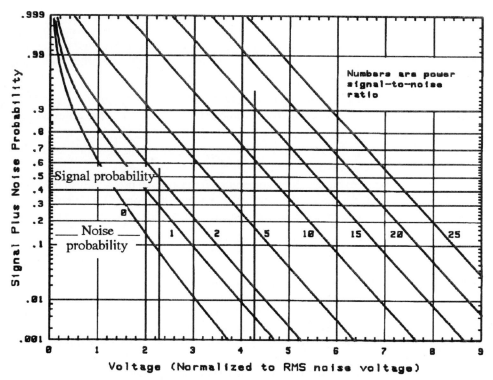

Figure 5-2. Modified Rayleigh (Ricean) Probability Function and Distribution for Signal Plus Noise.

5-2. Detection in Noise

variable is voltage, either noise alone or signal plus noise, normalized to the root-mean-square (rms) voltage of the noise.

The threshold is set for the desired noise probability P_n using Fig. 5-1. Selecting a threshold voltage of 2.3 times the rms noise voltage gives a noise probability of 0.085. Figure 5-2 shows that for a signal-to-noise ratio of 2 (3 dB), the signal probability is 0.5. This is a reasonable signal probability for radars but the noise probability is much too high.

The noise probability is improved (lowered) by increasing the threshold value. At a threshold of 4.3 times the RMS noise, the noise probability drops to 10^{-4}, which is better, but still not low enough for most radar applications. Improving the noise probability by raising the threshold to 4.3, however, also lowers signal probability to about 0.02. This tells us that for a noise probability of 10^{-4}, a signal-to-noise ratio of 3 dB is simply not sufficient for good detection. If the signal-to-noise ratio were raised to 10 (10 dB) with the threshold at 4.3, the signal probability becomes 0.62, which is adequate for most radar applications.

In summary, low noise probabilities require higher thresholds, which in turn require higher signal-to-noise ratios for reasonable detection probabilities.

All detection problems can be solved using figures similar to Figs. 5-1 and 5-2. Sometimes the information in the two figures is better displayed in the form shown in Fig. 5-3, known as Rice (or Ricean) curves. In this form, only one set of curves is necessary and the designer/evaluator does not need go through the step of finding a threshold voltage for false alarm probability and transferring it to the signal curves. On the other hand, the threshold voltage itself cannot be found from the Rice curves.

Example 5-2: A radar system requires a single-hit false alarm probability (noise probability) of 5×10^{-7} and a detection probability of 0.75. Find the required single-hit signal-to-noise ratio.

Solution: From Fig. 5-1, the threshold voltage needed for the specified noise probability is approximately 5.4 times the rms noise voltage.

From Fig. 5-2 (bottom) the signal-to-noise ratio giving a signal probability of 0.75 for a threshold of 5.4 times the rms noise voltage is about 17 (12.3 dB). Alternatively, Fig. 5-3 gives the answer directly, albeit with some interpolation required.

5-3. Signal Integration and Target Fluctuations

The preceding treatment is somewhat simpler than actual radar detection, since it is for single hits on non-fluctuating targets. The information presented in Figs. 5-1, 5-2, and 5-3 are for single hits on non-fluctuating targets. Several factors restrict the use of this material. First, radars seldom attempt to detect a single signal echo. Commonly, several echoes are processed together (integrated) with the processed composite applied to the threshold. This somewhat changes the distribution of Fig. 5-2. Second, real targets fluctuate (Ch. 4), which further modifies the probability distribution. Additionally, signal-plus-interference other than for thermal noise have different probability distributions.

Actual detection designs must thus allow for at least six factors: false alarm probability, detection probability, signal-to-interference ratio, interference type and statistics, target

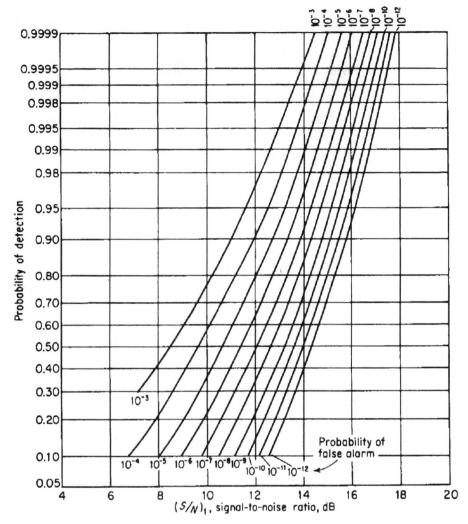

Figure 5-3. Rice Curves for Single Hit, Non-Fluctuating Target. (Reprinted from Skolnik [2]. © 1980 McGraw-Hill Book Company; used with permission)

fluctuations, and the number of hits integrated into a look. A detailed treatment of all of them is beyond the scope of this book. However, a comprehensive reference is available for signal detection in noise and the reader is encouraged to consult it for more details [1].

Signal integration can be non-coherent, where signal and interference phases are ignored, or coherent, where phase is considered. Diagrams showing the placement of the integration process in the signal flow are shown in Fig. 5-4.

Figure 5-4a is non-coherent integration, which takes place after envelope demodulation of the signal. This form of demodulation discards signal phase. Figure 5-4b shows integration taking place at the intermediate frequency and is coherent. Figure 5-4c shows

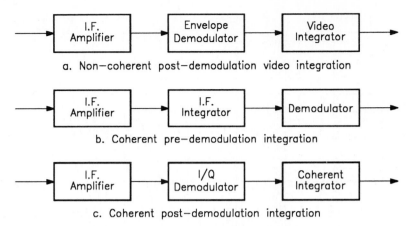

Figure 5-4. Signal Integration.

post-demodulation integration on I and Q signal components. It is coherent integration and is the most common type found in modern radars.

The signal-to-noise ratio with integration is raised by some function of the number of pulses integrated, as discussed in Ch. 3 and in the following equation.

$$S/N_N = S/N_i \, (N / L_i) \tag{5-7}$$

S/N_N = the signal-to-noise ratio resulting from a look of N hits

S/N_i = the average signal-to-noise ratio from a single hit

N = the number of pulses integrated in the process, called the integration number

L_i = the integration loss, a function of detection probability, false alarm probability, number of pulses integrated, and target fluctuation statistics

Coherent integration: Coherent integration is more effective than non-coherent, because more is known about the coherent signal. With it, the integration loss usually ranges from 1.0 to 1.7, and is rarely larger than 2.0, whereas with non-coherent integration, the loss can be as great as the square root of the integration number for very large integrations.

Meyer and Mayer [1] treat coherent integration by calculating an equivalent integrated signal-to-noise ratio and applying this S/N to single-hit process. Their equivalent signal-to-noise ratio, modified slightly (by inserting the 1/N factor) to fit the figures following, is

$$\overline{(S/N)_N} = (1/N) \sum_{k=1}^{N} (S/N)_k \tag{5-8}$$

$\overline{(S/N)_N}$ = the mean signal-to-noise ratio after integration

$(S/N)_k$ = the signal-to-noise ratio of the kth sample

k = the sample number

N = the number of samples in the look (the integration number)

The above relationship gives an average (mean) signal-to-noise ratio for the multiple echoes. The integrated S/N is thus N times the mean S/N of the individual hits.

$$(S/N)_N = N \overline{(S/N)_N} \qquad (5\text{-}9)$$

$(S/N)_N$ = the integrated signal-to-noise ratio

Coherent integration loss: Thus using Meyer and Mayer's model, an integration of N hits on a constant RCS target (case 0) produces an N-times increase in the signal-to-noise ratio, as indicated in Ch. 3. There are two losses which must be factored into this simplistic model. First is the loss caused by windowing the signal. Most coherent integrations are implemented as either Fourier transforms or correlations. To prevent leakage, where signals in one Doppler or range bin spread to other bins, the signals are windowed prior to the process.

Figure 5-5 shows the need for windowing. Two signals occurring simultaneously in time are hypothesized; one is a moving target, the other is clutter, with the clutter echo much larger than that of the moving target (the normal situation). Figure 5-5a is the actual spectrum of the two signals. Figure 5-5b is the spectrum calculated with a Fourier transform, but without windowing (actually with what is known as a rectangular window — see Ch. 11). The moving target is not detectable because of interference from the clutter. Figure 5-5c shows the calculated spectrum of the same data, but after a particular window is applied. Note that the moving target can now be discriminated from the clutter.

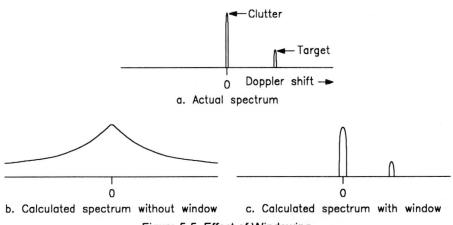

Figure 5-5. Effect of Windowing.

Windowing involves selectively discarding portions of the signal, and thus produces a loss in signal-to-noise ratio. It is discussed in Ch. 11, and Table 11-1 gives values for some representative windows. (This portion of the coherent integration loss is the column in Table 11-1 labeled "Equivalent Noise Bandwidth.") An examination of the table indicates that for commonly used window functions, the loss is less than 2.0.

The other portion of the coherent integration loss involves signal phase. Coherent integration of moving targets requires that the motion be removed before signal summation. This is done by subtracting from each consecutive hit a phase equivalent to the target's motion between hits (Fig. 5-6). The process is discussed in Sec. 11-3. If the motion compensation is perfect (Fig 5-6b), the integrated signal is the linear sum of the components and there is no loss. If not, the summation does not have the maximum possible value (Fig. 5-6c). This integration loss component is related to the scalloping loss, and its worst-case value for each of the windows treated is found in the "Scalloping Loss" column of Table 11-1. Other phase errors contribute to this loss, but for values of N greater than about four are usually negligible.

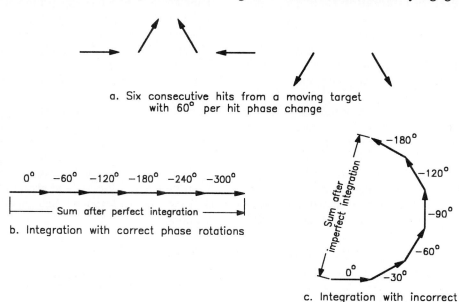

Figure 5-6. Coherent Integration and the Phase Loss.

The total coherent integration loss is thus

$$L_{i(C)} = L_W L_P \quad \text{(power ratios)} \quad (5\text{-}10a)$$
$$L_{i(C)} \text{ (dB)} = L_W \text{ (dB)} + L_P \text{ (dB)} \quad \text{(decibels)} \quad (5\text{-}10b)$$

$L_{i(C)}$ = the coherent integration loss
L_W = the signal energy loss caused by the window
L_P = the phase component of the loss, also related to the window

Some examples of the effect of coherent integration are given in Figs. 5-7 through 5-9. To use these figures, the radar equation and Eq. 5-8 are used to calculate an *equivalent average single-hit* signal-to-noise ratio, which is adjusted for the integration loss and used with the figures to find the detection probability.

Figure 5-7 is for eight pulses coherently integrated. Note that far more signal is needed for targets exhibiting large fluctuations (cases 1 and 2) than for steady targets. With coherent integration, the signal-to-noise ratio is a strong function of the amount of fluctuation. The trends are similar between false alarm probabilities of 10^{-4} and 10^{-9}, with more signal being required for the lower P_{fa}. The numbers used here are typical for low PRF search radars.

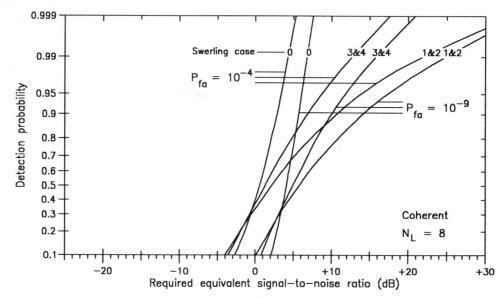

Figure 5-7. Detection Probability for Eight Pulses Coherently Integrated. (Data from Meyer and Mayer [1])

The following example illustrates the use of the equations and figures.

Example 5-3: A Swerling case 3 target is found to have the following S/N_k for eight consecutive hits of +9.0 dB, +8.0 dB, +12.4 dB, 10.0 dB, +9.8 dB, +11.9 dB, +8.9 dB, and +10.4 dB. A Dolph-Chebyshev ($\alpha = 3.0$) window is used. If the false alarm probability is 10^{-9}, find the detection probability. Use half the scalloping loss (in dB) as the phase loss.

Solution: First, the mean S/N must be found. The dB values are converted to power ratios and Eq. 5-8 used to find their mean. The values are 7.94, 6.31, 17.38, 10.00, 9.55, 15.49, 7.76, and 10.96. Their sum is 85.40, and the mean is 10.67, which is +10.28 dB. Table 11-1 gives the integration loss for this window as 1.51 (1.79 dB) and the scalloping loss as 1.44 dB. Using half the scalloping loss (0.72 dB) gives a total loss of 2.51 dB, for an equivalent S/N of 7.77 dB. Figure 5-7 gives the detection probability as about 0.65.

Figure 5-8 shows the result of integrating 64 pulses and specifying the same two false alarm probabilities. These data are applicable to medium PRF modes. Although a Pfa of 10^{-4} seems high, values are for medium PRF are often in this range. This is possible because typical MPRF modes examine several dwells before making the final detection decision, in a process called M of N detection (see Sec. 5-4). These multiple dwells are often a part of the deghosting processes for resolving range and Doppler ambiguities (Ch. 6). It is obvious from the figures and discussions that detections after integrating 64 pulses requires considerably less signal (about 9 dB) than integrating 8 hits.

Figure 5-9 shows the same data for 1024 integrated pulses, a typical value for high PRF (pulse Doppler). In these modes, multiple dwells are often not used, and the lower probability of false alarm is probably more useful.

5-3. Signal Integration and Target Fluctuations

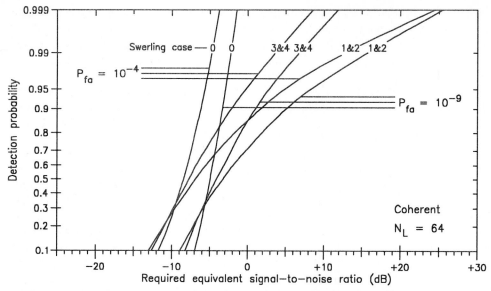

Figure 5-8. Detection Probability for 64 Pulses Coherently Integrated.
(Data from Meyer and Mayer [1])

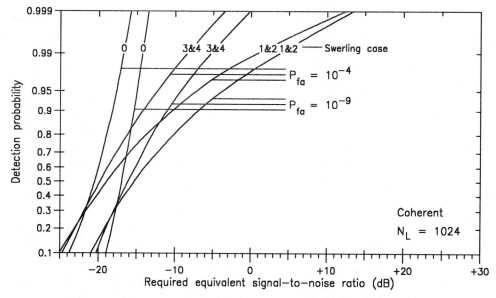

Figure 5-9. Detection Probability for 1024 Pulses Coherently Integrated.
(Data from Meyer and Mayer [1])

Target fluctuations and coherent integration: Target fluctuations in coherent integration are accounted for in Meyer and Mayer's coherent integration model (Figs. 5-7 through 5-9) and the equivalent S/N of Eq. 5-8. The difficulty in evaluating coherent detection is in modeling Eq. 5-8, a task beyond the scope of this book. Some comments are, however, in order.

Figures 5-7 through 5-9 show that the required equivalent S/N is a function of the amount of fluctuation and not its rapidity. In the calculation of the equivalent S/N, however, there is a marked difference between the integration effect on slow and rapid fluctuations.

The reason can be seen from the definitions of the fluctuation cases and RCS and a knowledge of the signal integration process. Targets which fluctuate are measured and described as having a radar cross-section which is the *average* of the fluctuations, with this average RCS used in the radar equation to predict single-hit signal-to-noise ratio. Conventionally, RCS reports of targets use the *median* form of the average (median RCS means that half the measurements reported RCS greater than the median and half the measurements were less). Case 1 and case 2 targets fluctuate over several orders of magnitude of radar cross-section. For example, a target which exhibits, for eight consecutive hits, radar cross-sections of 0.1, 0.4, 0.75, 0.99, 1.01, 1.33, 2.5, and 10 m^2 has a median RCS of 1 m^2 and would be Swerling case 2.

Signal integration, however, is a *summation* process, which returns a value related not to the median, but to the *mean* of the individual hits (Eq. 5-8). The mean value of the above target is 2.14 m^2. The radar equation in this example predicts detection based on an RCS of 1.0 m^2, but the actual detection is based on a 2.14 m^2 target. With rapidly fluctuating targets, where many different and divergent RCS values are present within a look (dwell), signal integration is likely to produce a consistently high effective RCS, and detection is assisted.

Slowly fluctuating targets are, on the other hand, more difficult to detect than those which are constant or which fluctuate rapidly. Consider a slowly fluctuating case 1 target of the same median RCS as given above (1.0 m^2). Four consecutive hits are likely to be the result of RCS values of perhaps 0.085, 0.095, 0.105, and 0.115 m^2 for one scan and perhaps 9.85, 9.95, 10.05, and 10.15 m^2 on the next. The median RCS is still 1.0 m^2, but the mean for the first scan is 0.1 m^2 and 10.0 m^2 for the second. If the probability of detection is to be high, the median RCS must be specified much higher with case 1 targets than with case 2. This is because in a look at a case 2 target, large RCS hits will be present to compensate for small RCS hits. In case 1, this is not true, since all hits in a scan are of approximately the same RCS.

Figure 5-10 shows the problem with slowly fluctuating targets. It shows that for moderate detection probabilities (0.5 − Fig. 5-10a), the penalty paid for detecting the case 1 target is small. As the detection probability increases (to 0.7 − Fig. 5-10b), the signal-to-noise requirement increases dramatically. The reader is encouraged to try this on a case 2 target − the required S/N increase to reach a P_D of 0.7 from 0.5 is much smaller than for case 1.

Thus because of their constancy of RCS across a dwell, slowly fluctuating targets require higher median signal-to-noise ratios for high detection probabilities than do rapidly fluctuating targets. The amount of fluctuation also influences detection, although to a somewhat lesser extent than rapidity. This will be discussed in the next part of this section for non-coherent integration (Figs. 5-13 and 5-14). It is emphasized again that the effect of

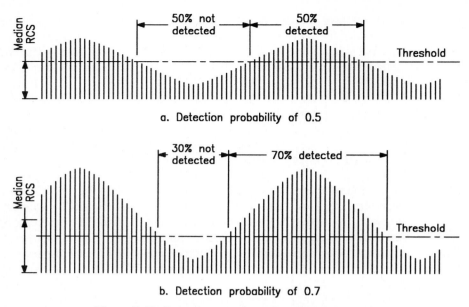

Figure 5-10. Detection of Slowly Fluctuating Targets.

fluctuation is more on finding the equivalent S/N than on applying it to the model, explaining the apparent inconsistency between the detection result and Figs. 5-7 through 5-9. Because RCS is defined as its median but used as its mean, the effective RCS with rapid fluctuation may be considerably greater than its specified value.

A second effect of rapid fluctuation is that when spectrum analysis is used for signal integration (common in coherent radars), large rapid fluctuations cause the signal spectrum to spread, leaving less signal in the Doppler bin containing the target. This effect is of considerably less importance than the mean/median effect discussed above.

The effect of fluctuation rapidity is illustrated in Figs. 5-13 and 5-14. Although the data are for non-coherent integration, the principle is evident. In the final analysis, fluctuation rapidity has more effect on detection than the amount of fluctuation.

Non-coherent integration and detection: The analysis is the detection of non-coherently integrated signals is considerably more difficult than with coherent integration. The analysis is extensive enough that a rigorous summary is not attempted here. The factors in the analysis are single-hit S/N, integration number, false alarm probability, target fluctuation statistics, and detection probability. The reader is referred to Meyer and Mayer [1] for further information.

There are about 400 plots, called Meyer plots, in the reference. They evaluate non-coherent detection for a wide variety of conditions. All five Swerling cases are covered, as are false alarm probabilities from 10^{-1} to 10^{-10}. An example is given as Fig. 5-11, this one for a Swerling Case 2 target and false alarm probability of 10^{-5}. Meyer plots use single-hit signal-to-noise ratios calculated from the radar equation using the target's median RCS. Unlike coherent integration, an effective S/N does not have to be calculated. The reader should be aware that there are fluctuation models other than Swerling. To use these models, the calculations would have to be extended to them.

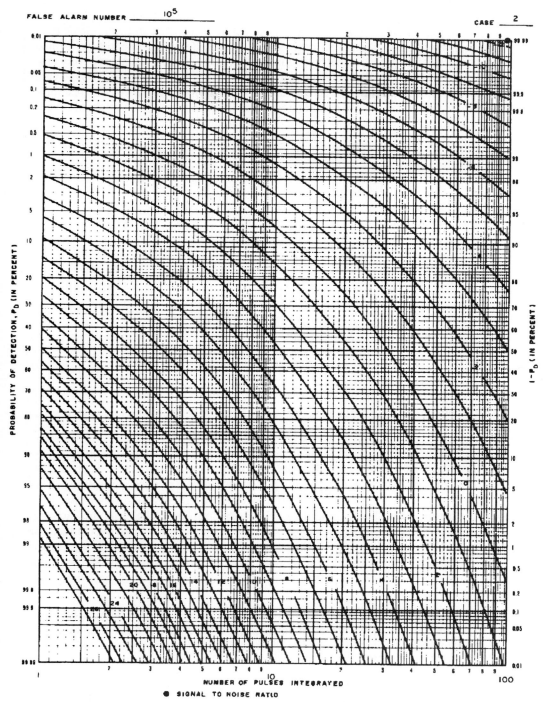

Figure 5-11. Meyer Plot Example. (Reprinted from Meyer and Mayer [1]. © Academic Press; used with permission)

5-3. Signal Integration and Target Fluctuations

Whereas coherent integration losses are generally small and only slightly influenced by the number of pulses integrated, non-coherent integration losses are strongly related to the integration number. Losses range from essentially unity for small integrations of non-fluctuating targets to approximately the square root of the integration number for very large integrations. Many non-coherent integrations take place in CRT phosphors. These integrations also have losses of about the square root of N. See Fig. 3-21 for an A-Scope example. PPI losses are about the same. A range of integration losses for various conditions are shown in Fig. 5-12.

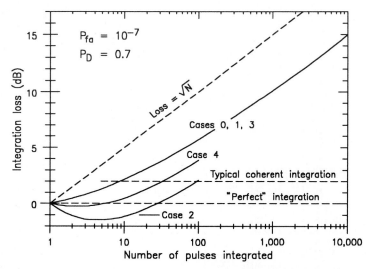

Figure 5-12. Integration Loss Examples. (Data from Meyer and Mayer [1])

Note that for some conditions (small integration numbers, rapid target fluctuations), the integration loss has negative decibel values. This means that the integration is better than "perfect" (for example, integrating eight pulses produces a process gain of 12). This apparent impossibility is again explained by the fact that the radar equation uses median RCS but the integration uses the mean. With rapidly fluctuating targets, the mean RCS across a look can be considerably larger than the median. For example, a target giving consecutive RCSs of 20, 1.0, and 0.05 m^2 would be expected to yield a certain signal-to-noise ratio, based on the median RCS of 1.0 and an integration number of 3. It actually yields S/N based on the mean RCS of 10.5, for an effective integration number of 10.3, over three times that expected.

Example 5-4: A radar system non-coherently integrates 10 pulses. The target to be detected is case 2, and the radar equation gives the signal-to-noise ratio for a single hit on the median RCS as 4 dB. The false alarm probability for the ten pulses is to be 10^{-5} (FAN is 10^{+5}). Find the probability of detection for the integrated signal.

Solution: Figure 5-11 fits the given data. It shows that for these conditions, the detection probability is about 0.66.

As with coherent integration, if targets fluctuates, additional signal-to-interference ratio is usually needed to produce the same probability of detection as that of constant (case 0) targets. Figures 5-13 and 5-14 show *examples* of the effect of target fluctuations on non-coherent signal detection for various probabilities of detection and two integration numbers (8 in Fig. 5-13 and 64 in 5-14), with false alarm probabilities of 10^{-4} and 10^{-9}. Note from

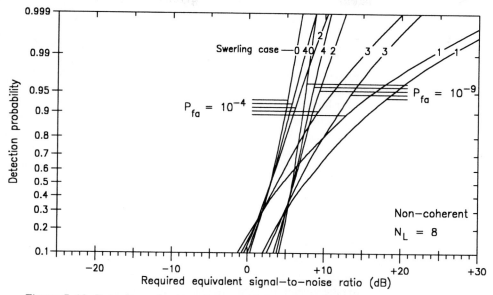

Figure 5-13. Detection after Non-Coherent Integration of Eight Hits. (Data from [1])

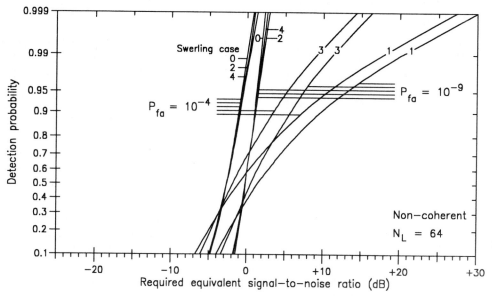

Figure 5-14. Detection after Non-Coherent Integration of 64 Hits. (Data from [1])

5-3. Signal Integration and Target Fluctuations

the figures that it is easier to detect rapidly fluctuating targets (cases 2 and 4) than those which fluctuate slowly (cases 1 and 3). The reasoning of Fig. 5-10 applies.

If a single hit were used for detection (uncommon), the rapidity of fluctuation ceases to be a factor (if only one hit is used, the RCS of the hits just previous and just after are of no consequence). In this case, the *amount* of fluctuation governs detection. This is implied in Figs. 5-7 through 5-9, which apply for single-hit detection if the S/N is adjusted.

Comments on detection: Target fluctuations and signal integration profoundly affect detection, which without them is relatively straightforward. Note that in the case of single-hit detections, the additional signal required is determined primarily by the *amount* of fluctuation. Case 1 and 2 targets require about the same amount of signal. Case 3 and 4 targets also require about the same signal increase, but as a group they require a lesser increase than do cases 1 and 2 (at least for high detection probabilities).

On the other hand, in detection of integrated signals fluctuation *rapidity* has the greater effect. Cases 1 and 3 (slow fluctuations) and case 0 (no fluctuations) behave about the same. Cases 2 and 4 (rapid fluctuations) affect integration in a much different way than the others, primarily again because target RCS reporting is *median*, whereas effective integration is based on *mean* RCS.

5-4. M of N Detection

In addition to the target and radar parameters given above, another factor exerts a strong influence on detection in modern radars. Because of range and Doppler ambiguities (Ch. 6), it is common for radars to take several independent looks at the same target space, usually at different PRFs and/or frequencies. Detection is enhanced by using these multiple independent looks in what is called *M-ary*, or *M of N* detection. For example, suppose eight independent looks were made in a particular resolution volume. Further suppose that the system would report a detection if a target were observed in the resolution cell (using the threshold as described above) for at least three of the eight looks. This would be a 3 of 8 detection scheme.

The statistics for M of N detection are described in Eq. 5-11 (detection) and 5-12 (false alarm).

$$P_D = \sum_{J=M}^{N} \frac{N!}{J!(N-J)!} P_S^J (1-P_S)^{N-J} \qquad (5\text{-}11)$$

P_D = the probability of detection (M of N)
M = the required number of successes for detection
N = the number of looks processed together
P_S = the probability of detecting the target on each look

$$P_{FA} = \sum_{J=M}^{N} \frac{N!}{J!(N-J)!} P_n^J (1-P_n)^{N-J} \qquad (5\text{-}12)$$

P_{FA} = the probability of false alarm (M of N)
M = the required number of successes for detection
N = the number of looks processed together
P_n = the probability of detecting interference on each look

Figure 5-15 shows the effect of several M of N schemes. The top figure gives the probability of false alarm for several M of N. Note that in all cases where the noise probability is reasonable, each of these M of N schemes results in an improved false alarm probability (the false alarm probability is *smaller* than the noise probability).

Figure 5-15 (bottom) shows detection probability *versus* signal probability for the same Ms of N. Note again that in most cases where the signal probability is reasonable, the detection probability is improved (it is *greater* than the signal probability). In both cases, signal detection and noise rejection, M of N usually helps. Its cost is that several looks must be made at the target instead. In systems where multiple looks are necessary for other reasons, such as ambiguity resolution, this cost of this detection improvement is some extra system complexity, which may be virtually entirely software.

Note that Figs. 5-15 (top) and 5-15 (bottom) display the same data. The vertical scale in Fig. 5-15 (top) is logarithmic, emphasizing the small expected values of P_{FA}. That of Fig. 5-15 (bottom) is a normal scale, emphasizing the mid-range expected values of P_D. Equations 5-11 and 5-12 also are of the same statistics, reflecting the fact that signal and interference are treated by the same process.

Example 5-5: An airborne radar operating at medium PRF (ambiguous in both range and Doppler — Ch, 6) examines a target space with 8 looks of 64 pulses each. If detections occur in 3 or more of the 8 looks, a target detection is declared (3 of 8) The radar is looking up (out of clutter) and noise is the primary interference. The detection threshold for each look is set at 4.0 times the RMS noise and the signal-to-noise ratio per hit is −6.3 dB. The total integration loss, including the signal processing window, is 2.3 dB (1.7 ratio). Assume the desired targets are case 0.

(a) Find the per-look detection and false alarm probabilities.

(b) Find the probability of detection (3 of 8) and probability of false alarm for the above conditions. Assume no target fluctuations (a questionable assumption under these circumstances).

Solution (a): The per-look signal-to-noise ratio is 9.5 dB, which is the input S/N of −6.3 dB, plus an integration number of 18.1 dB (64) less an integration loss of 2.3 dB (1.7). Figure 5-1 gives a per-look false alarm probability of 3.0×10^{-4}. Figure 5-2 gives a per-look detection probability of about 0.68.

Solution (b): Figure 5-15 (top) or Eq. 5-12 shows the false alarm probability for 3 of 8 and a 3×10^{-4} per-look probability to be about 2×10^{-9}, which is a considerable improvement. Figure 5-15 (bottom) or Eq. 5-11 gives the 3 of 8 detection probability of about 0.984, which is also a considerable improvement.

Comment: In the above example, both the detection and false alarm probabilities were improved (detection increased and false alarm decreased). This will often happen if the per-look numbers are fortuitous. The per-look false alarm probability

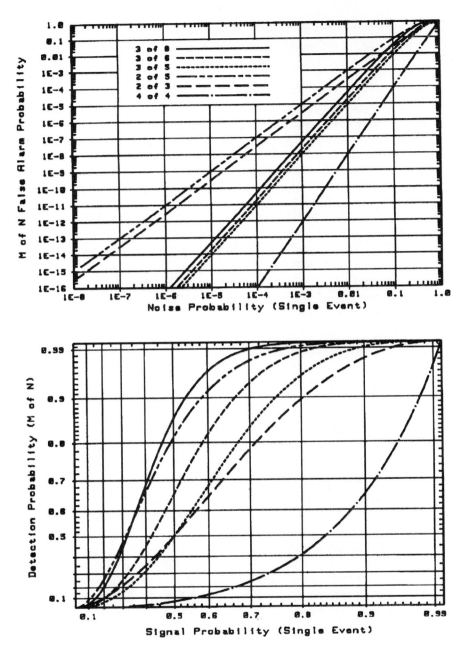

Figure 5-15. M of N Detection.

was small, and therefore the probability of getting three or more false alarms in eight tries is smaller. The per-look detection probability was one-half, but the overall detection probability was greater than one-half because there are more ways to succeed in 3 of 8 than to fail (0 of 8, 1 of 8, and 2 of 8 are failures — the other six conditions are successes). Therefore the M of N probability of success for a large per-look probability is usually increased.

M of N detection is a powerful method of dramatically improving detection statistics. The down side is that more inputs have to be processed, meaning that the radar's dwell (time-on-target) must be greater, slowing the search scan. This is compensated for by the fact that most M of N schemes involve either tracking radars, which only look at a single target and do not scan, or medium PRF modes, which can emit and process many samples quickly.

5-5. Threshold-Setting Concept — CFAR

Several methods are used to establish the detection threshold. The simplest is to use a manually operated potentiometer or a long time constant averaging circuit to set a constant threshold. This can be quite effective in some types of systems, but suffers where widely varying interference is encountered. Figure 5-16 shows the effect of using a constant threshold where the residual interfering signals vary widely in amplitude. In this case, the interference is clutter residue. There is no setting of this threshold which will detect both targets without also detecting clutter.

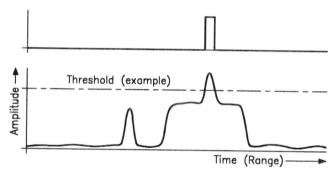

Figure 5-16. Detection with a Fixed Threshold.

More widely used schemes for setting the detection threshold sense the average interference level and set the threshold so that a relatively constant number of false alarms occur per unit of time. This method is called adaptive threshold, or constant false alarm rate detection (CFAR). It produces a threshold which "follows" the contour of the interference, as in Fig. 5-17.

Cell-averaging CFAR: Figure 5-18 shows one CFAR implementation. In it, the signal level in a few bins (range or Doppler) on each side of the bin being tested for a target are averaged together. It is assumed that these bins contain only interference. The average value of the signal in these bins is multiplied by a constant and the result is the threshold. Thus, the probability of false alarm is established, as in Fig. 5-1. The content of the center bin is

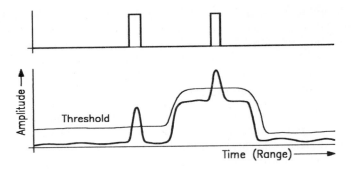

Figure 5-17. Detection with an Adaptive Threshold.

tested with the derived threshold to determine whether or not a detection occurs in that bin. The signals are then shifted one bin and the process repeated. Using this process, called cell averaging CFAR, each bin is tested against a threshold determined by the average signal level in a few bins on either side of it. Note that this technique is also effective against clutter and jamming.

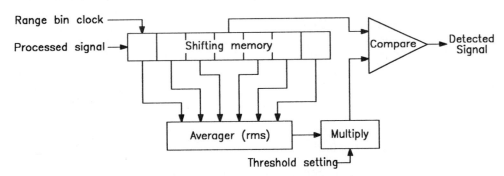

Figure 5-18. Cell-Averaging CFAR.

Figure 5-17 shows how the threshold behaves in a cell-averaging scheme. An analysis can be found in several sources, including [5]. The threshold developed by cell-averaging is

$$V_{TH} = [M_{TH}\ 1/(N-1)] \left[\sum_{n=-N/2+1}^{N/2-1} V(n) \right] \qquad (5\text{-}13)$$

V_{TH} = the detection threshold voltage (Volts)

M_{TH} = the threshold multiplier found from, for example, Fig. 5-1

N = the number of cells, including the cell being tested

n = the cell summation counter

$V(n)$ = the complex voltage in cell n (Volts)

In cell averaging, it is assumed that the probability distribution of the interference is known, and if noise it is usually Rayleigh. Clutter and certain types of ECM are not Rayleigh in character and the false alarm rate may not be as expected. In these situations, it may be necessary to sample the number of false alarms and modify the threshold based on this count. This results in what is known as a parametric or distribution-free detector [2].

Cell averaging as shown above is a function of range, and takes place on a range bin basis. Cell averaging using Doppler bins is also done, as is cell averaging using adjacent angle (azimuth) bins.

Guard-band CFAR: Another method of achieving CFAR is shown in Fig. 5-19. In it, the interference level is determined by examining the frequency bands adjacent to the band containing the signal. The threshold is then set based upon this interference level. This type of CFAR is effective against broadband interference, such as barrage jamming, where the interference level is, over some multiple of the target signal's bandwidth, independent of frequency. It is not effective against clutter, which falls within the signal bandpass, nor is it effective against spot jamming, for the same reason.

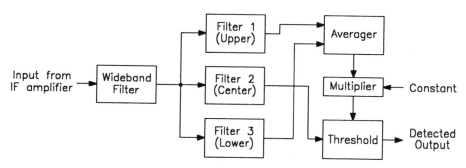

Figure 5-19. Guard-Band CFAR.

Clutter-map CFAR: Many modern search radars extend the cell-averaging scheme to the azimuth dimension as a method of achieving CFAR in clutter. The radar's space is divided into range and azimuth bins (Fig. 5-20). Over several scans, a moving average of the clutter residue in each range-azimuth cell is determined. The signal on each scan is compared to a threshold calculated from the moving average. Any abrupt changes in the processed signal in a cell are assumed to be moving targets. Since the averaging time constant is long, moving targets are assumed to change cells before they have time to materially influence the moving average.

Limiting CFAR (analog): Another CFAR method subjects the signal and broadband interference to hard-limiting, which rejects all amplitude information in both signal and interference. Signal-to-interference ratios can be improved through limiting because of a relationship in spectrum analysis where the amplitude of a waveform in one domain (time) equates to area in the other.

The example shown here is signal plus impulse jamming. The jamming waveform is a train of pulses which are much narrower than the signal pulses. The jamming thus has a

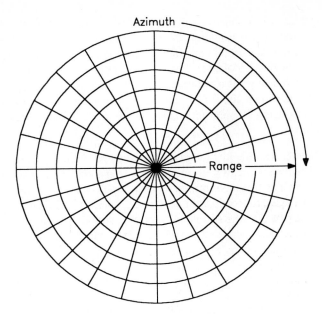

Figure 5-20. Clutter Map.

much wider bandwidth than the signal, and effectively acts as impulses to filters matched to the signal. If impulses are applied to the input of a filter, the output is the impulse response of the filter, which will have time components outside the limits of the impulse. This impulse response, which occupies considerable time, can interfere with a desired signal. See Fig. 5-21.

One analog implementation of this process is shown below. Figure 5-22 is its basic block diagram and Fig. 5-23 gives the envelopes of the waveforms. This process, called Dicke-fix, is actually performed at the radar receiver's intermediate frequency.

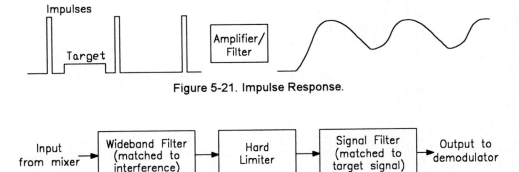

Figure 5-21. Impulse Response.

Figure 5-22. Analog Limiting CFAR (Dicke-fix).

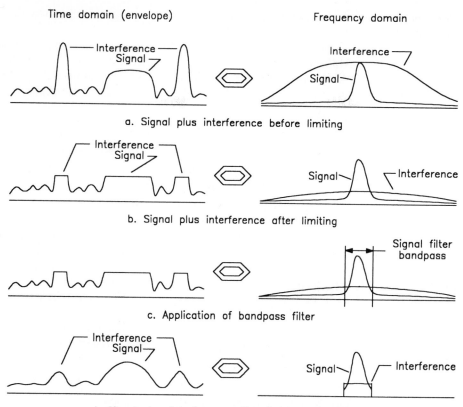

Figure 5-23. Analog Limiting CFAR Waves and Spectra.

Signal plus high amplitude, wideband interference is amplified in the broadband amplifier, the response of which is matched to the *interference* (Fig. 5-23a). The amplified signal plus interference is then limited, sometimes well into the system noise. After limiting, the signal and interference have the same amplitude in the time domain, and hence the same *area* in the frequency domain. Since the bandwidth of the interference is greater than that of the signal, however, the interference amplitude in the frequency domain is smaller than the echo's amplitude. Within the signal bandpass, there is less interfering energy than target energy. See Fig. 5-23b.

After limiting, the residual signal plus interference is passed through a filter which matches the *signal* (Fig. 5-23c, and 5-23d). Now the frequency domain area of the signal is greater than that of the interference, since the filtered bandwidths are the same, but signal amplitude is greater. In the time domain, the amplitude of the signal is thus greater than that of the interference (area in frequency yields amplitude in time).

Thus this and other limiting schemes work only if the bandwidth of the interference is much greater than that of the target echo signal, such as with wide-band noise jamming (also known as barrage noise) and with impulse jamming (very narrow pulses of jamming signal).

5-5. Threshold-Setting Concept - CFAR

Because of the extreme non-linearity which takes place in the limiting, it is difficult to resolve in Doppler after a scheme of this sort (Chs. 2 and 11).

Since signal is discarded along with interference, these methods degrade signal-to-noise ratios and are not effective where the dominant interference is receiver noise. If the signal-to-*noise* ratio is large, the improvement in signal-to-interference ratio is approximately the ratio of the bandwidth of the interference to the bandwidth of the signal.

$$G_{P(DF)} \approx B_{WB} / B_{NB} \qquad (5\text{-}13)$$

$G_{P(DF)}$ = the process gain to wideband interference with the Dicke-fix (analog limiting) process, assuming S/N and $I/N \gg 1$

B_{WB} = the bandwidth of the interference and the matched bandwidth of the wideband amplifier

B_{NB} = the signal and narrow band amplifier bandwidth

Digital limiting CFAR: Another form of limiting CFAR is sometimes found in radars which transmit digitally coded waveforms, usually as a binary code phase modulated onto the transmitted signal. Section 2-4 shows phase-coded transmit waveforms. Not all, or even most, phase-coded waveforms undergo limiting in their detection processes.

The echo signals can be limited by discarding all the information except the sign of the real and imaginary parts of the signal (the signs of I and of Q — see Ch. 6), as shown in Fig. 5-24. After limiting, the echoes are comprised of N binary bits of I and N binary bits of Q. Without interference, the echoes have the same binary code as the transmitted signal.

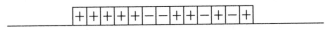

Figure 5-24. 13-bit Barker Sequence after Limiting.

Detection of limited phase-coded signals is described by the M of N process in Eqs. 5-11 and 5-12, where N is the number of bits in the code sequence. The probability of noise matching the code for each bit is exactly 1/2, since only two outcomes are possible. If, for example, the code sequence is 13 bits long and all 13 must match for a detection, the probability of noise meeting this criterion is 2^{-13} or 0.000122 (Eq. 5-12). In other words, on average each 8192 (2^{13}) pulses will produce one false detection The amplitude of the interference, because of limiting, has no effect on the false alarm rate, hence CFAR.

The example of the 13-bit sequence illustrates the biggest problem with digital hard-limiting CFAR. Unless there are a large number of bits in the code, the one-pulse false alarm rates are high. Clearly, the more bits in the code, the lower P_{fa}.

Example 5-6: A radar transmits a 31-bit pseudorandom phase code and the echo is hard limited. The per-pulse detection criterion is that 29 of 31 code segments must match the transmitted pattern. Find the single-pulse false alarm probability and the signal probability for each bit if the single-pulse detection probability must be 0.7. Then find the per-bit signal-to-noise ratio to give these values.

Solution: The first part is easy, merely applying Eq. 5-12 for M = 29, N = 31, and P_n = 0.5 (the probability that one bit of hard-limited noise will match the transmit code bit). The false alarm probability is 2.31×10^{-7}.

The second part, finding the per-bit signal probability for a 0.7 P_D for the pulse is not nearly so easy. Equation 5-11 finds the detection probability for a given signal probability, but there is no exact solution to go the other way, finding P_S given P_D. Therefore, trial-and-error is appropriate. If the per-bit signal probability is 0.9, the pulse detection probability is only 0.39. For a signal probability of 0.95, the detection probability is 0.80. For 0.94, it is 0.715. If the detection probability for one bit were 0.938, the detection probability for the whole pulse (29 of 31) would be about 0.70 (actually 0.699). The *Radar Handbook*, 1st ed. [6] gives a design method, but with a computer or programmable calculator, iteration is probably easier and is just as accurate.

Figure 5-1 shows that for a noise probability of 0.5, the threshold is 1.2 times the rms noise voltage. For a 0.938 signal probability at this threshold, Fig. 5-2 shows the required per-bit S/N to be about 3 (4.8 dB).

5-6. Further Information

Detection has been treated by many authors. Of the more recent works (if 1973 can be called recent), the classic is [1] As stated in the chapter, it treats non-coherent detection extensively and touches on coherent detection. Fortunately, as shown earlier in this chapter, coherent detection is much easier to estimate than non-coherent.

[1] D.P. Meyer and H.A. Mayer, *Radar Target Detection: Handbook of Theory and Practice*, New York: Academic Press, 1973.

Information on detection can be found in many sources, several of which are cited in the chapter and listed below. Much of the original work can be found in the writings of Marcum [7] and Swerling [8].

[2] M.I. Skolnik, *Introduction to Radar Systems*, 2nd ed., New York: McGraw-Hill, 1980.

[3] G.V. Trunk, Ch. 8 in M.I. Skolnik (ed.), *Radar Handbook*, 2nd ed., New York: McGraw-Hill, 1990.

[4] L.V. Blake, Ch. 2 in M.I. Skolnik (ed.), *Radar Handbook*, 2nd ed., New York: McGraw-Hill, 1990.

[5] J.W. Taylor, Ch. 3 in M.I. Skolnik (ed.), *Radar Handbook*, 2nd ed., New York: McGraw-Hill, 1990.

[6] J.W. Caspers, Ch. 15 in M.I. Skolnik (ed.), *Radar Handbook*, 1st ed., New York: McGraw-Hill, 1970.

[7] J.I. Marcum, "A Statistical Theory of Detection by Pulsed Radar, and Mathematical Appendix," *IRE Transactions*, vol. IT-6, April 1960, pp. 59-267.

[8] P. Swerling, "Probability of Detection for Fluctuating Targets," *RE Transactions*, vol. IT-6, April 1960, pp. 269-308.

6

TARGET ECHO INFORMATION EXTRACTION PART II:
Sampling, PRF Classes, Range, Doppler, Height-Finding

Introduction

In the last chapter, it was shown how radars detect the presence of targets. This chapter explores extraction of other target information, including range, Doppler, and height. The role sampling plays in the extraction of range and velocity of targets is investigated. PRF classes are explored further, and the usage of the different PRF classes is described.

6-1. Sampling

Before investigating target information extraction, sampling must be introduced. Sampling is a process where information is gathered at discrete times only and information at all other times is discarded. All systems which process data digitally must sample, and that includes most modern radars.

Sampling in pulsed and CW radars: Pulsed radars are inherently sampled data systems. Each time the illumination energy hits a target and a portion of the echo is captured by the radar, a *sample* of the target parameters is obtained. The only information available about the target — its position, signal amplitude, and signal phase — comes from these samples. Any change in the target's parameters taking place between samples is unknown to the radar. The radar can be said to continuously monitor the target *only* if position, amplitude, and phase changes are slow enough so that information between the samples can be filled in by interpolation.
  Sampling in CW radars is less obvious, but exists nonetheless in digitally processed CW radars, which most are. Sampling in these systems takes place in the signal analog signal to digital conversion process.

Sampling Criterion: The Nyquist sampling criterion states that in order for information to be fully recoverable after sampling, it must be sampled more than twice for each sinusoidal cycle of information present. The sample rate (f_s) must be greater than twice the highest frequency present (Eq. 6-1). Signals which are undersampled are not lost; they are folded or aliased into the spectrum after sampling, which is defined from $-f_s/2$ to $+f_s/2$. These aliased signals appear at incorrect spectral locations and can interfere with sufficiently sampled signals.

$$f_N > 2 f_{max} \qquad (6\text{-}1)$$

f_N = the Nyquist sample rate

f_{max} = the highest frequency present in the signal sampled

In the recovery of information from samples, *the lowest frequency which matches the samples is the one recovered.* Figure 6-1 shows a sinusoidal signal with sampling. In Fig. 6-1a there are more than two samples per cycle and the Nyquist criterion is met. The lowest frequency which matches the samples is that of the original signal. Thus the signal is fully recoverable from the samples alone.

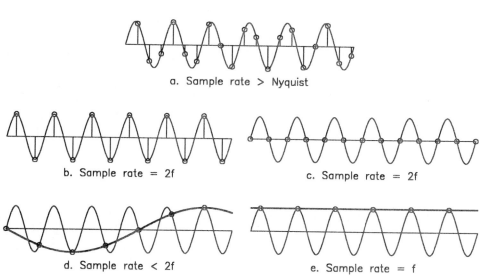

Figure 6-1. Sampling Fundamentals.

In Fig. 6-1b and 6-1c, the signal is sampled exactly twice per cycle. It is obvious that the signal cannot be recovered from the samples of Fig. 6-1c. This is one example of a blind phase. Although it appears that the signal is recoverable in Fig. 6-1b, in fact the Nyquist criterion is not met and the information cannot be totally recovered. The signal frequency is known from the samples, but whether it is a positive frequency (from an inbound target) or negative (from an outbound target) is lost in sampling.

In Fig. 6-1d, the sample rate produces less than two samples per signal cycle. This is known as undersampling. Note that a new sinusoid at a frequency lower than that of the original fits the samples. This new signal, called an alias, is the frequency reported by a spectrum analyzer using this sample rate. Undersampling results in the loss of information about the signal being sampled and may cause the introduction of unwanted interfering signals.

Finally, in Fig. 6-1e the signal frequency is an integer multiple of the sample rate. The samples indicate that the signal is constant, or DC. Targets whose Doppler shifts are an integer multiple of the sample rate (PRF) are interpreted as stationary. These are blind Dopplers and blind velocities.

Sampling and sample rates: Sampling in the time (oscilloscope) domain is a process of multiplying the wave to be sampled by a sampling waveform (Fig 6-2). The result in the frequency (spectrum analyzer) domain is that multiple copies of the signal spectrum appear. One is the spectrum of the original signal and the others are aliases (Fig. 6-3).

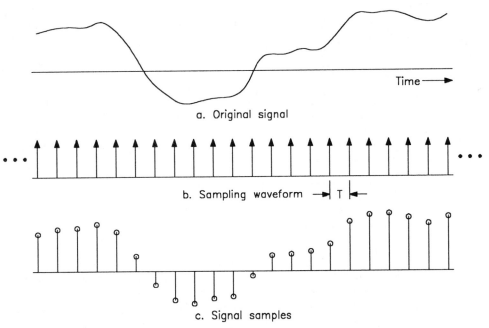

Figure 6-2. Sampling Process.

Multiplication in the time domain produces convolution in frequency, as explained in Ch. 11. The sampling waveform is ideally a uniform train of impulses, a mythical rectangular pulse with infinite amplitude, zero width, and unity area. In the discrete world, an impulse is simply a single sample with unity amplitude. An infinite uniform train of impulses separated by T in the time domain has a spectrum which is likewise an infinite uniform train of impulses separated by frequency $1/T$. A frequency domain impulse at frequency f represents a complex sinusoid $\cos(f) + j \sin(f)$.

Convolution with impulses is simply a matter of copying the wave to be convolved at the location of the impulse. Thus, the spectrum of a sampled wave is an infinite number of copies of the spectrum of the wave sampled, with aliases separated by $1/T = f_s$ (Fig. 6-3).

Changing the sample rate changes the spacing of the spectral aliases. Because of the reciprocal relationship between time samples and frequency samples, low sample rates place the aliases close to one another. *As long an the aliases do not overlap, the information is*

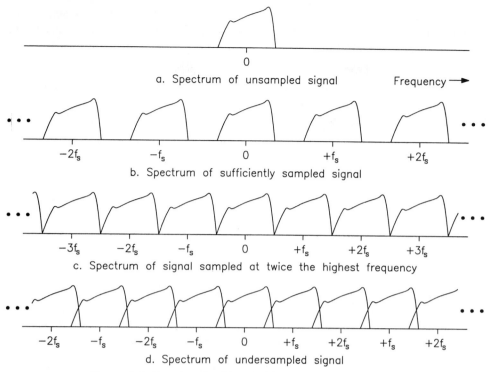

Figure 6-3. Spectra of a Signal without and with Sampling.

recoverable. Figure 6-3a shows the spectrum of an original unsampled signal. Figure 6-3b shows the spectrum of the same signal after sufficient sampling. Figure 6-3c shows the signal sampled at exactly twice the highest frequency present (*not* Nyquist sampled).

In Fig. 6-3d, the highest frequencies are undersampled, and the spectral copies overlap, preventing isolation by filters. Note that even though the lower frequencies in the signal are sufficiently sampled, the higher frequencies are not. The aliases of the undersampled higher frequencies will interfere with the sufficiently sampled lower frequencies, precluding recovery of the signal *even if the high frequencies are not of interest and not meant to be recovered*. The ability to recover the signal from the samples depends on *sufficiently sampling all components of the wave, not just those of interest*.

It is worth saying again: *Signals which are undersampled do not disappear. They remain, but with their information content distorted.* The practical result is that any signal which is undersampled remains in the system as an alias, and may, at its shifted frequency, interfere with other desired signals.

Figure 6-3 shows a broadband, continuous spectrum. Doppler analysis often deals with multiple discrete sinusoids. Consider Fig. 6-4. The signal under consideration, shown in Fig. 6-4a, is the superposition (sum) of three equal-amplitude sinusoids at frequencies of 700, 1300, and 2000 Hz. It is sampled at 3000 samples per second. Before sampling (Fig. 6-4b), the spectrum is composed of six distinct lines (the negative frequencies will be discussed in the I/Q sampling section of this chapter – don't worry about them for now, except to recognize that they are present). As shown earlier, a signal's spectrum after it is sampled is the spectrum before sampling repeated at all integer multiples of the sample frequency (Fig. 6-4c), including the negative frequencies.

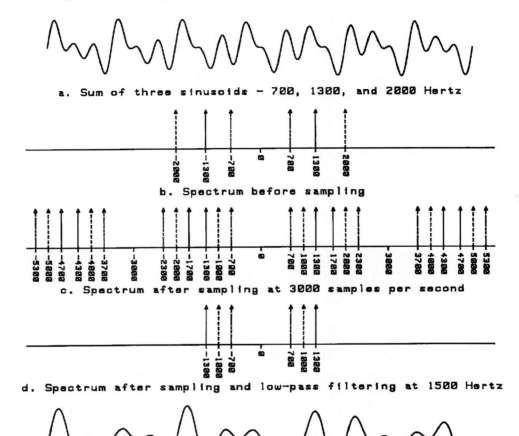

Figure 6-4. A Sampling Example.

The original signal is recovered by filtering the samples with a low-pass filter which cuts off at one-half of the sample rate. Figure 6-4d shows the result after all components above ±1500 Hz are removed by the filter. Three frequencies are left: 700 Hz, 1300 Hz, and *1000 Hz*. The 1000-Hz component, not present in the original signal, is the result of aliasing of the undersampled 2000-Hz component. The new frequency produced by undersampling is called an apparent frequency. The apparent frequency of a 2000 Hz signal sampled at 3000 sps is 1000 Hz. Figure 6-4e shows the time domain signal after sampling and recovery.

The apparent frequency after sampling is a function of the frequency being sampled and the sample rate, and can be found with the following algorithm.

$$\text{Remdr} = f_d \text{ MOD } f_s \qquad (6\text{-}2)$$

IF |Remdr| < $f_s/2$ THEN
    f_a = Remdr
ELSE
    IF Remdr > 0 THEN
        f_a = Remdr $- f_s$
    ELSE
        f_a = Remdr $+ f_s$
    END IF
END IF

f_d = the frequency being sampled (the Doppler shift — Hertz)
f_s = the sample frequency (the PRF — Hertz)
f_a = the apparent frequency of f_d after sampling (the apparent Doppler shift — Hertz)

The MOD operator is the remainder after the first argument is divided by the second. [For example, 8 MOD 3 = 2, and 9.3 MOD 2.5 = 1.8. The calculator equivalent is FRAC(A/B) x B, where FRAC is the fractional part of the quotient.]

$$X \text{ MOD } Y = \text{Fract}(X/Y) \times Y \qquad (6\text{-}3)$$

Fract(X/Y) = the fractional part of X/Y

Example 6-1: A signal consisting of five sinusoids is sampled at 5500 samples per second. The five sinusoids are at frequencies of +2400 Hz, −7500 Hz, +14300 Hz, +22900 Hz, and −31250 Hz. Find the frequencies recovered after sampling.

Solution: One can solve this problem either with algorithm (equation) 6-2, or by simply using common sense to remove multiples of the sample frequency until the remainder falls between $-f_s/2$ and $+f_s/2$, which are the limits of signals which are Nyquist sampled and between which all recovered frequencies must fall.

The +2400-Hz signal is Nyquist sampled and is recovered as +2400 Hz. The −7500-Hz signal is recovered as −2000 Hz (−7500 + 5500). The +14300 Hz signal is recovered as −2200 Hz (14300 − 5500 three times). The +22900-Hz wave is recovered as +900 Hz (22900 − 5500 four times). The −31250-Hz wave is recovered as +1750 Hz (−31250 + 5500 six times).

Information recovery from samples: Information is recovered from its samples by isolating the spectrum of the original signal through filtering. See Fig. 6-5. As long as the original spectrum can be isolated, the information is recoverable. If the aliases overlap the spectrum, recovery is not possible.

The information can theoretically be recovered if it was sampled at the Nyquist rate. However, sample rates higher than Nyquist (oversampling) make information recovery

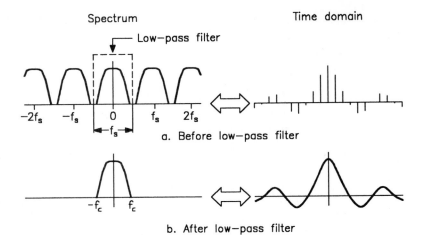

Figure 6-5. Information Recovery from Samples.

easier, since it can be done with simpler low-pass filters. Consider the compact disk (CD) audio recording system, spectra of which are given in Fig. 6-6. A sample rate of 44,100 sps is used on information containing frequencies up to 20,000 Hz. Figure 6-6a shows the sampled spectrum with the low-pass filter necessary to recover the signal. This filter must have a very low response at frequencies of 22,050 Hz or higher to prevent aliases from distorting the information. If the music is to be undistorted, the filter cannot start to roll off until above 20,000 Hz. To start rolling off at 20,000 Hz and attenuate perhaps 90 dB above 22,050 Hz requires very steep skirts and a very complex and expensive filter. The penalty for using a lesser filter design is that it attenuates music frequencies of less than 20,000 Hz and distorts the music.

In many CD players, a process called oversampling is implemented, whereby a mathematical algorithm effectively doubles the sample rate (see Sec. 11-10). With a new sample rate of 88,200 sps, the filter required is of a much less stringent design (Fig. 6-6b).

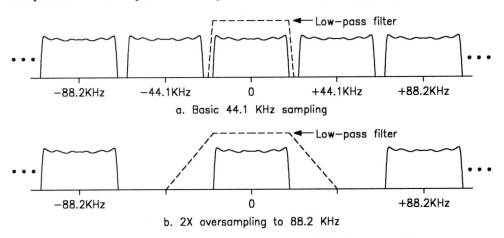

Figure 6-6. Effect of Oversampling on Information Recovery Filter.

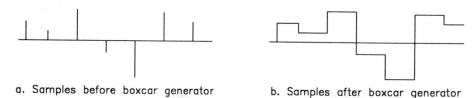

a. Samples before boxcar generator   b. Samples after boxcar generator

Figure 6-7. Boxcar Generation.

One problem with using a straight low-pass filter for information recovery from narrow samples is that the samples themselves contain little energy. Low-pass filtering *averages* the samples and produces an output signal with little amplitude. For example, 1.0 V, 1.0-μs pulses at 1000 pps average to only 0.001 V. Because of this, the samples are usually applied to a sample-and-hold circuit before the low-pass filter. This circuit, sometimes called a boxcar generator, stretches the samples to fill the entire inter-sample period, thus raising their average value and raising the amplitude of the filter output (Fig. 6-7). Boxcar generation is in itself a low-pass filtering technique and is used extensively in radar signal handling. See Ch. 7 for further uses.

Radar information rates: Radar information exists at several different rates, and two levels of sampling take place, shown in Fig. 6-8. First, individual targets are sampled at a rate consistent with the information about them which is to be recovered. Second, the search volume is sampled at a rate which resolves targets (in range) to the level desired. Usually the highest information rate and hence the highest sample rate involves range resolution.

The sample rate for individual targets is the PRF. The basic criterion for setting the PRF is the rate at which an individual target can change. The most rapid change for most targets is the phase shift from hit to hit caused by the Doppler shift. Radars which recover all information about the Doppler shift must sample at a rate equal to at least twice the highest Doppler frequency present. This form of sampling is called high PRF, or pulse Doppler. Later discussions in this chapter and Ch. 12 amplify high PRF definitions and uses.

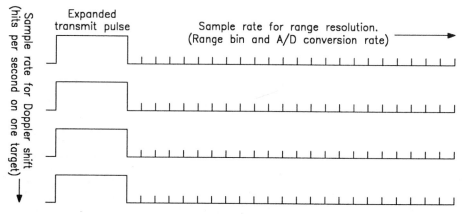

Figure 6-8. Radar Information Sampling Rates.

6-1. Sampling

Recall from Ch. 1 that PRF cannot be increased without limit to accommodate Nyquist sampling for Doppler. Range is also a consideration, and higher PRFs severely restrict the recovery of ranging information. The range/Doppler trade-off is addressed later in this chapter.

Note: There are techniques whereby the Doppler shift of undersampled targets can be determined, so long as the number of target echoes in a particular time bin is limited. The only way, however, to do *unrestricted* measurements of Doppler shift is to Nyquist sample the targets. See the discussions of medium PRF (MPRF) in Secs. 6-3 and 6-4.

For Doppler processing, only one range bin at a time is examined. For example, a Doppler process may involve 16 consecutive hits in range bin 13. Doppler processes for signals in range bins 12 and 14, and so forth, are entirely separate and independent of those for bin 13. Search radars, which usually process for all bins, may have to perform hundreds of Doppler processes for each look. Tracking radars most likely process only a few bins.

The second sampling rate is the rate at which information from the echo signal space is gathered and stored. In search radars and some tracking radars, this type of data gathering is done at the *range bin rate*, which usually matches the range resolution information rate instead of the Doppler shift. A rule of thumb is that range resolution is 150 m for every 1.0 μs of processed pulse width, corresponding to a bin rate and sample rate of 1.0 MHz. In radars which use digital signal processing, this is the rate at which the A/D converter operates.

Generally speaking, then, the PRF samples for Doppler, and the range bin rate samples for range resolution.

Example 6-2: A radar (no pulse compression) has a PRF of 3500 pps and divides its interpulse period into 512 equal range bins plus a dead time of 55 μs between the end of the last range bin and the start of the transmit pulse (neglect the transmit pulse width). Find:

a. The range resolution if the bin width matches the resolution
b. The probable pulse width
c. The approximate receiver bandwidth
d. The A/D conversion rate
e. The sample rate for range resolution
f. The sample rate for Doppler
g. The maximum Doppler which can be measured unambiguously

Solution: The PRF of 3500 gives an interpulse period of 285.7 μs. Subtracting 55 μs leaves 230.7 μs for range bins. Dividing by 512 gives a bin width of 450.6 ns.

a. If the range resolution matches the bin width of 450.6 ns (the most common practice), the range resolution is the range equivalent of 450.6 ns, or 68 m (Eq. 1-7).

b. The most probable transmit pulse width, without pulse compression, equals the bin width, or 450 ns.

c. The receiver bandwidth for a 450 ns pulse is, from Eq. 2-29, approximately 2.22 MHz. Receiver bandwidths are usually actually about 1.2 times the reciprocal of the pulse width, or 2.66 MHz.

d. There is one A/D conversion for each range bin. The conversion rate is therefore the reciprocal of the bin width or 2.22 MHz. There are actually two conversions per bin, one for I and one for Q. They are normally done in two separate channels, each with a rate of 2.22 MHz.

e. The sample rate for range resolution equals the bin rate, 2.22 MHz.

f. Doppler processing operates on signal from several consecutive occurrences of the same range bin. Therefore the sample rate for Doppler is the rate at which a specific range bin (e.g., bin 334) is generated. This is the PRF, 3500 samples per second.

g. The maximum Doppler which can be measured unambiguously is one-half the PRF, from the Nyquist sampling criterion, or 1750 Hz.

PRF classes: As shown in Ch. 1, there are three PRF classes, based on the sampling criteria for range and Doppler. Range and Doppler for the three classes is shown for a single target in Fig. 6-9. In range, the rectangular pulses represent transmissions and the shorter triangular pulses are the signal echoes from the target. The center line is the true range to the target. In Doppler, the taller trapezoidal pulses are aliases of the clutter echo spectrum. In stationary radars, these aliases occur at zero Doppler shift and at integer multiples of the PRF. In moving radars, the clutter spectrum is offset by the radial clutter velocity. Again, the shorter triangular shapes represent the Doppler shift of a target and the center line is the position of the target's true Doppler shift. The PRF classes are discussed later in this chapter.

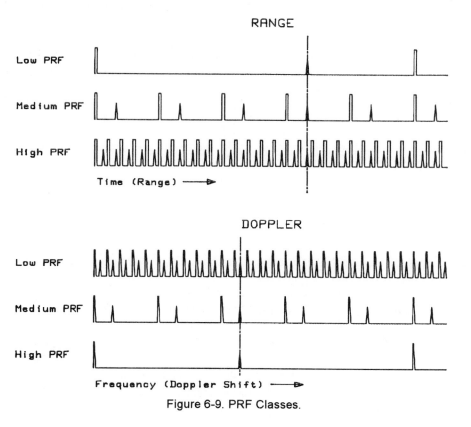

Figure 6-9. PRF Classes.

6-1. Sampling

— *Low PRF (LPRF):* In low PRF systems or modes, the transmitted pulses are far enough apart in time so that all target echoes return from the last pulse return before the transmitter fires again. Ranging is unambiguous.

Doppler shift is almost always undersampled at low PRF. As a result, few low PRF radars/modes can recover the target's Doppler shift. They may be able to sense the fact that the target was moving, but are unable to tell how fast.

A typical low PRF for a moderate range system is 500 pps. This PRF allows unambiguous ranging to 300 km. At X-Band, the maximum Doppler shift which is sufficiently sampled is 250 Hz, corresponding at X-Band to a target closing at 3.75 m/s, which is 8.4 mph or Mach 0.011.

— *High PRF (HPRF):* High PRF occurs when the sample rate is fast enough to sufficiently sample the Doppler shift of the target, in other words the PRF is greater than twice the highest possible target-induced Doppler shift. High PRF is also known as pulse Doppler.

In most cases, the high PRF pulses occur at such a short PRI that only very short range target echoes return before the transmitter fires again. Therefore range is highly ambiguous and recovery of target range at high PRF is problematical.

A typical high PRF at X-Band is 300,000 pps, which Nyquist samples target closing at a little over Mach 5. Its maximum unambiguous range is 500 m.

— *Medium PRF (MPRF):* A medium PRF mode undersamples in Doppler, but not badly so. It is too high for unambiguous ranging of most targets, but not as much so as high PRF. In other words, both range and Doppler are said to be "moderately ambiguous" to this PRF class.

It would seem that because of ambiguities in both range and Doppler, this PRF class would have relatively limited use. In fact, as radar data processors become more capable and it becomes possible to resolve the ambiguities (deghost), this PRF class takes on considerable usefulness. Along with pulse compression, it allows designers to solve the detection/range/Doppler/range resolution problem (detection requiring high energy, and hence high PRF and long pulses, ranging requiring low PRF, Doppler measurement requiring high PRF, and range resolution requiring short pulses). MPRF modes are now present in most modern military airborne radar systems and are beginning to show up in other applications as well. MPRF and deghosting are discussed later in this chapter.

A typical X-Band medium PRF is on the order of 15,000 pps, resulting in a maximum unambiguous range of 10 km and a maximum unambiguous closing velocity of 112 m/s (252 mph).

Example 6-3: A radar operates in the S-Band at 2.92 GHz. The design maximum range is 250 nmi and the design maximum closing velocity is Mach 4.0 at sea level. The range resolution is 100 m. Find:
  a. The minimum PRF for Nyquist sampling
  b. The maximum PRF for unambiguous range recovery
  c. The maximum unambiguous range at the high PRF
  d. The maximum unambiguous target radial velocity at the low PRF
  e. The A/D converter rate

Solution: The minimum high PRF must account for Nyquist sampling. The maximum low PRF must unambiguously range. The A/D converter rate is determined by the range resolution cell size.

a. Mach 1.0 at sea level is approximately 338 m/s. Mach 4.0 is therefore 1352 m/s. The Doppler shift for this velocity at 2.92 GHz is 26,320 Hz. The minimum sample rate is double this value, or 52,640 samples per second. (*Note:* In practice, targets are normally somewhat oversampled to account for imperfect filters. A factor of 1.28 is commonly used. With this added criterion, the sample rate (PRF) would be 67,400 samples per second.)

b. 250 nmi is 463,000 m (1852 m per nautical mile). The round-trip propagation time is 3.087 ms (3×10^8 m/s propagation velocity). This means that once the radar transmits, it cannot transmit again until at least 3.087 ms have elapsed. This gives a PRI > 3.087 ms, or a PRF < 324 pps.

c. A PRF of 52,640 pps corresponds to a PRI of 19.0 μs. The maximum unambiguous target, then, must have a propagation time of less than 19.0 μs. This corresponds to a range of 2,850 m.

d. A PRF of 324 pps can Nyquist sample a maximum data rate of 162 Hz. Using 162 Hz as the Doppler shift, the maximum unambiguous velocity is 8.32 m/s (corresponding to 18.6 mph — considerably slower than a good human sprinter).

e. If the range resolution is 100 m, the A/D converter must capture one sample set for each 100 m of range. The propagation time for 100 m is 0.667 μs, which is the time between A/D conversions. The rate is therefore 1.50 MHz.

6-2. Direct and I/Q Sampling

Doppler shift in pulsed radars manifests itself as the change in the phase of target echo signals from hit to hit (Secs. 1-3 and 1-5). It follows that in order to recover the Doppler shift, the system must not only sample the target at a sufficient rate, but must also measure and record the phase of each received echo. Two methods are available for gathering samples of the target's position and amplitude, direct sampling, and *I/Q* sampling.

Most non-radar and non-communication digital signal process direct samples signals in the time domain. Direct time sampling occurs when the signal amplitude is "frozen" at regular intervals and these sampled-and-held voltages are either digitized or processed in analog form.

The spectrum computed from direct samples is symmetrically two-sided, containing both positive and negative frequencies with a negative spectral component matching each positive component. If a sinusoidal wave is thought of as a rotating vector, it is not possible to discern the direction the vector rotates with direct sampling. Thus the *sense* of the Doppler shift is lost; that is, the information as to whether the target was closing or moving away is not available. The mathematical reasoning for this phenomenon is found in Euler's identities for cosine and sine.

$$\text{Cos } 2\pi f t = 1/2 \ e^{j2\pi f t} + 1/2 \ e^{-j2\pi f t} \qquad (6\text{-}4)$$

$$\text{Sin } 2\pi f t = -j/2 \ e^{j2\pi f t} + j/2 \ e^{-j2\pi f t} \qquad (6\text{-}5)$$

Each of the exponentials on the right side of the equations represent a rotating vector, completing *f* revolutions per second. The terms with the positive exponent rotate counterclockwise and the negative exponent terms rotate clockwise. The spectra of the cosine and sine are shown in Fig. 6-10. It can easily be seen that the vector sum of the cosine's components is totally real and the spectrum of the sine is totally imaginary. Thus in both cases, the direction of rotation is lost. This is the consequence of direct sampling either the real or imaginary amplitude of the signal.

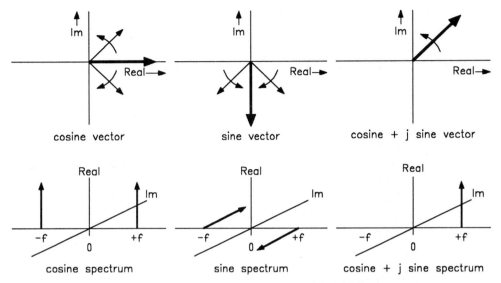

Figure 6-10. Cosine and Sine Identities and Spectra.

If the cosine and sine are combined as shown in Eq. 6-6, however, the spectrum is one-sided and a signal where both its real (cosine) and imaginary (sine) components are known has its direction of rotation (positive or negative frequency) determined (see Fig. 6-10, right).

$$\text{Cos } 2\pi ft + j \text{ Sin } 2\pi ft = e^{j2\pi ft} \tag{6-6}$$

Thus cosine $+j$ sine sampling allows us to determine the direction of rotation and separate inbound from outbound targets. The process of recovering both real and imaginary signal component is known as I/Q demodulation, or I/Q sampling. I stands for in-phase and is the cosine (real) component. Q stands for quadrature, and is the sine (imaginary) component.

In I/Q recovery the signal is multiplied by a reference wave of the same frequency and at the reference phase (the COHO – Sec. 2-1). Figure 6-11 is a block diagram of the process and Fig. 6-12 gives samples of the waveforms.

In the I channel, the echo signal is multiplied by the COHO, which at zero phase represents the cosine reference.

$$v_I(t) = V_S \text{Cos}[2\pi ft + \phi(t)] \times V_C \text{Cos}[2\pi ft] \tag{6-7}$$

$V_S \text{Cos}[2\pi ft + \phi(t)]$ = the target echo voltage

$V_C \text{Cos}[2\pi ft]$ = the COHO voltage – the phase of the COHO is by definition zero

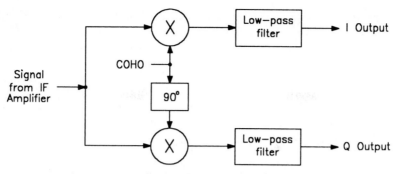

Figure 6-11. I/Q Demodulation Block Diagram.

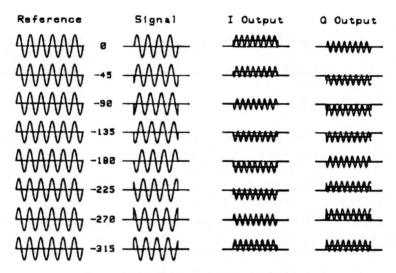

Figure 6-12. I/Q Demodulation Waveforms.

f = the frequency of both the echo and the COHO

$\phi(t)$ = the Doppler phase, the echo's instantaneous phase as a function of time – 360° phase shift is one Doppler cycle

By use of the trigonometric identity for the product of two cosines ($\text{Cos}A\ \text{Cos}B = 1/2\ \text{Cos}(A+B) + 1/2\ \text{Cos}(A-B)$, and setting V_C to 2, Eq. 6-7 becomes

$$v_I(t) = V_S \text{Cos}[4\pi ft + \phi(t)] + V_S \text{Cos}[\phi(t)] \qquad (6\text{-}8)$$

The first term of Eq. 6-8 is a sinusoid having double the frequency of the signal and COHO. The second term is the average (or DC) value of the product, and represents the signal voltage times the cosine of the signal phase angle. The first term is removed by the low-pass filter. The second term becomes the I component of the received echo.

6-2. Direct and I/Q Sampling

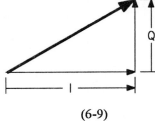

Figure 6-13. Phasor Recovery from *I* and *Q*.

$$v_I(t) = V_S \mathrm{Cos}[\phi(t)] \qquad (6\text{-}9)$$

In like manner, the signal echo is multiplied by the COHO shifted +90° (which is Sin($-2\pi ft$)). Applying the appropriate trigonometric identity [Cos$A$ Sin($-B$) = 1/2 Sin($A-B$) $-$ 1/2 Sin ($A+B$)] and filtering the double-frequency component, the Q signal is recovered, representing the signal voltage times the sine of the signal phase.

$$v_Q(t) = V_S \mathrm{Sin}[\phi(t)] \qquad (6\text{-}10)$$

The two signals, *I* and *Q* completely represent the echo signal, with *I* being the *real* component and *Q* being the *imaginary*. The signal processor, having the in-phase and quadrature signal components, can reconstruct the signal magnitude and phase (Fig. 6-13).

$$V_S^2 = V_I^2 + V_Q^2 \qquad (6\text{-}11)$$

$$\phi = \mathrm{Tan}^{-1}(V_Q / V_I) \qquad (6\text{-}12)$$

Note: The arctangent must accept both *I* and *Q* as arguments if the full 360° of phase is to be recovered. This is done by the FORTRAN or C++ ATAN2 function. Most conventional single-argument arctangent functions, such as are found in calculators and most computer languages produce results only from $-90°$ to $+90°$ and must be corrected.

The *I* and *Q* voltages for eight consecutive hits on an outbound target, with one Doppler cycle in the eight hits, are shown in Fig. 6-14. Note that by using Eqs. 6-11 and 6-12 on this data, a vector rotating clockwise can be recovered. This vector represents a negative frequency of one cycle in the time necessary for the eight samples. If, for example, the PRF were 1000 pps, the eight samples take 8 ms. One cycle in 8 ms is 125 Hz, which is the apparent Doppler shift for this data.

A spectrum computed from *I/Q* samples is one-sided, preserving the direction of rotation of a signal vector. The spectrum of an inbound target with I/Q would show only a positive Doppler shift, whereas with direct sampling, both positive and negative Doppler shifts show. Likewise, an outbound target with I/Q would show only a negative Doppler shift, instead of both positive and negative with direct sampling. See Fig. 6-15 for an example of spectra calculated from direct and *I/Q* sampling.

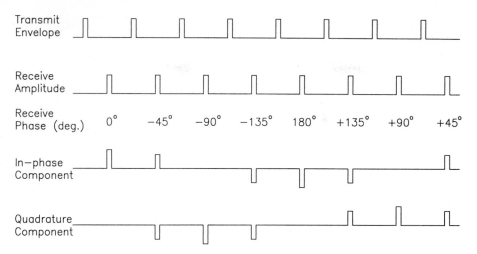

Figure 6-14. *I* and *Q* Signals for an Outbound Target (One Range Bin Only).

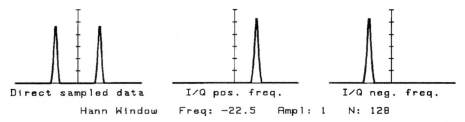

Figure 6-15. Spectra of Direct and *I/Q* Sampled Waves.

6-3. Ranging

Ranging determines how far from the radar the target is located. Range is found by measuring the time from a unique feature of the illumination waveform to the same feature on the echo. Since the velocity of propagation is known, range is found from

$$R = \frac{c\, T_p}{2} \qquad (6\text{-}13)$$

R = one-way range from radar to target
c = the velocity of propagation (see Table 1-2)
T_p = the total round-trip propagation time

With pulsed waveforms, the time reference is usually the centroid of the transmitted wave's envelope, and the propagation time ends at the centroid of the received echo. It is sometimes advantageous to range to the leading edge of the echo pulse, in cases where the target of interest separates inbound from a larger target (missile launch from an aircraft), and in certain types of deceptive ECM (leading-edge ranging – Ch. 7). Where targets of interest separate from larger vehicles outbound (e.g., instrumentation tracking of multi-stage vehicles), trailing-edge ranging is sometimes used.

Four range units are commonly used in radars. The two-way propagation time at sea level for each of the four units is given in Table 6-1. For other units, see Appendix A.

Table 6-1. Range Units and Time Approximations

Unit	Equivalence	Propagation time per unit
Kilometer	1000 m exactly (≈3280.84 ft)	6.6734 μs
Statute mile	5280 ft exactly (≈1609.34 m)	10.740 μs
Radar mile (also know as Data mile)	6000 ft exactly (≈1828.80 m)	12.204 μs
Nautical mile	1852 m exactly (≈6076.12 ft)	12.359 μs

Example 6-4: Measurement shows that the propagation time from the center of the transmission to the center of the echo pulse, corrected to the antenna's position, is 127.349 μs. The local propagation velocity is determined to be 299,714,000 m/s. Find the range from the radar to the target.

(*Note:* In all cases, the time of transmission and the time of echo detection must be corrected to the location of the antenna – propagation in the radar's internal plumbing may be substantial.)

Solution: From Eq. 6-13, the range is found to be 19,084.14 m. Using the data in Table 6-1, this range equals 11.858 statute miles, 10.435 radar miles (also called data miles), or 10.305 nautical miles.

Conditions of ranging: If the transmitted waveform is periodic in time, which is the usual case, three ranging conditions can occur, illustrated in Fig. 6-16. Two of these conditions can result in erroneous ranging. In all three cases in the figure, the arrows show the actual propagation time (range).

In the first condition (Fig. 6-16 top), the echo from a transmit waveform segment is received before the segment is repeated. This condition represents simple pulse-delay ranging and is unambiguous, *if* it is known that the echo is caused by the transmit pulse immediately preceding it. This is the assumed condition in many systems.

Received echoes which arrives at the same time the transmitter is operating are from targets at blind ranges (Fig. 6-16 center). Most monostatic radars cannot sense these targets. An infinite number of blind ranges exist, separated by the range equivalent of the time between transmit pulses.

To find true range, each echo must be paired with the illumination wave which caused it. If the pairing of illumination/echo is not known, ranging is ambiguous. Most ranging systems, in their initial measurement, recover only the propagation time from the last

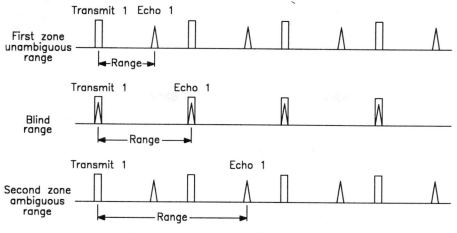

Figure 6-16. Ranging Conditions.

transmit wave, which is the apparent propagation time $- T_A$, which in turn gives a sometimes incorrect range called the apparent range $- R_A$. Figure 6-16 bottom shows this condition.

Range ambiguity and range zones: To identify which illumination pulse causes a particular target echo, the concept of a range zone is needed (Fig. 6-17). A target echo caused by the latest transmit wave and which arrives at the radar before the next transmit waveform period is said to be in range zone 1, where apparent and true ranges are equal. If more than one waveform period but less than two periods elapse before the signal echo arrives from the first waveform, the target is in range zone 2. If more than two periods but less than three elapse, it is in zone 3, and so forth.

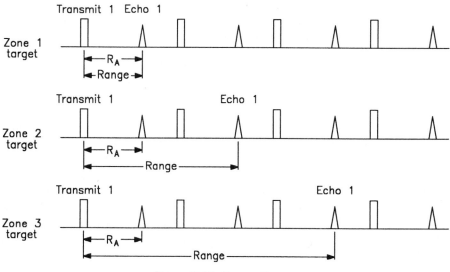

Figure 6-17. Range Zones.

6-3. Ranging

If the transmitted waveform is periodic, propagation times for zone 1 ranging cannot exceed the waveform period less the pulse width (T_U). In many radars, this is the maximum propagation time for unambiguous ranging. In zone 1, apparent range equals true range.

$$T_U = PRI - \tau_E \approx PRI \approx \frac{1}{PRF} \qquad (6\text{-}14)$$

T_U = the maximum zone 1 propagation time
PRI = the waveform period
PRF = the pulse repetition frequency
τ_E = the transmitted pulse width

$$R_u = \frac{cT_U}{2} = \frac{c(PRI - \tau_E)}{2} \approx \frac{c}{2\,PRF} \qquad (6\text{-}15)$$

R_U = the maximum unambiguous range

The following equations give the relationships between apparent propagation time and range, range zone, and true propagation time and range.

$$T_A = T_P \text{ MOD } PRI \qquad (6\text{-}16)$$
$$R_A = R \text{ MOD } [(c\,PRI)/2] \qquad (6\text{-}17)$$
$$T_P = T_A + (n_R - 1)\,PRI \qquad (6\text{-}18)$$
$$R = c\,T_P/2 \qquad (6\text{-}19)$$
$$R = c\,[T_A + (n_R - 1)\,PRI]/2 \qquad (6\text{-}20)$$
$$R = c\,[T_A + (n_R - 1)/PRF]/2 \qquad (6\text{-}21)$$
$$R = R_A + (c/2)\,[(n_R - 1)/PRF] \qquad (6\text{-}22)$$

T_A = the apparent propagation time
T_P = the actual (true) propagation time
R_A = the apparent range
R = the actual (true) range
n_R = the range zone (as defined above)
PRF = the repetition frequency of the illumination waveform
PRI = the interval (period) of the transmit waveform (the reciprocal of the PRF)

The MOD operator is defined in Sec. 6-1.

Range deghosting: Targets interpreted as being at incorrect ranges because of ambiguity are called range ghosts. The process of determining the true propagation time and range from the apparent propagation times and ranges is range deghosting. Equation 6-22 shows that a target which is outside range zone 1 shows different apparent ranges for different PRFs (Fig. 6-18). Most, but not all, radars which deghost in range use this phenomenon in their processing.

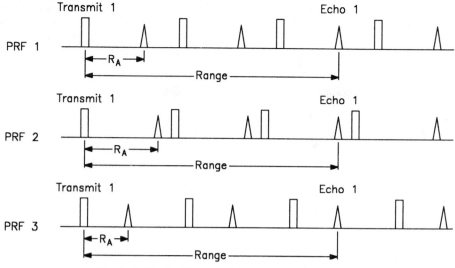

Figure 6-18. Ambiguous Range and PRF Change.

Figures 6-19 and 6-20 show a two-step method of range deghosting. The first step is to illuminate the target space using a number of different PRFs and test each range bin for detections (Fig. 6-19). In this example, the PRFs and range bins are chosen so that each range bin is 1.0 km wide and that there are an integer number of range bins in each waveform period.

The zone 1 target (target 1 at 7-km range) shows the same apparent range regardless of the PRF. Targets in other zones (target 2 at 26 km, and target 3 at 42 km) have different apparent ranges for different PRFs. In the example, eight pulses of each of five PRFs are used. Note that target 1, which is zone 1 to all PRFs, appears after the first pulse of each look. Echoes from targets 2 and 3 (both outside zone 1 at all PRFs) are not detected until after the transmitter has fired for the second, third, fourth, or fifth time, depending on the target and the PRF.

After the five looks of eight pulses each, decisions are made from the detections about which range bins contain targets. A 3 of 8 scheme is used here, and it finds all three targets for each PRF. A summary of the five PRFs is thus made. For example, a 3 of 8 scan of PRF 1 shows targets in bins 7, 11, and 12, for apparent ranges of 7, 11, and 12 km. Targets are detected using PRF 2 at apparent ranges of 2, 6, and 7 km, and so forth. At this point, the system does not know which target is in which range bin — it only knows that detections were made in several bins. In other words, it doesn't yet know target 1 from target 2 from target 3. Note that for PRF 5 (11,530 pps), target 2 is at a blind range, and that PRF 2 and PRF 3 give identical results even though two of the targets exchanged positions. The targets are identified by number only for the convenience of the reader.

To prevent having to wait for all echoes before starting the next PRF group, the illumination and receiver frequencies may be changed between groups. This also may help in Doppler deghosting, discussed later in this section.

In the second step (Fig. 6-20), the detection summaries from the first step are organized as shown. Those from one period are repeated enough times to extend the range from zero to the maximum instrumented range of the radar, placing the detections at each possible ambiguous range. The extended range bin system is then examined bin by bin from range zero to the maximum instrumented range. In this example, if detections are present in a particular range bin for three of the five PRFs (3 of 5 detection), a target is assumed to be in that bin. Targets are thus declared in bins 7, 26, and 42. If each bin is 1 km wide, there are three targets at 7-, 26-, and 42-km range.

This scheme works nicely if the number of detections is limited. As more detections occur, the probability increases that the deghosting criteria will be met erroneously, producing false targets at ranges where no targets exist. Even with our limited number of targets, several cases came within one detection of producing false alarms. The solution is to increase the number of PRFs, being very careful with their values. Computer simulations of various PRF combinations are helpful in this PRF design.

It is advantageous to change the transmitted frequency at the same time the PRF is changed, so that residual targets from one PRF will not contaminate the measurement made by the next sample rate.

Approached analytically, the process of ambiguity resolution involves writing Eq. 6-22 (or Eq. 6-18) for each PRF and solving simultaneously for true range. An examination of the equation shows a problem. The knowns in the equation set are the apparent range (measured for each PRF), the PRF, and the velocity of propagation. The unknowns are true range and the range zone. For one PRF there is one equation and two unknowns. For two PRFs, there are two equations and three unknowns (the true range and the two range zones). No matter how many PRFs are used, there is always one more unknown than equations.

According to classical algebra, this system is unsolvable and we can never find true range. This is not necessarily true. The range zones are natural numbers (integers greater than zero), and the true range is limited to the instrumented range of the radar. Within these restrictions, it is possible to choose PRFs which resolve ambiguities with a high probability of being correct. It is the task of the radar designer to assure that there are enough PRFs and that the PRF values are such that for a reasonable number of targets the probability of zone error is very low. The goal of PRF design is to have only one *reasonable* solution. For example, the data may give a target as being at 12 miles or 200,012 miles — only the 12 mile solution is reasonable and for all practical purposes it represents the correct range. Numerous targets and few PRFs greatly increase the probability that more than one reasonable solution will exist, and false alarms will occur. The analysis of this is beyond the scope of this book, but lends itself well to numeric solution using computer simulations.

Other methods of range deghosting exist. One much simpler scheme is for single target tracking radars is described later in this section. These radars and modes have a much easier time with ambiguities than search systems because they deal with only one target and a few range bins at a time.

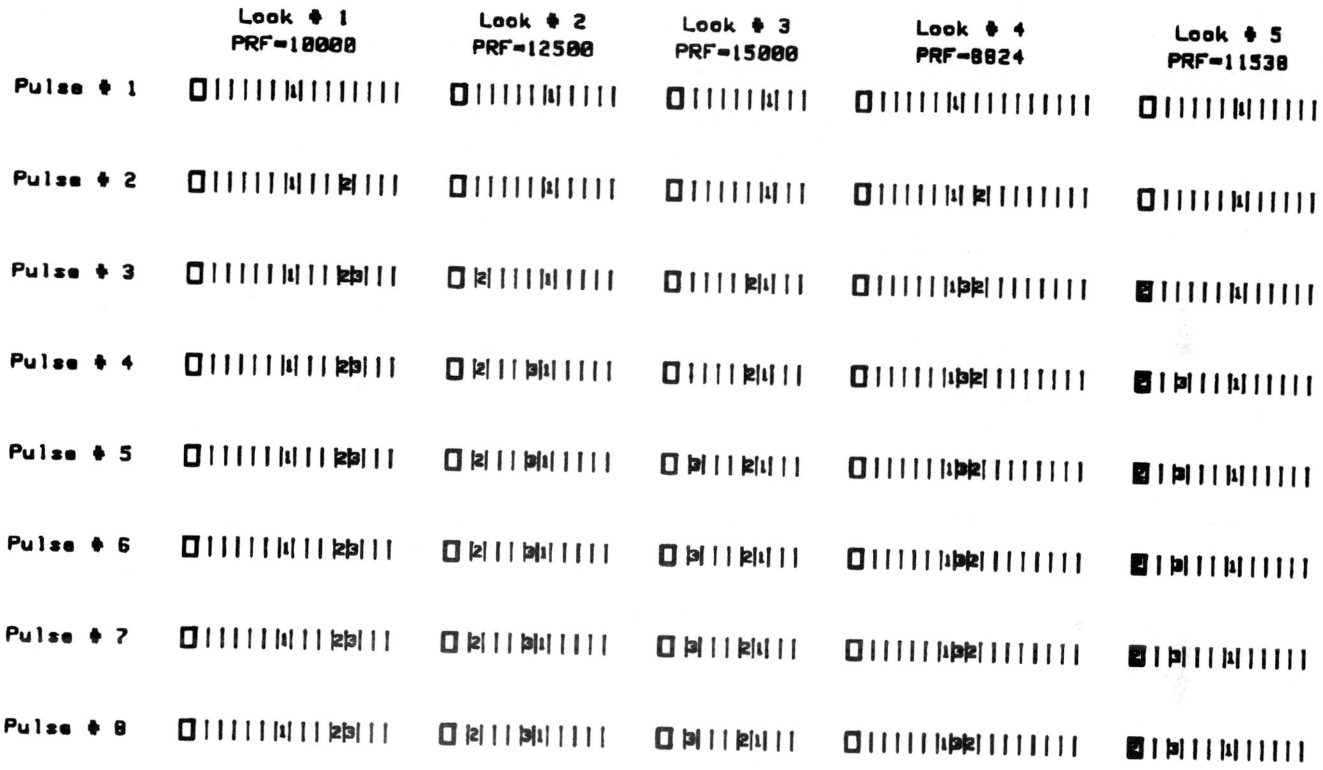

Figure 6-19. Range Deghosting Step 1.
The rectangles are transmit pulses. The range bins are 1 km wide. Target 1 is at 7 km range, target 2 at 26 km, and target 3 at 42 km.

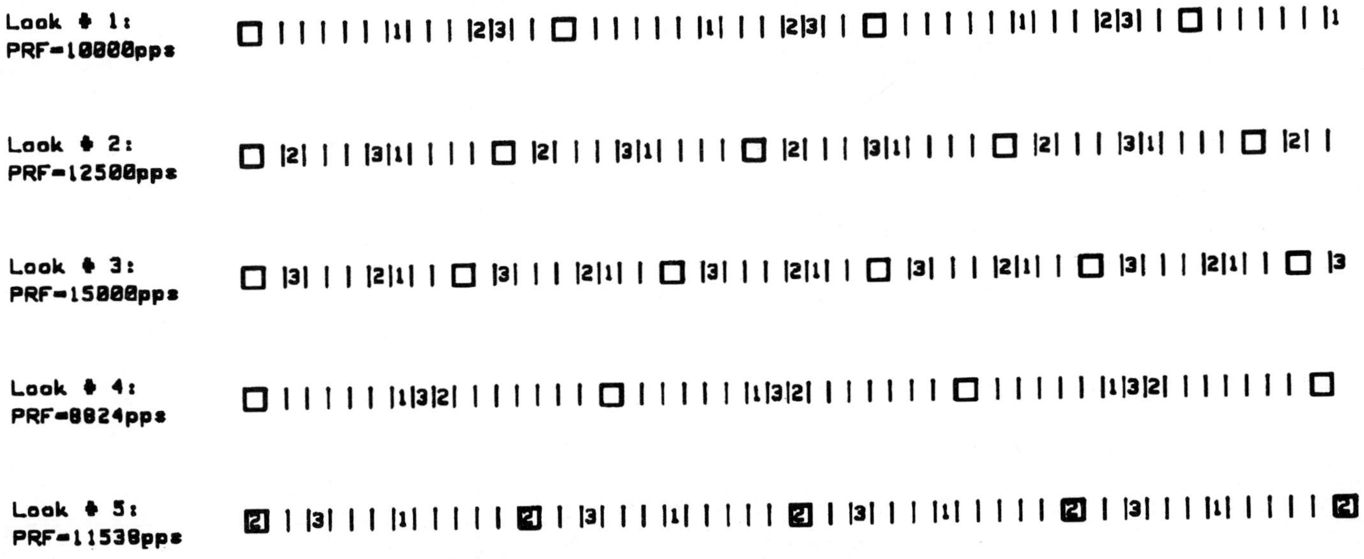

Figure 6-20. Range Deghosting Step 2.
The rectangles are transmit pulses. The range bins are 1 km wide. Target 1 is at 7 km range, target 2 at 26 km, and target 3 at 42 km.

Effective PRF for ranging: If multiple PRFs are used to find the true range of ambiguous targets, a first approximation of the effective PRF is the greatest common factor of the PRFs.

$$PRF_{RE} = \text{GCF}(PRF_1, PRF_2, PRF_3, \ldots) \tag{6-23}$$

PRF_{RE} = the effective PRF for ranging (the maximum range which the PRF combination can unambiguously determine is found from Eq. 6-3 using this PRF value)

Example 6-5: At a PRF of 1250 pps a radar observes a target at 104 km. At PRFs of 1000 pps and 1500 pps, the radar observes targets at 14 km and 64 km, respectively. If the target did not move significantly between observations, what is its shortest possible true range? Find also the greatest unambiguous range for these three PRFs.

Solution: A trial-and-error solution is used. A PRF of 1250 pps has 120.0 km between transmit pulses (range equivalent of 1/1250 s). Equation 6-22 shows that a target observed at 1250 pps with apparent range of 104 km can actually be at 104 km, 224 km (104 + 120), 344 km (104 + 2*120), 464 km, 584 km, 704 km, 824 km, and so forth. Likewise, a target with apparent range of 14 km at 1000 pps (150.0 km between transmit pulses) can actually be at 14 km, 164 km, 314 km, 464 km, 614 km, 764 km, and so forth. An apparent range of 64 km at 1500 pps (100.0 km between transmits) can actually be 64 km, 164 km, 264 km, 364 km, 464 km, 564 km, 664 km, 764 km, and so forth. The shortest range where the target could be for all three PRFs is 464 km.

Equation 6-23 is used to find the greatest unambiguous range of the three PRFs. The prime factors of 1250 are 2, 5, 5, 5, and 5. Those for 1000 are 2, 5, 5, 5, and 4. For 1500, they are 2, 5, 5, 5, 3, and 2. The greatest common factor is 2 x 5 x 5 x 5 = 250. The maximum unambiguous range corresponding to 250 pps is 600 km. Note that if the process used to find the range in the example is carried further, a match between the three PRFs and their target observations is also made at 1064 km (464 + 600). For many, but not all, radars this would be unreasonable, being too far to return enough signal for detection.

Blind ranges: If the arrival of a signal echo and a transmission occur simultaneously the echo is interpreted as being at zero range. In pulse radars, the receiver is blanked by the transmitter at this time. This is a range eclipse, or blind range, and occurs where:

$$R_B = \frac{n_R \, c \, PRI}{2} \tag{6-24}$$

R_B = a blind range

n_R = the range zone (any integer > 0)

Radars are blind for at least the pulse width. Since close-in clutter also masks targets, they are generally blind for longer times. Range deghosting will resolve some range eclipses (note target 2 with PRF 5).

Minimum range: To be seen by pulsed radars, targets must be beyond a certain minimum range, determined by the transmit pulse width (τ_E), the duplexer recovery time (τ_D), and the recovery time of other circuits (τ_M) such as antenna phase shifters (Sec. 9-11). With short pulse radars, the duplexer recovery usually limits, whereas pulse width sets minimum range in long pulse systems. Phased array radar minimum range is usually set by the phase shifters. The minimum range is expressed below.

$$R_{min} = c\,(\tau_E + \tau_{DM}) \tag{6-25}$$

R_{min} = the minimum range for target detection
τ_E = the transmit (expanded) pulse width
τ_{DM} = the greater of τ_D and τ_M
τ_D = the duplexer recovery time
τ_M = the recovery time of other circuits

Range deghosting with single-target track – a special case: If only one target is present, there is simpler method of range deghosting. The system, while maintaining a constant average PRF, transmits a unique PRF jitter pattern and looks for the same pattern on the echoes. This method is illustrated in Fig. 6-21, from a ranging machine used in versions of the AN/FPS-16 instrumentation radar. This radar is medium PRF in this mode; the principle applies to both medium and high PRF.

In this example, one set of pulses per PRF cycle (20 Hz in this case, corresponding to a maximum range of 7,500 km (4050 nmi) is coded. The code is that the first pulse is delayed from its normal position 2000 yd (12.2 μs). The second is delayed 4000 yd (24.4 μs) from its normal position, the third 6000 yd (36.3 μs), and the fourth 8000 yd (48.8 μs). In each range interval after the first delay, one range gate is developed at the normal position, and one delayed 2000 yd. A counter counts range intervals. The first detection in the 2000-yd delayed gate tentatively identifies the target's zone. This is called a "find." In the next three intervals, the delayed range gates are placed 4000, 6000, and 8000 yd from their normal position, respectively. If detections are made in all three of these delayed gates, the range zone identified by the "find" is confirmed (called a "verify"). If the target zone is found but not verified the process is repeated. Since the target must be detected in all four delayed gate positions before the range zone is verified, this is a 4 of 4 detection scheme (Ch. 5).

6-4. Target Velocity (Doppler Shift)

When a radio signal reflects from an object which is moving toward or away from the radar, its frequency is shifted, causing the received echo to be at a slightly different frequency from the transmitted energy. This is the Doppler shift, and a target's velocity can be measured by sensing the magnitude and direction of the Doppler shift.

Please refer to Fig. 6-22. The radar transmits on a frequency f_T. The target is moving toward the radar at radial velocity v_R. Radial velocity is the component of target velocity directly toward or away from the radar. By definition, closing velocity is positive velocity. The frequency which would be sensed by a frequency meter in the target is f_{tgt}. If the target echo is considered to be from a signal generator in the target, f_{tgt} is the signal generator's frequency. The derivation of the exact Doppler shift follows the figure.

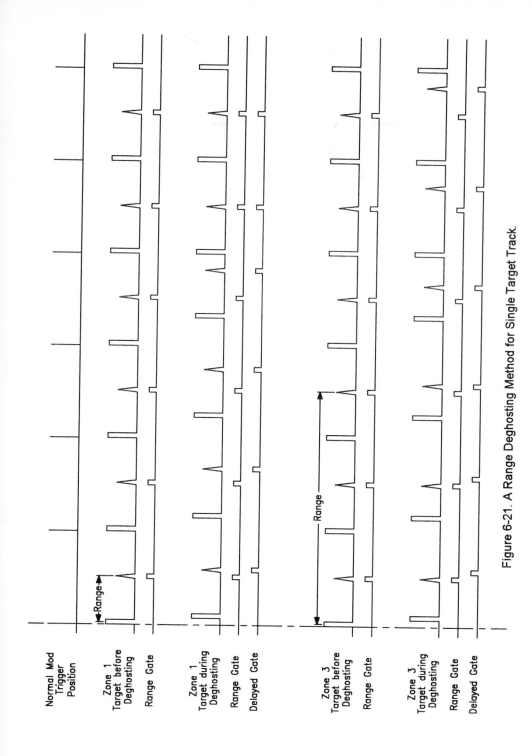

Figure 6-21. A Range Deghosting Method for Single Target Track.

303

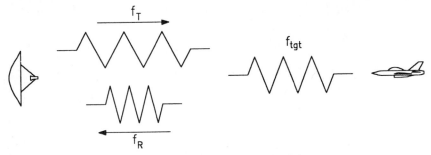

Figure 6-22. Doppler Shift Principle.

$$\Phi_T = 2\pi R / \lambda_T \tag{6-26}$$

Φ_T = the total one-way phase from radar to target (radians)
R = the one-way range to the target (meters)
λ_T = the illumination wavelength at the radar (meters)

If the target's range to the radar is changing with time, the illumination frequency as seen by a frequency meter in the target is

$$f_{tgt} = f_T + \Delta f_{tgt} \tag{6-27}$$

f_{tgt} = the frequency of the illumination seen by the target
f_T = the illumination frequency seen by the radar
Δf_{tgt} = the frequency change caused by the target's motion

The frequency change caused by the target's motion is the cycle equivalent of the time rate of change of total phase.

$$\Delta f_{tgt} = \frac{d\Phi_T}{dt} \frac{1}{2\pi} \tag{6-28}$$

The time rate of change of total phase, $d\Phi_T / dt$, is the derivative of Eq. 6-26.

$$\frac{d\Phi_T}{dt} = \frac{2\pi}{\lambda_T} \frac{dR}{dt} \tag{6-29}$$

But, dR/dt is radial velocity between target and radar.

$$v_R = dR / dt \tag{6-30}$$

Making the proper substitutions into Eq. 6-27 and converting wavelength to frequency ($f = c/\lambda$) gives the frequency of the illumination as seen by a frequency meter in the moving target.

$$f_{tgt} = f_T + f_T \frac{v_R}{c} \tag{6-31}$$

The target can now be treated as a transmitter emitting frequency f_{tgt}, moving at v_R to the radar. Using exactly the same reasoning as Eqs. 6-26 through 6-31, the frequency of the echo as seen by the radar can be found.

$$f_R = f_{tgt} + \Delta f_{rdr} \tag{6-32}$$

$$\Delta f_{rdr} = \Delta f_{tgt} = \frac{d\Phi_T}{dt}\frac{1}{2\pi} \tag{6-33}$$

After a bit of algebra, expressions for the echo frequency and the Doppler shift emerge.

$$f_R = f_T \frac{1 + v_R/c}{1 - v_R/c} \tag{6-34}$$

$$f_d = f_T \left[\frac{1 + v_R/c}{1 - v_R/c} - 1\right] \tag{6-35}$$

Equations 6-34 and 6-35 are the expressions for the exact received frequency and Doppler shift, respectively. Useful approximations result if the radial velocity is much less than the electromagnetic propagation velocity ($v_R/c \ll 1$). *Unless otherwise stated, the approximations will be used throughout this text.*

$$f_R \approx f_T(1 + 2v_R/c) \tag{6-36}$$

$$f_d \approx 2f_T v_R/c \tag{6-37}$$

Since propagation velocity is usually specified in meters per second, it is useful to have conversions for the other common velocity measurements. Table 6-2 gives *approximate* common conversions, with more complete conversion tables found in Appendix A.

Table 6-2. Approximate Velocity Conversions

To convert FROM:	TO:	Multiply by:
Statute miles per hour (smph)	Meters per second (m/s)	0.4470
Nautical mph (Knots)	Meters per second (m/s)	0.5144
Kilometers per hour (km/h)	Meters per second (m/s)	0.2778
Meters per second (m/s)	Statute miles per hour (mph)	2.2369
Meters per second (m/s)	Nautical mph (Knots)	1.9438
Meters per second (m/s)	Kilometers per hour (km/h)	3.6000

Mach 1 at sea level is approximately 338 m/s.

Remember that only the component of target velocity radial to the radar causes a Doppler shift. Note that if the target being observed is in an antenna sidelobe, the Doppler shift is the result of the target velocity radial the axis of that sidelobe. This phenomenon causes much difficulty, particularly with moving radars working against clutter. It is discussed in Ch. 5.

Example 6-6: An orbiting vehicle traveling 7900.000 m/s is tracked by a radar at 5.75000 GHz illumination frequency. Track data gives the horizontal angle between the radar's axis and the velocity vector as 34.774° and the vertical as 11.294°. Find the Doppler error from using the approximation of Eq. 6-37. Assume all numbers given are exact.

Solution: The radial velocity is 6363.463 m/s (Eqs. 1-4 and 1-5). The exact Doppler shift (Eq. 6-35, and c = 299,792,000 m/s) is 244,107.17 Hz. Using the approximation gives the Doppler shift as 244,101.99. The error, even for orbital velocities, is less then 0.0022 %.

Ambiguous and blind Dopplers: Figure 6-23 shows the different conditions found in Doppler measurements, as a function of the Doppler shift and the rate at which the target is sampled.

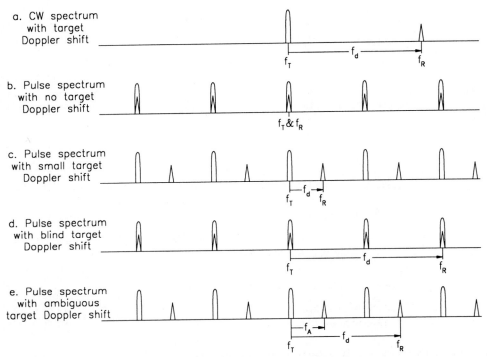

Figure 6-23. Doppler Shift Conditions.

In CW systems, Doppler shift is recovered by simply measuring the frequency displacement of the target echo spectrum from the transmit spectrum. Since the transmit spectrum is (ideally) a single line, the recovery of Doppler shift is relatively straightforward (Fig. 6-23a).

In pulsed systems the transmit spectrum is composed of an infinite number of spectral lines, separated by the sample rate (the PRF). Four spectral conditions are possible in pulsed systems (Fig. 6-23b through 6-23e).

The first condition (Fig. 6-23b) is a target whose range is not changing with time; it is either stationary, is moving at the same rate as the radar, or has a velocity vector perpendicular to the axis of a stationary radar. In this case, the echo and the transmitter are at the same frequency. The echo is at the same frequency as clutter in a stationary radar. In airborne radars, the primary clutter returns are at a different frequency from the transmitter, but the altitude return is at zero Doppler shift.

The second condition, Fig. 6-23c, results if the echo spectral lines are shifted less than half the PRF from the transmit lines, and the Nyquist sampling criterion is met. Recall that the frequency of a sampled wave is extracted as the *smallest frequency span from a received spectral line to the closest transmit line*.

In the third condition (Fig. 6-23d), the Doppler shift equals an integer multiple of the PRF, and the moving target's echo spectral lines fall at the same frequencies as transmit spectral lines and those from stationary target echoes. In most cases, clutter returns and transmitter leakage into the receiver will effectively mask the moving target echo. Also in many systems the Doppler filters attenuate echoes at zero Doppler shift and at integer multiples of the PRF for clutter rejection. Regardless, a moving target at these Doppler shifts will not be detected. This is a blind Doppler, and the associated radial velocity is a blind speed. It should be noted that blind Dopplers can also occur in CW systems which have digital signal processors. Here, the Doppler shift is blind if it is an integer multiple of the signal sample rate. Blind Dopplers and blind speeds are

$$f_B = n\, PRF \tag{6-38}$$

$$v_B = (c\, n\, PRF) / (2 f_T) \tag{6-39}$$

f_B = the blind Doppler shift
v_B = a blind radial velocity
n = any positive integer
PRF = the pulse repetition frequency (or the signal sample rate in CW radars)

Blind Doppler shifts and velocities are a serious problem for low PRF radars. Not only is the radar blind at exact multiples of the PRF, but also for a considerable frequency band around these multiples, so that the moving components of clutter can be rejected (Secs. 4-11 and 4-12). This clutter spectral width may occupy an appreciable fraction of the total Doppler spectrum, and is exacerbated by lower PRFs and higher transmitted frequencies. This is discussed further later in this section.

The final condition is where the moving target's Doppler shift is greater than half the PRF. The Doppler measurement still reports the displacement to the nearest transmit line. As shown in Fig. 6-23e, the Doppler measurement is now ambiguous, with the reported (apparent) Doppler shift being different from the true Doppler shift. The relationship between the apparent and true Doppler shifts is given below.

$$f_A = [(f_d\, MOD\, PRF) - PRF]\, MOD\, PRF \tag{6-40a}$$

or

$$f_A = [(f_d\, MOD\, PRF) + PRF]\, MOD\, PRF \tag{6-40b}$$

Whichever has the smaller absolute value

f_d = the true Doppler shift
f_A = the apparent Doppler shift from aliasing

Doppler deghosting: Doppler shift is measured ambiguously if the Nyquist sampling criterion is not met, as in medium and low PRF radars and modes, but Doppler ambiguities can sometimes be resolved. The technique, similar to that for range deghosting, is to observe the target at different PRFs, record the apparent Doppler shifts from each, and try to work out the ambiguities.

Doppler zones exist, just as do range zones. There is a Doppler zone number, which can be any integer (negative, zero, or positive), associated with each target. The true Doppler shift is

$$f_d = f_A + n_D \, PRF \qquad (6\text{-}41)$$

n_D = the Doppler zone number, an integer which can be negative, zero, or positive

An examination of Eq. 6-41 shows the same problem we had with range; no matter how many PRFs are used, there will always be one more unknown than equations. As with range, if there are enough correct PRFs and not too many targets, a unique solution of all *reasonable* true Doppler shifts can be found.

Figures 6-24 and 6-25 illustrate a trial-and-error technique similar to that used to deghost range. In this example, there are three targets having Doppler shifts of +1500 Hz, +5500 Hz, and −4000 Hz, and the PRFs are 3500 pps, 4500 pps, 5500 pps, 6000 pps, and 6500 pps. Assume for all PRFs that the Doppler bins are 500 Hz wide. (Chapter 11 will show that setting the same Doppler bin width for all PRFs is more difficult than having the same range bin widths.)

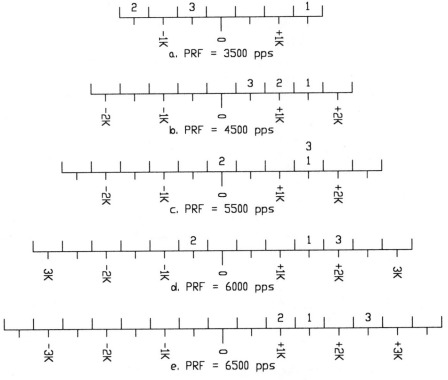

Figure 6-24. Doppler Deghosting — Step 1.

Table 6-3 shows the apparent Doppler shifts at each of the PRFs. The sampling is I/Q and the sign of the frequency is preserved. Figure 6-24 shows the results after detections are attempted in each bin of each PRF. Again, the targets are numbered to make visualizing the problem easier — in the actual analysis it is not known at this point which target produces

Table 6-3. Apparent Doppler Shifts for Deghosting Example.

Actual Doppler (Hz)	Apparent Doppler Shift at PRF (Hz):				
	3500 pps	4500 pps	5500 pps	6000 pps	6500 pps
+1500	+1500	+1500	+1500	+1500	+1500
+5500	−1500	+1000	0	−500	−1000
−4000	−500	+500	+1500	+2000	+2500

which detection. As with range, the frequency may also be changed between PRFs to prevent echoes from one PRF from contaminating the measurements from the next PRF.

Figure 6-25 shows how the results of the spectrum analyses are arranged to resolve ambiguities. Each spectrum is repeated to the highest Doppler frequency analyzed. In this figure, all possible Doppler shifts are shown for each identified apparent Doppler. A vertical

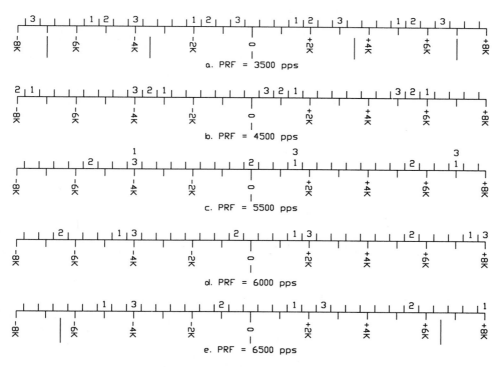

Figure 6-25. Doppler Deghosting — Step 2.

6-4. Target Velocity (Doppler Shift)

five statistics. It must be remembered that, as with range, there will always be one more unknown than equations (PRFs) and that ambiguities still exist. It is up to the radar designer to assure that for a reasonable number of targets, only one *reasonable* solution set exists.

An alternative method is to vary the transmit frequency instead of the PRF, producing apparent Doppler shifts which vary with transmitted frequency rather than sample rate. The disadvantage of this method is that range and Doppler deghosting cannot be done simultaneously, as they can when the PRF is varied. As stated earlier, in many systems both frequency and PRF are changed, first to prevent echoes from one PRF from contaminating those of the next PRF, and second to make Doppler bin widths more convenient.

Frequency agility and Doppler: In some radar systems, the transmit frequency is changed (hopped) from pulse to pulse, look to look, or in some other pattern. Detecting rapidly fluctuating targets is easier than detecting those which fluctuate slowly. The RCS of most targets can be made to fluctuate more rapidly if the illumination frequency is changed rapidly (see Ch. 4). Also, an agile radar is harder to jam than one which operates on a single frequency. If a jammer can "lock" to the radar's frequency, it can concentrate all its jamming power into the radar's bandpass (spot jamming). A randomly agile radar forces most jammers to spread their energy over all the frequencies the radar uses, with the result that less jamming power is within the bandpass of the radar at any instant in time.

Frequency agility, however, complicates the analysis of target Doppler shift. The reason is that the Doppler shift changes with transmitted frequency, and becomes a function of range where the illumination frequency changes within a look. In order to recover Doppler shift, the range may need to be known to an unattainable accuracy. The following development gives the Doppler shift in the presence of frequency agility. *Note:* the relationship developed leads to the approximation of Doppler shift of Eq. 6-37, and requires that the radial velocity be much less than the electromagnetic velocity of propagation. The development of exact relationships is left as an exercise.

$$\Phi_{T2} = 2 \frac{2\pi R}{\lambda_T} = \frac{4\pi f_T R}{c} \quad (6\text{-}42)$$

Φ_{T2} = the round-trip phase from radar to target
R = range
λ_T = illumination wavelength
c = velocity of propagation
f_T = transmit frequency

If illumination frequency is constant, time differentiating Eq. 6-42 gives:

$$\frac{d\Phi_T}{dt} = \frac{4\pi f_T}{c} \frac{dR}{dt} \quad (6\text{-}43)$$

Making the substitutions from the previous derivation ($d\Phi_T/dt = 2\pi f_d$, and $dR/dt = v_R$), Eq. 6-43 becomes identical to Eq. 6-37.

If both range and the illumination frequency vary with time (frequency agility), differentiation of Eq. 6-42 with respect to time and making the appropriate substitutions yields

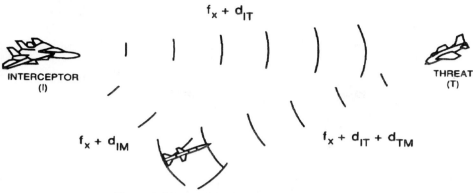

Figure 6-26. Semi-Active Guidance Geometry.
(Reprinted from Eichblatt [3]; © 1989 AIAA, used with permission.)

$$f_d \approx 2 f_T \frac{v_R}{c} + \frac{2R}{c} \frac{df_T}{dt} \qquad (6\text{-}44)$$

In order to find the Doppler shift with frequency agility, range must be known to extraordinary accuracy, greater than is normally the case. Therefore with frequency agility, the Doppler measurement becomes difficult. The solution is to use look-to-look, rather than pulse-to-pulse frequency agility.

Doppler in semi-active radar homing (SARH): The closing rate of a semi-active radar guided missile and its target is found by comparing the frequency of the echo signal with that of the rear-reference. See Sec. 1-6 and Fig. 6-26. The following derivation of the SARH Doppler shift is adapted from Eichblatt [3].

The starting point is the basic equation for exact received frequency (Eq. 6-35), modified somewhat to account for both source (transmitter) and observer (receiver) motion, and given as Eq. 6-45. Again, only the velocity components radial to the missile and target produce Doppler shift. Note that the velocity conventions are different from those used previously. Figure 6-27 shows the velocity conventions used in this derivation.

$$f_O = f_S \frac{c + v_O}{c - v_S} \qquad (6\text{-}45)$$

f_O = the frequency seen by the observer

f_S = the source (illumination) frequency from the perspective of the illuminator

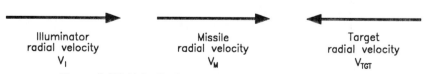

Figure 6-27. Velocity Conventions for SARH Developments
(Arrows Indicate Positive Velocity).

6-4. Target Velocity (Doppler Shift)

v_O = the radial velocity of the observer (positive is the direction of source propagation)

v_S = the radial velocity of the source (positive is the direction of source propagation)

c = the velocity of propagation

The target echo frequency as seen by the missile's front receiver is

$$f_{MF} = f_{TGT} \frac{c + v_M}{c + v_{TGT}} \qquad (6\text{-}46)$$

f_{MF} = the frequency at the missile's front receiver
f_{TGT} = the frequency of the target's echo
v_M = the missile's radial velocity
v_{TGT} = the target's radial velocity

$$\text{but, } f_{TGT} = f_I \frac{c + v_{TGT}}{c - v_I} \qquad (6\text{-}47)$$

f_I = the illumination (source) frequency
v_I = the illuminator's radial velocity

Combining Eqs. 6-46 and 6-47 gives the missile's front receive frequency in terms of the illuminator's velocity and frequency, the target's velocity, and the missile's velocity.

$$f_{MF} = f_I \frac{c + v_M}{c - v_I} \frac{c + v_M}{c - v_{TGT}} \qquad (6\text{-}48)$$

The frequency at the missile's rear reference receiver is

$$f_{MR} = f_I \frac{c - v_M}{c - v_I} \qquad (6\text{-}49)$$

f_{MR} = the frequency at the missile's rear reference receiver

In Doppler shift terms, the above equations can be expressed as

$$f_{TGT} = f_I + f_{IT} \qquad (6\text{-}50)$$
$$f_{MF} = f_T + f_{IT} + f_{TM} \qquad (6\text{-}51)$$
$$f_{MR} = f_I + f_{IM} \qquad (6\text{-}52)$$

f_{IT} = the Doppler shift from illuminator to target
f_{TM} = the Doppler shift from target to missile
f_{IM} = the Doppler shift from illuminator to missile

Subtracting Eq. 6-57 from Eq. 6-56 yields the Doppler difference between the front and rear receivers in the missile.

$$f_{MF} - f_{MR} = f_{IT} + f_{TM} - f_{IM} \qquad (6\text{-}53)$$

Substituting the values of f_{MF} and f_{MR} from Eqs. 6-48 and 6-49 gives the Doppler shift recovered by the missile in terms of the source frequency and the radial velocities.

$$f_{MF} - f_{MR} = \frac{c + v_{TGT}}{c - v_I} \frac{c + v_M}{c - v_{TGT}} - f_I \frac{c - v_M}{c - v_I} \qquad (6\text{-}54)$$

$$f_{MF} - f_{MR} = \frac{(c + v_{TGT})(c + v_M) - (c - v_M)(c - v_{TGT})}{(c - v_I)(c - v_{TGT})} \qquad (6\text{-}55)$$

$$f_{MF} - f_{MR} = \frac{2c(v_M + v_{TGT})}{(c - v_I)(c - v_{TGT})} \qquad (6\text{-}56)$$

Doppler filters and clutter notches: One of the principal uses of the Doppler shift in radar systems is to separate moving targets from clutter. The total signal, consisting of moving target echoes plus clutter, is filtered to remove the clutter. This is possible only if clutter echoes have different Doppler shifts from moving target echoes. It is possible for moving targets to be masked by and confused with clutter, as was shown in our discussion of blind Dopplers.

Filters must be implemented to pass the moving target signals and reject clutter. They are usually in the form of periodic band-stop filters. They must be periodic because stationary clutter (ignoring moving radars for now) appears not only at zero Doppler shift, but also at all integer multiples of the PRF.

Clutter itself has finite spectral width (Secs. 4-11 through 4-13), and for that reason the clutter reject filter notches cannot have zero width. This causes major problems in low PRF radars and modes. Figure 6-28 shows a typical spectrum of clutter plus a moving target, as seen by a pulsed or sampled CW radar.

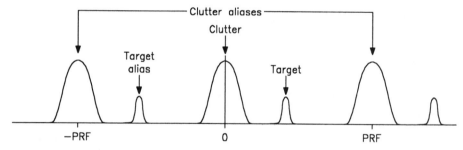

Figure 6-28. Typical Clutter Spectrum with One Moving Target.

Figure 6-29 shows an idealized clutter filter's response. Assume that the radar operates at 10.0 GHz and that the clutter must be removed to a velocity of plus and minus 7.3 knots, corresponding to a clutter Doppler shift of plus and minus 250 Hz. The notches thus must be 500 Hz wide and occur at zero and integer multiples of the PRF. Now assume a PRF of 1000 pps. In Fig. 6-29a, fully half of the Doppler spectrum is taken up by the notches, leaving only half for the detection of moving targets. If a moving target is in a notch, the radar cannot see it (blind Doppler). There are ways around this all-or-nothing response, involving PRF agility and discussed in Ch. 12, but the loss caused by the notches being a large fraction of the total available spectrum is real and irreducible.

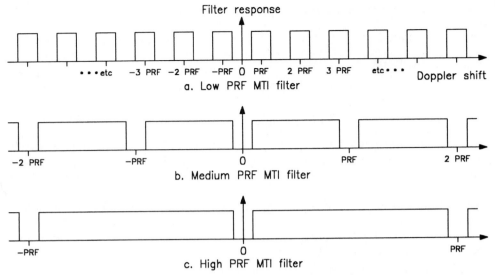

Figure 6-29. Doppler Filter Clutter Notches.

It is easy to see that if the notches were made a little wider or the PRF a little lower, the filter's response to moving targets would disappear entirely. This is a major problem in suppressing clutter at low PRFs, made virtually impossible in airborne radars where rejection of clutter to ± 80 knots is often needed.

Figure 6-29b shows what happens when the PRF is raised. Notches the same width as before occupy a smaller fraction of the total spectrum. In Fig. 6-29c, the PRF is higher yet, and the notches have even less detrimental effect. With increasing PRF, the moving target loss caused by the notch width is reduced. Conversely, at higher PRFs the notches can be made wider without increasing the loss.

The bottom line is that Doppler recovery, both from the Nyquist sampling standpoint and that of the clutter notch, works better at higher PRFs. Newer radars generally trend toward higher PRFs for these reasons, plus illumination energy levels are higher with the same transmitted peak power. Deghosting and high PRF/CW techniques, discussed in Secs. 6-3 and 6-5, address the worsening range ambiguities brought on by higher PRFs.

Sidelobe Doppler: Another source of difficulty with Doppler information recovery, unique to moving radars, is that clutter has a net Doppler shift and clutter viewed through antenna sidelobes has an unpredictable Doppler shift. The radar compensates for the velocity of clutter viewed by the antenna's main lobe (main beam clutter), which is a function of the radar's velocity over the clutter, the antenna angles from the radar's velocity vector, and the transmitted frequency. Since the angles of the sidelobes are different from those of the main lobe, sidelobe clutter may appear as a moving target. See Ch. 4 for more information on sidelobe clutter.

6-5. Range and Velocity with CW and Pulse Doppler Waveforms

An entirely different set of problems occurs when the radar's transmitted waveform is CW or pulse Doppler (high PRF). There are no time-variant features in a CW waveform other than the cycles themselves, and their period is much too short for most ranging purposes. Pulse Doppler PRFs are likewise too high for pulse-delay ranging (measuring the time delay between transmit and echo).

In both CW and pulse Doppler, ranging is usually accomplished by modulating the RF carrier wave and processing the echo modulation to recover range. Three schemes are commonly used: frequency modulating the carrier with a sinusoidal waveform, frequency modulating the carrier with a triangular waveform, and phase-coding the carrier wave. In each case, the period of the modulation now sets the maximum range, rather than PRF.

Sinusoidal FM CW: The simplest modulation form is sinusoidal FM, where the RF carrier is frequency modulated with a sinusoid (Fig. 6-30). The range to a target is found by measuring the phase shift between the transmitted modulation wave and the modulation recovered from the echo. The propagation time is

$$T_p = \frac{\delta\phi}{360 f_M} \qquad (6\text{-}57)$$

T_p = the propagation time from transmit to echo

$\delta\phi$ = the phase of the echo modulation referenced to the transmitted modulation, in degrees

f_M = the modulation frequency

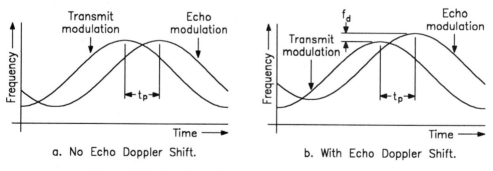

Figure 6-30. Ranging with Sinusoidally Modulated FMCW.

Figure 6-30a shows sinusoidal FM ranging to a target with zero Doppler shift. If Doppler is present, as in Fig. 6-30b, range is still found from the modulation phase shift (Eq. 6-57), and the Doppler is taken from the DC value of the comparison between transmit and echo modulation.

The biggest drawback to sinusoidal FM ranging is that multiple targets are not resolved. Each demodulated echo is a sinusoid at the FM modulation frequency, with a phase shift corresponding to its range, and a DC offset corresponding to its Doppler shift. The demodulated echo from two targets is the sum of the demodulated echoes from the individual targets. The vector sum of two DC-offset sinusoids at the same frequency is a third DC-offset sinusoid, which appears to this system as a single target with range and Doppler shift between those of the individual echoes. Sinusoidal FM ranging is most commonly used in single-target track radars, where only one target is present.

Example 6-7: An X-Band CW radar (10.4 GHz) uses sinusoidal FM for ranging. Its modulation frequency is 90 Hz. The phase difference between the received modulation and that transmitted is 3.5°. Find the target's range.

Solution: The period of a 90 Hz wave is 0.01111 s. Each degree of phase is 1/360 of this, or 30.86 μs. 3.5° takes 108.0 μs, which is the propagation time (Eq. 6-57 gives the same answer). The target is thus 16,204 m, or 8.75 nmi from the radar.

Triangular FM CW: Figure 6-31 shows the principle of triangular carrier FM ranging, in this case, on a single target with no Doppler shift. The range information is contained in the instantaneous frequency difference between the signal echo at the radar and the radar's present transmit frequency. In the absence of Doppler, this frequency difference is the same for the up and down sweeps and is proportional to the time delay between transmission of a segment of the wave and reception of the echo from that same segment.

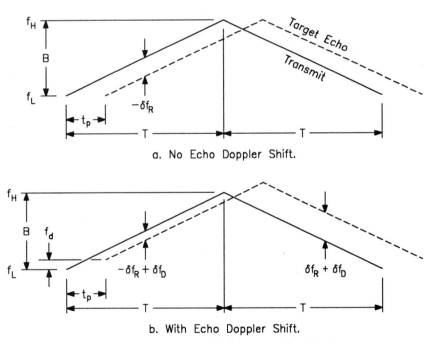

Figure 6-31. Ranging with Triangular FMCW.

Equation 6-58, derived from similar triangles in Fig. 6-31a, gives the relationship between transmit frequency, sweep bandwidth, transmit/receive difference frequency and target range.

$$T_p = \frac{T \, \delta f_R}{B} \tag{6-58}$$

$$B = f_H - f_L \tag{6-59}$$

$$R = \frac{c \, T \, \delta f_R}{2 \, B} \tag{6-60}$$

T_p = the round-trip propagation time
T = the period of the modulation wave
δf_R = the frequency difference between receive signal and transmit signal caused by target range
f_H = the highest frequency in the transmit wave
f_L = the lowest frequency in the transmit wave
R = range
c = electromagnetic propagation velocity

A Doppler shifted target modifies the received wave as shown in Fig. 6-31b. Now there are two difference frequencies; one for the illumination up-sweep and one for the down-sweep. These two frequencies, δf_{US} and δf_{DS}, contain both the Doppler shift frequency and the frequency shift caused by target range. Assuming the Doppler shift is greater than the range shift, δf_{US} represents the difference between the Doppler and range frequencies and δf_{DS} represents the sum of the two frequencies. Assuming the slopes of the two sweeps are additive inverses (equal magnitudes, different signs), it is a simple matter to separate the two using Eqs. 6-61 and 6-62.

$$f_d = (\delta f_{DS} + \delta f_{US}) / 2 \tag{6-61}$$

$$\delta f_R = (\delta f_D - \delta f_U) / 2 \tag{6-62}$$

f_d = the Doppler frequency shift
δf_R = the frequency shift caused by range

Example 6-8: A CW radar has a center frequency of 10.0 GHz and triangularly sweeps a bandwidth of 2.0 MHz at a 200 Hz rate. The frequency difference between transmit and receive on the up-sweep is 65,510 Hz and on the down-sweep it is 82,650 Hz. What is the target's range and radial velocity?

Solution: By Eqs. 6-62 and 6-62, the range frequency is 74,080 Hz and the Doppler shift is 8,570 Hz. Eq. 6-58 gives a propagation time of 92.6 μs, which translates to 13,890 m, or 7.50 nmi. The Doppler shift translates to a radial velocity of 131.2 m/s, or 255 Kt.

Resolving multiple targets: Triangular FM allows resolution of multiple targets. If two targets are present, they produce two echo signals at the radar, each with different instantaneous frequency differences with the illumination. The composite signal is the sum of the echoes. A spectrum analysis of the up-sweep frequency differences and one of the down-sweep differences gives the individual targets in different frequency bins, assuming they are resolvable. Figure 6-32 shows typical spectra. The problem is that there is no

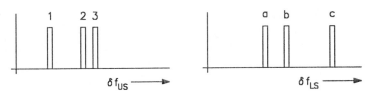

Figure 6-32. Typical Spectra of Up- and Down-Sweep.

information whereby an up-sweep difference frequency (1, 2, and 3) can be tied to a down-sweep difference (a, b, and c), and thus targets viewed this way are "scrambled."

The answer is to add a third waveform segment which has no modulation (Fig. 6-33). A spectrum analysis of echoes in this spectrum gives the Doppler shifts of the multiple targets. It then becomes a matter of trial-and-error pairing of the up-sweep difference with the down-sweep differences to see which pairings produce the same Doppler shifts as the unswept signal segment. Another method is to use multiple sweeps with different slopes.

$$f_d = \delta f_{LS} \tag{6-63}$$

f_d = the Doppler frequency shift

δf_{LS} = the difference frequency of the level (constant frequency) waveform segment

Example 6-9: A airborne pulse Doppler radar illumination waveform is frequency modulated as shown in Fig. 6-33. Each sweep segment is 10.0 ms long. The low frequency is exactly 10.0 GHz and the sweep is 2.00 MHz. The spectrum analysis of the up-sweep shows three echoes of $-38,655$ Hz, $-50,878$ Hz, and $-85,487$ Hz. The down-sweep has echoes at $+74,513$ Hz, $+86,665$ Hz, and $+95,802$ Hz. The level sweep shows echoes of 24,005 Hz, 22,462 Hz, and $-5,487$ Hz. Find the range and closing velocity of the three targets.

Solution: It is known from the level waveform segment that three Doppler shifts are present: 24,005 Hz, 22,462 Hz, and $-5,487$ Hz. Assume the up-sweep frequency of $-38,655$ Hz pairs with the down-sweep frequency of 74,513 Hz. By Eq. 6-61, the Doppler shift of this pairing is 17,929 Hz. This does not match one of the known Doppler shifts and therefore this pairing must be incorrect. Now pair the $-38,655$ Hz up-sweep with the 86,665 Hz down-sweep. This gives a Doppler shift

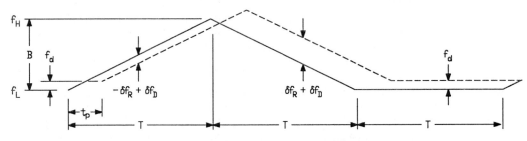

Figure 6-33. Three-Segment Linear FM Ranging Waveform.

of 24,005 Hz, which is one of the known Dopplers. This pairing could be correct. As a check, pairing the −38,655 up-sweep with the 95,802 down-sweep gives a Doppler shift of 28,574, which is not one of the identified Dopplers. Thus the pairing of −38,655 Hz up-sweep and 86,665 down-sweep is probably correct, and by Eq. 6-62 gives a range frequency of 62,660 Hz. This target's range (Eq. 6-60) is approximately 47,000 m (25.4 nmi), and its closing velocity, from Eq. 6-37, is 360 m/s (700 Knots).

By like reasoning, the −50,878 Hz up-sweep frequency is found to pair with the 95,802 Hz down-sweep, giving a target with a range of 55,000 m (29.7 nmi) and a closing velocity of 337 m/s (655 Kt).

The −85,487 Hz up-sweep frequency pairs with the 74,513 Hz down-sweep, giving a target at a range of 60,000 m (32.4 nmi) with a closure of −82.3 m/s (−160 Kt). This target, with its negative Doppler shift and velocity, is tail-aspect and moving away from the radar.

Resolution performance with FMCW waveforms is the same as with pulsed. Frequency resolution is the reciprocal of the dwell time.

$$\Delta f_S = 1/T \qquad (6\text{-}64)$$

Δf_S = the frequency resolution of each waveform segment
T = the time extent of the segment

Doppler resolution is a direct result of the segment frequency resolution. Thus Doppler resolution is as found for pulsed waveforms in Sec. 1-6.

$$\Delta f_d = \Delta f_S \qquad (6\text{-}65)$$

Δf_d = the Doppler resolution

Range resolution is the range equivalent of the frequency resolution. Substituting frequency resolution (Eq. 6-64) into the range/frequency relationship (Eq. 6-60) gives the same range resolution as for pulsed waveforms (Sec. 1-3).

$$\Delta R = \frac{cT(1/T)}{2B} = \frac{c}{2B} \qquad (6\text{-}66)$$

ΔR = the range resolution

Since Doppler shift is a function of illumination frequency, a concern is that the Doppler shift may change significantly over the sweep. Sweep bandwidths used in most radars are generally narrow enough so that this is not a problem. If a sweep bandwidth of 2.0 MHz is used at 10.0 GHz average carrier frequency against a Mach 5.5 target, the Doppler shift at the lowest transmitted frequency (9.9990 GHz) is 123,921 Hz, and at the highest frequency (10.0010 GHz) it is 123,946 Hz. If the total Doppler spectrum is divided into 1024 bins of 300 Hz each, the Doppler shift differs by only 0.08 bin across the sweep. Faster targets and wider sweep bandwidths may make this a problem.

Where only zero Doppler shift is present (as in some diagnostic radars), the sweep can be unidirectional, and a single spectrum analysis gives a range profile of the target space.

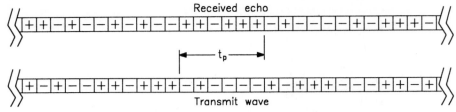

Figure 6-34. Ranging with Phase-Coded CW.

Phase-Coded FM CW: Phase-coded waves are CW signals with digital phase modulation applied. Ranging is simply a matter of measuring the time from the transmission of a code segment to the time of reception of the same segment. The process used is generally correlation of the received signal with a delayed copy of that transmitted (see Ch. 11). The delay corresponding to the correlation function maximum is the propagation time. Figure 6-34 shows the principle.

The ability to range accurately and unambiguously requires that a code be selected which has only one peak in its autocorrelation function over the range of possible propagation times. The code should also have small partial sums at alignments different from the target's propagation time. These requirements are examined in Chs. 11 and 13.

Estimation of range from a fly-by with a CW radar: If a radar which measures Doppler shift but not range flies by a target, a method exists for estimating the radar-to-target range at the closest point. This is useful in certain instrumentation tasks, such as miss-distance indicators, and for fuzing warheads. Figure 6-35 shows a fly-by example and the Doppler shift at various stages of the scenario. Approaching the target, the radar sees up-Doppler. If it does not hit the target, at one point in the flight path the radar sees zero Doppler. As it flies away, it sees down-Doppler. The range from radar to target at its closest approach is related to the velocities and the slope of the Doppler history, which is in turn determined by the line-of-sight rate. Steeper slopes indicate nearer misses; if the missile hits the target, the slope is infinite. The derivation of range from the various velocities and the slope of the Doppler history is left as an exercise.

Ranging in pulse Doppler modes: Ranging in pulse Doppler systems can be accomplished by one of two principal means. First, the pulse Doppler system can be treated as though it were CW, with the carrier frequency modulated. The modulation can be sinusoidal, triangular, or any of several other waveforms.

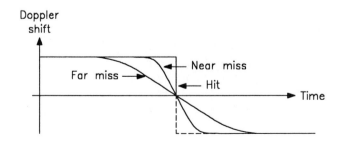

Figure 6-35. Doppler History for Fly-By.

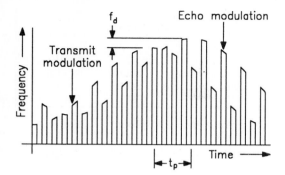

Figure 6-36. Ranging with Sinusoidally Modulated FM Pulse Doppler.

Pulse Doppler FM ranging works exactly the same as CW, except that the transmitted and echo signals are broken up into pulses. The instantaneous echo carrier frequency still differs from that of the illumination by a function of the range to the target. The phase of the modulation still contains the target's range, and the DC value of the modulation contains its Doppler shift. Figure 6-36 shows the waveforms with pulse Doppler.

The second way is to treat pulse Doppler as a pulsed system and deghost using multiple PRFs, illustrated in Figs. 6-19 and 6-20 and in the discussion associated with them. The problem here is that the ambiguities are so extreme that very large numbers of PRFs must be used. Also, there are few range bins in pulse Doppler systems (sometimes only one), which restricts the use of this method. Just as the carrier frequency can be continuously varied, so also can the PRF, which gives another method of pulse Doppler ranging.

Table 6-4 summarizes the strengths and weaknesses of each of the various CW and pulse Doppler ranging methods.

Range in semi-active guidance: Pure semi-active guidance, using CW or pulse Doppler illumination without modulation or a data link with the illuminating vehicle, does not directly produce the range from missile to target. If no further information is available, guidance consists of keeping the missile pointed at the target in order to assure intercept. Figure 6-37 shows what happens. In many cases, the required missile maneuver in the last stages of flight may exceed its capability and increase the probability of a miss.

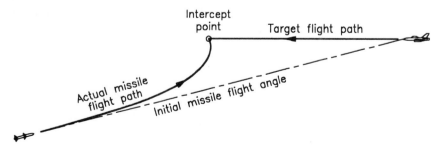

Figure 6-37. Flight Path with Missile Always Pointed at Target.

6-5. Range and Velocity with CW and Pulse Doppler

Table 6-4. Comparison of Pulse Doppler Ranging Methods.
(after Mooney and Skillman [4])

Parameter	Multiple discrete PRFs	Continuously variable PRFs	Linear carrier FM	Sinusoidal carrier FM
Range accuracy	Excellent	Poor	Poor	Fair
Rejection of spurious detections from PRF harmonics	Excellent	Very poor	Excellent	Excellent
Clutter rejection	Excellent	Good	Good	Good
Target detection near clutter	Excellent	Good	Poor	Good
Number of targets generated by the process when multiple targets are:				
Different velocity	None	None	Few	None
Same velocity	Many	Few	Few	Few
Peak power required	High	Low	Low	Low
Range performance	Good	Excellent	Good	Excellent
Is the speed of measurement suitable for search modes?	Yes	No	Yes	No
Performance of track through eclipse	Excellent	Excellent	Poor	Poor
Number of range-gated receivers required:				
Track	≥1	≥1	...	≥1
Search	Many	...	≥1	...
Auxiliary hardware required	Some	Much	Little	Little

If an estimate of range is available, the missile can compute an intercept point and avoid the violent terminal maneuver. Unless a data link with the launch vehicle is present, however, range is not usually known. In this case, an intercept technique which does not require range, called proportional guidance is often used (Fig. 6-38). The missile flies a path such that the angle between the missile's flight path and the target, called the line-of-sight angle, is kept constant, rather than the missile being pointed at the target. This angle is available from the front antenna, which tracks the target.

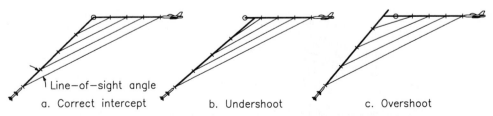

a. Correct intercept   b. Undershoot   c. Overshoot

Figure 6-38. Flight Path with Proportional Guidance.

Figure 6-38a shows a correct intercept using proportional guidance. Note the constant angle. Figure 6-38b shows an undershoot, where the missile will fly behind the target. Note that the angle is decreasing with time. Figure 6-38c shows an overshoot, where the missile will fly in front of the target. Now the angle increases with time. The rate at which the angle changes is called the line-of-sight rate, and for a correct intercept should be zero.

The principle shown in the figure is somewhat simplified in that the velocities are assumed constant and the target does not maneuver, but it shows the concept.

6-6. Evaluation of the PRF Classes

After discussions of range and range ambiguities, and Doppler and Doppler ambiguities, it is now appropriate to look more closely at the PRF classes. This will be done in outline form, adapted from and inspired by similar discussions in Stimson [1]. Although somewhat oriented toward airborne radars, the discussions below apply equally to ground based and shipboard systems.

First, some definitions are required. Look-down means that the radar looks at the target and clutter simultaneously, and the two are not normally resolvable in range. The various aspects for air-to-air operations are shown in Fig. 6-39 and defined below.

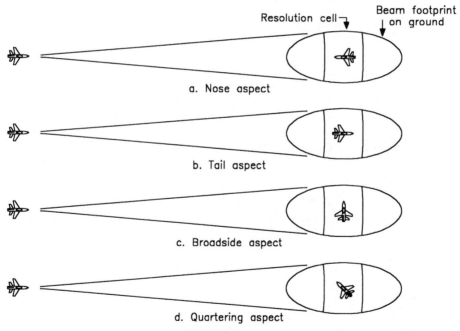

Figure 6-39. Radar / Target Aspects.

- *Nose aspect* is where the airborne radar and the target are flying almost directly toward one another. Almost all motion is radial and the radial velocity is approximately the sum the two vehicles' velocities. The target's Doppler shift is greater than that of the clutter.

- *Tail aspect* is where the radar chases the target, with the two moving in the same direction. Most motion is radial, and the radial velocity is the difference between the radar's velocity and that of the target. The target's Doppler shift is less than that of the clutter.
- *Broadside aspect* is where the target's velocity vector is approximately perpendicular to that of the radar. The radial velocity between the radar and target is approximately the radar's velocity. The target's Doppler shift is approximately equal to that of the clutter.
- *Quartering aspect* is where the target's motion is partially radial to or from the radar and partly tangential. The Doppler shift can be greater than that of the clutter is the aspect is nose quartering (shown), or less than that of the clutter in tail quartering.

LOW PRF: Low PRF is where all targets are zone 1 in range and are highly ambiguous in Doppler. It allows simple pulse-delay ranging. It is a low energy waveform because of its low duty cycle. High energy requires high peak power.

Low PRF advantages:
- Low PRF is good for look-up (out of clutter) at limited ranges by land based, shipboard, and airborne radars. It is also good for real-beam ground mapping from aircraft.
- It is good for precise range measurement, and simple pulse delay ranging is supported.
- It supports good range resolution.
- In air-to-air operations, the normal sidelobe (altitude) return can be rejected with range resolution and range gating.

Low PRF disadvantages:
- Low PRF is poor for air-to-air look-down and in any application with high levels of clutter. The clutter filter notch widths necessary to reject the clutter spectrum may be larger than the PRF, leaving no room for moving targets. At best, the notches are a very large fraction of the total Doppler spectrum. The alternative is a narrow notch which allows part of the clutter spectrum to be treated as moving.
- Ground moving targets can be a problem in air-to-air look-down and in any other application with high clutter levels, because their rejection may require notches so wide as to preclude the reception of moving targets.
- Because of the severe undersampling, Doppler ambiguities are generally too severe to be resolved, precluding accurate velocity measurements and discrimination among targets with different velocities. Doppler shift can be sensed, but its measurement is difficult. Single-target velocities can be obtained by measuring the rate at which range changes. Multiple targets with the same range/azimuth/elevation bin cannot be resolved this way.
- Transmit duty cycles are low, hindering target detection. To obtain the high energy required for long range detection, low PRF requires higher peak power or larger amounts of pulse compression than do other PRF classes. This is basically a short-range PRF class compared to the others.

— This is the easiest PRF class to intercept, because of the required high peak powers. Coded waveforms somewhat alleviate this problem.

HIGH PRF: High PRF is where all targets are Nyquist sampled in Doppler and are in Doppler zone 0. Doppler ambiguities do not exist. Range is highly ambiguous. It supports the simplest recovery of Doppler shift. It is a high energy waveform because of its high duty cycle. High energy can be obtained with modest peak power. High PRF is inherently the longest range PRF class.

High PRF advantages:
— High PRF has good nose-aspect capability, where high closure rates assure that the target Doppler shift will be greater than any clutter Doppler shift. Sidelobe clutter is not a problem with this PRF class.
— High average power and energy are available because of the high duty cycle associated with high PRF. It is the best PRF in this aspect for long-range target detection. In most radars which have multiple PRF class capability, the longest range modes, both search and track, are high PRF.
— Main lobe clutter can be rejected without also rejecting target echoes, because the target Doppler shifts are always greater than that of the clutter.
— The altitude return is rejected because of the target's Doppler being greater than that of the altitude return.
— High PRF works well on closing targets in clutter from shipboard and land-based radars, such as sea-skimming missiles and low-altitude cruise missiles and aircraft. This is because the closing targets have much greater Doppler shifts than the clutter.

High PRF disadvantages:
— Detection of low closure rate targets, such as tail aspect where the radar and the target are moving at about the same velocity, is poor. This is primarily because sidelobe clutter can have the same Doppler shift as the target. Because of its susceptibility to sidelobe clutter (Ch. 5), high PRF performs poorly in the presence of any target closure rate which results in a Doppler shift less than that of the nose-aspect clutter.
— Range ambiguities are great enough to often preclude accurate ranging to multiple targets. Likewise, range resolution can be poor in the presence of large numbers of targets. The fault lies with the methods used to discriminate between many targets in range and Doppler (Sec. 6-5).
— Zero closing rate targets, including tail aspect in air-to-air, and crossing targets in stationary systems, are particularly difficult at high PRF, because of poor range performance. They may be effectively masked in Doppler by the altitude return, transmitter spillover, and clutter.

MEDIUM PRF: Medium PRF (MPRF) is ambiguous in both range and Doppler, but the number of zones containing possible targets is limited in both dimensions. The limited number of zones normally means that both range and Doppler ambiguities can be resolved. This is the only PRF class where both range and Doppler of large numbers of targets are normally found. It is a medium-energy class, and modes using medium PRF usually function at medium and short ranges.

Medium PRF advantages:
- Medium PRF has good capabilities in all target aspects. With the ability to find both true range and true Doppler for multiple targets, medium PRF can cope satisfactorily with main lobe clutter, sidelobe clutter, and the altitude return.
- Because of wide spectral expanse between clutter notches and because of range gating capability, which is not available in high PRF, ground moving targets can be readily rejected.
- Pulse delay ranging is possible, with resolution of range ambiguities.
- Zero closure rate targets can be handled because the altitude return can be resolved and rejected in range.

Medium PRF disadvantages:
- Detection range against both low and high closure rate targets can be limited by sidelobe clutter. However, since this PRF class can find both true range and true Doppler for multiple targets, it has the best chance of dealing with sidelobe clutter.
- Medium PRF is ambiguous in both range and Doppler. Both ambiguities must be resolved. To do this, the total number of simultaneous targets must be limited.
- Special measures, such as elaborate STC or guard antennas and receivers are needed to reject sidelobe returns from strong ground targets (see Ch. 9).

6-7. Radar Height-Finding

Three-dimensional search radars/modes and tracking radars/modes locate targets in range, azimuth, and elevation. Additionally, a class of radars known as the 2-D height-finding radars located targets in range and elevation. To determine a target's altitude above sea level, the target's elevation angle, its range, the earth's curvature parameters, and atmospheric bending of the electromagnetic waves (refraction) must be processed together. Three basic kinds of height-finding are done: flat earth height-finding, spherical earth height-finding with no correction for beam refraction, and height-finding with corrections for a non-spherical earth and beam refraction. This topic is discussed further in Sec. 14-2.

Flat earth height-finding assumes the earth is flat and the transmit and echo signal paths are straight lines. The target parameter found is the height above the plane of the radar (the so-called Z dimension).

$$Z = R \sin \theta_{EL} \qquad (6\text{-}67)$$

Z = the target's height above the plane of the radar

R = the range

θ_{EL} = the elevation angle from radar to target

Round-earth height-finding is done by solving the triangle, shown in Fig. 6-40, which extends from the center of the earth to the radar to the target and back to the earth's center. Two sides of the triangle are known: the distance from the center of the earth to the radar ($E_0 + H_r$), and the distance from the radar to the target (the slant range R). The angle included between these two sides is also known; it is the elevation angle plus the right angle between the horizontal plane of the radar and an earth's radius through the radar. The law of cosines gives a solution for the third side.

$$(E_0+H_t)^2 = (E_0+H_r)^2 + R^2 - 2(E_0+H_r) R \cos(\theta_{EL}+90°) \qquad (6\text{-}68)$$

E_0 = the earth's radius
H_t = the height of the target above mean sea level (MSL)
H_r = the elevation of the radar above MSL
R = the range (slant range) from radar to target
θ_{EL} = the radar's elevation angle (degrees)

Microwave beams do not travel in straight lines in the atmosphere, with denser air at the surface bending the beam downward, and producing a reported target altitude higher than the actual altitude (Fig. 6-41). This bending of the electromagnetic beam is caused by refraction.

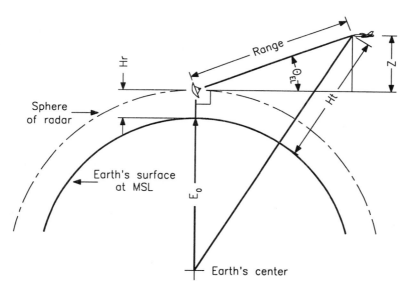

Figure 6-40. Radar Height-Finding Geometry.

Refraction is determined by the density gradient of the atmosphere with altitude, dn/dh, where n is the index of refraction and h is the altitude above mean sea level (MSL). To find target altitude accurately, dn/dh must be known to greater accuracy than is normally possible, particularly if the radar scans a large volume of space. Accurate determination of dn/dh has in the past involved carefully probing the atmosphere with instrument packages attached to rockets or balloons, an expensive and continuous task. Recently declassified Strategic Defense Initiative (SDI) studies have suggested that by observing extraterrestrial objects at known locations, atmospheric bending can be accurately determined and continuously compensated for. These studies have been in the optical frequency bands, but may have applications in the radar bands. For an excellent overview, see Ref. [5].

6-7. Radar Height-Finding and Coverage

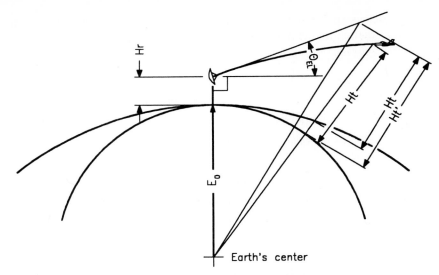

Figure 6-41. Height-Finding with Refraction.

Once the refractive path of the beam is determined, one convenient way of correcting this bending is to assume the earth's radius to be somewhat larger than its actual value.

$$E_0' = k E_0 \qquad (6\text{-}69)$$

E_0' = the adjusted earth's radius, for use in Eq. 6-68

k = a factor for adjusting the earth's radius

The factor k can be modeled from dn/dh profiles, radar altitude, range, and elevation angle. In many systems, an average number is used, sometimes fixed and sometimes adjusted for local surface atmospheric conditions. For the entire earth, all seasons, and all times, a value of $k = 4/3$ is about right, giving the so-called 4/3 earth model of refraction and its correction.

If k is a constant, the assumption is that the refractive index decreases linearly with altitude (dn/dh is constant). More realistic models recognize various curvatures for dn/dh, the most common of which is exponential, expressed as follows [6].

$$n(h) = 1 + (N_{MSL} \times 10^{-6}) \exp(-\gamma h) \qquad (6\text{-}70)$$

$n(h)$ = the index of refraction as a function of h

N_{MSL} = the refractivity at the surface, projected to MSL

γ = a factor governing the exponential rule used

h = the altitude above MSL (meters)

The various CRPL [7] models treat the refractive index as in Eq. 6-70. Common values are $N_{MSL} = 313$ and $\gamma = 0.00014386$, resulting in an atmosphere whose refractivity is half the MSL value at 4818 m (15,809 ft). The correction for target altitude is found from Snell's Law, and requires numerically solving a differential equation describing beam position as a function of propagation time and is beyond the scope of this book (see [8]).

Reflections off the earth's surface, either from the radar antenna's main lobe or sidelobes, also affects height-finding accuracy. Further study is beyond the scope of this text and is left to the reader. A starting point is [7].

Radar coverage diagrams: For each search radar, a radar coverage diagram, such as Fig. 1-53, is developed. It shows the altitude versus range of targets which can be seen by the radar. Coverage diagrams recognize refraction, usually with the 4/3 earth model or one of the CRPL models.

Radar horizon: Radars can see only those targets which are above their horizon. Figures 6-42 and 6-43 are nomographs linking the maximum range at which a radar above a smooth spherical earth (as over the oceans) can see targets at various altitudes. Figure 6-42 is for high altitude targets and Fig. 6-43 is for low (sea-skimmers). Both ignore refraction and ducting.

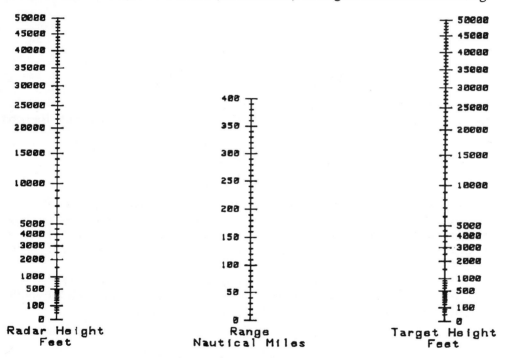

Figure 6-42. High-Altitude Radar Horizon Nomograph.

6-8. Further Information

General references: The following two books give information on airborne operations, medium PRF, deghosting, and ranging at high PRF.

[1] G.W. Stimson, *Introduction to Airborne Radar*, El Segundo, CA: Hughes Aircraft, 1983.

[2] G.V. Morris, *Airborne Pulsed Doppler Radar*, Norwood, MA: Artech House, 1988.

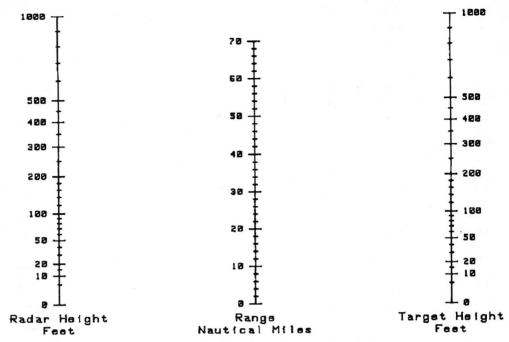

Figure 6-43. Low-Altitude Radar Horizon Nomograph.

Cited references:

[3] E.J. Eichblatt, Jr. (ed.), *Test and Evaluation of the Tactical Missile*, Vol. 119 of *Progress in Astronautics and Aeronautics* series, Washington, DC: American Institute of Aeronautics and Astronautics, 1989.

[4] D.H. Mooney and W.A. Skillman, Ch. 19 in M.I. Skolnik (ed.), *Radar Handbook*, 1st ed., New York: McGraw-Hill, 1970.

[5] C.S. Powel, "Mirroring the Cosmos," *Scientific American*, vol. 265, no. 5, November 1991, pp. 112-123.

[6] The National Bureau of Standards, Fort Collins Colorado, has done extensive atmospheric modeling and evaluations of radio refraction in the atmosphere. Although no specific references are given here, anyone doing work on radar height-finding should obtain their publications list.

[7] L.V. Blake, Ch. 2 in M.I. Skolnik (ed.), *Radar Handbook*, 2nd ed., New York: McGraw-Hill, 1990, p. 2.35.

[8] D.K. Barton and H.R. Ward, *Handbook of Radar Measurement*, Norwood, MA: Artech House, 1984, Appendix D.

7

TRACKING RADAR

7-1. Tracking Introduction

Tracking is the process whereby the radar follows the position of one or more objects in space, ignoring the content of the space not occupied by the target(s) being tracked. It has the advantage that one (or more) target(s) can be located with much greater accuracy than with search radars and modes. The target(s) being tracked are more or less continuously monitored so that their position is known regardless of maneuvering, as opposed to search systems and modes where target positions are only samples every few seconds, with no knowledge of target maneuvering between scans. For true tracking to take place, the target(s) must be sampled at the Nyquist rate (Ch. 6) for the track servo bandwidths and target maneuvering bandwidths. This usually requires that the target(s) be sampled from 10 to 20 times per second or more (Fig. 7-1).

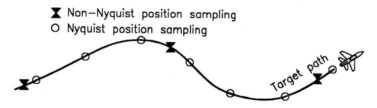

Figure 7-1. Track Sampling.

Track types: Four distinct types of tracking are recognized.

— *Single target track (STT)*, where the radar follows a single object and ignores all others. The rate at which STT radars and modes sample the target is the PRF.

— *Spotlight track*, where the radar follows one object for a period of time, switches to a second object, which it follows for a period of time, switches perhaps to other targets, following them for the prescribed time, and finally switches back to the first target to repeat the cycle. Spotlight tracking is less accurate than single-target tracking because while the radar spotlights one target, the positions of the others are unknown.

— *Multi-target track*, in which the radar continuously monitoring the position of several targets, with each target sampled many times per second. Effectively, one radar performs the function of as many tracking radars as there are targets being tracked. Multi-target tracking requires that the antenna's beam position be changeable essentially instantaneously, and this is normally possible only with electronically scanned antennas. For true multi-target tracking, each target must be sampled at the radar servo and target maneuvering bandwidths' Nyquist rates, requiring that each be sampled 10 to 20 times per second.

— *Track-while-scan (TWS)* is the process whereby a search radar samples the position of several targets once per scan and uses sophisticated smoothing and extrapolation algorithms to estimate the position of the targets between scan samples. Strictly speaking, it is not really tracking, since the sample rate (which equals the search antenna's scan frequency), is less than Nyquist for the servos and target maneuvering. Typical target update rates are between once per second and once per 15 seconds. Target maneuvering in particular between scans can result in significantly lower accuracy position estimates than those obtained from true Nyquist sampled tracking.

Tracking radar modes: Two distinct modes are present in all tracking: *acquisition* and *track*. In the acquisition mode, the radar is designated (pointed) to a location in space which is the best available external estimate of the target's position. Designation sources can be:

— Other tracking radars, which give azimuth, elevation, and range data
— 3-D search radars, giving azimuth, elevation, and range data
— 2-D search radars, furnishing azimuth and range only, requiring the tracking radar to scan in elevation
— Optical trackers, giving azimuth and elevation position only, with the track radar scanning in range
— Computer estimates of target position
— Data from the radar itself operated in a search mode
— Manual operator-controlled searches

In automatic track mode, the systems use information derived in the antenna, receiver, and tracking circuits to move the antenna beam so that its axis points in the direction of the target. Most also use timing circuitry to follow the object in time-delay (range tracking). Some radars also track velocity. A block diagram of a tracking radar is shown in Fig. 7-2.

Angular errors must be developed from only one target at a time, otherwise the system will not accurately track the desired target. In cases where more than one target can be present in the antenna beam, the desired target must be selected and other targets rejected based on some criterion. This process, called gating, is shown in Fig. 7-3. Usually the selection is made based either on the time of arrival of the signal (range gating) or on the Doppler shift of the signal (Doppler gating). If the signal is gated, a mechanism must be present to generate and modify the gates. Time selection is done in a range tracker or ranging machine, described in Sec. 7-5. Frequency selection is accomplished in a Doppler tracker, or velocity tracker, shown in Sec. 7-6.

Track methods are summarized in Sec. 1-6. As stated there, tracking radars are classified by the method used to derive angle errors, with the principle types being conical scan, lobing, amplitude comparison monopulse, and phase comparison monopulse. Track-while-scan derives errors using centroid detection. These methods are described in later sections in this chapter.

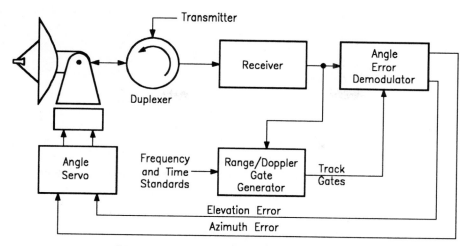

Figure 7-2. Tracking System Block Diagram.

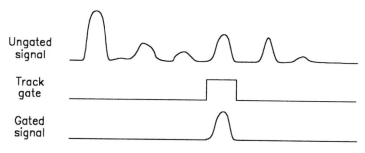

Figure 7-3. Track Gating.

7-2. Tracking System Properties and Parameters

Two parameters must be defined before tracking can discussed further: error gradient, and depth-of-null. They measure the tracking system's ability to convert angular track errors into signals, which can be applied to servos to move the antenna or beam to the location of the target. Although discussed as angle tracking parameters, they apply equally to range and Doppler tracking. Before describing the parameters, however, some knowledge of tracking antenna patterns is necessary.

Tracking antenna patterns: Tracking antennas have three patterns: sum, azimuth error, and elevation error. The sum pattern is the normal antenna pattern, as seen in Ch. 2. The error patterns represent the relationship between angular error and error signal amplitude. Ideally, the error pattern has no response (zero gain) at the same angle at which the sum pattern has maximum gain. One of the two main lobes of an error pattern produces a signal which is in-phase with the signal from the sum pattern and the other is 180° phase to the sum. Ideal sum and difference patterns are shown in Fig. 7-4.

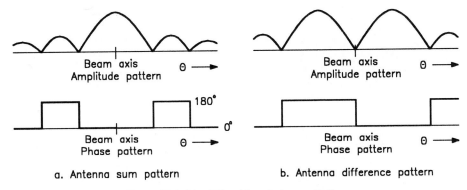

a. Antenna sum pattern b. Antenna difference pattern

Figure 7-4. Ideal Tracking Antenna Pattern.

Error gradient: All tracking system have a parameter known as error gradient, which is the slope of the servo error signal *versus* angular error (Fig. 7-5). It is usually established at some standard test point in the angle error system and is measured in volts of error signal per mil (or degree) of angular error. It is one of the factors that determine the tracking servo parameters. To check error gradient requires a simulated target at a fixed location, such as furnished by a boresight tower (see Sec. 10-10). Error gradient measurements are usually taken with considerably larger angular errors than would be encountered in normal tracks. Thus it does not convey what is happening down in the null itself. That is the job of the next parameter discussed, depth-of-null.

Depth-of-null: Another important parameter in all tracking systems measures the cancellation of signal in the error channels when the angular track error is zero and the error channel is nulled. Ideally, with zero track error, the error voltages should be zero. An antenna pattern taken at the azimuth (or elevation) output of the system should look like Fig. 7-4b, with complete cancellation in the null and an abrupt change in error polarity from plus to minus, corresponding to effective phases in the error antenna pattern of 0° and 180°.

In real systems, the antenna patterns are as shown in Fig. 7-6, with a shallower, rounded error null and the error phase gradually changing from 0° to 180°. The depth-of-null parameter is defined as the ratio of error channel pattern peak gain to its gain in the null. Specifications of 30 to 40 dB are common.

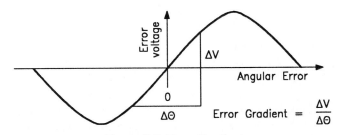

Figure 7-5. Error Gradient.

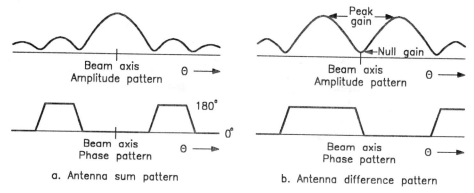

a. Antenna sum pattern b. Antenna difference pattern

Figure 7-6. Real Tracking Antenna Pattern.

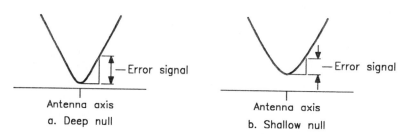

a. Deep null b. Shallow null

Figure 7-7. Error Pattern Null Detail.

$$DON = G_{E(Pk)} / G_{E(Null)} \qquad (7\text{-}1)$$

DON = the depth-of-null

$G_{E(Pk)}$ = the peak antenna gain of the error channel

$G_{E(Null)}$ = the error channel antenna gain in the null

The result of poor depth-of-null is shown in Fig. 7-7. In Fig. 7-7a, the depth-of-null is large, and in the null a small angular error produces a large error voltage (the null error gradient is large).

7-3. Conical Scan Angle Tracking

Conical scan systems use an antenna beam which is offset slightly (*squinted*) from the antenna's axis, and the squinted beam is scanned rapidly in a circular path around the axis. If the target is not located on the axis of the antenna, the amplitude of the signal echo varies with the antenna's scan position, because different gains are pointed at the target at different points in the scan. The tracking system senses the echo signal amplitude as a function of scan position to derive the target's location from the scan axis. This track error is used to automatically position the antenna axis to coincide with the target location. The target position is derived sequentially, and several target echoes are required to locate the target.

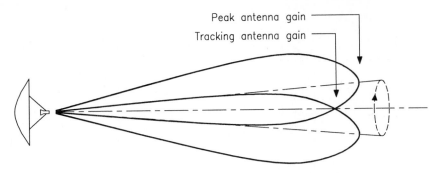

Figure 7-8. Conical Scan Beam Configuration (3-dB Crossover).

Many conical scan systems scan both the transmit and receive beams. Others scan only the receive antenna beam. This latter type of tracking, called COnical Scan on Receive Only, or COSRO, is much less susceptible to a type of ECM called gain inversion jamming, described later in this section. Some other systems must scan the transmitter in order to function properly.

Conical scan angular error development: Angular errors are developed from the amplitude modulation of the echo signal caused by the antenna's scan. The target being tracked is (usually) edited by a range gate or Doppler gate so that only the signal from the target of interest is applied to the tracking circuits. Figure 7-8 shows a typical beam configuration for conical scan tracking.

For true tracking, the scan frequency must be at the Nyquist rate or greater. In a pulsed radar, there must be at least four samples (pulses) per scan: two for sampling in azimuth, and two for elevation.

The beam is squinted and scanned either mechanically by offsetting the feed and rotating or wobbling it, or electronically with phase shifters. If the feed is rotated, the polarization of the antenna changes with scan position. If it is wobbled, also called nutated, the polarization remains constant (see Fig. 7-9).

The amount of squint is usually such that the gain of the beam on the antenna axis is reduced from peak gain by somewhere between 3 dB and 6 dB. If both the transmit and receive beams are scanned, the two-way gain loss at crossover is double (in dB) that of the one-way loss. To achieve a 3-dB crossover (one-way, 6-dB two-way) requires that the squint angle be about half the 3-dB beamwidth. Figure 7-8 shows a 3-dB squint.

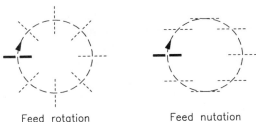

Figure 7-9. Feed Rotation and Nutation.

Feed rotation   Feed nutation

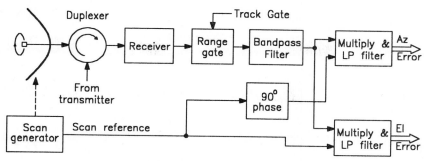

Figure 7-10. Conical Scan Error System Block Diagram.

There is a squint loss (Ch. 3) caused by not tracking at the peak of the antenna gain. This loss is determined by the squint angle and by whether or not the transmit as well as the receive beam is scanned. A 6-dB squint has a larger error gradient at the expense of a higher squint loss.

Figures 7-10 is block diagram of the conical scan error system. The scan generator causes the squinted beam to scan in its circular pattern, and at the same time produces a sinusoidal signal (the scan reference) which varies at the scan rate and whose phase is locked to the scan. In the example shown here, the positive peak of the reference occurs when the squinted beam is above the axis, the negative zero crossing when the beam is to the right, the negative peak when the beam is down, and the positive zero crossing when the beam is to the left. The scan reference is used to demodulate the errors.

Figure 7-11 shows the cross-section of a typical scan, with the letters indicating various possible target positions in the scan. Figure 7-12 shows the signal echo patterns, called the signal history, received from a non-fluctuating target at each of the positions of Fig. 7-11. If the target is located at position O, on the axis of the antenna, there is no track error and its amplitude does not vary because of the scan. The signal history of targets at any other position exhibits sinusoidal amplitude modulation. The amplitude of the modulation gives the amount of track error; the modulation phase with respect to the scan reference gives the direction. If the target is scintillating, this amplitude variation interferes with the tracking process and ultimately limits the accuracy of this type of tracking (Sec. 7-10).

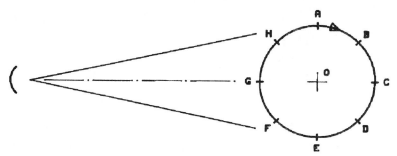

Figure 7-11. Scan Cross-Section and Target Positions.

7-3. Conical Scan Angle Tracking

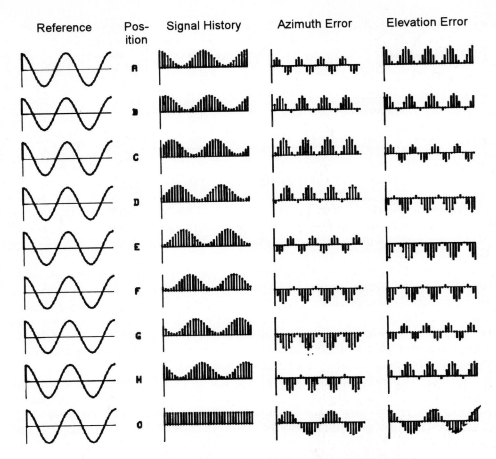

Figure 7-12. Waveforms in Conical Scan Error Development.

Consider target position A. The elevation error is the product of the signal history and the scan reference, both of which vary sinusoidally at the same frequency. The product of two sinusoids of the same frequency is a sinusoid with double the frequency of the two inputs and a DC component which is determined by the amplitude of the signals and the phase difference between them (Eq. 7-2).

$$V_{EL} = V_{SE} \cos(2\pi f_{SC} t + \phi_{SE}) \, V_{SC} \cos(2\pi f_{SC} t) \qquad (7\text{-}2)$$

V_{EL} = the elevation error voltage

V_{SE} = the envelope of the signal history voltage

f_{SC} = the scan frequency

ϕ_{SE} = the phase of the signal history envelope with respect to the scan reference

V_{SC} = the scan reference voltage (henceforth normalized to 1)

$$V_{EL} = 1/2 V_{SE} V_{SC} [\cos(\phi_{SE}) + \cos(4\pi f_{SC} t + \phi_{SE})] \qquad (7\text{-}3)$$

Chap. 7. Tracking Radar

The second term of the product is removed in a low-pass filter and the first (DC) term is applied to the tracking circuits to reposition the antenna beam. A target in position A gives a large positive average value in its elevation error.

$$V_{EL} = 1/2 V_{SE} V_{SC} [\cos(\phi_{SE})] \qquad (7\text{-}4)$$

A target in position E produces a signal history modulated at 180° phase from the scan reference. This product has a negative DC component. In this tracking scheme, a positive elevation error drives the beam upward, and a negative error drives it downward.

Targets in positions C and G produce signal histories modulated at 90° phase from the reference. The product of two equal frequency sinusoids at 90° has a DC component of zero. Thus targets in these positions produce zero elevation error, which is correct.

The azimuth error is the product of the signal history and the reference phase-shifted 90° (Eqs. 7-5, 7-6, and 7-7).

$$V_{AZ} = V_{SE} \cos(2\pi f_{SC} t + \phi_{SE}) V_{SC} \sin(2\pi f_{SC} t) \qquad (7\text{-}5)$$

V_{AZ} = the azimuth error voltage
V_{SE} = the signal history voltage envelope
f_{SC} = the scan frequency
ϕ_{SE} = the phase of the signal history envelope with respect to the scan reference
V_{SC} = the scan reference voltage (henceforth normalized to 1)

$$V_{AZ} = 1/2 V_{SE} V_{SC} [\sin(\phi_{SE}) + \sin(4\pi f_{SC} t)] \qquad (7\text{-}6)$$

After low-pass filtering, the azimuth error becomes

$$V_{AZ} = 1/2 V_{SE} V_{SC} [\sin(\phi_{SE})] \qquad (7\text{-}7)$$

Most tracking receivers have automatic gain control (AGC). The function of AGC (Ch. 2) is to hold the receiver output constant, regardless of the amplitude of the signal input. AGC operates with a long time constant, and only adjusts for slow amplitude fluctuations. Thus, the AGC holds the average signal level constant, while preserving the rapid variations caused by the scan. AGC makes the tracking errors a function of only the amount of target angular displacement from the beam axis, and not a function of RCS or range.

If conventional low-pass filters were used for the functions of Eqs. 7-4 and 7-7, the short duty cycle of the error pulses would cause the result to have a very small amplitude. Therefore, as a first step in the filtering process, the pulses are stretched over the PRI in a process called boxcar generation, shown in Fig. 7-13. Usually, boxcar generation is performed before error demodulation.

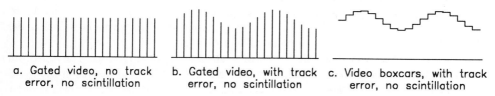

a. Gated video, no track error, no scintillation
b. Gated video, with track error, no scintillation
c. Video boxcars, with track error, no scintillation

Figure 7-13. Conical Scan Gated Video.

7-3. Conical Scan Angle Tracking

The track signal is further filtered in a bandpass filter centered at the scan frequency with a bandwidth equal to twice the servo bandwidth (Fig. 7-10). Its purpose is to remove any target echo amplitude changes at frequencies other than that of the scan. This reduces the effect of target scintillation and also of amplitude modulated (AM) jamming at other than the scan rate. AM jamming at the scan rate interferes with the track and is difficult to suppress, giving good reason to hide the scan rate, if possible. This bandpass filter is usually applied prior to multiplying the filtered signal history by the scan reference.

After the multiplication of the filtered signal history with the reference, the low-pass filter to remove the double-frequency component of the product is applied. Its cutoff frequency equals or exceeds the servo bandwidth (in many cases, this filter sets the servo bandwidth). Its action is shown in Fig. 7-14.

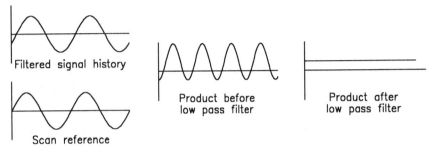

Figure 7-14. Error Generation and Filtering.

Most systems which use conical scan do so because it is less complex and expensive compared to monopulse. A few, such as the U.S. Navy's Automated Carrier Landing System (ACLS) must use it to make the system function.

Scan rates: The rate of the conical scan is determined primarily by sampling considerations and by the mechanism used to generate the scan. At least four target echoes must be received for each scan cycle, two for azimuth error determination and two for elevation. The maximum scan rate, therefore, is the one-fourth of the PRF. Beyond this Nyquist criterion, the scan rate is set by the scan mechanism. Mechanical scans generally require more time than electronic scans, and rates in the order of 30 scans per second have proven popular.

Conical scan vulnerabilities: Conical scan is vulnerable to amplitude modulated jamming, in particular to a self-protection technique called gain inversion (Fig. 7-15), where a false target signal is created with amplitude variations which are out of phase with the tracker's scan. This produces errors of the wrong sense and drives the beam off target, rather than centering the antenna axis at the target's position. This vulnerability is exacerbated when the conical scan is present on the transmitted signal, since the jammer can lock to the scan and produce exact out-of-scan-phase ECM.

The radar's transmitted signal is received and repeated by the jammer. The repetition of the signal is assumed to be instantaneous so that the ECM signal arrives at the radar simultaneously with the echo of the target carrying the jammer. If the target is off the radar antenna's axis, the jammer receiver sees the radar's transmit signal as varying sinusoidally. The phase of the variation is the same as the phase of the variation of the echo received by the radar. The jammer demodulates the transmit amplitude variation, inverts it, and sends it back to the radar. The radar receives its *lowest*, rather than highest, signal when the antenna beam is pointed closest to the target (Fig. 7-15b). If the jammer amplitude is much greater than that of the echo, the radar tracks the jammer. With inverted angle errors, the radar will be unable to track.

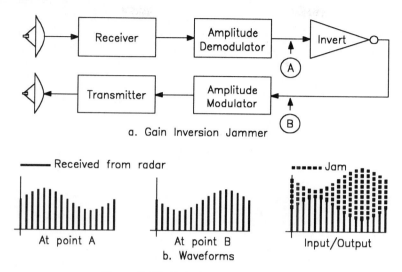

Figure 7-15. Gain Inversion Jamming.

7-4. Lobing Angle Tracking

Lobing radars generate angular errors by sequentially placing squinted beams in discrete angular positions around the antenna's axis. This is similar to conical scan, except that the beam occupies discrete locations instead of being continuously scanned. As with conical scan, target signal amplitude contains the track error information, and the target's location is determined sequentially, with several echoes being required.

Modern lobing systems usually squint electronically. Since the beam position can be changed very rapidly, they seldom scan the transmitted beam, performing lobe on receive only, or LORO. They can also use virtually any scan pattern (not necessarily A to B to C to D as in Fig. 1-57) and do not have to use the same pattern each time. For these two reasons, lobing is much less affected by amplitude modulated jamming than is conical scan.

LORO is commonly used in track modes of multi-mode radars and is a common tracking scheme for SARH (semi-active radar homing) missiles. It is, in most cases, however, being gradually replaced by monopulse tracking.

Figure 7-16 depicts a typical electronically lobed antenna. It is divided into quadrants, with each quadrant being fed through a single phase shifter. On transmit, the phase shifters are all set to the same value, producing a boresight beam. On receive, the phase shifters are set to provide a *squinted* beam, with the squint angle usually being a constant and the squint direction changed for each pulse. Four squint directions are usually used, as shown in Fig. 7-17.

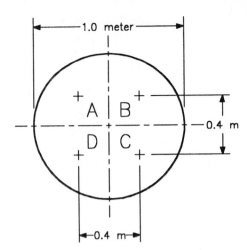

Figure 7-16. Lobing Antenna.

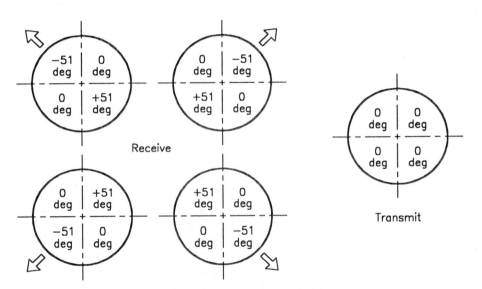

Figure 7-17. Lobing Beam-Steering Example.

To explain the principle, consider that the antenna of Fig. 7-16 receives target echoes on a frequency of 10.00 GHz (wavelength is 3.0 cm) and that the effective centers of the four quadrants are as shown in the figure. Equation 7-8, derived in Ch. 9, describes beam steering in an array. It indicates that if adjacent quadrants (which behave as elements in the four-element array) are fed 51° from one another, the phase steering (squint) angle is about 0.61°. Figure 7-17 shows the four phase combinations required to squint the beam to four discrete locations, in this case up and left, up and right, down and right, and down and left. If the antenna also transmits, a fifth phase pattern is used, with all phase shifters set to the same value. This produces an on-axis transmit beam.

$$\Delta\phi = \frac{360 \, S \, \sin\theta}{\lambda} \qquad (7\text{-}8)$$

$\Delta\phi$ = the phase difference between array elements (the quadrants, in this case)

S = the space between elements (centers of radiation of the quadrants, in this case)

θ = the beam steering angle from the antenna's axis

λ = the wavelength

Most modern lobing antennas are planar slot arrays. Figure 7-18 is a schematic diagram of an antenna of this type with phase shifters. The phase shifts and switches in the figure depict the phases necessary for lobing. There are four phase shifters, each of which can be switched to three phase values.

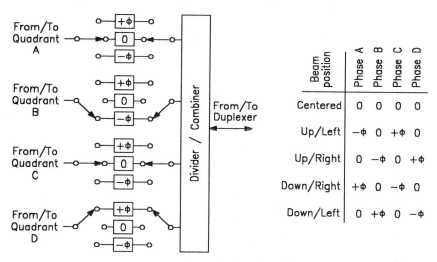

Figure 7-18. Lobing Beam-Steering.

7-5. Amplitude-Comparison Monopulse Angle Tracking

This method of generating angular track errors is similar to lobing, in that a target is examined with multiple squinted antenna beams and the relative amplitude of the echoes in each beam determines the angular error. The difference is that the beams are produced simultaneously instead of sequentially. This is the origin of the name monopulse, meaning that a full angle error set can be derived from a single pulse. In addition to other advantages, such as higher error rates, monopulse is immune to gain inversion and AM jamming. There are many variations of monopulse tracking, only some of which use four beams. Please see the Further Information section for references.

The concept of monopulse tracking is easiest explained with reflector antennas, but applies equally well to arrays. The initial discussions will be of reflectors, with arrays left until some concepts are established.

7-5. Amplitude-Comparison Monopulse Angle Tracking

Consider Fig. 7-19. Energy arriving at the antenna from the direction of the antenna's axis focuses to the parabola's focal point. A compound feed of at least four antennas is placed at the parabolic focus, as seen in Fig. 7-20, and this on-axis energy divides equally among the four feeds. Energy arriving from off-axis focuses to a point other than the focal point and divides such that some feed antennas receive more energy than others. The pattern of energy amplitudes in the four feeds is determined by the amount and direction of the signal angular offset from the reflector's axis. Considered another way, the four feeds individually form the four beams, with feed A generating a beam down and to the right, feed B generating the beam pointed down and to the left, and so forth.

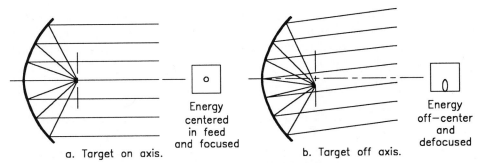

Figure 7-19. Monopulse Concept.

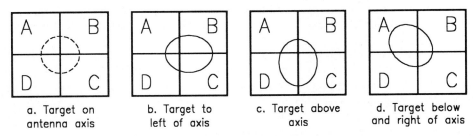

Figure 7-20. Monopulse Feed.

The distances between horns are small and the phases of the four signals A, B, C, and D are within a few degrees of one another. For this discussion, it will be assumed that the phases are identical.

The four feeds are connected in a microwave circuit called a monopulse comparator, shown in Fig. 7-21. The feeds are treated as pairs. In this case, the pairs are arranged vertically, but horizontal pairing works just as well. The signals from the two feeds in a pair are combined so that their sum and difference are formed. The four new signals are $A + D$, $A - D$, $B + C$, and $B - C$. Note that if $A > D$, then $A - D$ is positive, and if $A < D$, $A - D$ is negative. Positive and negative in microwave signals refers to phase. Positive signals are 0° to the reference and negative signals are 180° to the reference. Monopulse tracking is self-referencing, and the phase reference is the sum signal ($A + D$), not the transmit phase, as is the case in virtually all other parts of the radar.

The two pairs of sums and differences are themselves applied to sum and difference circuits. The second-level sum of first-level sums is $A + B + C + D$, and is called the sum

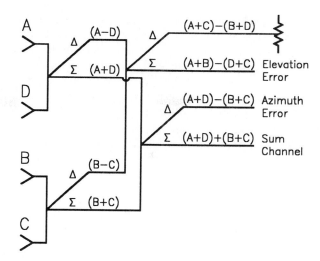

Figure 7-21. Monopulse Comparator Action.

signal. It is used for range and Doppler tracking, for displays, and serves as the phase reference for the error signals.

The second-level difference of first-level sums is, from Fig. 7-21, the azimuth error. Again, its polarity indicates phase, with reference to the sum signal.

The second-level sum of first-level differences is the elevation error, with the phase reference still being the sum signal. The purpose of the second-level difference of first-level differences is to terminate signals which would otherwise cause the azimuth and elevation channels to cross-talk.

It is important in this type monopulse that the angular errors be small enough so that the signals arriving at the four feeds be in-phase with one another. It helps that the surface of the feed is physically small, a few wavelengths across at most. In arrays, where the antenna surface is very large and signals arriving from angles off-axis present different phases to the segments into which the array is divided, the phase beam-steering shifters must equalize all the segment phases before the monopulse errors are developed.

The transmit energy is handled differently in different systems. In the four-element feed described above, the comparator sum channel port connects to the duplexer and the transmit signal is applied here. In others, a separate feed is used for the transmit and sum channel. In still others, the transmit feed is separate from receive. This is difficult to implement properly, but eliminates the need for a duplexer. For examples, see the feeds in Fig. 7-22. Many other configurations are possible. See [1] and [4].

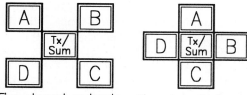

Five-element rectangle   Five-element diamond

Figure 7-22. Some Other Monopulse Feed Configurations.

7-5. Amplitude-Comparison Monopulse Angle Tracking

The echo signals exiting the three active ports of the monopulse comparator are at the radio frequency and very low levels. A separate receiver is required for each, which is the principal drawback in implementing monopulse. The stringent requirements on the three receivers are that the gains and phase shifts of the three error channels be tightly controlled. Figure 7-23 shows two simplified block diagrams of receiver schemes, one of which uses time-division multiplexing so that the two error signals share one receiver. Since half the error signal energy is lost in the multiplexed receiver, it is used only in applications where size and weight outweigh the need for tracking sensitivity. With modern circuitry, the switches are often heavier and bulkier than the third receiver, making the three-receiver scheme preferable even in applications where size and weight are major design factors. Detailed block diagrams of three-channel receivers are found in Ch. 10.

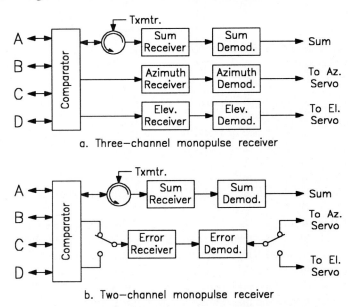

Figure 7-23. Monopulse Receiver Configurations.

Because the monopulse errors are derived from the amplitude and phase of the comparator output signals, the receiver channels must be matched in gain and in phase at all gains for successful tracking to occur.

Monopulse tracking with array antennas: Amplitude comparison monopulse tracking with array antennas is more complex than with reflectors. At least five beams must be produced. They are the transmit beam and the four squinted error-detecting beams. With the four- or five-horn (or more) reflector feeds, each feed produced a squinted beam. This is not true with arrays. If the array were divided into four quadrants (Fig. 7-24), each sub-array would simply produce a wider beam on-axis and no track error would result.

A clue as to what is required is found in the discussion of lobing radars. The squinted beams were formed sequentially using switchable phase shifters. To create four squinted beams simultaneously requires that the signals from the four quadrants be divided into four

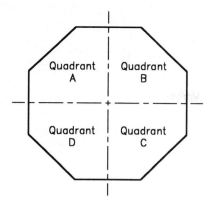

Figure 7-24. Array Divided for Monopulse Tracking.

parts, which in turn feed four beam-forming networks. These networks, each with the phase shifters necessary to produce one beam, synthesize the four squinted beams (upper left, upper right, lower right, and lower left). These four beams are fed to the sum-and-difference circuitry of a monopulse comparator to produce the sum, azimuth error, and elevation error signals necessary for tracking. Figure 7-25 shows the concept and Fig. 7-26 shows the dividers, phase shifters, and combiners necessary to make it work. Since the phase shifts — at least over a narrow bandwidth — are fixed, they can be implemented as lengths of waveguide. Planar slot array antennas which do monopulse tracking are easy to identify by the extremely complex manifold on the back of the plate. Much of this waveguide complexity is in the dividers, combiners, and phase shifts to implement monopulse tracking.

The above discussion does not apply to space-fed lens arrays (see Ch. 9). They behave, for tracking purposes, exactly like reflectors and can use the relatively simple feeds and comparators of reflectors.

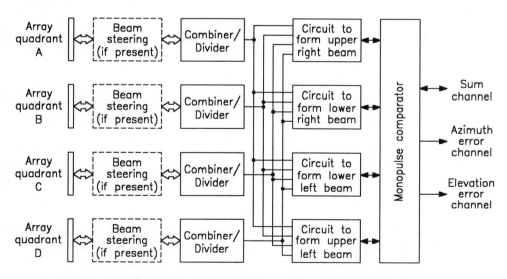

Figure 7-25. Monopulse Tracking Array Simplified Block Diagram.

7-5. Amplitude-Comparison Monopulse Angle Tracking

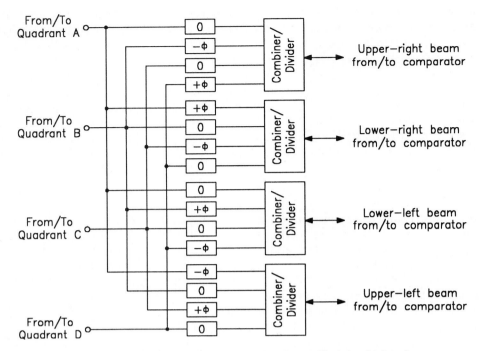

Figure 7-26. Array Monopulse Tracking Beam-Forming Networks.

Arrays can be divided into more than four segments for tracking. Although a discussion of these implementations is beyond the scope of this book, array tracking with up to 16 segments is done and has certain advantages. The reader is referred to the Further Information section of this chapter.

Mechanically scanned arrays, such as the planar slot array of Fig. 1-3, are pointed at the target such that the energy arriving at each quadrant is in-phase with that of the other quadrants. Tracking and mechanical movement of the antenna maintain this phase relationship throughout the track. Phased arrays can have energy arriving at the quadrants which is out of phase with the others. Before the four quadrant signals can be divided, shifted, and combined for monopulse tracking, their phases must be equalized. This is a natural result of the beam-steering phase shifters and requires no additional circuitry. Note in Fig. 7-25, however, that the beam-steering phase shifts must occur before the squinted beams for tracking can be formed.

7-6. Phase-Comparison Monopulse Angle Tracking

Phase monopulse differs from amplitude monopulse in that the phase of the signal received in the different antenna elements determines angle errors (in amplitude monopulse, a requirement was that the phases of the signals at the elements be the same). Phase monopulse uses an array of at least two antennas spaced some distance from one another. Separate arrays are required for azimuth and elevation or for north/south and east/west, with a complete phase monopulse tracking system requiring at least four antennas. The phase of the signals received by the elements are compared. If the antenna's axis is pointed at the target, the phases are equal; if not, they differ. The amount and direction of the phase difference is the magnitude and direction of the error and is used to drive the antenna. Figure 7-27 shows one dimension of phase monopulse tracking.

Phase monopulse systems are normally large with respect to the wavelength, because widely spaced elements yield larger phase shifts for a given angular error. This, however, leads to ambiguities, as seen in Fig. 7-28. Here, an off-axis signal produces identical phases at the two outer elements. The ambiguities are resolved by adding antennas closer to the axis with the inner pair resolving ambiguities and the outer pair giving the system high error sensitivity. In very large systems, such as the Very Long Baseline Arrays (VLBA) used in radio astronomy, three or more antennas pairs are required for each axis.

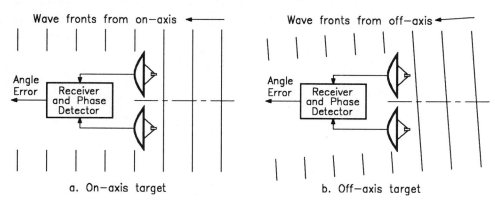

Figure 7-27. Phase Monopulse Principle.

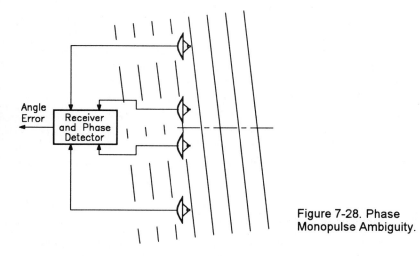

Figure 7-28. Phase Monopulse Ambiguity.

7-6. Phase-Comparison Monopulse Angle Tracking

7-7. Wideband Monopulse Tracking

Under certain conditions, it is possible to derive range and cross-range information about several scatterers simultaneously by tracking one of the objects in one range bin and using the monopulse errors in the other bins to derive their cross-range position with respect to the object being tracked [4]. In order for this technique to work, all the objects being analyzed must be within the antenna's beam simultaneously, and each must occupy a different range resolution cell. Note that this is not imaging, where many scatterers can occupy a single range bin and still be resolved.

Figure 7-29 shows the principle. Each member of the cluster of objects shown is simultaneously within the antenna's beam and each is resolved in range. The object labeled "A" is tracked by the radar, and its track error is zero (or corrected with the monopulse errors to zero). The presence of targets in other range bins gives their range position with respect to the tracked object, and the monopulse tracking errors in these other range bins indicate their relative cross-range position. The individual target positions can be found in three dimensions: range, azimuth cross-range, and elevation cross-range.

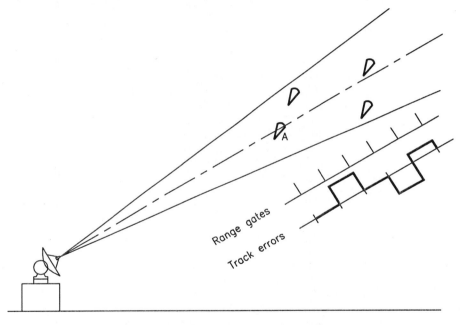

Figure 7-29. Wideband Monopulse Concept.

The accuracy of this technique depends on several factors.

— Track errors are proportional to the ratio of the video amplitude in the error channel to that of the sum channel. Thus the amplitude of the signals from each object other than the one nulled by the antenna tracking mechanism must be accurately measured in all three channels: sum, azimuth, and elevation. Alternatively, each range bin can be served by its own automatic gain control (AGC) voltage, causing the sum channel signal amplitudes to be constant within each range bin regardless of target RCS.

— The ratio of the error channel amplitudes to sum channel amplitude must be known functions of angle errors. These quantities are functions of antenna patterns (sum and difference) and receiver gain linearities. Both must be known over the angular span used and for all frequencies used [5].

— Azimuth and elevation tracking errors must be independent of one another and must not cross-talk.

An example of results obtained by this method is shown in Fig. 7-30.

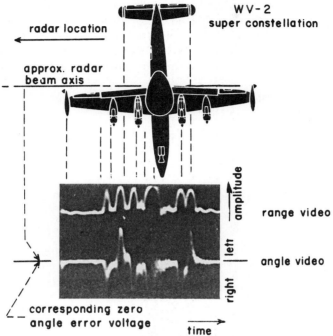

Figure 7-30. Wideband Monopulse Example. (Reprinted from D.D. Howard [6]. ©1990 McGraw-Hill Book Company, used with permission)

7-8. Track-While-Scan Principles

Multiple targets can be followed simultaneously by search radars in a process called track-while-scan, or TWS. As the radar scans by each target of interest, its position is reported to a tracking program in the radar's data processor. After several consecutive reports, the data processor can smooth the target positions and predict each target's location for the next scan. Because the position updates are separated widely in time (usually several seconds) track-while-scan is not, strictly speaking, tracking. The accuracy of this method is lower than true tracking systems, where the target's maneuvering capability is Nyquist sampled, that is, sampled at a sufficient rate to give a true representation of the target's flight path — typically a minimum of 20 samples per second. It does, however, allow many targets to be simultaneously monitored by a single search radar.

Track-while-scan systems must track the location of one or more targets as the antenna searches. Because of this, the tracking circuits cannot move the antenna to center its axis on targets. The result is that, although monopulse techniques can be and are used in TWS, the most common method of angular error development is centroid detection (Fig. 7-31). This is a method by which, as the antenna scans by a particular target, a signal amplitude versus antenna position pattern is recorded and an algorithm locates its centroid. The centroid is the value of the independent variable (time) where the area under the figure to the right of this value equals the area to the left of the value. Other systems locate targets precisely with antenna difference patterns, as in monopulse tracking.

The system then places a gate around each target position and tracks the signal within the gate (Fig. 7-31c). The gates are usually in azimuth and range. On initial acquisition, the system does not know the target's direction or speed and the gate must be large enough so that any reasonable target will be within it on the next scan. As the target is observed for several scans, its motion becomes better known and the gates are made smaller. After many scans, the gate size is determined only by the radar's measurement errors and possible target maneuvering.

Algorithms for track-while-scan data smoothing and extrapolation are introduced in Sec 7-11.

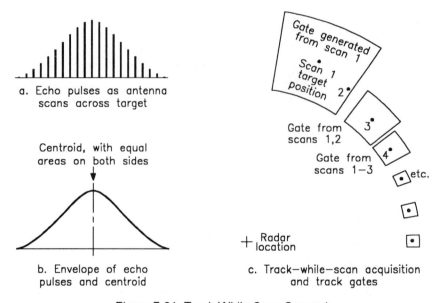

a. Echo pulses as antenna scans across target

b. Envelope of echo pulses and centroid

c. Track-while-scan acquisition and track gates

Figure 7-31. Track-While-Scan Concept.

7-9. Range Tracking

Range tracking is used to find accurate target range information and to develop the time (range) gates to exclude signals which are not part of the target tracked (Fig. 7-3). Range errors are usually developed with a split-gate method shown in Fig. 7-32. A description of a split-gate range tracker is given in Fig. 7-33, and its timing given in Fig. 7-34. This system is called a *range machine* although the days of mechanical range tracking are long gone.

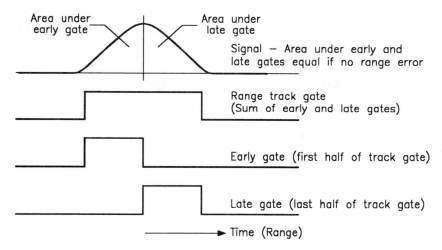

Figure 7-32. Split-Gate Range Error Development.

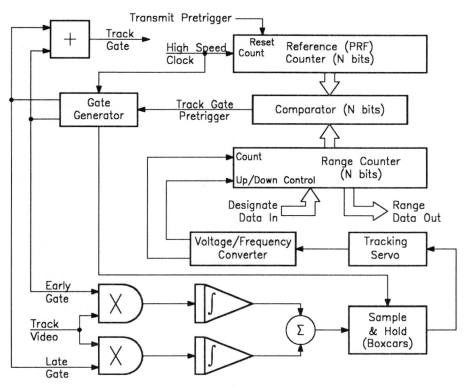

Figure 7-33. Split-Gate Digital Ranging Machine.

7-9. Range Tracking

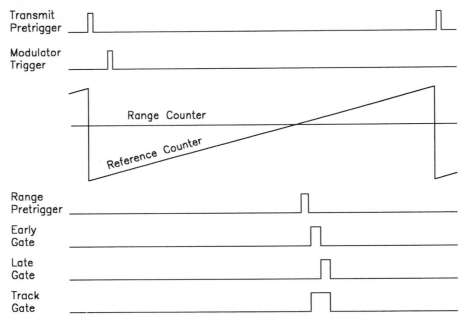

Figure 7-34. Range Machine Timing.

At some predetermined time before the transmitter is scheduled to fire, a counter known as the reference counter is set to its zero state. This counter is sometimes called the PRF counter because in many systems it also develops the transmitter timing. It then counts a high speed clock, producing the ramp count shown in Fig. 7-34. The range (time) from the transmit output to the track gate is stored in the range counter. At the time when the two counts match, a trigger is generated which, after some predetermined delay, causes the early, late, and track gates. The delay between the trigger which sets the reference counter (the transmit pre-trigger) and the delay between the coincidence pulse from the comparator (the range pre-trigger) and the crossover of the early and late gates is the same, adjusted for cable and transmission line delays within the radar.

Target echo signal is examined by the early and late gates, and the *area* under the signal in each gate is measured by the integrators shown in Fig. 7-33. If the areas are not equal, a range error voltage is generated. After processing in the range servo, this error voltage is converted to pulses which modify the content of the range counter such that the error is minimized. Thus the gates "follow" the target in time, or range.

In a digital ranging machine, the motion of the gates in time is not continuous, but occurs in small steps. The granularity of the steps is a function of the frequency of the clock counted by the reference counter. Equations 7-9 through 7-11 describe this relationship.

$$\Delta t = 1/f_{CL} \qquad (7\text{-}9)$$

$$\Delta Rg = c\,\Delta t/2 \qquad (7\text{-}10)$$

$$\Delta Rg = c/(2f_{CL}) \qquad (7\text{-}11)$$

Δt = the time step granularity of the gates

ΔRg = the range granularity, the size steps the gates take
f_{CL} = the clock frequency
c = the velocity of propagation in the medium

Thus a system using a 75 MHz clock, for example, has a range granularity of 2 m. Range machines can actually have finer granularity than the clock rate indicates by using a multi-phase clock generated in a tapped delay line. If in the above example ten equally spaced clock phases were available, the range granularity would drop to 0.2 m.

Prior to the initiation of track, the range machine must be set to the approximate range of the target, such that the target lies somewhere within the track gates. This process, called designation, is accomplished by simply presetting the range counter to the estimated position of the target.

7-10. Velocity Tracking

Some systems track targets in velocity (Doppler), both for data recovery and to provide angle track gates if angle error data is Doppler processed. The Doppler tracking servo can function in one of a number of ways; the Doppler tracking machine shown in Fig. 7-35 is typical.

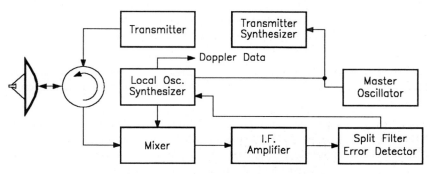

Figure 7-35. Doppler Tracking Machine.

The Doppler error is produced at the intermediate frequency in a discriminator or a split-filter system very similar to the split-gate range error discriminator. The split-filter error generator is shown in Fig. 7-36. The Doppler track error is represented by the difference between the target intermediate frequency and the system's nominal intermediate frequency. After the error is filtered and amplified in a servo, it is used to change the receiver local oscillator frequency until the Doppler-shifted signal is at the nominal IF. The Doppler shift is the amount the local oscillator has to be "pulled" from its nominal value. If the local oscillator is lower in frequency than the transmit frequency, the Doppler shift is

$$f_d = f_T - f_C - f_{LO} \qquad (7\text{-}12)$$

f_d = the Doppler shift (Hertz)
f_T = the transmit frequency (Hertz)
f_C = the COHO frequency, which is the nominal IF (Hertz)
f_{LO} = the local oscillator frequency when the track error is zero (Hertz)

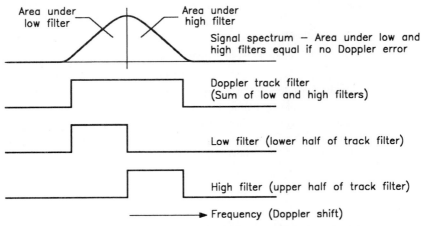

Figure 7-36. Doppler Split-Filter Error Development.

7-11. Tracking Accuracy

Introduction and concepts: The accuracy with which a radar tracks targets determines, to a great extent, its ability to fulfill its mission. Tracking accuracy is a function of the radar, the target, and the propagation medium. There are several sources of errors in tracking, described below.

— *Bias errors* are primarily equipment mis-calibrations. Examples are antenna pedestal encoders improperly set and inadequate compensation in range for internal cable lengths and lengths of signal transmission lines.

— *Propagation medium errors* are from a lack of knowledge of the signal propagation path to and from the target, or misapplication of that knowledge. Refraction of the signal as it passes through the atmosphere causes range and elevation angle errors. Range errors caused by uncertainties in the refractive index are small, but elevation errors can be large. See Table 1-2 and Sec. 6-7.

Elevation errors, discussed in Sec. 6-7, can be large — up to thousands of feet in target altitude. In over-the-horizon radars, uncompensated motion of the ionosphere causes target Doppler shift errors.

— *Servo lag errors* are caused by tracking servos requiring small errors in order to generate the voltages needed to move the antenna pedestal or range machine. For example, a Type I servo (Sec. 7-10) requires an error to maintain a constant pedestal angular velocity. The relationships which describe servo lag are

$$K_V \equiv \frac{\text{Servo drive velocity}}{\text{Lag error needed to cause that drive velocity}} \qquad (7\text{-}13)$$

$$K_A \equiv \frac{\text{Servo drive acceleration}}{\text{Lag error needed to cause drive acceleration}} \qquad (7\text{-}14)$$

$$K_J \equiv \frac{\text{Servo drive jerk}}{\text{Lag error needed to cause drive jerk}} \qquad (7\text{-}15)$$

And so forth (higher order constants)

K_V = the servo velocity constant (1/s)
K_A = the servo acceleration constant (1/s²)
K_J = the servo jerk constant (1/s³)

Velocity, acceleration, and so forth are in the appropriate units (mils/s, °/s, m/s, mils/s², ...)

Lag is in the same distance units (mils, degrees, meters, ...). Positive lags are behind the target, negative are ahead.

Example 7-1: A target being tracked flies a path causing a pedestal angular velocity of 24 mils/s and an angular acceleration of -3.4 mils/s². For this servo, K_V is 100 and K_A is 30 (typical numbers for modern servos are somewhat higher than these). Find the lag error.

Solution: Equation 7-13 gives the lag error caused by the angular velocity as +0.24 mil. Equation 7-14 gives the acceleration lag error as -0.11 mil. The total lag is the sum, +0.13 mil, with positive indicating that the pedestal/beam is pointing behind the target.

— *Noise errors* are the result of the finite signal-to-noise ratio, and the effect of noise on range is shown in Fig. 7-36. Tracking usually takes place at the centroid of the target's signal; noisy signals have uncertain centroids, as shown in the figure. Noise errors are inversely proportional to the voltage signal-to-noise ratio, which is the square root of the power signal-to-noise ratio.

— *Target-induced errors* are those which are caused by the target, not the radar. The principal example of this error is target glint, discussed in Ch. 4. The effect of glint on track accuracy is covered in the angle accuracy discussion.

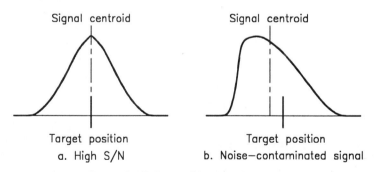

Figure 7-37. Noise Effect on Accuracy.

Range accuracy: Range accuracy is the measure of (1) how accurately the radar determines the time-of-arrival of the received signal echo, and (2) how accurately the time delay from transmit to the located echo is measured. The accuracy of signal echo time-of-arrival is, in turn, determined primarily by how precisely the radar can locate an identifiable feature of the echo.

7-11. Tracking Accuracy

Most radar systems and modes locate the time of arrival of the target at the *centroid* of the echo signal envelope, but some others measure the time-of-arrival to the *leading edge* of the signal envelope. In general, centroid detection results in the more accurate ranging. Leading edge tracking is used in special applications such as in radar altimeters, and is used in conventional radars in the presence of separating targets (such as an aircraft launching a missile) and with certain types of jamming. It is easier to jam down-range (at a greater range than the target echo) than to jam up-range. In both these cases, leading edge target location will tend to stay with the signal having the shorter range, which is likely to be the desired target.

Range errors occur by three mechanisms (see Fig. 1-22). Time-of-arrival, or "noise" errors are errors in the measurement of the time-of-arrival of the signal, caused primarily by interference contaminating the echo pulse. This results in a signal centroid occurring at a time other than that of the actual center of the echo pulse (Fig. 7-37). Time-base errors are caused by erroneous time bases and mis-calibration of the system, such as zero range set at some time other than the center of the transmit pulse at the antenna. Servo lag errors occur in tracking systems where the range servo does not "keep up" with the target, with the result that the reported position lags the true position. Since most modern range machines are electronic and massless, range servo lag is much less of a problem than in the angular dimensions.

Time-of-arrival ("Noise") errors: The main contributing parameters to signal echo time-of-arrival determination are the pulse width and the signal-to-interference ratio of the processed signal. Figure 7-38 shows the relationship between the signal plus noise and the time-of-arrival error.

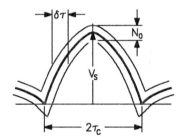

Figure 7-38. Range Noise Error.

If one simplifies the signal shape of Fig. 7-38 to triangular, or takes the time error $\delta\tau$ at the average slope between zero and the signal peak, the relationship between signal, noise, and time error becomes a simple proportion [7].

$$\frac{\delta\tau}{N_0} = \frac{2\tau_C}{V_S} \qquad (7\text{-}16)$$

$\delta\tau$ = the RMS time-of-arrival error

N_0 = the RMS noise voltage

V_S = the peak signal voltage

τ_C = the compressed pulse width if compression used, the pulse width otherwise

Solving the proportion for time-of-arrival error and substituting the power signal-to-noise ratio for the ratio of the voltages give an expression for range noise error.

$$S/N = 1/2 \, (V_S / N_0)^2 \tag{7-17}$$

S/N = the power signal-to-noise ratio

$$\delta\tau = \frac{2\tau_C}{\sqrt{2\,S/N}} \tag{7-18}$$

The simplistic approach taken in the above derivation is refined by Barton and Ward [8]. Their relationship for time-of-arrival accuracy from a single pulse is

$$\delta\tau = \frac{\tau_C}{2\sqrt{S/N}} \quad \text{(single pulse)} \tag{7-19}$$

$$\delta\tau = \frac{1}{2B\sqrt{S/N}} \quad \text{(single pulse)} \tag{7-20}$$

B = the matched bandwidth to the pulse τ_C

Range measurements are almost never made with a single pulse. For multiple pulses, the time-of-arrival error given by Barton and Ward is

$$\delta\tau = \frac{1}{2B\sqrt{PRF\,T_d\,S/N}} \quad \text{(multiple pulses)} \tag{7-21}$$

$$\delta\tau = \frac{\tau_C}{2\sqrt{PRF\,T_d\,S/N}} \quad \text{(multiple pulses)} \tag{7-22}$$

PRF = the pulse repetition frequency of the pulsed radar

T_d = the look, or dwell time (time over which data is gathered for the measurement).

Range error is found from time of arrival error by

$$\delta R = v_p \, \delta\tau / 2 \tag{7-23}$$

δR = the RMS range error

v_p = the velocity of propagation

Note that the relationships given in Eqs. 7-21 and 7-22 assume that no ranging error results from target motion during the look. This can be true only if the range measurement has *a priori* knowledge of the target's motion parameters. In the general case where this knowledge does not exist, long look times degrade rather than enhance range accuracy.

Time delay measurement errors: Errors in time delay measurement are from two sources. First are bias errors, caused by various system calibration errors, including mistiming of the transmit pulse. Second are time base errors, or errors in the "clock" which reads the target time of arrival. Bias errors are generally independent of target range. Clock errors cause a ranging error which is a linear function of target range. These errors are summarized in Eq. 7-25.

$$\Delta T_{p(td)} = \Delta T_B + \Delta T_C R \tag{7-24}$$

$\Delta T_{p(td)}$ = the total propagation time error (seconds)
ΔT_B = the time bias error (seconds)
ΔT_C = the clock error (seconds per meter)
R = the target range (meters)

Once the total time error is determined, the range error is found as

$$\delta R = \frac{c \, \delta T_p}{2} \qquad (7\text{-}25)$$

δR = range error (meters)
c = the velocity of propagation (meters per second)
δT_p = the total round-trip propagation time error (seconds)

Angle accuracy: Several factors determine the accuracy to which a radar can locate a target in the angular dimensions: (1) antenna beamwidth, with narrower antenna beams locating targets more accurately; (2) signal-to-interference ratio (higher ratios result in more accurate positioning); (3) target amplitude fluctuations (scintillation); (4) target phase fluctuations (glint), and (5) in the case of tracking radars, servo system noise.

Figure 7-39a shows the contribution of each of these factors in angular signal location for conical scan and lobing. Figure 7-39b shows the same for monopulse. Note that, whereas conical scan and lobing accuracy are affected by target scintillation, monopulse is not.

Signal-to-noise ratio is the first factor affecting accuracy. In Fig. 7-39, the line labeled "noise" represents the relative random noise interference generated in the receiver as a function of range. For a given RCS target, the effect of noise increases with target range. The error attributed to noise is inversely proportional to the voltage signal-to-noise ratio, derived below.

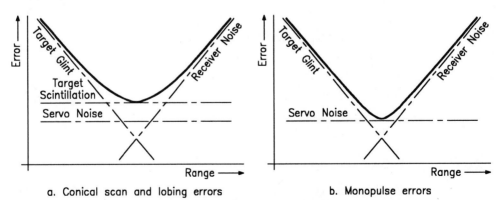

Figure 7-39. Angle Error Components.

The derivation of angle noise error is on the same principle as that for time-of-arrival error (after Toomay [7]). Figure 7-40 shows the concept. Assuming the beam shape is triangular or that the error slope is taken half way from null to peak gain (approximately the 3-dB beamwidth), a proportion exists as

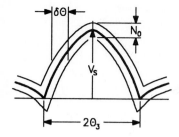

Figure 7-40. Angle Noise Error.

$$\frac{\delta\theta}{N_0} = \frac{2\theta_3}{V_S} \qquad (7\text{-}26)$$

$\delta\theta$ = the RMS angle error
θ_3 = the 3-dB beamwidth
N_0 = the RMS noise voltage
V_S = the peak signal voltage

Solving the proportion for $\delta\theta$ and substituting the power signal-to-noise ratio (Eq. 7-18) gives

$$\delta\theta = \frac{2\theta_3}{\sqrt{2\ S/N}} \qquad (7\text{-}27)$$

As with range, Barton and Ward [8] have refined the above relationship as

$$\delta\theta = \frac{\theta_3}{K_{AM}\sqrt{2\ \tau\ B\ S/N}} \qquad (7\text{-}28)$$

$\delta\theta$ = the RMS angular error caused by noise
θ_3 = the 3-dB beamwidth of the antenna
K_{AM} = a constant whose value depends on the type of angular measurement (monopulse or conical scan) [8]
τ = the pulse width
B = the equivalent noise bandwidth
S/N = the power signal-to-noise ratio (valid for > 6 dB)

The error caused by noise is proportional to range^4 for a constant RCS target.

Target-induced angle errors: These errors are caused mainly by target phase fluctuations, or glint (Ch. 4). Glint errors are linear errors; that is, the amount of error is expressed in meters rather than in degrees. The linear glint error is solely a function of target characteristics and frequency and is not a function of range. As can be seen in Fig. 7-41, the angular glint error corresponding to a constant linear error is inversely proportional to range; the angular glint error *increases* as the target nears the radar.

$$\delta\theta_G = \frac{K_G}{R} \qquad (7\text{-}29)$$

7-11. Tracking Accuracy

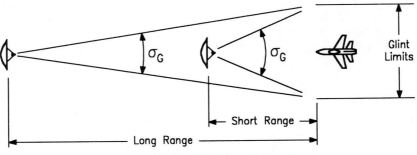

Figure 7-41. Glint Error Concept.

$\delta\theta_G$ = the angular error caused by target glint

K_G = a constant whose value depends on target span and target glint parameters

R = the range to the target

Target amplitude fluctuations cause tracking errors in tracking radars which depend on sequential target amplitude variations to develop their tracking information: the conical scan and lobing systems. Scintillation track errors are very sensitive to the frequency of the target scintillation; the closer it is to the scan or lobing rate, the greater the error. On the other hand, this frequency sensitivity can be used to reduce these errors by using rapid frequency agility, making them vary so rapidly that the angle servos average them. See Sec. 4-6.

Angular error caused by multipath: Multipath occurs when there is more than one propagation path from radar to target and back, and is primarily a problem in the elevation dimension. It is most severe for targets at low elevation angles. An example of multipath error is given in Fig. 7-42.

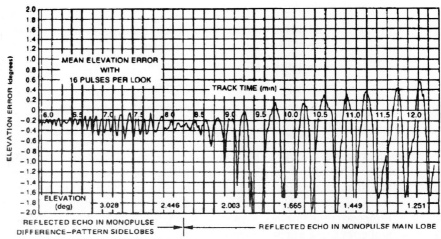

Figure 7-42. Elevation Error Caused by Multipath, S-Band Radar. (Reprinted from D. D. Howard [9]. © 1990 McGraw-Hill Book Company, used with permission)

The defense against multipath is to prevent the reflection point on ground or water from being in the main response of the antenna. Some systems where this is a particular problem implement antennas which are much larger in elevation than azimuth to produce a narrow elevation beam and exclude the reflection point from the beam. This is the technique used in the U.S. Navy's Phalanx Close-In-Weapon System (CIWS, pronounced "sea-whiz") radar, which tracks inbound threats and directs a gun toward them (it also tracks the outbound bullets). Elevation errors cause the bullets to miss the target.

Total angular error example: Figure 7-43 gives the error estimates for the AN/FPS-16(V) instrumentation tracking radar for a 5m² target at elevation angles greater than 6°.

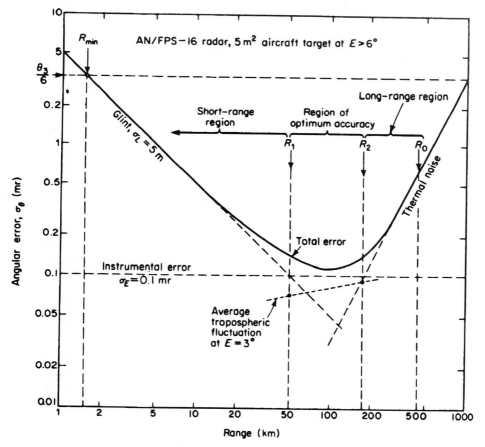

Figure 7-43. Estimated Errors for the AN/FPS-16(V) with Specified Target. (From Barton & Ward [8]. © 1984 Artech House, used with permission)

7-11. Tracking Accuracy

7-12. Tracking Servos

The tracking servos are the circuits and mechanical components which take the track errors from the error demodulators and convert them into the azimuth and elevation pedestal motions which keep the antenna pointed at the target. Or, in phased arrays, the servos move the beams to keep them on the target. A detailed treatment of servomechanisms is beyond our purpose, but we will look at some of the processes special to radar servos.

Figure 7-44 gives the block diagram of a basic electromechanical servo. Its input is a desired action. After filtering and amplification, an actuator is driven, causing an output action. The actual action is compared to the desired action, generating an action error signal, which is what drives the output action. Servos come in many types, but all of them contain this desired action/actual action/error feedback mechanism.

Radars almost always use two feedback loops in each electromechanical servo: one outer loop controlling the overall system and one inner loop controlling the velocity at which the action occurs. Figure 7-45 shows this inner loop.

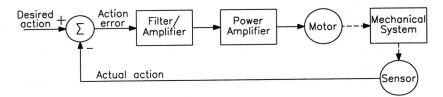

Figure 7-44. Basic Servomechanism.

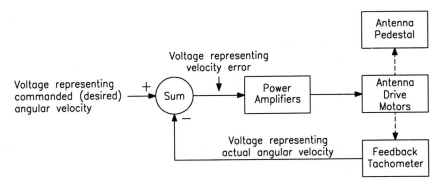

Figure 7-45. Velocity Feedback Loop.

At some point in the servomechanism, a voltage or in the case of computed servos, a digital word is generated representing the commanded velocity. An error is generated by comparing this command velocity to the actual velocity. This error is amplified and drives the actuator, in this case pedestal drive motors. The actual velocity is measured and fed back to correct the motor drive. This loop prevents the rate at which the pedestal moves from being dependent on external influences, such as wind.

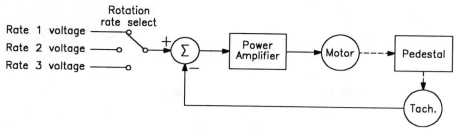

Figure 7-46. Search Radar Servo.

The velocity feedback loop may comprise the entire servo in a search radar, which must keep the pedestal moving at the desired rate. Figure 7-46 shows a simple search radar servo with velocity feedback.

Servomechanisms are classified according to the way they transfer input signals to the outputs. Analog servos are given a type number, with the name indicating the number of integrations take place between the input and output. For example, a Type 0 servo has no integrations. If the output is position, the input must be position. The Type 0 servo has little use in radar servos.

Type 1 and Type 2 servos are shown in Fig. 7-47. With the switch closed, the servo is Type 1 and has a single integration in the velocity feedback loop. Its characteristic is that *a constant input produces a changing output at a constant velocity*. In antenna servo terms, input represents velocity and a constant input produces a constantly changing pedestal position. Zero input produces a constant position at the output.

$$p_1(t) = \int v_I(t)\, dt + p(0) \tag{7-30}$$

$p_1(t) = $ the Type 1 output position

$v_I(t) = $ the input voltage

$p(0) = $ the position at time 0 (the constant of integration)

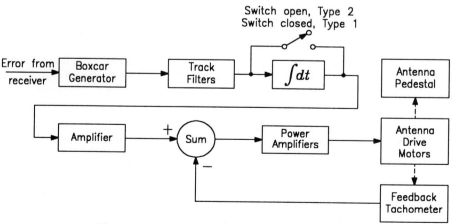

Figure 7-47. Type 1 and Type 2 Pedestal Servo.

7-12. Tracking Servos

The primary disadvantage of a Type 1 servo is that tracking a moving target requires an error. This problem is solved with the Type 2 servo, shown in Fig. 7-47 with the switch open. Here the input represents acceleration and the output is position. The output of the integrator circuit is velocity, and the constant of integration provides a constant velocity with zero input. The result is a more accurate track than with the Type 1.

$$p_2(t) = \int\int v_I(t)\, dt\, dt + \int V(0)\, dt + p(0) \qquad (7\text{-}31)$$

$p_2(t)$ = the Type 2 output position

$v_I(t)$ = the input voltage

$V(0)$ = velocity at time zero (first constant of integration)

$p(0)$ = position at time zero (second constant of integration)

Type 3 servos have two integrators plus the velocity loop and treat inputs as jerk, with output being position. They allow yet more accurate tracks but are more difficult to stabilize. Type 2 is the most common analog pedestal servo.

Over the past two decades, servomechanisms have evolved from analog to digital. In these computed servos, the tracking errors and feedback tachometers are digitized and a digital drive signal is computed. This drive signal is converted to analog and amplified, and provides power to the actuator. Hybrid servos also exist wherein part of the implementation is digital and part analog. Figure 7-48 is a simplified block diagram of a computed servo.

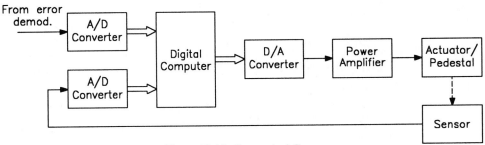

Figure 7-48. Computed Servo.

The algorithm which generates the drive signal can have many forms [10]. Its purpose is twofold. The first is to smooth the data, interpolating between error samples. The second is to predict where the target should be for the next sample. One major advantage of the computed servos is that they can be used where the target position sample rate is very low, as in track-while-scan systems. A well-known algorithm is the *Kalman Filter*. Its detailed examination is beyond the scope of this book, but its principles are quite simple.

The basis of the computed servo is the computation of a running average, formulated by Bayes in the eighteenth century.

$$X(n) = X(n-1) + [O(n) - X(n-1)]/n \qquad (7\text{-}32)$$

$X(n)$ = the average after n observations (the running average).

$X(n-1)$ = the average after $n-1$ observations.

$O(n)$ = the value of the nth observation.

This principle can be used to find a better estimate of a present value and to extrapolate the value to a predicted position at the $(n+1)$th observation. The relationship for upgrading the present value estimate (smoothing) is

$$X_S(n) = X_P(n) + \alpha[O(n) - X_P(n)] \qquad (7\text{-}33)$$

$X_S(n)$ = the smoothed estimate of X after n observations.
$X_P(n)$ = the predicted value of X after $n-1$ observations.
$O(n)$ = the value of the nth observation
α = a weighting function

The weighting function takes into account an estimate of the accuracy of observation n. If the measurement is estimated to be totally accurate, α is set to unity and the smoothed value equals the observed value. If the observation is deemed totally unreliable, the smoothed value is left equal to that predicted from $n-1$ observations.

If the value observed is position, the same smoothing principle can be applied to the positional change, or its velocity.

$$V_S(n) = V_S(n-1) + \beta[O(n) - X_P(n)]/T \qquad (7\text{-}34)$$

$V_S(n)$ = the smoothed velocity after the nth observation.
$V_S(n-1)$ = the smoothed velocity after the $(n-1)$th observation.
β = another weighting factor.
T = the time between measurements.

Note that if the estimate of the exactness of the present observation is high, β is set to unity and the smoothed velocity is corrected for the change in observed position from the prediction. If the measurement is considered unreliable, the present observation is discarded and the previous smoothed velocity becomes the present smoothed velocity ($\beta = 0$).

Knowing the present smoothed position and the present smoothed velocity allows the next predicted (extrapolated) position to be calculated as

$$X_P(n+1) = X_S(n) + V_S(n)T \qquad (7\text{-}35)$$

$X_P(n+1)$ = the predicted position for time $n+1$.

Equations 7-34 and 7-35 are the well-known α-β filter, and form the basis for tracking and track-while-scan algorithms. The remaining tasks are to calculate α and β to yield optimal tracks for various conditions.

The Kalman filter optimizes the α-β filter for straight-line motion. It can be shown [11] that the α-β filter is a Kalman filter if

$$\alpha = \frac{2(2n-1)}{n(n+1)} \qquad (7\text{-}36)$$

$$\beta = \frac{6}{n(n+1)} \qquad (7\text{-}37)$$

It can be seen from the above equations that the weighting functions α and β approach zero after a large number of observations, which minimizes the effect of later observations. This is optimal for straight-line motion, but gives unsatisfactory results on objects whose motion changes unpredictably with time, such as maneuvering radar targets.

7-12. Tracking Servos

Just as the Kalman filter optimizes α and β for straight-line motion, other optimizations have been calculated for various maneuvering targets. A detailed development is beyond the scope of this text, but it can be shown [12] that one satisfactory maneuvering relationship is established if α and β are related as

$$\beta = \frac{\alpha^2}{2 - \alpha} \tag{7-38}$$

Radar targets sometimes travel in straight lines, in which case it is desirable to approach the Kalman filter, with some restriction on the number of observations used to set the weighting factors. In other situations, their flight paths may be curved but highly predictable. In yet others, they may undergo totally unpredictable high-G maneuvers. It is often desired to adapt the tracking filter to the target conditions. One method used in track-while-scan is to use multiple tracking gates, shown in Fig. 7-49.

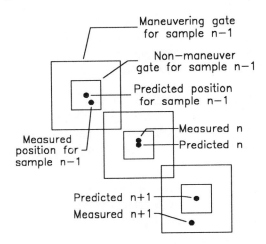

Figure 7-49. Maneuvering and Non-Maneuvering Gates.

The inner gate in the figure is formed and extrapolated to accommodate straight-line or predictably curved motions, with maneuvers of 1-G or less [13]. As long as the target stays within this non-maneuvering gate, the filter gains (α and β) are low and the gate motion is heavily smoothed. The outer gate detects unpredicted target maneuvers and causes the filter gains to be increased. This allows the tracking system to follow high-G maneuvers, but degrades the track in a noisy environment.

Some provision must also be made for target fades and false alarms. Trunk [10] suggests if a detection occurs in the maneuvering gate without one in the track gate, that two tracks be maintained, one assuming the target was not detected on the last scan and continuing the old non-maneuvering track, and one assuming the maneuvering detection was valid. Data from subsequent scans are used to resolve which track is valid. At times when the target fades below the detection threshold, the system must sense this and maintain gate motion based on the last available data (called coasting).

The α-β filter principle can, of course, be expanded to accommodate acceleration, jerk, and higher order motions.

The α-β filter can be implemented in any coordinate system the radar accommodates. In track radars and modes, spherical coordinates may be advantageous, since they are most closely related to the servos they control. In track-while-scan, it is often desirable to implement tracks in local Cartesian coordinates. In both cases, radar platform motions will probably need to be removed before sending target reports to the track algorithm.

The translation of the above relationships and remarks to radar parameters, including the signal-to-noise ratio's effect, is left to more specialized texts [14] and [15].

Secant compensation: As the elevation angle of a track increases, the required azimuth motion for a given target motion increases. The azimuth track gain must thus be increased at high elevation angles. The azimuth tracking gain relationship is

$$G_{AZ} = G_{AZ}(0) \operatorname{Sec}(\theta_{EL}) \tag{7-39}$$

G_{AZ} = the azimuth servo gain.

$G_{AZ}(0)$ = the azimuth servo gain at zero elevation.

θ_{EL} = the elevation angle of the track.

Elevation plunge: In many tracking systems, the antenna pedestal can move in elevation beyond the zenith and can track at angles around 180°. This capability, called plunge, is primarily for calibration and is rarely used for targets which go "over the top."

Visualizing an antenna pedestal tracking a target near the horizon with the pedestal plunged shows that the azimuth errors are reversed; a target which is to the right of the antenna's axis is corrected with a right-error with the pedestal at zero elevation, but with a left-error with the pedestal plunged. Thus when the elevation angle exceeds 90°, the azimuth tracking errors must be reversed.

7-13. Track Acquisition

Before automatic track can be initiated, the target must be within the antenna's beam and within some time and/or Doppler gates. Placing the radar's beam and gates in proximity to the desired target is the process of track acquisition. The process of physically moving the beam and gates to the target's estimated position is called designation and the source of this information is the designation source.

Designation sources can supply information in any or all of the radar's coordinates. The most common sources are from 2-D search radars, supplying azimuth and range; 3-D search radars supplying azimuth, elevation, and range; optical systems supplying azimuth and elevation; and manual acquisitions, where the target's position is known only roughly.

The designation data can be in synchro, DC voltage, or digital form. In the case of manual acquisitions, it is usually in the form of voice commands.

The accuracy of this information must be sufficient to place the tracking radar within its antenna beamwidth of the target, and within its acquisition range and Doppler gates. If the designation data do not meet these criteria, scans must be initiated. The type of scan used depends on the designation data available. For designation from 3-D search radars and optics, the antenna scans can be quite small. From 2-D search radars, the scan can be small in azimuth but needs to cover large elevation angles. Figure 7-50 shows some possible scans.

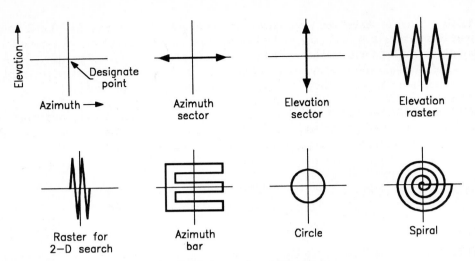

Figure 7-50. Some Acquisition Scans.

7-14. Track Anomalies

Tracking systems are subject to a wide variety of abnormal operations, some caused by the radar itself, some by the target, some by the environment, and some by active interference. In this section, some anomalous behaviors are examined.

Sidelobe tracks: It is possible to acquire a target through the radar's sidelobe rather than its main beam. This is particularly true with a beacon track, where the radar is tracking a signal from a transponder in the target vehicle, rather than the echo from the target itself (see Sec. 3-17 for a general description of beacons). This is because of the one-way usage of the antenna for both beacon interrogation and reply causing the sidelobe signals to be attenuated by the one-way sidelobes. Primary radar (skin) tracks have less problem since the antenna's two-way sidelobes have much less gain than one-way. In case of beacons in particular, sidelobe tracks are difficult to avoid and hard to identify, unless the operator can simultaneously view the beacon return and the primary target echo. Generally, only when the target is viewed by the antenna's main lobe is there enough signal to display the skin echo.

Sidelobe tracks have a very high target position uncertainty. A three-dimensional representation of sidelobes would show that a track at the peak sidelobe gain would yield a position which lies on a circle around the true target location, at least for close-in sidelobes from antennas many wavelengths above the ground or sea. In many antenna patterns, the odd-order sidelobes (first, third, and so forth) invert tracking errors, thus contributing to even more position uncertainty.

Multipath and tracking: Figure 7-42 showed that when two signal paths to the target are present, elevation position uncertainty is high. Discussions in Ch. 4 show that there is also a target fluctuation component associated with multipath which can cause the signal to drop to very low levels and may result in loss of track. One defense against multipath is to exclude the ground or sea reflection with narrow antenna beams. Another is to disengage the elevation servo during multipath conditions and track in azimuth and range only. Most tracking radars with operator intervention capabilities have this feature. The elevation data during this semi-automatic track is unreliable, but it is better than losing track.

Track drops because of low signal: At extreme ranges and with low RCS targets, obviously, tracks are less accurate and reliable than with high signals. Many tracking radars require about 20 dB of single-hit S/N for full accuracy tracks. Lower accuracy tracks can usually be maintained down to about 3- to 6-dB S/N, below which the servos will not follow the target.

Track drops because of clutter: Clutter is a problem with tracking because the servos may lock to the clutter residue instead of to the target echo. Any action which improves signal-to-clutter ratio, including narrowing pulse width (thus improving resolution in range and rejecting clutter) and activating Doppler processing will help.

Track drops because of target fluctuations: Slow target fluctuations (Ch. 4) may lower the S/N to below the required tracking level for long enough for the servos to "lose" the target, even though there is enough median RCS at the target's range for reliable track. In this case, the servo must "coast" until the signal is sufficient to track. Computed servos are good at this, and will usually keep the gates and antenna in the correct position for reacquisiton unless the target maneuvers during the signal fade. Any action which causes the fluctuations to become rapid, including frequency agility, alleviates this problem.

Track anomalies because of ECM: Many types of ECM are aimed at disrupting radar tracks ([16], [17], [18], and [19]). All axes, range, angles, and Doppler can be attacked. Disruptive techniques vary from noise jamming, which effectively lowers the signal-to-noise ratio, to false target generators which place "echoes" in positions and at Doppler shifts where none actually exist.

Many self-protection techniques are based on the augmentor shown in Fig. 7-51. In it, illumination is received from the radar, amplified, modified, and returned to the radar. If its equivalent RCS (Sec. 3-14) is much greater than the echo from the target carrying it, the radar tracks the augmentor rather than the echo. In range-gate pull-off (RGPO) [13], the modifier is a variable delay causing the return from the augmentor to appear at a different range from the target (Fig. 7-52). RGPO can be down-range with a simple delay. To deceive up-range, the delay must be lengthened to occupy the next pulse period. If the radar's transmit pulses or ranging modulations cannot be anticipated, up-range jamming is not possible.

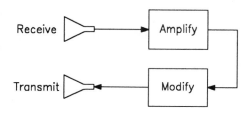

Figure 7-51. Augmentor.

7-14. Track Anomalies

A defense against range-gate pull-off is to track the leading edge of the target instead of the centroid (Fig. 7-53). If the ECM is not too much greater in amplitude than the echo, the tracker will stay with the echo. Sometime leading-edge tracking is done on log video, which compresses the amplitude difference between the ECM and echo signal.

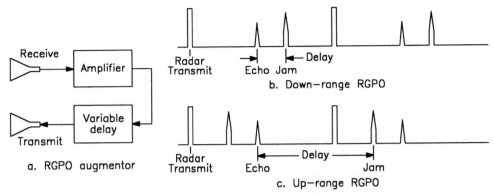

Figure 7-52. Range-Gate Pull-Off.

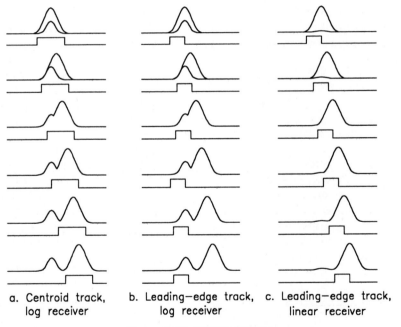

Figure 7-53. RGPO Actions.

Velocity-gate pull-off deceives Doppler trackers by modifying the augmented signal's frequency. This can be done with a single-sideband modulator or with a phase-shifter and sawtooth oscillator (Fig. 7-54). With the SSB modulator, the radar's illumination is shifted by the frequency of the augmentor's offset oscillator. This offset is added to the Doppler shift

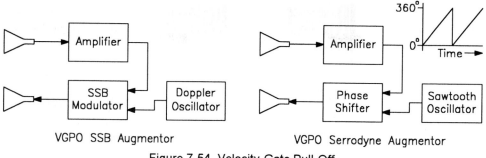

Figure 7-54. Velocity-Gate Pull-Off.

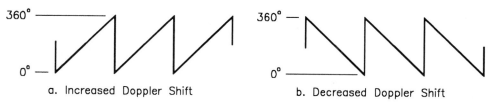

Figure 7-55. Serrodyne Waveforms.

of the target carrying the augmentor. If sending the true Doppler to the radar is desired, the offset is zero. If the radar is to be deceived into measuring the target as having a higher than actual velocity, the offset is a positive frequency (Ch. 11). If the apparent velocity is to be reduced, the offset is a negative frequency.

The phase-shifter and sawtooth oscillator implementation is called a serrodyne jammer, after the shape of the modifying wave (Fig. 7-55). The phase shift is calibrated to be 0° at the sawtooth's lowest voltage and 360° at the highest. If the wave is as shown in Fig. 7-55a, one sinusoidal cycle is *added* to the wave returned to the radar by each sawtooth cycle, and the apparent Doppler shift is increased. If the sawtooth is inverted as shown in Fig. 7-55b, sinusoidal cycles are *subtracted* and the apparent Doppler shift is decreased.

Cross-eye jamming (Fig. 7-56), consisting of two augmentors, deceives in angles by increasing the target's glint. The augmentors may or may not modify the signal in other ways, such as delay and frequency. Several techniques fall under the heading cross-eye. Figure 7-56 shows one of them.

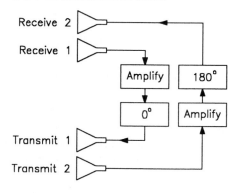

Figure 7-56. Cross-Eye.

7-14. Track Anomalies

7-15. Further Information

General information: The following three books contain extensive information on tracking and tracking performance.

[1]   D. D. Howard, Ch. 18 in M.I. Skolnik (ed.), *Radar Handbook*, 2nd ed., New York: McGraw-Hill, 1990.

[2]   D.K. Barton (ed.), *Monopulse Radar*, Volume 1 of the Artech House series, *Radar*, Norwood, MA: Artech House, 1974.

[3]   D.K. Barton and H.R. Wade, *Handbook of Radar Measurements*, Norwood, MA: Artech House, 1984.

Cited references:

[4]   D.D. Howard, "Single Aperture Monopulse Radar Multi-mode Antenna Feed and Homing Device," *IEEE Intern. Conv. Mil. Electron. Conf. Proc.*, 1964

[5]   D.D. Howard, "High Range-Resolution Monopulse Radar," *IEEE Transactions on Aerospace and Electronic Systems*, vol. AES-11, September 1975, pp. 749-755.

[6]   D.D. Howard, Ch. 18 in M.I. Skolnik (ed.), *Radar Handbook*, 2nd ed., New York: McGraw-Hill, 1990, p. 18.32.

[7]   J.C. Toomay, *Radar Principles for the Non-Specialist*, 2nd ed., New York: Van Nostrand Reinhold, 1989, pp. 57-60.

[8]   D.K. Barton and H.R. Ward, *Handbook of Radar Measurements*, Norwood, MA: Artech House, 1984, Chs. 2 through 6, Ch. 8.

[9]   D.D. Howard, Ch. 18 in M.I. Skolnik (ed.), *Radar Handbook*, 2nd ed., New York: McGraw-Hill, 1990, p. 18.47.

[10] G.V. Trunk, Ch. 8 in M.I. Skolnik (ed.), *Radar Handbook*, 2nd ed., New York: McGraw-Hill, 1990.

[11] A.L. Quigley, "Tracking and Associated Problems," *IEE Conf. Radar — Present and Future*, London: 1973, pp. 352-357.

[12] T.R. Benedict and G.W. Bordner, "Synthesis of an Optimal Set of Radar Track-While-Scan Smoothing Equations," *IRE Transactions*, vol. AC-7, July 1962, pp. 27-32.

[13] G.V. Trunk, "Survey of Radar ADT," *Naval Research Laboratory Report 8698*, June 1983.

[14] S.C. Bozic, *Digital and Kalman Filtering*, London: Edward Arnold Ltd., 1979

[15] A. Farina and F.A Studer, *Radar Data Processing*, New York: John Wiley & Sons, Inc., 1985 (two volumes).

The next reference is a two-volume set, is typewritten, is of marginal copy quality, and is frightfully expensive. It also contains virtually every openly discussed technique for ECM and ECCM. It starts with a theory section, and then presents ECM and ECCM techniques in encyclopedia form. It is an invaluable source for radar engineers.

[16] L. B. Van Brunt, *Applied ECM*, Dunn Loring, VA: EW Engineering, 1978.

Another EW source is given below. This publisher has several EW books.

[17] D. C. Schlenher, *Electronic Warfare*, Norwood, MA: Artech house, 1986.

The following two books are published either annually (*International Countermeasures Handbook*) or biannually (*Jane's Radar and EW Systems*) and are excellent sources of EW information.

[18] *International Countermeasures Handbook*, Palo Alto, CA: EW Communications, Inc., published annually.

[19] B. Blake, *Jane's Radar and Electronic Warfare Systems*, 2nd ed., Coulsdon, Surrey, UK: Jane's Information Group, 1990 (published biannually).

8

RADAR TRANSMITTERS AND MICROWAVE COMPONENTS

8-1. Transmitter Introduction and Functions

The function of the radar transmitter is to generate the electromagnetic signal with which the target volume is illuminated. There are two fundamentally different types of radar transmitters: coherent amplifying and non-coherent. Coherent transmitters produce a signal generated from internal oscillators whose phase is known prior to transmission (Ch. 2). Coherent-on-receive and non-coherent transmitters produce a signal whose phase is unknown prior to the start of the transmit output, and whose phase, if used, must be measured after the transmission starts.

The block diagram of a typical coherent transmitter is shown in Fig. 8-1, and the major blocks and parameters are described in Ch. 2. The configuration described below is typical, but the reader should realize that many configurations are possible, and a comprehensive treatment of them is beyond the scope of this book.

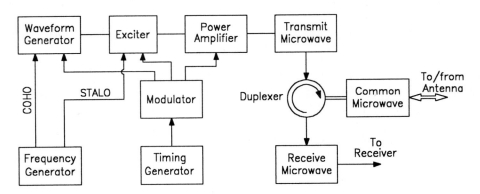

Figure 8-1. Radar Transmitter Generic Diagram.

Transmitter types: Figure 8-2 shows the "family tree" of radar transmitters, classed according to their final power amplifier and divided into three groups, each of which has applications in specific radar types.

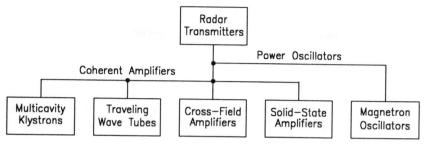

Figure 8-2. Transmitter Types.

- *O-Type* - linear beam - microwave amplifying tubes, including multi-cavity klystrons, traveling wave tubes, and twystrons
- *M-Type* - crossed-field - microwave amplifying tubes, including crossed-field amplifiers (CFAs), and magnetrons
- *Solid-state* microwave amplifiers

O-Type amplifiers are characterized by their high gain, high power capability, extremely low noise, and relatively large size and weight for their power rating. Klystrons are narrow-bandwidth devices and traveling-wave tubes (TWTs) are wideband. In applications where extremely low transmit noise is needed, including those where MTI Improvement factors must be large, O-Type tubes prevail, with TWTs favored where frequency agility is needed.

M-type crossed-field amplifiers have low gain, moderate- to high-power capability, moderate noise, wide bandwidth, and are small. They have a unique property in that if no DC power is applied, they function as low-loss transmission lines, and drive power feeds through them. They are used in applications where very high MTI-I is not needed, such as ground-based and shipboard PRF radars.

M-type magnetrons are essentially oscillating CFAs. They have moderate- to high-power capability, moderate to high noise, and are physically small and light. They are the device of choice if Doppler is not a consideration, or if low MTI-I is satisfactory. The majority of all radars in the world have magnetron transmitters, if small-boat and aircraft radars are included in the count. The trend in high-performance military and air-traffic control radars is away from this type. Another use of magnetrons is in microwave ovens — most home ovens use a CW magnetron operating at 2.54 GHz.

Table 8-1 summarizes the characteristics of the transmitter types.

8-1. Transmitter Introduction and Functions

Table 8-1. Transmitter Characteristics
(after Weil [7])

	Klystron	TWT	Linear CFA	Reentrant CFA	Solid State	Magnetron
Gain	High	High	Very Low	Low	Moderate	[a]
Bandwidth	Narrow	Wide	Wide	Wide	Wide	[a]
Noise output	Very low	Very low	Moderate	Moderate	Low	Moderate / high
DC Voltage	High	High	Moderate	Moderate	Low	Moderate
Efficiency	Low	Low	Moderate	Moderate	Moderate	High
X-Rays	High	High	Low	Low	None	Low
Relative size	Big	Medium	Medium	Small	Medium[b]	Small
Rel. weight	Heavy	Medium[c]	Medium	Light	Medium[b]	Light

[a] A magnetron is an oscillator, and as such has neither gain nor bandwidth.

[b] Solid state devices and circuits are themselves small and light. Because a large number is required to achieve high RF power, solid state transmitters tend to be large. Exceptions are active arrays. See Sec. 8-6.

[c] Gridded TWTs, often found in aircraft, are smaller and lighter than conventional TWTs, because the modulator for gridded tubes is simple and small.

8-2. Klystron Transmitters

Multi-cavity amplifying klystrons are widely used as radar transmitting amplifiers. The other type, reflex klystrons, are used in radar receivers (Ch. 10). A schematic diagram of a multi-cavity klystron is given in Fig. 8-3.

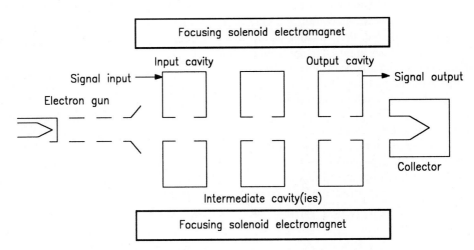

Figure 8-3. Multi-Cavity Klystron Schematic Diagram.

378 Chap. 8. Transmitters and Microwave Components

The electron gun includes the cathode, heater, anode, and control and focusing electrodes. Electrons are emitted from the hot cathode and are controlled, focused, and accelerated through a few tens of kilovolts. The output of the gun is a compact beam of electrons of typically a few tens of amperes, with all electrons moving at essentially the same velocity. The electron gun in the picture tube of a TV set works in much the same fashion, except on a much smaller scale.

The electron beam is introduced into the gap in a microwave cavity resonator. This cavity is the input or buncher cavity. Figure 8-4a shows the interaction of the cavities and beam during one of the signal half-cycles and Fig. 8-4b shows the other half-cycle.

Electrons exit the electron gun all traveling at the same velocity. When they pass the gap in the buncher cavity, they interact with the electric field of the input signal in the cavity. During one of the signal half-cycles (Fig. 8-4a) they flow into a positive field and are accelerated. During the other half-cycle (Fig. 8-4b), the electric field is reversed and the electrons are slowed.

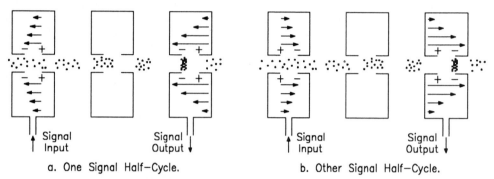

a. One Signal Half-Cycle.   b. Other Signal Half-Cycle.

Figure 8-4. Interaction of Cavities and Electron Beam.

The electrons then drift, during which time the accelerated electrons "catch up with" those which were decelerated. The result is that the electrons bunch, forming areas of high charge density alternated with areas of low charge density.

As they pass through the output cavity, these charge bunches induce electromagnetic waves in the cavity. If the bunching is complete enough, the fields in the output cavity exceed those in the input cavity which started the bunching process and gain results.

In the catcher cavity, the electrons are de-bunched, giving up their microwave energy to the cavity and its load. Once the electrons leave the output cavity, they drift to the collector. The reason klystrons have relatively low efficiencies is that significant energy remains in the beam after the catcher cavity. This energy is converted to heat (and to a lesser extent, X-rays) at the collector. Most high-power klystrons are liquid cooled, with most of the cooling needed at the collector.

One of the major factors determining gain is the physical length of the tube from buncher cavity to catcher cavity. Less velocity difference between fast and slow electrons is needed to achieve full bunching if the electrons drift for a long distance. Thus a smaller input signal to a long tube produces the same output that would require a larger input signal in a short tube, and the gain of the longer tube is greater.

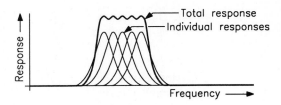

Figure 8-5. Stagger-Tuning Effect.

Only the buncher and catcher cavities are necessary for klystron action. The intermediate cavity(ies) simply make the tube function better, giving it more gain and the opportunity for wider bandwidth. Klystrons, with their resonant cavities, are inherently narrow bandwidth tubes. The addition of the intermediate cavity(ies) makes stagger-tuning of the many resonant circuits (tuning each to a slightly different frequency) possible, widening the overall bandwidth (Fig. 8-5).

The bunching process is displayed graphically in an Applegate diagram (Fig. 8-6). Its vertical axis is distance across the tube and the horizontal axis is time. The lines represent individual electron trajectories. The slope of the line thus represents electron velocity.

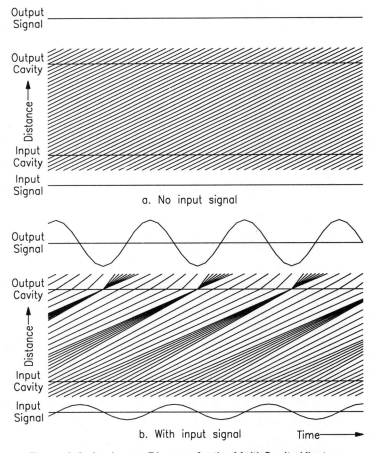

Figure 8-6. Applegate Diagram for the Multi-Cavity Klystron.

The focusing solenoid electromagnet causes a DC magnetic field longitudinal to the electron beam. The electrons are all of the same charge and polarity and repel one another. The beam tends to diverge, and as electrons move perpendicular to the beam, they cross the focusing magnetic lines of force and are directed back into the beam.

Electrons which stray from the beam and strike the internal structure of the tube provide a way of testing the focusing's effectiveness. Since the drift tube is usually connected to its body of the klystron, these stray electrons cause current to flow from the body to ground. The focusing electromagnetic field is simply adjusted to minimize this body current. The DC biases and connections necessary to measure the body current are shown (for a traveling wave tube, which is bias-wise almost identical to a multi-cavity klystron) in Fig. 8-13.

The most common focusing is electromagnetic. Focusing is also possible with permanent magnets, which generally makes smaller and lighter tubes, but sacrifices the opportunity to adjust the focusing to minimize body current. A tube focused this way is called permanent magnet focused klystron — PMFK.

Electrostatically focused klystrons — ESFKs — are also used. In these tubes, the electron beam periodically passes through a ring-like structure called a "lens" which has on it a high negative voltage. The negative voltage repels the electrons, "squeezing" them back into a tight beam. Unlike magnetic focusing, electrostatic focusing is not applied continuously and the electron beam spreads between lenses. Thus the lenses must be numerous and close together.

8-3. Traveling Wave Tube (TWT) Transmitters

The traveling wave tube transmitter is characterized by high gain, broad bandwidth, very low noise, and in some varieties, easy controllability with low-level modulation. Figure 8-7 is a simplified schematic diagram of a TWT.

Note its structural similarity to the multi-cavity klystron. The only significant difference is the signal structure, which uses a slow-wave structure (SWS) in place of the cavities. A SWS is a transmission line where the forward signal propagation is slower than normal. A

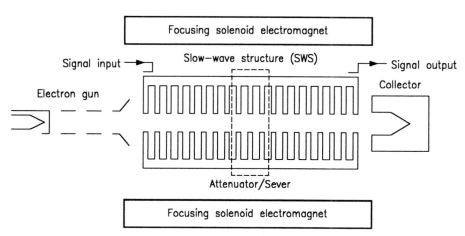

Figure 8-7. Simplified Schematic Diagram of Traveling Wave Tube (TWT).

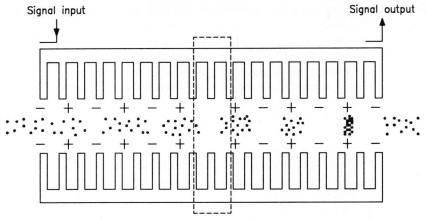

Figure 8-8. Slow-Wave Structure / Electron Beam Action in a TWT.

simple wire helix, used as the SWS in many low-powered TWTs, has this property. The propagation of signal along the helix wire is governed by its transmission parameters and is close to that of a conventional line. The simple fact that the signal travels much farther around the line than its forward progress along the line slows the effective propagation velocity. *The SWS propagation velocity and the velocity of the electron beam must be carefully matched, because effective TWT amplification requires that electrons interact with the same signal segment for the length of the tube.*

TWTs, like klystrons, amplify by velocity modulating an electron beam such that a small input signal causes bunching of the electrons. Like the klystron, the length of the electron drift space affects gain, and the electron beam must be focused. The mechanism of amplification by a TWT differs from that of a klystron in that the interaction between the signal and the electron beam is continuous, rather than occurring at just a few points (the cavities). Figure 8-8 shows the interaction of the SWS and the electron beam in TWT amplification.

An electron beam is generated by an electron gun, as in a klystron. Signal is introduced onto the SWS at the RF input port. The signal energy on the SWS is coupled into the electron beam, reducing the RF energy on the SWS. See Fig. 8-9 for relative RF energy levels on the SWS and in the beam.

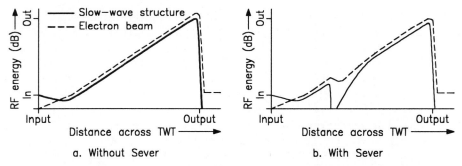

Figure 8-9. RF Energy Levels on the Slow-Wave Structure and Electron Beam.

The SWS propagation velocity and the axial velocity of the electron beam are the same, so that each electron interacts with the same signal point for the entire trip through the tube. Electrons are attracted to positive signal nodes on the SWS and are repelled from negative nodes. Thus some electrons are slowed down and some are speeded up (velocity modulation).

For the present discussion, disregard the attenuator in Fig. 8-8 and consider only Fig. 8-9a. As the faster electrons catch up with the slower ones and bunching begins, the RF energy in the beam quickly exceeds that on the SWS. After this crossover, which is quite close to the input, energy is coupled from the electron beam to the SWS. As bunching increases, more and more of the beam's DC energy is converted to RF and the RF level in the beam increases. Through coupling, the RF energy on the SWS also increases. At the output point, most of the RF energy is taken from the SWS, and the beam debunches. Thus only a small fraction of the total RF energy remains when the spent beam enters the collector.

The gain of the above process can be quite large. The gain of many moderate power TWTs is high enough so that the entire driver amplifier chain can be solid-state. The bandwidth can also be large, since the slow-wave structure is non-resonant (in some TWTs it is in fact resonant — more about this later).

The slow-wave structure introduces increased bandwidth as compared to cavities. Unfortunately, it also introduces a feedback path from output to input which does not exist in klystrons. Any signals reflected by the output load travel backward on the SWS to the input. If there is sufficient reflection of this backward wave at the input, oscillations can occur. This is the reason for the attenuator or sever.

In any system, oscillation occurs if the following conditions are met *simultaneously* (see Fig. 8-10): *loop gain greater than unity AND loop phase equals zero*. The loop gain is the gain from one point in the system — point A in Fig 8-10 — around a loop, and back to the same point. The loop phase is the phase shift from a point, around the loop, and back to the same point. Integer multiples of 360° are the same as zero.

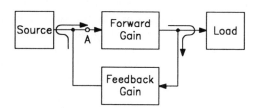

Figure 8-10. Loop Gain and Phase in a Feedback Amplifier.

The zero loop phase criterion is met in all TWTs. The SWS is many wavelengths long, and there will be a frequency at which the loop phase is zero at the same time the forward TWT gain is large. Thus the only way oscillation can be prevented is by restricting the loop gain to less than one. (This is just the opposite of the conditions found in low frequency feedback circuits, such as operational amplifiers. In these circuits, the loop gain is almost always much greater than one and the only way to prevent oscillation is to prevent loop phase from becoming zero at any frequency where forward gain is high.)

Since the forward gain of a TWT is large, the only way to prevent the loop gain from meeting the oscillation criterion is to introduce large losses in the feedback portion of the loop. This is accomplished in three ways. First, the load return loss must be high so that little output energy is reflected. The TWT loads must be well matched to their amplifiers, a requirement which does not exist in klystrons. Second, the signal which does reflect and

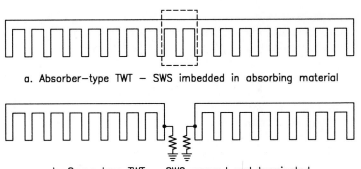

a. Absorber-type TWT — SWS imbedded in absorbing material

b. Sever-type TWT — SWS severed and terminated

Figure 8-11. Attenuator and Sever TWT Types.

make its way to the input must not be reflected back in the forward direction. This requires that the source return loss be high, so that the source absorbs most of the reflected energy. Third, even after the return losses at both ends are made large, most TWTs have so much gain that they require a signal attenuator be placed in the slow-wave structure.

The sum of the load return loss (dB), the source return loss (dB), and the SWS attenuation (dB) must exceed the forward TWT gain if the loop gain is to be less than unity. This SWS attenuation can be accomplished by placing a physical attenuator, usually some form of RF absorber, around the SWS. This form is shown in Fig. 8-11a and is the configuration of many low-power TWTs. In higher power tubes, the slow-wave structure is usually broken and both ends terminated in well-matched loads. This form of attenuation (Fig. 8-11b) is known as a sever.

Figure 8-9b shows the energy-related result of the attenuator/sever. Signal on the SWS is attenuated to near zero and the only forward signal path across the sever is in the electron beam. Forward RF energy lost here somewhat reduces the overall gain of the tube, but greatly reduces the requirement for impractical return losses at the TWT load and source. Some tubes have enough forward gain that the SWS must be severed more than once to prevent oscillations from occurring in sections of the tubes. Figure 8-12 shows relative signals at various parts of the TWT.

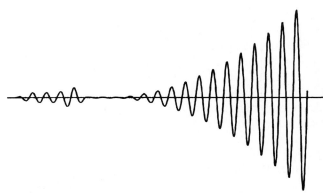

Figure 8-12. Signal Amplitude across Single-Sever TWT.

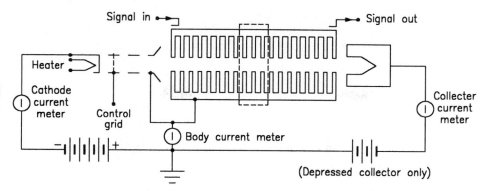

Figure 8-13. TWT Bias Connections and Body Current Meter.

As with klystrons, the focusing magnetic field can be adjusted by measuring and minimizing the body current. Figure 8-13 shows the DC bias connections for a TWT and the location of the body current meter.

TWT types: TWTs are usually classified by their slow-wave structure type. Low-power TWTs often use a simple wire helix as the SWS. Slow-wave structures in higher power tubes can be of the basic helical form, but must use heavier material to conduct the heat buildup caused by electrons striking the SWS. Figure 8-14 shows part of a higher power helical TWT structure. Helices can also be wound as shown in Fig. 8-15. This is a bi-filar helical TWT.

Figure 8-16 shows a ring-bar TWT, with its relatively massive slow-wave structure. Many high power TWTs are of this construction.

A configuration which exhibits good heat dissipation and vibration resistance is the coupled-cavity [9] slow-wave structure of Fig. 8-17. It is often found in radars where the tubes must be physically small and where vibration is high, such as in aircraft.

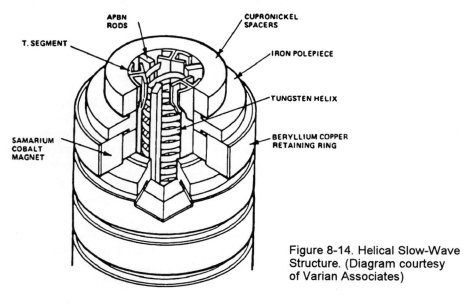

Figure 8-14. Helical Slow-Wave Structure. (Diagram courtesy of Varian Associates)

8-3. Traveling Wave Tube Transmitters

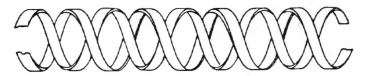

Figure 8-15. Bi-filar Helix Slow-Wave Structure.
(Reprinted from Mendel [8]. ©IEEE. Used with permission)

Figure 8-16. Ring-Bar Slow-Wave Structure.
(Reprinted from Mendel [8]. ©IEEE. Used with permission)

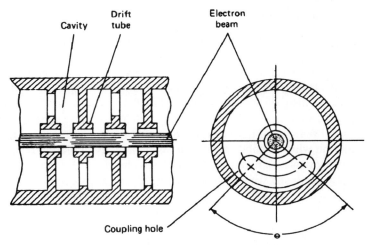

Figure 8-17. Coupled-Cavity Slow-Wave Structure.
(Reprinted from Mendel [8]. ©IEEE. Used with permission)

The cavities introduce resonant components into traveling wave tubes, but the effect on bandwidth is not as severe as it would seem. First, the cavities are resistively loaded, which broadens their bandwidth. Second, adjacent cavities are extremely closely coupled, as evidenced by the large coupling ports (called kidneys). Resonant circuits by themselves and multiple loosely-coupled resonant circuits have narrow bandwidths. As coupling increases, however, the bandwidth broadens (Fig. 8-18). With the tight coupling of modern TWTs, untuned bandwidths of 15% are readily attainable, sufficient for most radar applications. Most modern airborne radars use coupled-cavity TWTs.

TWT modulation and control grids: To pulse a TWT, the electron beam must be modulated on and off. If no signal were applied while the beam was on, the tube would put out large amounts of noise and dissipate much heat. As stated earlier, pulse modulation can be applied in one of two ways. A high-level modulator provides the operating DC voltage

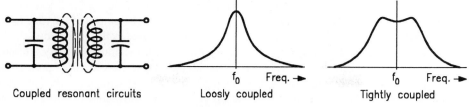

Figure 8-18. Effect of Coupling on Resonant Circuits.

and current to the TWT cathode in the form of a negative pulse of tens of thousands of volts and tens of amperes of current. The hardware required for high-level modulation is impractical in certain environments. Section 8-7 discusses high-level modulation.

Low-level modulation is accomplished by applying a control voltage to a control electrode (grid) in the tube. A grid voltage which is negative with respect to the cathode repels electrons back to the cathode and prevents the flow of beam current. A positive grid voltage allows the electron current to flow.

A simple wire grid, such as shown in Fig. 8-19a, suffices in low-power vacuum tubes because the current is small and the heat generated by electrons colliding with the grid wires is readily dissipated. High power tubes, with large electron beam currents, cannot be controlled in this manner. The grid structure simply cannot dissipate the heat generated by collisions.

The solution is to use a shadow grid to protect the control grid (Fig. 8-19b). The shadow grid is shaped and placed so that the control grid is in its electron shadow and few collisions occur between electrons and the control grid structure. The shadow grid itself is protected because it is relatively massive, and it is at the same potential as the cathode. The electrons which strike it have little velocity and hence little energy. The invention of the shadow grid was the step which allowed TWT transmitter amplifiers, with their high gain, wide bandwidth, and very low noise, to be used in airborne radars. This same control scheme could be used in klystrons.

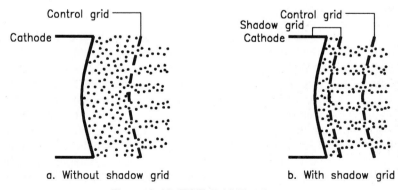

Figure 8-19. TWT Grid Structures.

8-3. Traveling Wave Tube Transmitters

TWT efficiency and depressed collector TWTs: The efficiency of TWT amplifiers is relatively low because much of the beam's energy remains at the output of the tube to be dissipated at the collector. The electrons' DC velocity is determined by the anode-to-cathode potential. In some tubes, the collector is at the same potential as the anode (usually ground) and has little effect on the electron velocity. If the electrons were slowed after leaving the slow-wave structure and before striking the collector, the energy dissipated in the collector would decrease and the tube's efficiency would rise. Depressed collectors function by operating the collector at a negative potential with respect to the anode and tube body (Fig. 8-20). This technique decreases the energy dissipated at the collector, and thus the cooling load.

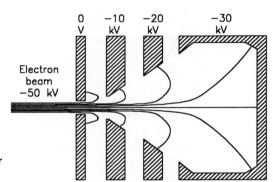

Figure 8-20. Depressed Collector and Electron Trajectories.

Twystrons: Some radars (the AN/TPS-43, for example) use a tube which is a hybrid of a multi-cavity klystron driver section and a TWT final power amplifier (FPA) in one vacuum envelope, called a twystron. It has a very high power capability and relatively wide bandwidth, achieved by stagger tuning several cavity stages.

8-4. Crossed-Field Amplifier (CFA) Transmitters

Crossed-field amplifiers are so named because their operation depends on the interactive effects of a DC electric field and a DC magnetic field at right angles to one another. Figure 8-21 shows a basic linear-format CFA. The cathode (also known as the sole) emits electrons which are drawn by the electric field toward the anode. A magnetic field in the drift path causes the electrons to curve such that their basic motion is horizontal. See reference [11].

The signal is introduced onto the slow-wave structure in much the same way as in a traveling-wave tube, and the signal and electrons propagate from left to right at the same velocity. As with a TWT, each electron interacts with the same signal element for the length of the tube and velocity modulation occurs. Unlike the TWT, the modulation is perpendicular to the electron beam motion. Thus instead of bunching, the electron beam forms a wave (Fig. 8-22), with electrons opposite positive SWS nodes accelerated upward and those opposite negative nodes accelerated downward. Gain results when coupling of RF energy from the beam to the SWS exceeds coupling from SWS to beam.

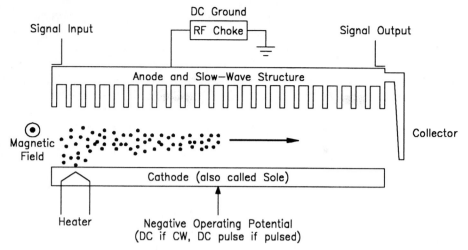

Figure 8-21. Linear CFA Concept (without signal).

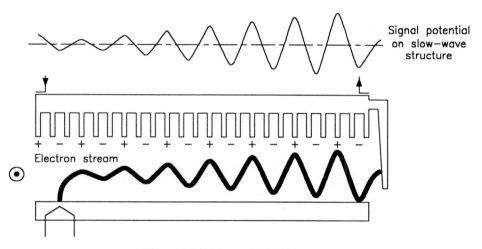

Figure 8-22. Linear CFA With Signal.

CFAs, particularly linear CFAs, have less gain than TWTs because of the relatively loose coupling between the SWS and beam in the CFA. Attenuators and severs are in many cases not necessary because of the low forward gain.

The lack of gain is partially alleviated by providing positive feedback within some crossed-field amplifiers. Figure 8-23 shows two re-entrant, or circular format, CFAs (the difference between the two amplifiers is described in a later paragraph in this section). The amplification process is identical to that for linear format tubes, except that the electron beam moves in a circular path, and the electron "wave" develops in the circular path. In the CFA shown, a portion of the electron beam continues to circulate after the output, carrying some of the signal back to the input.

8-4. Crossed-Field Amplifier Transmitters

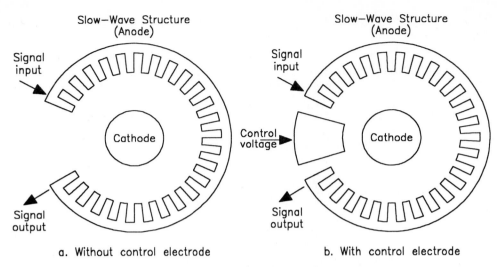

Figure 8-23. Circular Format (Re-entrant) CFAs.

Since CFAs are usually quite efficient (50 percent or so), most of the RF energy in the electron beam is coupled to the output, causing the electron stream to "debunch" at the output, similar to TWTs and klystrons.

The positive feedback and the lack of an attenuator/sever in most circular CFAs causes there to be four distinct operating modes, summarized in Table 8-2. The passive mode (DC off, RF on) allows this type CFA to act as waveguide (albeit expensive waveguide) and simply pass the RF drive to the output. The fourth mode (DC on, RF off) is the magnetron mode, discussed in Sec. 8-5.

Table 8-2. Reentrant CFA Modes.

Mode Number	DC Power	RF Drive	CFA Action — Output
I	Off	Off	No output
II	On	On	Amplified RF drive
III	Off	On	RF drive (acts as waveguide)
IV	On	Off	Full power noise

Some radar transmitters take advantage of the passive mode to produce two distinct power output levels. This process, called power programming, allows the transmitter to emit high power when needed, such as when the antenna beam is pointed at the horizon and where very long range targets can exist. When beams are pointed at high elevation angles, targets in the atmosphere are close to the radar and require much less illumination power. Thus the transmitter's power supplies and cooling can be saved for use when truly needed.

The tube in Fig. 8-23a must be high-level modulated. Some CFAs have control electrodes, similar in action to the TWT's grid, which allow the electron beam to be shut off between transmitter emissions. A schematic diagram of this type CFA is shown in Fig. 8-23b.

A cross-section of a typical circular format crossed-field amplifier is given in Fig. 8-24, showing the placement of the tube's major features. As with most crossed-field devices, this tube uses a permanent magnet, which contributes to its relatively small size and light weight. The tube shown is liquid-cooled. Because of their high efficiency, many crossed-field devices are air-cooled, eliminating the requirement for elaborate plumbing systems and heat exchangers.

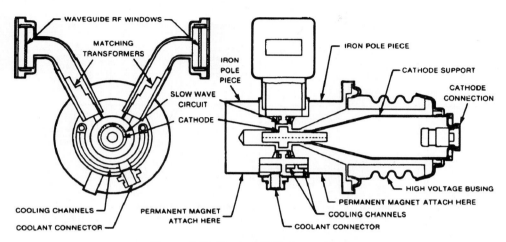

Figure 8-24. Typical CFA Cross-Section.
(Diagram courtesy of Varian Associates)

CFA types: Many different crossed-field devices exist (Fig. 8-25), including amplifiers and oscillators. They are classified by how the electron beam is generated (injected from an electron gun or emitted from the cathode), whether their format is linear or circular, whether the signal propagates in the same direction as the electron beam (forward wave) or in the opposite direction (backward wave), and whether they normally amplify or oscillate.

8-5. Magnetron Transmitters

Reentrant crossed-field amplifiers can be constructed with resonant slow-wave structures and operated in the fourth mode in Table 8-2, with DC input and no RF input. Instead of producing full power noise, the output is a sinusoid at the frequency at which the slow-wave structure resonates. This magnetron, invented in its present form by H.A.H. Boot and J.T. Randall in England in 1939, was the major breakthrough that led to practical microwave radars. Magnetrons are sometimes known as standing-wave crossed-field devices, because of the appearance of an apparent standing wave in the electron stream.

The slow-wave structure shown in Fig. 8-26 is an early version, the multi-cavity magnetron. It oscillates at the frequency of the cavities. Since the cavities cannot be tuned to exactly the same frequency, the output is more like narrow band noise than a pure sinusoid.

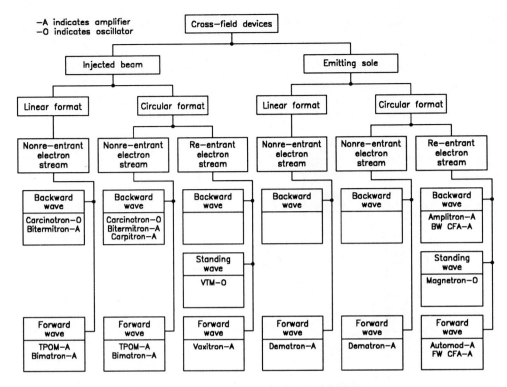

Figure 8-25. Crossed-field Devices Family Tree.
(After D. Fink [10])

The straps connecting alternate cavities are to prevent moding, which is an out-of-sync oscillation at some frequency other than the design frequency. It is associated with some rather spectacular failure modes in magnetrons. Many other slow-wave structures were used in early magnetrons, but few survive except for the one described below.

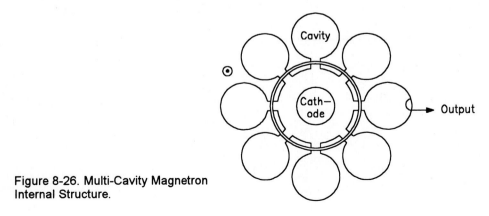

Figure 8-26. Multi-Cavity Magnetron Internal Structure.

A device which outputs a much purer sinusoid than multi-cavity magnetrons is the coaxial magnetron [12], which uses a single cavity (Fig. 8-27). Most magnetrons made today, including those in home microwave ovens (at 2.45 GHz), are coaxial. Coaxial magnetrons also have the advantage that they are easily tuned by simply changing the size or characteristics of single cavity.

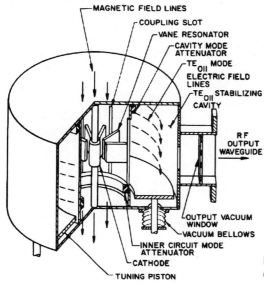

Figure 8-27. Coaxial Magnetron.
(Diagram courtesy of Varian Associates)

Magnetron frequency: The frequency on which a magnetron oscillates depends on three factors: the cavity resonant frequency, the output power, and reflections from the load. The natural frequency of the magnetron is its cavity resonant frequency with nominal output power and no load reflection. If the magnetron's frequency is changed by changing the input DC voltage, often measured as the output RF power, the magnetron is said to be *pushed*. If its frequency is changed by the output load, the magnetron is *pulled*. Each magnetron type has a characteristic plot called a Rieke diagram which describes the output frequency as a function of RF power out and load reflections (Fig. 8-28).

Many radar applications require that the transmitter be tunable during operation for frequency agility. This is in addition to normal tuning for single-frequency operation by the bellows in Fig. 8-27. Amplifying transmitters are best suited to this technique, but some magnetrons can be tuned rapidly enough to provide useful agility. Techniques for rapid tuning include mechanically moving an internal tuning device through magnetic coupling (the so-called spin-tuned magnetrons), and the use of internal compressible rings in the cavity (ring-tuned magnetron). Several proprietary schemes for rapid tuning are available, including Varian's Accusweep™ system. In magnetron frequency agile operation, the transmit pulse must be sampled to set the frequency of the receiver's local oscillator (see automatic frequency control in Ch. 10).

8-5. Magnetron Transmitters

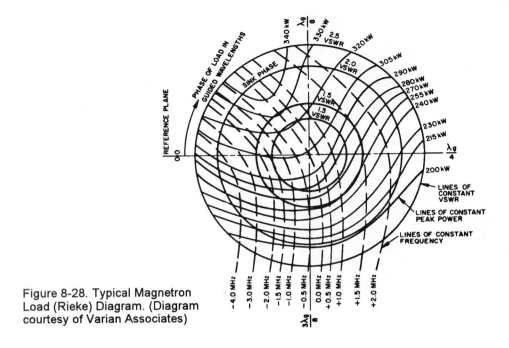

Figure 8-28. Typical Magnetron Load (Rieke) Diagram. (Diagram courtesy of Varian Associates)

8-6. Solid-State Transmitters

The advantages of solid-state devices over vacuum tubes are well known and include lower voltage requirements and high reliability. Their primary disadvantages are low power output per device and relatively poor operation at higher microwave frequencies. Radar transmitters made up of solid-state devices thus are often reserved for the lower frequency bands (with exceptions) and contain many devices operating in parallel. One 110-kW L-Band transmitter built in the 1980s used approximately 250,000 junction transistors grouped 16 per package [13].

Many different types of solid state devices are capable of producing microwave power, both as amplifiers and as oscillators. They include:

— *Bipolar junction transistors (BJTs)* are conventional junction (NPN and PNP) transistors used primarily as amplifiers. They have high gain per device, but are usable only at lower frequencies. Their power output per device is low.

— *Field-effect transistors (FETs)* are primarily amplifiers. They have lower gain per device than BJTs but are usable to frequencies in the millimeter bands. Their power output per device is low. The noise produced by them can be very low. In the microwave bands, FET technology is considerably more advanced than BJT and they find wide use in radars as driver and power amplifiers.

— *Transferred-electron devices (TEDs)* are forward-biased semiconductor devices, usually in diode form, which have oscillating and amplifying modes. They operate well into the millimeter wave bands. Because they operate in forward bias, the voltage across them is low, restricting their power output per device and their efficiency. Their primary use in radars is as receiver local oscillators, receiver parametric amplifier pumps (Ch. 10), and test signal generators. The most commonly found TED is the *Gunn diode*.

– *Avalanche transit-time (ATT)* devices are complex diodes operating in reverse breakdown. They are primarily oscillators, with some having amplifying modes. They operate well into the millimeter bands, and because of their reverse bias operation can put out substantial power (100s of W CW). Their primary use is as pulsed and CW sources in the higher bands. ATT devices include *Read diodes*, *IMPATTs*, *TRAPATTs*, *BARITTs*, and other similar devices.

Solid-state devices can make up the entire transmitter in radars. Examples include the AN/FPS-115 PAVE PAWS early warning radar, the AN/TPS-59 tactical 3-D search, and the AN/FPS-117. They are also used as driver amplifiers in many transmitters which use tubes as their final power amplifier. An example is the "Super" AN/FPS-16, where FET amplifiers bring the signal level to about 10 W of peak power, after which a TWT driver and klystron FPA boost it to 3.0 MW.

Solid-state transmitter configurations: Because of the generally low power per device, solid-state radar transmitters are characterized by parallelism. One method of achieving this parallelism is by dividing the drive power among many amplifying modules and combining the resulting power onto a single transmission line. Transmitters which are configured this way fit in systems the same as conventional vacuum tube transmitters, with the full power concentrated in a single transmission line and applied to the duplexer. This concept is shown in simplified form in Fig 8-29a.

The other principal configuration is the so-called *active array*, in which each transmit amplifier feeds one element of an array antenna. Each element also has its own receive low-noise amplifier (LNA) and phase shifter. A simplified diagram of an active array is shown in Fig. 8-29b. A plus for this configuration is that duplexing takes place at low power, eliminating the need for a large and expensive component. Examples are the AN/TPS-59 and AN/FPS-117.

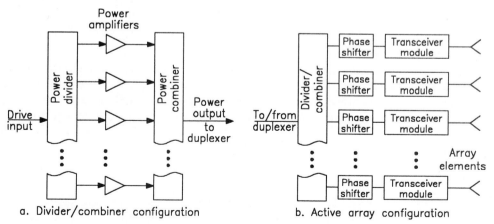

Figure 8-29. Solid-State Transmitter Configurations.

8-6. Solid-State Transmitters

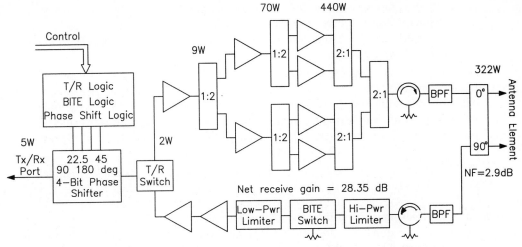

Figure 8-30. A Typical Transceiver Module (modeled after PAVE PAWS).

The block diagram of a transmit/receive active array module patterned after PAVE PAWS [14] is shown in Fig. 8-30. The powers shown on the module are total transmit power to that point. In this radar, each array is composed of 56 sub-arrays, each containing 32 modules, for a total of 1982 modules per face and 3584 modules per radar. It is estimated that the mean time between failure (MTBF) of each modules is 150,000 to 200,000 hours. The radar can operate for about 1.5 years without requiring module replacements, since a certain percentage of the modules can fail without materially affecting the radar's performance (in a process known as *graceful degradation*).

Reference [15] lists and describes a large number of solid-state transmitters found in various radar systems.

8-7. Modulators

Modulators provide control for transmitters, turning them on and off for pulsed operation. In this section, pulsed operation with both high- and low-level modulators is discussed.

Low-level modulators: Low-level modulators form a pulse of the proper width and level at the right time, apply the necessary DC biases, and send it to the control electrode of a transmit amplifier. Typical control pulses are a few hundred volts at an ampere or so. Figure 8-31 is of a simplified low-level modulator. The pulse is formed in the pulse generator, where the timing and width are set. It is amplified and passed through a pulse

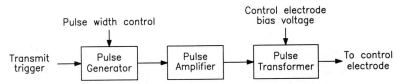

Figure 8-31. Low-Level Modulator Simplified Block Diagram.

Figure 8-32. Pulse Generator Examples.

transformer, where its voltage is adjusted (stepped up) and the transmit tube control electrode bias is applied. The resulting pulse is applied to the transmit tube's control electrode.

Pulse generators: The pulse generator sets the pulse time and width. Several types exist, two of which are shown in Fig. 8-32. In the first, the transmit trigger (sometimes called the modulation trigger, or mod trigger) sets a flip-flop. The "Q" output of the flip-flop is the pulse generator's output. This output is also applied to a delay line. After propagating through the delay line, it resets the flip-flop. Thus the delay line sets the pulse width and switching the delay line accomplishes pulse width diversity.

The second type uses a monostable (one-shot) multivibrator as the pulse generator. The trigger starts the pulse, and the capacitor and resistor set its width. This type allows easy pulse width diversity, either by switching the capacitor or resistor, or by simply switching several one-shots, one for each of the desired pulse widths.

Pulse amplifiers: Pulse amplifiers are simply video amplifiers which have the capability of outputting pulses of several hundred volts. They may use vacuum tubes or FETs, and may or may not have pulse transformers.

Pulse transformers: Pulse transformers in low-level modulators are used to insert the control electrode DC bias onto the pulse. In high level modulators, they also add the transmitting amplifier's filament voltage. Their action (in the high level configuration) is shown in Fig. 8-33.

High-level modulators – Hard tube: High-level modulators are divided into two types, soft tube and hard tube. Despite their names, these modulators can be implemented with either vacuum tubes or solid-state devices, or a combination. Hard tube modulators are simply low-level modulators with additional amplification to bring the pulse to a level high enough to provide the FPA's DC drive. The transmit energy comes from the high voltage power supply capacitors, and the pulse length is set by the width of a drive pulse from a pulse generator. The pulse width is easily changed, a major advantage to this configuration.

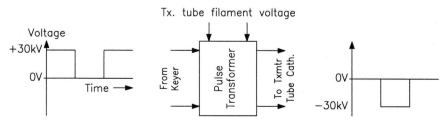

Figure 8-33. Action of Pulse Transformer.

8-7. Modulators

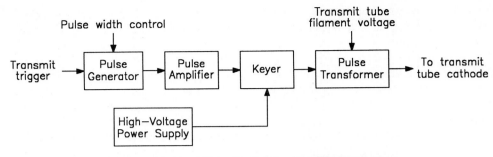

Figure 8-34. Hard Tube Modulator Simplified Diagram.

A block diagram of a hard tube modulator is shown in Fig. 8-34. The pulse generator forms the pulse and the driver amplifies it similarly to the low level modulator. The additional amplifier, called a keyer, brings the pulse to the level required by the transmitter. The pulse is applied to the pulse transformer, where the pulse voltage is adjusted and the transmit tube's bias and filament voltages are added. The pulse out of the transformer is usually negative and applied to the cathode of the transmit tube (see Fig 8-33).

Keyers are high-output pulse amplifiers which provide the DC drive for the output tube. They can be either vacuum tube or solid-state; the choice depends on the power output needed.

Some transmitters have circuits which sense arcs in the transmitter and waveguide. Upon detection, it is possible in hard tube modulators to instantly shut off the transmit pulse, thus protecting the transmitter. When the arc extinguishes, transmitters usually are ready to function again, missing only a few pulses, or even just a fraction of one pulse. This is another major advantage to the hard tube configuration.

Hard tube modulators – floating deck: This special configuration is shown in Fig. 8-35. It is useful in cases where very large voltage swings (to 200 kV or more) are needed. It can produce either a positive pulse for modulating the anode of the transmitter tube or a negative pulse for the cathode. To turn a positive pulse on, tube 1 is on and tube 2 is off. To turn it off, the opposite conditions apply. To produce a negative pulse, the above conditions are reversed.

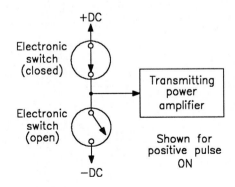

Figure 8-35. Floating-Deck Modulator.

398 Chap. 8. Transmitters and Microwave Components

Soft tube modulators: Soft tube modulators use thyratrons or silicon control rectifiers (SCRs) to switch energy taken from the high-voltage power supply and stored in a pulse forming network (PFN). Unlike the hard tube modulator, in this configuration the switch does not control the pulse width. The pulse width, set by the PFN, is independent of the trigger (within limits). A characteristic of thyratrons and SCRs is that a control electrode closes the switch, but the current through the device must drop below a certain level before they can be shut off. Thus in a soft tube modulator, the active device only starts the pulse; the PFN shuts it off.

Figure 8-36 shows the basic circuit. The high-voltage power supply charges the PFN through the charging circuit. The transmit trigger causes the switch to short-circuit, starting the pulse. The PFN transfers the stored energy to the transmitter for a period of time set by the PFN circuit values. When the PFN is discharged, it abruptly terminates the pulse. At some later time, the switch opens and the charging network begins transferring energy to the PFN for the next pulse. The action of the pulse transformer is the same as in a hard tube modulator.

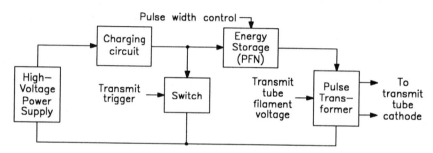

Figure 8-36. Soft Tube Modulator Simplified Block Diagram.

Pulse forming network: The pulse forming network is a simulated transmission line which behaves as illustrated in Fig. 8-37 (this is not the actual configuration, but shows the principle). Because of the use of the transmission line-type circuit, these modulators are also called line modulators.

A transmission line is shorted at one end and is connected at the other end through a source resistor and switch to a DC source. Assume the battery is 20 V and that $Z_S = Z_0 = 50\ \Omega$. The output voltage is zero if the switch has been open for a long time. When the switch closes, 10 V appears across the line at its input (10 V across the line, 10 across the source impedance

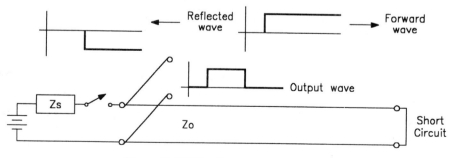

Figure 8-37. Line Modulator Principle.

8-7. Modulators

Z_S). A forward wave of 10 V and 200 mA (10 V ÷ 200 mA = 50 Ω) propagates from left to right on the line. When this forward wave reaches the short circuit, a reflection of −10 V and 200 ma occurs, which propagates right to left as the reflected wave. The voltage at any point on the line is the sum of the forward and reflected voltages. When the reflected wave reaches the input, it sums with the 10 V forward wave for a total of zero. This is the shut-off of the pulse. The pulse width is simply the time necessary for the wave to propagate down the line to the short and back to the input.

$$PW = 2L / v_G \qquad (8\text{-}1)$$

PW = the pulse width (seconds)

L = the length of the line (meters)

v_G = the propagation velocity on the line (meters per second)

A line modulator's pulse-forming network (PFN) is a lumped constant equivalent of the transmission line. Adding a parallel switch and a non-resistive (non-dissipating) charging circuit completes the modulator. Fig. 8-38 is a simplified schematic diagram.

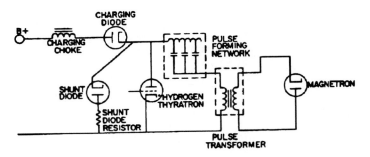

LINE TYPE MODULATOR

Figure 8-38. Line Modulator Simplified Schematic Diagram.
(Diagram courtesy of Varian Associates)

The PFN charges from the power supply through the charging choke and charging diode. When fully charged, the left side of the PFN is at the power supply voltage (it can actually be twice this voltage — see discussion of charging below) and the bottom of the PFN is grounded. The transmit trigger fires the thyratron, closing the switch and forcing the left side of the PFN to near zero voltage. This drops the bottom of the PFN to the negative of the power supply voltage (or twice this value). When the PFN has discharged, with the time set by the length of the equivalent transmission line, the pulse stops and the switch opens (thyratrons and SCRs open when the current through them drops below a threshold value). Afterward, the PFN recharges. The shunt diode and shunt resistor kill any reflections from a mismatched transmit tube load. There is often a diode across the pulse transformer to prevent the cathode of the transmit tube from going positive.

The relationships for line modulators follows.

$$Z_{OL} = \sqrt{L_N / C_N} \qquad (8\text{-}2)$$

$$\tau = \sqrt{L_N C_N} \qquad (8\text{-}3)$$

Z_{OL} = the PFN characteristic impedance
τ = the pulse width
L_N = the total PFN inductance (henry)
C_N = the total PFN capacitance (farad)

In most line modulators, the charging choke forms a resonant circuit with the PFN capacitance and the PFN inductance. Since the charging choke inductance is much greater than that of the PFN, the PFN inductance is negligible. The response of the charging choke and PFN capacitance after the switch opens is a damped sinusoid whose peak-to-peak voltage for the first cycle is very nearly twice the HVPS voltage. If the PRF is chosen as one-half this resonant frequency, the next pulse occurs one-half cycle after the previous pulse, when the PFN is at its peak voltage. If the charging resonant frequency is greater than twice the PRF, the charging diode prevents the PFN from discharging back into the choke. Figure 8-39 shows PFN voltages for these two resonant conditions.

For the PRF equal to twice the charging resonant frequency, the value of the charging choke can be found from

$$L_C = \frac{1}{\pi C_N PRF^2} \qquad (8\text{-}4)$$

L_C = the charging choke inductance (henry)

The dependence of the pulse width on the PFN and the PRF on the charging network makes this modulator less flexible than the hard tube modulator. Although small PRF changes are tolerable, inductors and capacitors must be switched to support pulse width and wide PRF diversity.

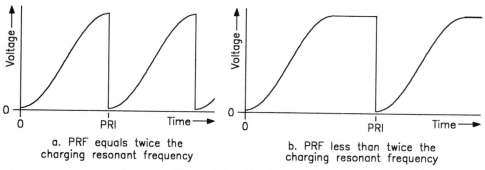

a. PRF equals twice the charging resonant frequency

b. PRF less than twice the charging resonant frequency

Figure 8-39. PFN Voltage.

8-8. High-Voltage Power Supplies

The high-voltage power supplies (HVPS) used in radars supply the high voltage either to the transmit tube, or to the high-level modulator, which in turn supplies it to the transmit tube. Most radar HVPS are three-phase full wave rectifiers at the frequency of the radar's prime power. The reason for using three-phase power is that the ripple frequency is three times that of a single-phase supply at the same power frequency, making the filter capacitor more effective.

The rectifiers can be vacuum tube or semiconductor, with modern designs favoring solid-state. One single vacuum tube can rectify the voltages required in radar HVPS, but solid-state diodes can't. They are therefore used in stacks, as shown in Fig. 8-40. The resistors pass a current several times the leakage current of the diodes and equalize the reverse voltage across the diodes. The configuration of the rectifiers is in a bridge in most modern power supplies, although a full-wave tapped transformer configuration is sometimes used.

Filtering is generally an L network of a filter inductor and filter capacitor. The filter capacitor must block approximately twice the HVPS voltage, and must of necessity be of rather small capacitance, making the three-phase power necessary. Inadequate filtering either causes excessive output RF amplitude changes from pulse to pulse (power supply ripple), or causes the output pulse to droop (voltage decrease across the pulse).

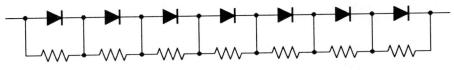

Figure 8-40. Rectifier Stack.

8-9. Transmitter Vacuum and Cooling Systems

Vacuum: Transmitting tubes must operate in a hard vacuum. Residual gases inside the tube form positive ions on collision with electrons in the beam. These massive ions accelerate toward and strike the cathode (negatively charged), and can damage it severely. Most tubes are evacuated during manufacture and have placed inside them a getter, a chemical which reacts with gas molecules remaining in the tube and removes them from the gaseous state. Once the getter is used up, it cannot remove gases and the tube is useless.

In tubes where residual gases are a problem, a vacuum pump is attached, and it runs during the tube's operation. The pump gathers up and ejects gas molecules from small leaks and those which are released from the tube's components (in a process called outgassing). These are gases, found primarily in higher power devices, which would overwhelm a getter. The pump often used is an ion pump, effective only at vacuums harder than about 10^{-2} Torr (1 Torr is approximately one atmosphere). Thus these pumps function well only in tubes with small leaks and limited outgassing.

Cooling: Excess heat from transmitters can be removed either by air or by liquid. In general, all but the lowest power linear-beam tubes (klystrons and TWTs) must be liquid cooled. This is because the heat is concentrated in the collector and there simply is not enough surface area for the relatively low heat capacity of air to effectively cool them.

Air cooling is more common (although by no means universal) in crossed-field tubes, because their primary heat source is their anode, which is the largest component of the tube. More information on vacuum and cooling can be found in reference [2].

8-10. Transmitter Monitoring and Testing

Many transmit parameters must be monitored during operation and tested during maintenance. Most radar transmitters have built-in capability for supplying the necessary signals for testing and monitoring. Figure 8-41 shows a typical transmitter test scheme. *See Appendix B for transmitter safety information.*

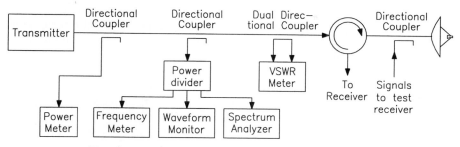

Figure 8-41. Typical Transmitter Test and Monitoring.

Power measurement: Transmit power is measured by sampling a fraction of the power output and using an RF power meter. The directional coupler which samples the power is usually a fairly high (numerically) ratio coupler and couples only a small fraction of the power (50 to 60 dB is typical). Because amplitude accuracy is important, a dedicated coupler is often used for this measurement.

Several corrections must be applied to the power meter reading to get the transmitter's true power. First is the coupler ratio itself. Second is the loss in the cable from the coupler to the transmitter power monitoring port. In most systems, the ratio given for the monitoring port includes both the coupler and this cable. Any cables from the port to the meter must be accounted for by the measurer. A way to avoid cable problems is to attach the power meter sensor directly to the transmitter's power monitoring port.

Finally, the power must be corrected for the transmitter's duty cycle. Most power meters read *average power*, but pulse transmitters are specified in *peak power*. The difference is the duty cycle correction factor, from Ch. 2. A few meters read peak power directly. The transmitter's power output relates to the power meter reading by the following equation:

$$P_T(\text{dBm}) = P_M(\text{dBm}) + dccf(\text{dB}) + L_C(\text{dB}) \qquad (8\text{-}5)$$

P_T = the transmitter's peak power (dBm)

P_M = the (average) power meter reading (dBm)

$dccf$ = the duty cycle correction factor (dB – see Ch. 3)

L_C = the loss from the monitoring port to the meter sensor (dB)

Frequency measurements: Frequency measurements, like power, are made more difficult by pulsing. The most accurate frequency meters are counters, which with few exceptions must have CW signal to operate accurately. Absorption wavemeters can measure the frequency of pulsed signals, but are less accurate and are harder to use than counters. A wavemeter is basically nothing more than a transmission line resonant circuit with a diode and DC meter arranged such that either a peak or dip is shown when the resonant circuit is tuned to the frequency of the input signal.

Pulse width and shape: Waveform monitors sample RF from the transmitter, demodulate it, and display the envelope on an oscilloscope. The choice of detector (demodulator) type is based on availability and power level. One class of detector which is often used is the so-called square-law detector, whose envelope output is proportional to the RF input squared. It has the advantage that RF *power* out of the transmitter is shown as envelope *voltage* on the oscilloscope, since power is proportional to voltage squared. If pulse width is to be measured at the pulse's half-power point, it shows on the oscilloscope as half-voltage.

To get the square-law effect, the detector must be terminated. The termination should be placed at the oscilloscope end of the cable to prevent pulse "ringing" in the test setup. Note that the termination is not necessarily the characteristic impedance of the cable — it should be matched to the detector. The manufacturer's specifications for the detector give the proper termination value.

The following discussion of spectrum analysis includes another pulse width measurement technique.

Spectrum: The spectrum of the transmitter's waveform tells much about how well the transmitter is operating. The spectrum depends on the wave being transmitted. For example, a pulsed transmitter producing a gated CW wave with a rectangular envelope has a spectrum whose envelope is a sinc function. One generating a linear FM chirp waveform has an approximately rectangular spectrum. Chapter 2 gives the spectra of the various waveforms.

An example of a proper spectrum (gated CW, rectangular envelope) is shown in Fig. 8-42a. It shows that the maximum spectral sidelobes are 13.5 dB below the peak of the main lobe and are balanced. Figure 8-42b shows an abnormal spectrum for the same waveform. This could result in a magnetron transmitter from excessive modulator droop, which pushes the magnetron and changes the frequency across the pulse. Simultaneous amplitude and frequency modulation at the same rate produces a characteristic unbalanced spectrum. Other magnetron abnormalities and those of amplifying transmitters will be found in their operation and maintenance manuals.

The pulse width of gated CW waveforms can be measured with a spectrum analyzer. Recall from Ch. 3 that the bandwidth between the spectral nulls surrounding the main spectral lobe is twice the reciprocal of the pulse width. This technique is often more accurate than using an oscilloscope. Unfortunately, this method works only with the gated CW waveform.

$$\tau = 2 / B_{N-N} \qquad (8\text{-}6)$$

τ = the gated CW pulse width (seconds)

B_{N-N} = null-to-null bandwidth of the main spectral lobe (Hertz)

VSWR and Return Loss: Reflections are measured using a dual directional coupler, with one coupled port giving an attenuated sample of the forward power and the other sampling the reflected wave. Finding the peak envelope voltage of each and taking their ratio gives the reflection coefficient, from which the VSWR can be calculated. If a standard VSWR test set is used, the detector characteristics are accounted for. If the measurement is made with laboratory detectors and oscilloscopes, the measurement must be corrected for the detector characteristics. If the detectors are linear, the ratio is the voltage reflection coefficient. If they are square-law (and properly terminated), the ratio measured is the *power* reflection coefficient.

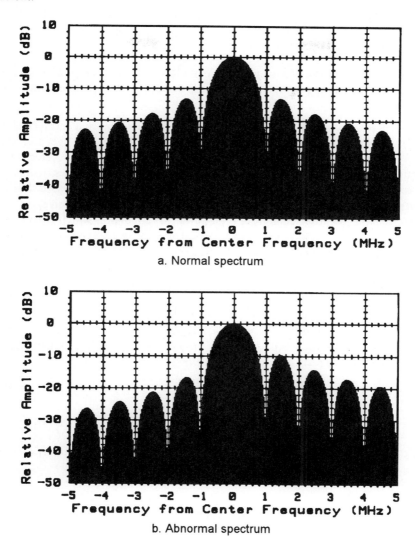

Figure 8-42. Gated CW Spectra.

Transmitter, Modulator, and HVPS DC and pulse monitoring: Most transmitters have built-in test points to be used for monitoring and testing DC and DC pulse voltages and currents. It is important that these test points be used, because they are generally situated at the bottom of voltage/current dividers and provide protection to the operator and test equipment. Any measurements made within the transmitter/modulator/HVPS itself can be extremely hazardous. One point to remember is that circuits must be broken to measure current. Doing so, even in parts of the transmitter where the normal voltages are low (the ground at the bottom of the pulse transformer, for example, Fig. 8-38) can result in lethal voltages being generated.

8-11. Microwave Components

The microwave components of a radar consist of the duplexer, transmission lines, isolators, couplers, and other ancillary components needed to make the system function and to monitor its performance.

Duplexer: Radars which share the antenna between transmitter and receiver require a method of alternately connecting the antenna to the transmitter and receiver and isolating the other system. The duplexer performs this role. It also protects the receiver from the transmitted power. Since the switching must take place in very short times (nanoseconds) and since large RF power levels can be involved, electronic switches are used. Most radar duplexers are one of two types: circulator, and TR tube (for *transmit-receive*).

The circulator duplexer contains a high-power RF circulator, which is composed of signal couplers and phase shifters such that a signal entering one port has a low attenuation path only to the next port in a particular direction (Fig. 8-43). The low attenuation paths in the circulator shown are ports 1-2, 2-3, 3-4, and 4-1. The remaining paths are high attenuation.

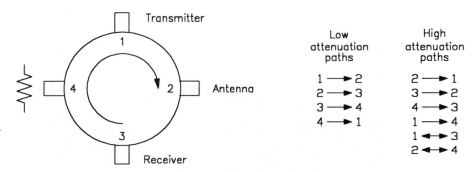

Figure 8-43. Circulator Duplexer.

The leakage (high attenuation) paths couple transmit power into the receiver, as do reflections from the antenna and components between the duplexer and the antenna. For this reason, the circulator duplexer may require extra protection for the receiver in the form of a signal limiter, discussed later.

Radar duplexers require only three ports: transmitter, antenna, and receiver. However, because of the relatively high leakage between the transmitter and receiver ports, four port circulators are commonly used. The fourth port is between the transmitter and receiver and is terminated, increasing the path loss. A high-power four-port circulator is shown in Fig. 8-44. Magic tees are common devices in radars and are discussed later in this section. After reading the section on magic tees, the reader should be able to explain the operation of this duplexer.

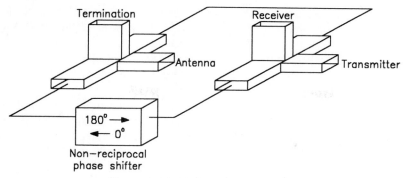

Figure 8-44. High-Power Circulator.

    The circulator is the only duplexer type which can be used in single antenna systems where the transmitter and receiver must be active simultaneously. Examples are CW systems and some pulse radars where tracking and separate missile illumination must occur simultaneously. However, there is considerable leakage of the transmitter into the receiver because of reflections by the antenna. Circulators are also used in receivers in negative resistance low-noise amplifiers (LNAs). See Ch. 10 for details.

    A TR duplexer is shown in one of its forms in Fig. 8-45, with this particular one being a dual TR duplexer. When the transmitter is active, the high power microwave signal causes gas-filled waveguide sections (the TR tubes) to ionize, short-circuiting the waveguide and causing power in it to be reflected. This directs transmit power to the antenna and protects

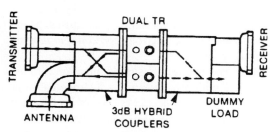

Balanced duplexer with Dual TR. Shown in the high level (transmit) condition.

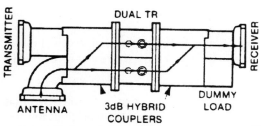

Balanced duplexer with Dual TR. Shown in the low level (receive) condition.

Figure 8-45. Dual TR Duplexer.
(Diagram courtesy of Varian Associates)

8-11. Microwave Components

the receiver. When the transmitter is off, the TR tubes act as normal waveguide, directing echo power from the antenna to the receiver. Leakage through the TR tubes on transmit is directed to an RF termination because the two leakage paths are of equal length to the termination port. The leakage paths to the receiver port are out-of-phase. The TR tubes must be matched to one another, since good receiver protection depends on the them leaking the same amount.

The isolation between the transmitter and receiver is better in this type duplexer than with a circulator, and some systems require no further receiver protection. It does not, however, allow simultaneous transmit and receive. It also has components which have a finite operating life, including the TR tubes and their power supplies. Most modern radars use circulators in one form or another.

TR tubes are sections of waveguide sealed at both ends with RF-transparent seals and filled with a gas mixture. The mixture is formulated for two properties (Fig. 8-46). The first is to cause the tube to ionize quickly and minimize the leakage before complete ionization (the *spike*). The second is to cause the ionization to be very complete and reduce the leakage after ionization (the *flat*).

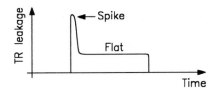

Figure 8-46. TR Leakage.

Two types of TR tubes exist, characterized by the method they use to shorten the spike. The keep-alive TR has a pointed electrode which is kept at a few hundred volts negative from the body of the tube. The point concentrates electric fields in its vicinity and causes the gas mixture in this area to partially ionize. The resulting free electrons speed the ionization process when RF power from the transmitter is absorbed.

Another method of maintaining a supply of free electrons to speed ionization is to use a radioactive gas which emits electrons (a beta emitter). Tritium, a hydrogen isotope consisting of one proton, one electron, and two neutrons, decays by one of its neutrons splitting into a proton and an electron, forming an isotope of Helium. The electrons released assist the ionization by the transmitter RF.

Both TR types have maintenance problems and finite lives. The keep-alive tube must have a keep-alive power supply, subject to failure. Radioactive TRs suffer from the fact that tritium has a half-life of 12.5 years, which causes them to have a finite life whether operating or on the shelf.

Some older radars use an arrangement of TR and ATR (anti transmit receive) tubes. Since these duplexers are highly individualistic in their operation, the reader is referred to their operations manuals.

Signal limiter: To protect the receiver from the transmitter, a signal limiter is often required in the path from the circulator to the receiver. One type, a solid-state diode limiter, is shown in Fig. 8-47. Signals entering the upper left port are divided in the hybrid into two equal parts, phased 90° apart, and placed on two transmission lines. If the peak-to-peak signal is less than 1.4 V_{p-p} (silicon diodes), the diodes are open circuits and do not interfere

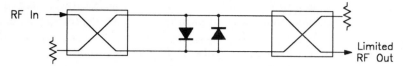

Figure 8-47. Diode Signal Limiter.

with the transmission. The signals combine at the output hybrid and output at the lower right port. To signals components greater than 1.4 V_{p-p}, the diodes short circuit, reflecting these components with a phase shift of 180°. The reflected components are directed to the lower left port of the input hybrid, which is terminated.

Magic tee: This device is a waveguide junction which has many uses in radar, particularly in transmitters, duplexers, and antennas (Fig. 8-48). Signals entering port H (the H-Plane junction) are divided, and equal in-phase parts appear at ports 1 and 2. Signals entering port E (the E-Plane junction) divide and appear at ports 1 and 2 in a 180° phase relationship. Signals entering ports 1 and 2 combine such that their vector sum is directed to port H and their vector difference to port E. The principle of operation is evident if the waveguides are considered to be parallel-wire transmission line. The H port is in parallel with ports 1 and 2 and the E port is in series. A full explanation is left as an exercise.

One application of magic tees is in the sum-and-difference circuitry needed for monopulse tracking, explained in Ch. 7. Another is the switching of a single transmitter between two antennas, shown in Fig. 8-49. If FPA-1 (final power amplifier 1) and FPA-2 feed the magic tee in-phase, the output goes to antenna 1. If the two FPA outputs are 180° phase, the output goes to antenna 2.

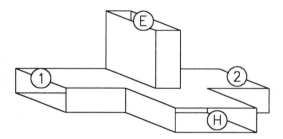

Figure 8-48. Waveguide Magic Tee.

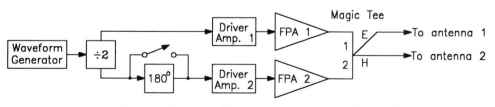

Figure 8-49. Magic Tee as a Transmitter Switch.

8-11. Microwave Components

Rotary joints: Waveguide rotary joints are necessary to transfer the transmitted signal and the receiver local oscillators from the fixed radar equipment to the rotating antenna. The rotating part of the joint must be either a circular waveguide or a coaxial line. Most are coaxial, since the transitions from rectangular to circular waveguide are bulky.

Many rotary joints couple more than one signal. Figure 8-50 shows a dual coupling, which in this case transfers transmitted power and the receiver local oscillator. Triple rotary joints also exist.

Most rotary joints use choke couplers for their sliding portions, so there does not have to be metal-to-metal contact. Dielectric seals between the sliding components also provide seals for waveguide pressurization (Sec. 8-13).

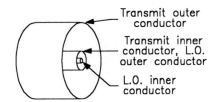

Figure 8-50. Dual Rotary Joint.

Other microwave components: Radars make extensive use of many microwave devices, such as isolators and directional couplers. The reader is referred to microwave texts for information about these devices ([3] and [16]).

8-12. Waveguide

Waveguide is commonly used in radars where high microwave power is present and where low losses are required. It is a hollow conducting "pipe" which propagates electromagnetic waves bounded by the walls of the guide. Waveguides can have many configurations, including rectangular, elliptical, circular, ridged, and double ridged.

The most commonly used type in radars is TE_{10} rectangular guide. It is the smallest possible guide for a given frequency. Its cross-section has two dimensions, universally called "a" (the longer dimension) and "b" (the shorter). The TE in its name stands for *transverse electric*, meaning that no component of the electric fields is in the direction of propagation. The lack of an "M" in the name indicates that magnetic field components are found in the direction of propagation. TM waveguide has transverse magnetic fields but not electric, and in TEM guide — most coaxial cables are TEM — neither the electric nor the magnetic fields have components in the direction of propagation.

The subscript 10 indicates that in the "a" direction there is one half-wave of the electric field, and in the "b" there is no electric field variation. Figure 8-51 shows the cross-section of a TE_{10} guide with the net electric fields.

Propagation through the guide is shown in Fig. 8-52. Electric fields arrange themselves at an angle across the guide such that the wavelength in the dielectric (often air) is as shown in the figure. Propagation, at the velocity of the dielectric, is at right angles to the fields. Where a field touches the side wall, it reflects as a field of the opposite polarity. The vectors in Fig 8-56 are the sums at a cross-sections where the peak electric fields are maximum. The

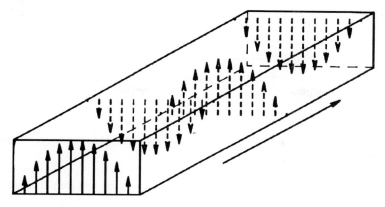

Figure 8-51. TE_{10} Waveguide Cross-Section with Net Electric Fields.

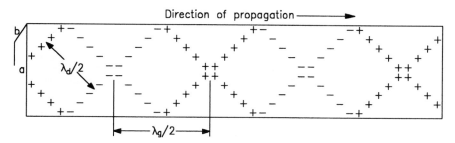

a. Lower frequency (longer wavelength)

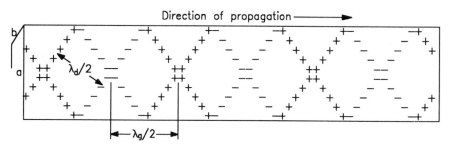

b. Higher frequency (shorter wavelength)

Figure 8-52. TE_{10} Propagation Electric Fields.

angles are a function of the frequency (wavelength). Part a of Fig. 8-52 shows a lower frequency (longer wavelength) and part b shows a higher frequency.

Figure 8-52 shows that the shortest wavelength which can be supported is where one-half wavelength in the dielectric is the "a" width of the guide. At this wavelength, the propagation is perpendicular to the axis of the guide and the signal simply bounces back and forth between the side walls. Thus the low frequency cutoff wavelength, is twice "a." The cutoff frequency is

$$\lambda_C = 2a \tag{8-7}$$

$$f_C = v_D / \lambda_C = v_D / 2a \tag{8-8}$$

λ_C = the cutoff wavelength
a = the waveguide inside broadwall dimension
f_C = the cutoff frequency
v_D = the dielectric velocity of propagation

TE_{10} is supported only where the dielectric wavelength is less than the "a" dimension. Thus this mode is theoretically supported over an octave bandwidth; in practice, satisfactory propagation occurs when the dielectric wavelength is from about 25 percent greater than the cutoff wavelength to about 5 percent less than "a." "b" must be less than half the dielectric wavelength for TE_{10} propagation.

The velocity at which energy propagates down the guide is called the group velocity. It is related to the dielectric velocity by the geometry of Fig 8-52.

$$v_G = v_D \sqrt{1 - (f_C/f)^2} \tag{8-9}$$

v_G = the group velocity
v_D = the velocity of the electromagnetic waves in the dielectric

The wavelength in the guide (the so-called guide wavelength) is the distance between two positive nodes along the guide.

$$\lambda_G = \lambda_D / \sqrt{1 - (f_C/f)^2} \tag{8-10}$$

λ_G = the guide wavelength
λ_D = the dielectric wavelength

Within the bounds of TE_{10} propagation, the guide wavelength must be greater than the dielectric wavelength. If the dielectric is air, with a propagation velocity very nearly the speed of light, this longer wavelength implies that the signal is propagating in the waveguide at greater than the speed of light. This phase velocity matches the guide wavelength. See slot antennas in Ch. 9.

$$v_P = v_D / \sqrt{1 - (f_C/f)^2} \tag{8-11}$$

v_P = the phase velocity ($> c$)

The propagation velocity (group) in a TE_{10} waveguide is a function of frequency, making the waveguide dispersive. For very wideband signals (for very high range resolution), waveguide is unsuitable as a transmission line. In these cases, a non-dispersive TEM mode line must be used. Unfortunately, the greatest advantage of waveguide is its low loss and high power capability, and one of the worst problems with coaxial cable is high loss and low power capability.

The characteristic impedance of waveguide is also a function of frequency and is given below.

$$Z_0 = Z_D / \sqrt{1 - (f_C/f)^2} \tag{8-12}$$

Z_0 = the characteristic impedance of the guide
Z_D = the impedance of the dielectric (120π Ω, in the case of air)

Waveguide pressurization: Waveguides are normally pressurized for two reasons. First is to prevent the external environment from contaminating the interior of the guide. If the waveguide leaks, it is highly desirable that it leak outward, so that water and other contaminants do not enter. When a radar is to be inoperative for long periods of time, it is especially important that the waveguide pressure be maintained.

Second, many gasses under high pressure are dielectrically stronger than the same gasses at low pressure, supporting higher voltages before arcing occurs. Pressure, then, increases the power-handling capability of the waveguide. Dry air is usually the gas used for pressurization unless the power-handling requirements exceed those available at reasonable pressures. In these cases, other gasses, such as sulfur hexafluoride (SF_6) are used.

Since the systems used to pressurize waveguide are electromechanical, they are subject to failure. At many radar sites, dry nitrogen is available so that the radar can continue to operate while the dehydrator/compressor is maintained. One must be careful to select the correct gas at these times. Oxygen is highly dangerous and must not be used unless one wants fires inside the waveguide. Helium, which is available at many radar sites for balloons to lift calibration targets, is a very weak dielectric and unsuitable for this purpose. See Table 8-3.

Table 8-3. Waveguide pressurization. (From Blake [17])
Power-handling capability (dry air at 1.0 atm = 1)

Pressure	Dry air	SF_6
1.0	1.0	4.0
2.0	2.8	15.5
3.0	4.0	21.0
4.0	5.8	
5.0	7.6	
6.0	9.6	

Waveguide table: The critical parameters of commonly used waveguide sizes are tabulated in Table 8-4. Not all waveguide sizes are given; only those which cover the commonly used radar bands. Only aluminum waveguides are shown. Attenuations are higher than shown for brass waveguides and lower for silver.

Table 8-4. Waveguide Parameters
(Hewlett-Packard [18])

Freq. Radar Band	Range GHz.	EIA WR-	JAN UG-[a] or MIL-W-85/	Width (a) cm.	Height (b) cm.	Cutoff Freq. GHz	Attn. low to high freq. dB/100m	CW Pwr. low to high freq. kW[b]
L	1.12-1.70	650	103	16.51	8.255	0.908	0.820-0.543	11900-17200
S	2.60-3.95	284	75[a]	7.214	3.404	2.080	3.074-2.105	2180-3100
C	4.90-7.05	159	-	4.039	2.019	3.71	6.622-4.980	754-983
X	8.20-12.40	90	67[a]	2.286	1.016	6.56	17.91-12.39	206-293
Ku	12.40-18.00	62	107[a]	1.580	0.790	9.49	26.71-19.63	119-157
K	18.00-26.50	42	66[a]	1.067	0.432	14.1	57.11-41.95	43-58
Ka	26.50-40.00	28	3-106	0.711	0.356	21.1	94.88-64.98	23-32
Q	33.00-50.00	22	3-010	0.569	0.284	26.35	101.2-67.78	14-20
U	40.00-60.00	19	3-013	0.478	0.2388	31.4	127.3-89.25	10-14
V	50.00-75.00	15	-	0.3759	0.1879	39.9	188.0-128.4	6-9
E	60.00-90.00	12	-	0.3099	0.1549	48.4	256.9-172.3	4-6
W	75.00-110.00	10	-	0.2540	0.1270	59.0	329.7-231.9	3-4

[a] Numbers marked with [a] are JAN UG- designations (eg. JAN UG-75). Those not marked are MIL-W-85/ designations.

[b] Assumes sea level with a breakdown of air dielectric at 15,000 volts per centimeter and a safety factor of 2.

8-13. Further Information

General references:

The following book is a good overview of transmitter design. Although it is out of print, many libraries have copies.

[1] G.W. Ewell, *Radar Transmitters*, New York: McGraw-Hill, 1981.

[2] A. S. Gilmour, Jr., *Microwave Tubes*, Norwood, MA: Artech House, 1986.

[3] S.Y. Liao, *Microwave Devices and Circuits*, Englewood Cliffs, NJ: Prentice Hall, 1980.

[4] D. Roddy, *Microwave Technology*, Englewood Cliffs, NJ: Prentice Hall, 1986.

Much useful information on all aspects of RF and microwaves (except waveguide) can be found in the following book.

[5] *The ARRL Handbook for the Radio Amateur*, Newington, CT: American Radio Relay League, published annually (67th ed,. 1990).

[6] T.A. Weil, Ch. 4 in M.I. Skolnik (ed.), *Radar Handbook*, 2nd ed., New York: McGraw-Hill, 1990.

Cited references:

[7] T.A. Weil, Ch. 7 in M.I. Skolnik (ed.), *Radar Handbook*, 1st ed., New York: McGraw-Hill, 1970.

[8] J.T. Mendel, "Helix and Coupled-Cavity Traveling-Wave Tubes," *Proceedings of the IEEE*, vol. 61, no. 3, March 1973, pp. 280, 293.

[9] R.B. Berry, *Coupled-Cavity Traveling Wave Tubes*, Palo Alto CA: Varian Associates, not dated.

[10] D.G. Fink, *Electronics Engineers' Handbook*, New York: McGraw-Hill, 1975, p. 9.50.

[11] *Introduction to Pulsed Crossed-Field Amplifiers*, Beverly MA: Varian Associates, not dated.

[12] *Introduction to Coaxial Magnetrons*, Beverly MA: Varian Associates, not dated.

[13] ITT Gilfillan, Van Nuys, CA.

[14] D.G. Laighton, Ch. 4 in E. Brookner (ed.), *Aspects of Modern Radar*, Norwood MA: 1988, pp. 279-281.

[15] M.T. Borkowski, Ch. 4 in M.I. Skolnik (ed.), *Radar Handbook*, 2nd ed., New York: McGraw-Hill, 1990.

[16] J. Helszajn, *Passive and Active Microwave Circuits*, New York: John Wiley & Sons, 1978.

[17] L.V. Blake, *Transmission Lines and Waveguide*, New York: John Wiley and Sons, 1969.

[18] Hewlett-Packard, *Microwave Test Accessories Catalog*, 1994 edition, Pp. 84-85.

9

RADAR ANTENNAS

9-1. Antenna Principles

The characteristics and principal parameters of radar antennas were discussed in Ch. 2. To review, the four principal antenna functions are

— To concentrate the transmitter's output into a narrow (angle) beam, thereby increasing the power density within the beam. The antenna property which accomplishes this is its gain.
— To match the impedance of the radar's transmission lines with that of the propagation medium, usually the atmosphere.
— To intercept and capture echo energy from targets. This capture is directional in the sense that the antenna captures energy more effectively from the direction the antenna's axis is pointing. This property is the antenna's effective aperture, and is related to its gain.
— To cause the transmit and receive beams to be steered in a preferred direction.

Reciprocity: Antennas are reciprocal. All characteristics and parameters apply equally to transmit and receive.

Development of the antenna beam: An antenna's pattern is its response as a function of the azimuth and elevation angle from the axis of the principal beam. To understand the beam's development, consider that all antennas, regardless of type or size, can be considered as an array of a large number of small elements. The far-field response of the array at any angle is simply the vector sum of the responses of the elements at the same angle. Figure 9-1 shows such an array as a transmit antenna (reciprocity assures that on receive it behaves exactly the same). An incremental signal is radiated from each element, in this case at equal phases. In the far field, the fields sum to form the response (fields sum voltages, not power — gain is in power terms). On the antenna's axis, signals from all the elements travel equal distances (within 1/16 wavelength — Ch. 2) and the incremental signals arrive in-phase with one another, producing the maximum response. Off-axis, the signals sum in a different fashion, producing the main beam shape, nulls, and sidelobes.

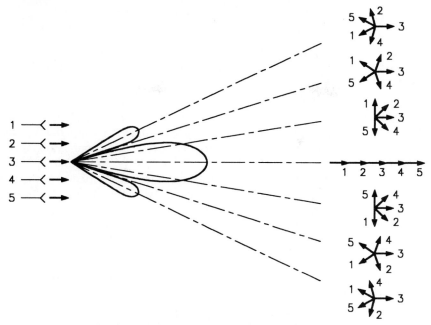

Figure 9-1. Antenna Beam Development.

By extending this argument, it can be shown that the far-field pattern of an antenna is the Fourier transform of its near-field illumination (Fig. 9-2).

$$E(u,v) = FT[i(x,y)] \qquad (9\text{-}1)$$
$$u = \sin(\theta_{AZ}) \qquad (9\text{-}2)$$
$$v = \sin(\theta_{EL}) \qquad (9\text{-}3)$$

$E(u,v)$ = the electric field beam pattern as a function of u and v

FT = the Fourier transform

$i(x,y)$ = the illumination as a function of the position on the antenna surface (in x, y coordinates). It can be in voltage or current terms.

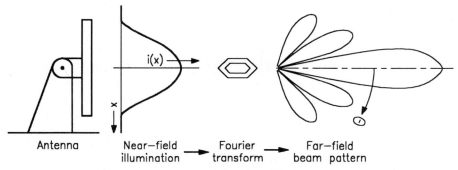

Figure 9-2. Fourier Transform Relationship in Antennas.

9-1. Antenna Principles

θ_{AZ} = the azimuth angle from the beam axis

θ_{EL} = the elevation angle from the axis

For continuously illuminated antennas such as reflectors and horns, the continuous Fourier transform (CFT) transforms the near-field illumination to far-field pattern. For discretely illuminated antennas (arrays), the discrete Fourier transform (DFT) is operative.

Almost all radar antennas use tapered illumination. Transmitted energy is not distributed uniformly over the antenna surface, and received energy striking various parts of the antenna is not used equally at its output. Most antennas have a higher response to energy striking their centers than to that striking the edges. This illumination taper is for sidelobe control.

The price paid for lowered sidelobes is that tapered antennas are effectively shorter than those uniformly illuminated. Thus tapered antennas have wider beamwidths, lower effective apertures, and less gain. Because sidelobes are system entry points for clutter and jamming, radar designers will virtually always chose low sidelobes over narrow beams and high gain.

An antenna's illumination function [$i(x,y)$] is the result of four factors:

— The *distribution of energy* across the surface of the antenna
— The *projected shape of the antenna* (the shape seen looking along the antenna's axis from the far-field)
— The *blockage of the aperture* (a blockage is an area of the antenna's surface which is not electromagnetically "visible" from the far-field)
— The *phase distribution* of the signals across the surface of the antenna

Figure 9-3 shows the principle of energy distribution across the surface of an antenna and its effect on the far-field beam pattern. The function $I(x)$ is a cross-section of the current distribution across the antenna aperture.

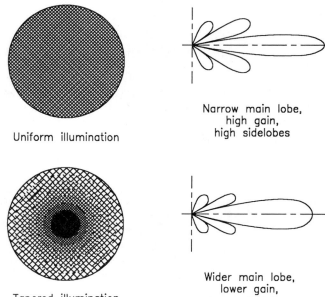

Figure 9-3. Effect of Antenna Energy Distribution.

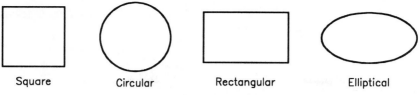

Figure 9-4. Antenna Shapes.

Figure 9-4 gives typical aperture shapes. The effective magnitude of the illumination at the aperture surface is the product of the current distribution and the antenna shape factor. For example, a circular aperture with uniform current distribution behaves as though it were circularly distributed. A circular aperture with uniform current distribution and a like-sized square aperture with circular current distribution will produce identical beam patterns.

Aperture blockage (Fig. 9-5) has the effect of reducing the length and area of an antenna, and distorting its illumination function. The net result is a widening of the beam (from length reduction), a loss of gain (from area reduction), and an increase in sidelobe levels (from illumination function changes).

Figure 9-6 shows the patterns of aperture blockage. Two illumination functions are produced: the unblocked aperture's illumination and the blockage illumination. The total illumination is their difference. Since the Fourier transform is linear (meaning that the transform of the sum of two components is the sum of their individual transforms), the resulting beam pattern is the difference of the two patterns (Fig. 9-7). In the development of blockage sidelobes, the phase of the sidelobe and blockage signals is important. This is why the patterns show positive (0°) and negative (180°) values.

The blocking feature can be part of the antenna, such as the feed assembly and its supporting struts on a reflector antenna. It can also be remote from the antenna, such as a mast which the antenna "looks through" at some azimuth angles, or a guard rail around the antenna. Any structure within the antenna's beam in its near field is an aperture blockage. At any time the antenna is looking at an external blockage, its beam pattern (particularly

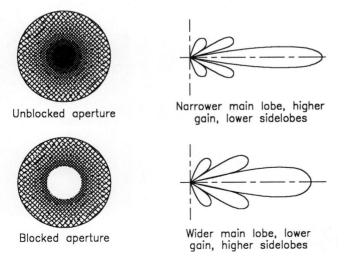

Figure 9-5. Aperture Blockage Effects.

9-1. Antenna Principles 419

sidelobe level) is different from the pattern when it is looking away from the blockage. The illumination of a blocked aperture is shown in Fig. 9-8.

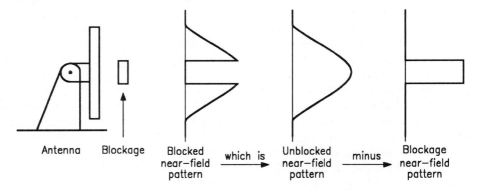

Figure 9-6. Blocked-Aperture Pattern Development.

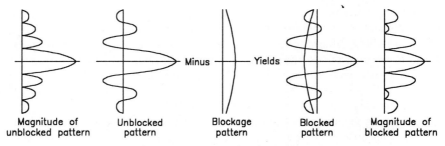

Figure 9-7. Aperture Blockage Patterns.

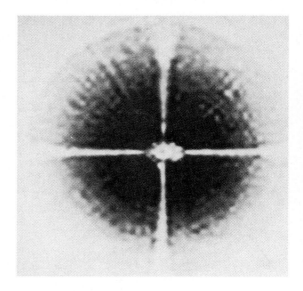

Figure 9-8. Blocked Aperture.
(Reprinted from Payne [5].
© IEEE, used with permission)

The phase of the antenna's illumination is important for two reasons, illustrated in Fig. 9-9. First, one of the tasks of an antenna is to produce a beam containing planar waves, which result in the greatest gain and effective aperture. The process by which antennas produce planar waves is called collimation (The term collimation also describes the alignment process which assures that the waves are planar.) To produce planar waves, all parts of the antenna must radiate such that in the far field these incremental signals sum in phase. If parts of the antenna radiate erroneous phases, the summation in the far field is less perfect, and the beam is wider, the gain is lower, and the sidelobes are higher than if the phases were correct.

Phase is also important because if there is a constant phase slope (or tilt) across the antenna, its maximum response will be in a direction other than along the antenna's axis. This is the principle behind phased array antennas.

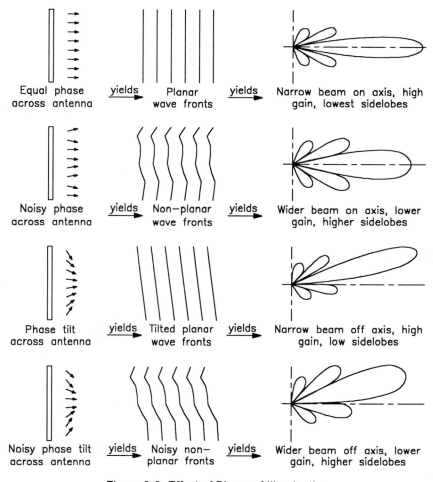

Figure 9-9. Effect of Phase of Illumination.

9-1. Antenna Principles

Antenna patterns for different illuminations: As we have said, many far-field antenna parameters depend on the total illumination in the near field. The total illumination is the product of the current (or voltage) distribution across the surface of the antenna and the antenna's shape factor. Table 9-1 gives the properties of several shapes and illumination functions. For a description of the functions themselves, see Ch. 11. The mathematics describing signal processing windows and antenna tapers are the same — the Fourier transform.

Table 9-1. Antenna beam properties.

Illumination function	Beamwidth (degrees); multiples of λ/D	Effective aperture; multiples of area	Highest sidelobe (dB below main lobe)	Sidelobe rolloff (dB per doubling of $\sin \theta$)
Rectangular	50.8	1.0	−13.5	−6
Triangular	73.4	0.50	−27	−12
Cosine	68.8	0.64	−23	−12
Cosine^2 (Hann)	82.5	0.50	−32	−18
Parabolic	66.5	0.74	−22	
Dolph-Chebyshev 3.0	82.5	0.48	−60	0
Kaiser-Bessel 2.5	90.0	0.44	−57	−6

9-2. Arrays of Discrete Elements — Principles

An array is a number of antennas, called elements, working together to form a single composite antenna. An understanding of the fundamentals of arrays helps explain many antenna phenomena, including the behavior of antennas near the ground or sea. In addition, many antennas encountered in communication, radar, and EW are arrays. The elements of an array can be organized in one of several ways, including those listed below.

- *Line array:* This is a one-dimensional array with multiple elements along its length, but is only one element high. All elements lie on a single line.
- *Planar array:* A planar array is two dimensional, with multiple elements in both length and height. All elements lie on the same plane.
- *Periodic array:* A periodic array is one where the element spacing is constant over the entire antenna.
- *Random array:* Random arrays have element spacings which are aperiodic.
- *Conformal array:* This array *conforms* to the shape of the object on which it is mounted, for example, the nose of an aircraft or missile.
- *Thinned array:* An array (usually, but not always, periodic) where not all element positions are filled. In other words, some elements are inert, or inactive.

Array-continuous antenna equivalence: Arrays behave identically (beamwidth, effective aperture, and gain) to continuous antennas, provided the spacing of the elements does not exceed one-half wavelength. In most circumstances, spacings up to one wavelength

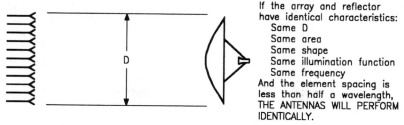

Figure 9-10. Array/Continuous Antenna Equivalence.

are allowed. If the spacing condition is met, the antenna relationships developed in Sec. 2-5 hold for the array. See Fig. 9-10.

The beam pattern of an array depends on both the array itself and the pattern of the elements. The pattern produced by an array of isotropic elements is called the array factor. The total pattern is the product of the array factor and the pattern of the elements, assuming they are all the same. This principle is known as the array multiplication principle, and is illustrated in Fig. 9-11.

The effect in an array of angular offset from the antenna's axis is shown in Fig. 9-12 and in the following equations. Using the center element as the reference, assuming the spacing between all elements is uniform (the array is periodic) and that the observation point is in the antenna's far field, where all elements are equidistant along the antenna's axis, the difference in distance to the observation point is

$$\Delta r = S \sin \theta \qquad (9\text{-}4)$$

Δr = the difference in distance between adjacent elements from the plane of the array to the observation point (meters)

S = the spacing between adjacent elements (meters)

θ = the angle from the antenna's axis to the observation axis

Expressed as a one-way phase shift, the distance from Eq. 9-4 is

$$\Delta \phi = \frac{2 \pi S}{\lambda} \sin \theta \quad \text{(radians)} \qquad (9\text{-}5)$$

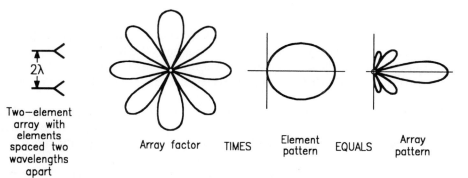

Figure 9-11. Array Multiplication Principle.

9-2. Arrays of Discrete Elements — Principles

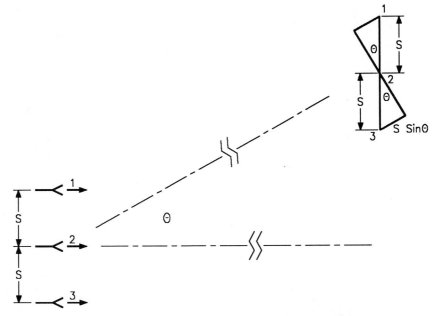

Figure 9-12. Array Beam Formation — Equal Phased Elements.

$$\Delta\phi = \frac{360\,S}{\lambda}\,\sin\theta \quad \text{(degrees)} \tag{9-6}$$

λ = the wavelength (meters)

Example 9-1: A three-element array, such as in Fig. 9-12, operates at 3.10 GHz with a uniform element spacing of 7.35 cm. All elements are fed at the same phase. Find the angle of the first null in the beam pattern.

Solution: To produce a null, the far-field elemental signals must cancel. Thus they must be distributed uniformly around a circle. With three elements, the elemental phasors must thus be 120° apart. Solving Eq. 9-6 for θ with a spacing of 7.35 cm and a wavelength of 9.68 cm gives an angle θ of 26.0°. This is the angle of the first null.

It was assumed above that the phase of each element was the same. If the phase is changed between elements by a constant amount, the array is said to have a linear phase tilt. This tilt causes the maximum summation to occur at an angle different from that of the array's axis, as shown in Fig. 9-13. This is the principle of phase steering, and is the basis of "phased array" antennas.

The steering angle is determined from the geometry shown in the figure inset, which is elements 3 and 4 expanded. The maximum summation (the main beam) will form in the direction where the elemental signals have the same phase. In the figure, the signal from element 4 leads that of element 3. Thus the signal from element 4 must travel enough farther than that of element 3 to give them the same phase. That extra distance is where S Sinθ equals the phase difference between elements 4 and 3. The steering angle is expressed in Eqs. 9-7 and 9-8, which are simply Eqs. 9-5 and 9-6 solved for sin θ.

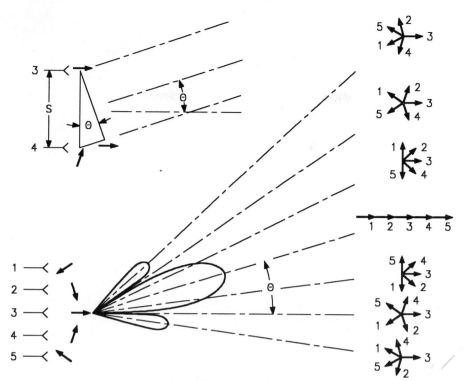

Figure 9-13. Effect of Phase Tilt on Beam Formation.

$$\sin \theta = \frac{\lambda \Delta\phi}{2\pi S} \quad (\Delta\phi \text{ in radians}) \qquad (9\text{-}7)$$

$$\sin \theta = \frac{\lambda \Delta\phi}{360 S} \quad (\Delta\phi \text{ in degrees}) \qquad (9\text{-}8)$$

θ = the beam steering angle from antenna axis

$\Delta\phi$ = the phase difference between adjacent elements

S = the element spacing (same units as wavelength)

λ = the wavelength

The above discussion considers arrays as transmitters. As receivers, they work exactly the same way, except that instead of splitting the signal among the elements, the elemental signals are summed (Fig. 9-14). Because of antenna reciprocity, the receive steering uses the same phase angles as transmit steering.

At steering angles other than zero (beam steered to the antenna's axis, the antenna is slightly shorter than if viewed from on-axis. Thus the beamwidth of an off-axis beam is slightly wider than on-axis and the gain and effective aperture are slightly less. The principle is shown in Fig. 9-15 and the off-axis beam parameters are

9-2. Arrays of Discrete Elements — Principles

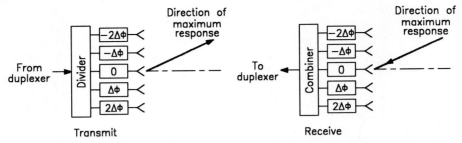

Figure 9-14. Comparison of Array Transmit and Receive.

$$\theta_{3(OFF)} = \frac{\theta_{3(ON)}}{\cos\theta_{AZ}\cos\theta_{EL}} \quad (9\text{-}9)$$

$$A_{E(OFF)} = \frac{A_{E(ON)}}{\cos\theta_{AZ}\cos\theta_{EL}} \quad (9\text{-}10)$$

$$G_{(OFF)} = \frac{G_{(ON)}}{\cos\theta_{AZ}\cos\theta_{EL}} \quad (9\text{-}11)$$

$A_{E(OFF)}$ = the off-axis effective aperture
$A_{E(ON)}$ = the on-axis effective aperture
$G_{(OFF)}$ = the off-axis gain
$G_{(ON)}$ = the on-axis gain
$\theta_{3(OFF)}$ = the off-axis 3-dB beamwidth
$\theta_{3(ON)}$ = the on-axis 3-dB beamwidth
θ_{AZ} = the azimuth steering angle
θ_{EL} = the elevation steering angle

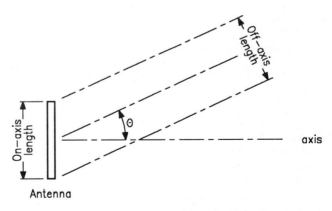

Figure 9-15. Beam Degradation with Steering Angle.

426   Chap. 9. Radar Antennas

9-3. Radar Antenna Configurations

Radar antennas come in many types and configurations. Figure 9-16 shows some of the more common varieties. All the reflectors except the cylindrical paraboloid are parabolas of rotation (paraboloids) and a cross-section is shown.

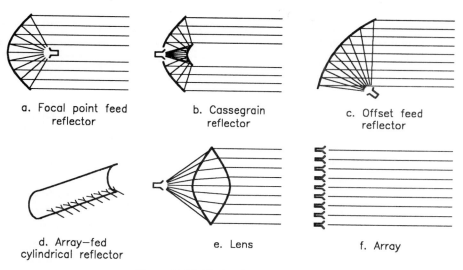

a. Focal point feed reflector
b. Cassegrain reflector
c. Offset feed reflector
d. Array—fed cylindrical reflector
e. Lens
f. Array

Figure 9-16. Radar Antenna Types.

— The *focal point feed reflector* antenna uses as a reflector a parabola of rotation (paraboloid) with the feed at the paraboloid's focus. Its illumination is set by the feed antenna's beam pattern only and is difficult to control. Aperture blockage is by the feed and its supports.

— The *cassegrain* reflector system uses two reflectors, the primary being a paraboloid and the secondary being an hyperboloid (hyperbola of rotation). The feed is near the primary reflector and faces in the direction of propagation. Illumination control is difficult and aperture blockage (from the secondary reflector and its supports) can be very large (for an exception, see the twist reflector cassegrain in the next section.

— The *offset feed reflector* is a focal point feed antenna with part of the reflector removed. The purpose is to reduce or eliminate aperture blockage. Illumination control is difficult.

— The *cylindrical parabolic reflector* antenna uses a reflector which is curved parabolically in one direction and is not curved in the other. It cannot be fed from a single point and is usually fed with a line array. This gives much more control over the illumination than is possible with a focal point feed antenna, and thus lower sidelobes.

— *Lens* antennas avoid the aperture blockage problem altogether, but illumination control is by feed antenna pattern and is difficult. This configuration is not used as much as it was in the past.

— *Array* antenna illumination is determined by the distribution of energy to and from the elements and can be very tightly controlled. This, plus no aperture blockage, gives arrays the potential of having very low sidelobes. If phase control of the elements is implemented, arrays can electronically steer beams without physically moving the antenna. Arrays are used in many radars at present and will be seen in more in the future.

Table 9-2 summarizes the characteristics of the different radar antenna configurations and gives their strong and weak points.

Table 9-2.
Characteristics of the Antenna Configurations

Antenna Configuration	Illumination Control	Aperture Blockage	Sidelobe Levels
Focal-point feed reflector	Poor	Moderate	High
Cassegrain reflector	Fair	Much	Moderate
Polarization twist Cassegrain	Fair	Moderate	Moderate
Offset-feed reflector	Poor	Little	Moderate
Array-fed cylindrical reflector	Good	Moderate to little	Low
Lens	Poor	Little	Moderate
Planar slot array	Good	Very little	Very low
Frequency steered array	Good	Very little	Low to moderate
Phase steered array	Good	Very little	Low to moderate

9-4. Sidelobe Suppression Techniques

It has been noted that antenna sidelobes cause problems in radar systems. They provide a path for jamming and EMI to enter the radar. In airborne systems, sidelobe clutter interferes with tail-aspect targets. Three methods are used to counteract the effects of sidelobes; low sidelobe antenna designs, sidelobe blanking (SLB), and coherent sidelobe cancellation (CSLC).

Low sidelobe antennas: Low sidelobe antenna designs are a result of careful attention to several factors, in addition to careful manufacturing and handling. Only the principles are discussed here. Detailed consideration of low sidelobe designs can be found in specialized texts.

The choice of the illumination function is critical. With reflector antennas, the designer can only control the illumination function in the design of the feed. With arrays, newer modeling techniques allow good control of illumination. This is one reason most low-sidelobe antennas are arrays.

Aperture blockage must be avoided, both in the antenna design and in its placement. Near-field blockages external to the antenna, such as masts, poles, guard railings, and so on, have the same effect on sidelobes as blockages attached to the antenna, such as feeds and struts. Again, because of their low blockage, array designs are favored.

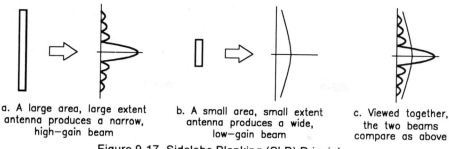

a. A large area, large extent antenna produces a narrow, high-gain beam

b. A small area, small extent antenna produces a wide, low-gain beam

c. Viewed together, the two beams compare as above

Figure 9-17. Sidelobe Blanking (SLB) Principle.

Surface smoothness in the case of reflectors and phase control in arrays affects sidelobe levels. In the discussion of phase-steered antennas it will be shown that phase control granularity has a major role in determining sidelobes. For this reason, newer phase-steered arrays tend toward finer control of phase-shifters. Changing from 4-bit phase control to 7-bit changes the phase control granularity from 22.5° to 2.8125°.

Sidelobe blanking (SLB): Radars which have sidelobe blanking, which includes most airborne and many shipboard military systems, have two antennas. One is the radar antenna, with its main lobe and sidelobes. The second is the sidelobe blanking antenna (also know as the "guard" antenna), whose gain is less than that of the main lobe of the radar antenna, but greater (ideally at all angles) than the radar antenna's sidelobes. See Fig. 9-17.

The received output from each antenna is fed to a receiver, with the two receivers gain-matched to one another. Outputs from the two receivers are compared on a range-bin or Doppler-bin basis (Fig. 9-18). If the output of the radar receiver exceeds that of the sidelobe blanking (guard) receiver, the signal must have come from the main lobe of the radar antenna. This is because the main lobe is the only part of the radar antenna's pattern where the gain exceeds that of the guard antenna. This signal is processed through the system. If the guard receiver's output is greater than that of the radar receiver, the signal must have come from a radar antenna sidelobe, and it is blanked.

The danger in this method of sidelobe suppression is that a main lobe target echo may arrive simultaneously with high level sidelobe interference. The legitimate main lobe target may be cancelled.

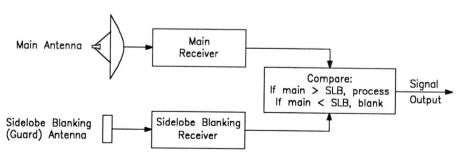

Figure 9-18. Sidelobe Blanking System.

9-4. Sidelobe Suppression Techniques

Coherent sidelobe cancellation (CSLC): The principle of CSLC is shown in Figs. 9-19 and 9-20. The idea is to generate an auxiliary antenna whose sidelobe gain and structure matches that of the radar antenna, but whose main lobe gain is smaller. The responses of the two antennas are subtracted, thus cancelling the response of the radar antenna's sidelobes but not that of its main lobe.

Before proceeding, two antenna principles must be reviewed. First is that the gain of an antenna's main lobe is proportional to the *area* of the antenna. Second is that the lobe structure and width of the main lobe and sidelobes is a function of the length of the antenna. The trick, then, is to produce an auxiliary antenna whose length is the same as that of the radar antenna (producing the same lobe structure) but whose area is much less than that of the radar antenna (giving it a lower main lobe gain).

Although it is theoretically possible to cancel all sidelobes with this technique, in practice this is impractical. Most CSLC efforts are aimed at cancelling the antenna's response in a particular direction (perhaps the direction of a jammer). This is what is implied in Fig. 9-19. As the radar beam's position changes with scanning or tracking, the CSLC array must be adjusted so that its response exactly matches that of the radar antenna in the direction of the interference. This is a complex antenna control problem. This is accomplished with the phase-shifters and attenuators in the CSLC elements.

Actually, a number of null responses can be created and a number of interfering sources suppressed. The maximum number of nulls which can be created and steered is the same as the number of elements in the CSLC array. The process of creating and steering antenna nulls for ECCM (Electronic Counter-Counter Measures) is also known as adaptive nulling. The performance of a system with SLB and CSLC is shown in Fig. 9-21.

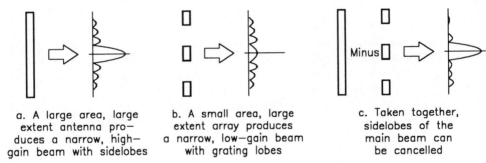

a. A large area, large extent antenna produces a narrow, high-gain beam with sidelobes

b. A small area, large extent array produces a narrow, low-gain beam with grating lobes

c. Taken together, sidelobes of the main beam can be cancelled

Figure 9-19. Coherent Sidelobe Cancellation (CSLC) Principle.

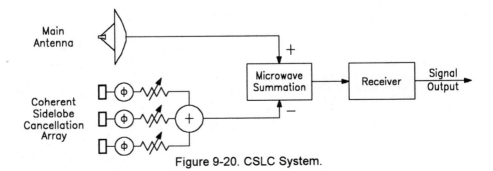

Figure 9-20. CSLC System.

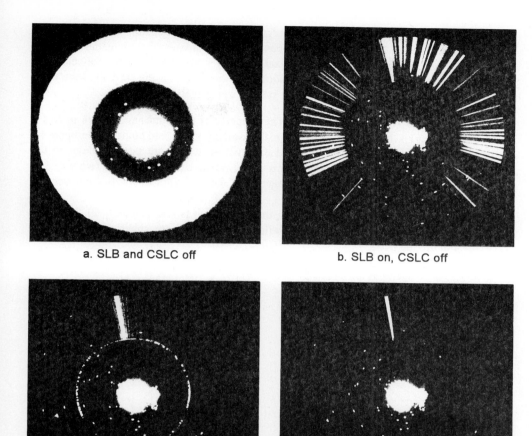

Figure 9-21. SLB and CSLC Effect.
(Reprinted from Johnson and Stoner [6]. © 1976 IEEE, used with permission)

9-5. Reflector Antennas

Reflector antennas consist of a reflector and a feed. The feed, placed at the focus of the reflector, is the source of transmitted energy and the collection point of received echoes. The reflector reflects the transmitted energy in the direction of its axis and reflects power coming from the axis direction to the focal point. Many configurations exist, three of which are used commonly in radars. First is the focal point feed reflector antenna (Fig. 9-16a). The reflector is a paraboloid (a parabola of rotation) and the feed is concentrated to a single point at the paraboloid's focus. The offset feed reflector (Fig. 9-16c) is a variant of the focal point feed reflector. Second is the cassegrain system, consisting of two reflectors (Fig 9-16b). Third is the cylindrical paraboloid reflector, which is parabolic in one dimension and cylindrical in the other (see Fig. 9-16d). In this case, the focus is a line rather than a point, and the feed must conform to this line. In most cases, the feed is a line array. The primary reflector is a paraboloid and the secondary is a hyperboloid, with the feed near the surface of the main reflector and pointed in the direction of the system's propagation.

Focal point feed reflector antenna: The property of the paraboloid which makes it useful as a reflector is that energy emitted from the focal point (the feed point) reflects parallel to the axis of the reflector. Conversely, energy whose waves are perpendicular to the reflector's axis are directed to the focus. This phenomenon can be viewed in two ways: as rays (lines parallel to the direction of propagation of the waves) or as waves (representing points of equal phase in the RF energy).

Figure 9-22 shows both representations. In the antenna's far-field (see Ch. 2), the waves are planar surfaces perpendicular to the reflector's axis, and are spherical between the reflector and the feed.

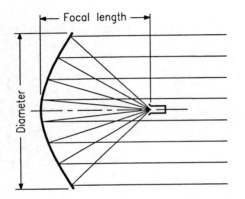 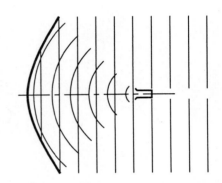

a. Ray diagram: Rays from focus reflect parallel to axis. Rays parallel to axis reflect to focus.

b. Wave diagram: Spherical waves from focus reflect planar (collimated). Plane waves reflect spherical.

Figure 9-22. Focal Point Feed Reflector Antenna.

Received energy from the axis direction concentrates to a small spot at the focus (Fig. 9-23). Energy from a direction other than the axis is focused to a point displaced from the focus and is diffused, or defocused.

The design parameters of a focal point feed antenna are the reflector's size and focal length, the feed's beam pattern, and the blockage of the aperture by the feed. The size of the reflector and its illumination pattern determine the antenna's beamwidth, gain, and (along with other factors) its sidelobes.

The reflector's focal length is the distance from its surface at the intersection of the axis to the focus. In general, the longer the focal length, the more the concentrated energy is displaced from the focus for a given signal angular displacement from the reflector's axis. This makes long focal length reflectors useful in tracking radars, since with them a given angular target error produces more error at the feed than with short focal length paraboloids. On the other hand, long focal length reflectors require more support for the feed and hence contribute to greater aperture blockage.

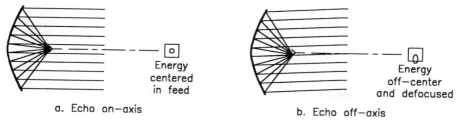

Figure 9-23. Reflector Focusing.

The pattern of the feed determines the illumination of the reflector. In general, the feed antenna must have a beamwidth which matches the angle seen from the focal point to the edges of the reflector, known as the subtended angle. Figure 9-24 defines subtended angle and Eq. 9-12 gives the relationship between the ratio of the reflector's focal length to its diameter (called the f-over-d, or f/d, ratio) to the subtended angle. This relationship is expressed graphically as Fig. 9-25.

$$\theta_{sub} = 4 \tan^{-1} [1 / (4 \text{ f/d})] \tag{9-12}$$

θ_{sub} is the feed subtended angle for a focal point reflector

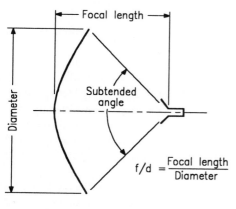

Figure 9-24. Feed Pattern: Subtended Angle Definition.

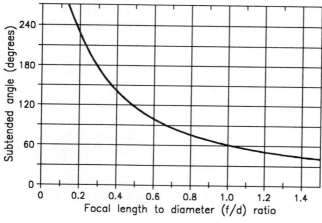

Figure 9-25. F/D Ratio and Subtended Angle.

9-5. Reflector Antennas

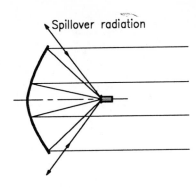

Figure 9-26. Spillover.

Several rules of thumb exist for relating feed 3-dB beamwidth to subtended angle, ranging from making the 3-dB beamwidth about 0.9 of the subtended angle to making the null-to-null beamwidth (about twice the 3-dB beamwidth for most illumination functions) equal to the subtended angle.

If the feed beamwidth is excessive, spillover occurs (Fig. 9-26), causing an undesired antenna response in the direction of the spillover. If the feed angle is too small, only a portion of the reflector is illuminated. Since only that portion of the reflector which is illuminated is effective, too narrow a feed beam results in a total antenna which has wider beams and less gain than predicted by the reflector's physical size and wavelength. In most cases, it is assumed that the feed matches the reflector and the length efficiency relationships of Ch. 2 hold, but occasionally a system is designed to use only part of the reflector. One such system is discussed later in this section.

Many two-dimensional search radars with reflector antennas use multiple feeds. The U.S. Federal Aviation Administration's (FAA's) Airport Surveillance Radar model 9 (ASR-9) is shown in Fig. 1-54. Its feed assembly is shown in Fig. 9-27. The two feeds produce two beams, one aimed at the horizon (the lower beam, upper feed) and one aimed upward (the upper beam, lower feed). The upper radar feed is matched to the reflector and illuminates its entire surface. It is at the reflector's focus, and produces a beam of maximum gain and minimum beamwidth which is pointed approximately horizontally (Fig. 9-28). This beam both transmits and receives.

The lower feed is below the reflector focus and illuminates most of the reflector. It creates a beam which is slightly lower in gain and pointed slightly upward, so that its lower edge does not touch the ground. It only receives. At short ranges where clutter is likely, the upper beam (lower feed) is fed to the receiver. At longer ranges, the lower beam is used (Fig. 9-28). This is a clutter-suppression technique found in many 2-D search radars.

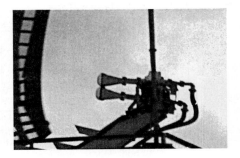

Figure 9-27. ASR-9 Feeds.
(Photograph by author)

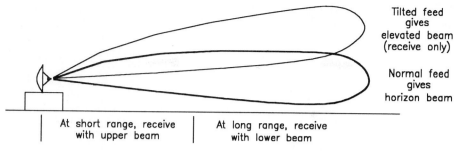

Figure 9-28. Elevation Beam Relationships. See Text.

The FAA's Route Surveillance Radar model 3 (ARSR-3) and several others use a focal point feed reflector with a feed similar to that shown schematically in Fig. 9-29. The feed has three horns, two for the radar and one for the secondary surveillance radar (SSR) which performs Air Traffic Control Radar Beacon System (ATCRBS) and Identification Friend or Foe (IFF) functions.

The two radar feeds produce upper and lower beams as described above and are used for clutter suppression. The other feed, labeled "beacon feed," is for the Secondary Surveillance Radar (SSR). Note that it is considerably larger than the radar feeds. Since the SSR frequencies (1.030 GHz interrogate and 1.090 GHz reply) are similar to the radar frequencies (about 1.25 GHz), the beacon *feed* has a considerably narrower beamwidth than the radar *feeds* and thus illuminates only a portion of the reflector (Fig. 9-30). The *antenna system* therefore has considerably lower gain and wider beams to the beacon than the radar. Also, the beacon feed is offset in azimuth from the radar, causing the beacon beam to point at a different azimuth angle than the radar beams. This is done so that the radar pulse, with several megawatts of power, will not arrive at the aircraft beacon (transponder) at the same time as the beacon interrogation, with about 2 kW of power. If they did arrive simultaneously, the radar pulse would prevent the beacon interrogation from being recognized.

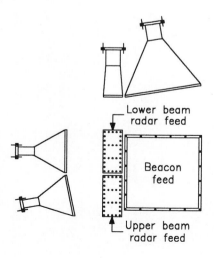

Figure 9-29. ARSR-3 Feeds.

9-5. Reflector Antennas

435

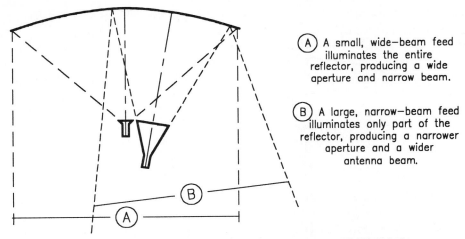

Figure 9-30. Comparison of Radar and Beacon Uses of ARSR-3 Antenna.

A close examination of the reflector in most 2-D air search radars shows excess curvature at the bottom compared to the top. This is to implement the shaped-beam patterns necessary to detect high-angle short-range targets without sacrificing gain at low elevation angles (see Sec. 2-5). A commonly used pattern is cosecant-squared, where echo amplitude at the radar is a function only of target RCS and altitude, and is independent of range.

Three-dimensional search radars resolve in elevation as well as azimuth and require steering of the beam in both directions. One implementation uses multiple feeds and switches between feeds for different elevation angles. These antennas usually have reflectors which produce roughly equal azimuth and elevation beamwidths. An example is the AN/TPS-43, shown in Ch. 1 (Fig. 1-56). These stacked beam antennas are gradually being replaced with arrays; the AN/TPS-43 is being replaced with the AN/TPS-75 (Fig. 9-31).

Figure 9-31. AN/TPS-75 Three-Dimensional Air Search Radar.

(Photograph courtesy of Westinghouse Electric Corporation)

Cylindrical paraboloid: One major problem with the focal point feed reflector system is that illumination control is difficult. Another is that electronic steering of the beam (for 3-D operation, for example) is awkward. Both these problems are resolved by using an array-fed cylindrical paraboloid reflector antenna. An example of this type of antenna is the U.S. Marine Corps and Air Force AN/TPS-63 (Fig. 9-32). This antenna is oriented vertically and uses a vertical line array as its feed.

Figure 9-32. AN/TPS-63 Cylindrical Paraboloid.

(Photograph courtesy of Westinghouse Electric Corporation)

Offset feed reflectors: One of the major causes of sidelobes is aperture blockage and one of the major causes of blockage is the feed and its supports. Some radar antennas reduce aperture blockage by placing the feed outside the aperture (Fig. 9-16c). This is often done with fan-beam search antennas. A few pencil-beam antennas also use this technique, including the AN/SPG-51 tracking radar and the AEGIS Mk-99 illuminator (Fig. 9-33).

Cassegrain reflectors: Long focal length paraboloids require that the feed supports by quite large, with their attendant aperture blockage. Also, focal point feeds do not allow much freedom in designing the reflector's illumination. The cassegrain system (Fig 9-16b) somewhat alleviates these problems by folding the optics to a shorter length, at the cost of the aperture blockage of the secondary reflector. By definition, cassegrain optics place the feed at or near the paraboloid main reflector and use a hyperboloid as the secondary reflector. Figure 9-34 shows a cassegrain antenna, the U.S. Navy's AN/SPG-60 tracking radar.

9-5. Reflector Antennas

Figure 9-33. Two AEGIS Illuminator Offset Feed Antennas. (Photograph courtesy of Bath Iron Works)

Figure 9-34. AN/SPG-60 Cassegrain Reflector Antenna.

(Photograph courtesy of Lockheed-Sanders)

The aperture blockage caused by the secondary reflector can be quite large. A technique for reducing its sidelobe-causing effect is the polarization twist cassegrain (Figs. 9-35 through 9-37).

In the example of the figures, the feed is horizontally polarized and the secondary reflector is made up of horizontal wires imbedded in a fiberglass shell. The wires are insulated from one another. Horizontal electric fields cannot be supported on the horizontal conductors and reflect with a 180° phase shift. Vertical electric fields can be supported, since the wires are insulated from one another. Thus, the structure is a reflector to horizontal polarization and is transparent to vertical polarization.

The primary reflector is a compound reflector, with an inner layer of wires insulated from one another and oriented 45° to the polarization of the feed (Fig. 9-35). The outer layer is a solid reflector 1/4 wavelength (in the material of the shell) from the inner layer. The action of this compound reflector is to reflect the wave and twist its polarization by 90°. The mechanism is shown in Fig. 9-36.

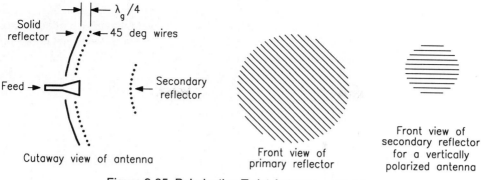

Figure 9-35. Polarization Twist Antenna — Part 1.

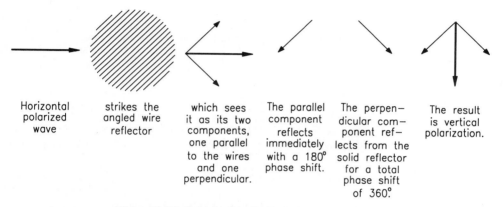

Figure 9-36. Polarization Twist Antenna — Part 2.

The action of the entire antenna is shown in Fig. 9-37. The transmitted wave is emitted from the feed horizontally polarized. It strikes the secondary reflector and is reflected. It then strikes the main reflector and is reflected with its polarization twisted to vertical. The vertical wave passes through the secondary reflector. Vertically polarized echo waves pass through the secondary reflector and strike the main reflector, where they are reflected and twisted to horizontal polarization. They then strike the secondary reflector and are directed to the feed. Therefore the secondary reflector is not an aperture blockage to either transmit or receive. The only blockage is the feed itself. Several radar antennas are based on this principle.

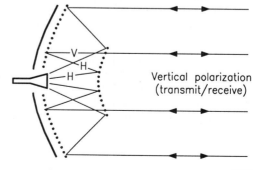

Figure 9-37. Polarization Twist Antenna — Result.

9-5. Reflector Antennas

A variant of this antenna is the inverse cassegrain shown in Fig. 9-38. In it, a horizontal spherical wave from the feed is collimated by a parabolic reflector composed of horizontal wires and directed back toward the feed. A flat compound plate reflects (retaining the collimation) and twists the wave to vertical polarization, which passes through the parabolic reflector unimpeded.

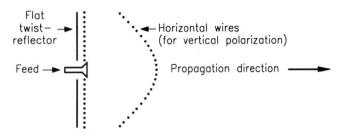

Figure 9-38. Inverse Cassegrain Antenna with Polarization Twist.

9-6. Lens Antennas

Microwave lens antennas (Fig. 9-16e) are yet another way to reduce aperture blockage and thus reduce sidelobes. Microwave lenses can be made of dielectrics (like optical lenses) or of metal slats.

Because focal lengths in radar antennas tend to be quite short by optical standards (photographers would kill for lenses with f/d ratios found in radar antennas), ordinary lenses would be impractically thick. For this reason, most microwave lens antennas use Fresnel lenses (also known as zoned lenses) in which the curvature matches that of common lenses, but in which the thickness is controlled by periodic wavelength (in the lens material) steps. See Fig. 9-39.

Other types of antennas are called "lenses." Examples are the bootlace lens antennas (Rotman and Archer lenses), and lens arrays. Each of these is actually a phased-steered array, and will be discussed in Sec. 9-9.

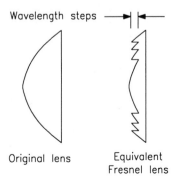

Figure 9-39. Fresnel (Zoned) Lens.

9-7. Mechanically Scanned Arrays

Array antennas have many uses in radars. When thinking of arrays, most people associate with the phase-steered array antennas such as AEGIS, but many other radar arrays exist. The properties of arrays which make them attractive as radar antennas are:

- There is excellent control of the illumination function in mechanically scanned arrays, resulting in low sidelobes.
- There is virtually no aperture blockage in arrays, resulting in low sidelobes.
- Arrays can be controlled easily and small amounts of electronic steering can be readily implemented, making arrays attractive as tracking antennas.

Mechanically scanned arrays are those in which there is no electronic beam steering, or where electronic steering is small and incidental to other antenna functions. There exist several basic types, including horn arrays, broadwall slot arrays, edgewall slot arrays, and microstrip arrays.

An example of a mechanically scanned array is the U.S. Navy's Target Acquisition System (TAS) air search radar, shown in Fig. 9-40. The antenna consists of a number of horns arranged in a horizontal line array. The SSR antenna associated with this radar points in the opposite direction from the radar antenna to prevent the radar's illumination from interfering with the SSR interrogation.

Figure 9-40. TAS Array.
(Photograph courtesy of Hughes Aircraft Company)

Broadwall slot arrays: Slot antennas are holes cut in conducting surfaces on which are RF waves. The mechanism by which slots radiate is through the interruption of current in these conducting surfaces. The slot antennas used in radars are cut in waveguides, hence they are called waveguide slot antennas.

Broadwall slots are cut in the wide wall of the waveguide, as shown in Fig. 9-41. Currents flow in TE_{10} waveguide walls as shown in the figure. The currents are shown frozen at a particular instant, although the currents and the voltage nodes propagate down the guide at its natural propagation velocity. The slots, which are 1/2 wavelength long and are spaced one-half the guide wavelength apart, interrupt the flow of current. The polarization of the radiation is at right angles to the slots (slot radiators behave exactly as dipoles rotated 90°). Slot positions must alternate across the center of the guide so that they all interrupt current flowing in the same direction and the incremental waves are in phase with one another in the far field.

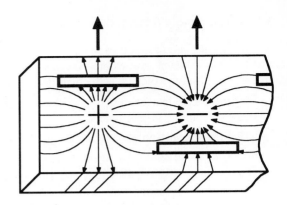

Figure 9-41. Waveguide Broadwall Slot Radiators.

The further the slots are from the center, the more net current they interrupt and the greater fraction of the RF power in the guide do they radiate (Fig. 9-42). Only slots which interrupt net current radiate. Slots cut in the center of the guide interrupt no net current and do not radiate. The resonant length of the slots is a function of frequency, slot position with respect to the waveguide centerline, and slot width. Slot width determines bandwidth. Slot design data can be found in the writings of Richardson and Yee ([8], [9], and [10], for example), and other sources.

Setting slot positions from centerline are one way to taper the array's illumination (Fig. 9-43). Taper is also achieved by compartmentalizing the waveguide run and varying the feed to each section of waveguide.

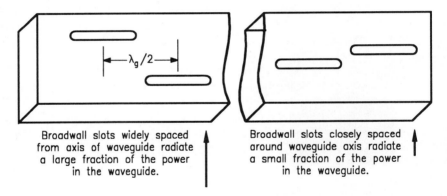

Broadwall slots widely spaced from axis of waveguide radiate a large fraction of the power in the waveguide.

Broadwall slots closely spaced around waveguide axis radiate a small fraction of the power in the waveguide.

Figure 9-42. Broadwall Slot Radiation Intensity.

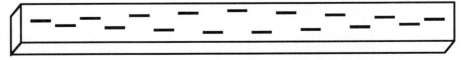

In broadwall slot arrays, illumination taper is achieved by varying the spacing of the slots from the center of the waveguide wall.

Figure 9-43. Broadwall Slot Array Illumination Taper.

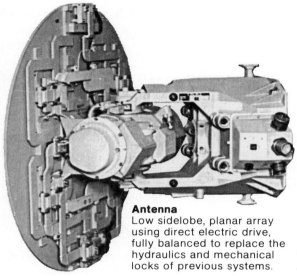

Antenna
Low sidelobe, planar array using direct electric drive, fully balanced to replace the hydraulics and mechanical locks of previous systems.

Figure 9-44. Broadwall Waveguide Slot Array Antenna (AN/APG-65).

(Photograph courtesy of Hughes Aircraft Radar Systems Group)

The line slot arrays, described above, are usually stacked to form a planar slot array. This antenna type is used in most modern missiles and airborne radars. It is compact and light weight, and has good pattern and sidelobe characteristics. Figure 9-44 shows a typical example (AN/APG-65 on the F/A-18).

There are two distinct kinds of slot arrays: standing wave arrays, and traveling wave arrays (Fig. 9-45) In many of these antennas, the waveguide runs are divided into chambers, in some cases to provide more flexibility in illumination taper, and in others for special applications such as monopulse tracking.

If the waveguide dividers are short circuits, as they are in most broadwall arrays, the array is standing wave (Fig. 9-45a). When excited from the manifold (transmit) or the slots (receive), a standing wave is set up in the waveguide segments. The power is roughly constant within each chamber. Illumination in a standing wave array is thus set by the distribution of power into the chambers from the manifold (the waveguide on the back of the antenna in Fig. 9-44) and the slot spacings from the waveguide axis.

In the traveling wave array, the guide may or may not be broken up into segments. In either case, the waveguide ends are terminated, rather than short-circuited (Fig. 9-45b). The illumination of a traveling wave array is set by three factors. First, like the standing wave

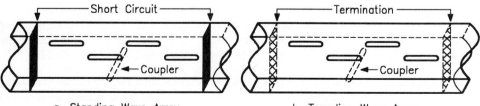

a. Standing Wave Array   b. Traveling Wave Array

Figure 9-45. Slot Array Types.

9-7. Mechanically Scanned Arrays

array, are the manifold coupling ratios. Second is the placement of the slots with respect to the waveguide centerline. The third factor, which is not present in standing wave arrays, is the distribution of energy within a waveguide segment. Many edgewall slot arrays (described later in this section) are of the traveling wave type.

Because of the terminations, the energy across a segment is not constant. Consider a waveguide segment fed from the manifold at its center. At the feed point (the center), the power level is highest. As power is radiated by the center slots, the level in the guide falls. At the ends there is less waveguide power than in the center. To achieve the proper illumination, the end slots must radiate a larger percentage of the waveguide power than they would in a standing wave array, because the power within the waveguide at this point is less than it would be in a standing wave array. Received signals, which strike each slot with equal amounts of power, enter the waveguide according to slot placement. The signals entering slots near the waveguide ends have a greater percentage absorbed by the terminations than at the center. Thus the same factors which determine illumination on transmit determine illumination on receive (in accordance with the reciprocity principle).

Element spacing: The spacing of elements in broadwall slot arrays is dictated by the waveguides themselves. For this discussion, assume an array with horizontal waveguide runs (vertically polarized). The horizontal spacing is one-half of a guide wavelength, which from Ch. 8 is known to be somewhat greater than half of the free-space wavelength. Ordinarily, spacings greater than one-half wavelength produce at least minor responses in the plane of the antenna (90° to the main lobe). Spacings of one or more wavelengths produce multiple main lobes called *grating lobes*. The array multiplication principle comes to the rescue in many antennas, however, since the elements do not respond appreciably in that direction.

The vertical spacing is dictated by the inside waveguide "a" dimension (Ch. 8), which must be greater than one-half and less than one dielectric wavelength. In addition, the waveguide wall thickness must be included. Thus the vertical spacing must be greater than the ideal of <1/2 wavelength. Again, because of the wide spacing, a response perpendicular to the axis is expected, but is minor because of the multiplication principle.

Element interactions: Elements in an array do not function independent of one another. One of the major difficulties in designing arrays is to account for the interactions between elements and their effects on the beam pattern. A discussion of this topic is beyond the purpose of this book, but information can be found in volumes such as the *Antenna Engineering Handbook* [1].

Bandwidth: The bandwidth of slot arrays is somewhat limited by the fact that the radiating elements are resonant and by the necessity for spacing them 1/2 guide-wavelength apart. Consequences of operating these antennas at frequencies other than the design center include increased VSWR, higher sidelobes, and a "smearing" of the beam caused by frequency steering across the antenna surface. Further information can be found in [10].

Manifold: The manifold distributes power to the waveguide runs and segments. Manifolds for non-tracking antennas, can be quite simple. Much more complex manifolds are required for lobing and particularly for monopulse tracking antennas (Sec. 7-4). A sure way of identifying a monopulse tracking planar slot array antenna is by the extremely complex manifold on its back.

The purpose of the manifold is to distribute transmitted power to the elements, and to sum the element receive responses. The following discussion summarizes these functions.

Used as a transmit antenna, power is directed to the waveguide manifold, from which it is distributed into the waveguide line arrays. The microwave signal introduced into each waveguide segment is set by the coupling ratios in that part of the manifold. Part of the illumination function of the antenna comes from the distribution of power from the manifold into the segments. Within each segment, power radiates through the slots according to their placement with respect to the centerline. These two actions set the illumination function.

As a receiving antenna, equal amounts of power impinge on each part of the array. The offset of the slots from the waveguide centerline determines what percent of this power enters the waveguide segments. This is part of the receive illumination function. The manifold couplers determine the rest of the illumination function. The incremental signals are summed at the manifold port. In the absence of ferrite phase shifters (which may have different phase shifts to transmit and received signals), the receive illumination is identical to that for transmit, resulting in a receive beam which identical to the transmit beam.

Edgewall slot arrays: Edgewall slots are cut in one of the shorter walls of the waveguide, and are slanted as shown in Fig. 9-46. In this part of the waveguide, currents flow between the broad walls in roughly straight paths. The slots are cut at an angle to interrupt the current and the polarization of the radiation is, as with the broadwall configuration, at right angles to the slot. Slots are paired, each at the same slant angle but in the opposite direction to its mate. The slots are spaced at 1/2 guide wavelength from each other. The far-field radiation is thus as shown in the figure, with the vertical electric field components of each slot pair cancelling. The net radiation from a horizontal waveguide run with edgewall slots is horizontally polarized.

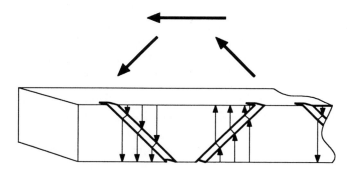

Figure 9-46. Edgewall Slot Radiator.

The fraction of the waveguide signal power which radiates, and the fraction of echo power striking the slots which is conducted into the waveguide, depends on the angle of the slots (Fig. 9-47). Slots cut directly across the edgewall from to bottom interrupt little current and radiate little. Slots cut parallel to the broadwalls radiate the maximum possible fraction, but the polarization is alternating vertical and there is no far-field broadside radiation.

Illumination taper is achieved in an edgewall slot array by varying the angles of the slot pair, as shown in Fig. 9-48. Information about edgewall slot design is found in many of the same references as for broadwall slots [7].

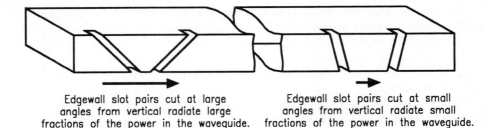

Edgewall slot pairs cut at large angles from vertical radiate large fractions of the power in the waveguide.

Edgewall slot pairs cut at small angles from vertical radiate small fractions of the power in the waveguide.

Figure 9-47. Edgewall Slot Radiation Intensity.

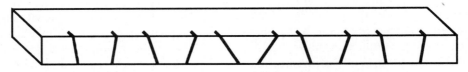

Illumination taper is achieved in sidewall slot arrays by varying the angle at which the slot pairs are cut.

Figure 9-48. Edgewall Slot Array Illumination Taper.

Edgewall slot arrays are often of the traveling wave variety. Most use waveguide segments which are much longer than those of standing wave broadwall slots, in many cases the full waveguide length. They can be fed at the center of the waveguide (center-fed arrays) or at the ends (end-fed arrays). End-fed arrays exhibit some squinting (steering) of the beam when operated at other than their design center frequency. Center-fed arrays are not steered by frequency changes, but their beamwidths are widened. See Sec. 9-10.

The illumination function of an edgewall slot array, if of the traveling wave variety, is controlled by the same factors that control broadwall traveling wave arrays: manifold distribution, slot positions (angles), and distribution of power within the waveguide.

Edgewall arrays exhibit very good illumination control and are used in applications where low sidelobes are of prime importance. The principal disadvantage of this configuration compared to broadwall slots is that these antennas are physically larger and heavier. Their advantages over broadwall arrays are better illumination control and the ability to place the elements closer together, because the "b" dimension of the waveguide, which must be less than 1/2 wavelength, determines the spacing between line arrays.

Antennas of this configuration are used in ground-based, shipboard, and airborne radars. Unlike broadwall slot arrays, most edgewall arrays of this configuration incorporate some form of electronic beam steering in one dimension (usually elevation). Examples include the AN/APY-1 and AN/APY-2 on the E-3A Airborne Warning and Control System — AWACS which have elevation phase steering (Figs. 9-49 and 9-50), and the AN/SPS-48 (Fig. 1-1) and AN/SPS-52 3-D search radars (Fig. 9-76), which frequency-steer in elevation.

Figure 9-49. Edgewall Slot Array — AN/APY-2 on E-3D Aircraft.
(Photograph courtesy of The Boeing Company)

Figure 9-50. AN/APY-2 Ultra-Low Sidelobe Waveguide Array.

(Photograph courtesy of Westinghouse Electric Corporation)

9-7. Mechanically Scanned Arrays

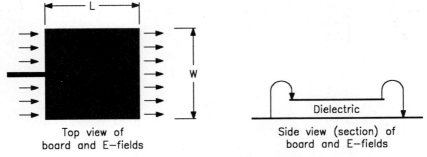

Figure 9-51. Radiation from a Microstrip Patch.

Microstrip array antennas: Microstrip is the simplest of the technologies of microwave printed circuits. It consists of a two-sided printed circuit board, one side of which is left whole as a ground plane with circuits etched on the other. Figure 9-51 shows the circuit side of an antenna etched onto a microstrip board. The length L of the antenna patch is 1/2 wavelength on the board (slightly less than the free-space wavelength). Radiation is from the edges of the patch normal to the feed point. Each patch thus acts as a two-element array radiating in-phase. The elements are the two edges and they are in-phase because of the half-wavelength separation of the edges.

The width W determines the beam pattern normal to the polarization and the input impedance of the patch (impedance matching is easy in microstrip, since transmission lines of virtually any impedance needed are available — set by the board parameters and the width of the trace).

Microstrip, because of its ease of construction and low cost, is gaining use in radar antennas, particularly where transmit power per element is relatively low.

A great advantage to microstrip is that power divisions and combination and impedance matching are very easy to implement. Figure 9-52 shows a simple microstrip array. Assuming the feed point is 50 Ω and the patches are 128 Ω, the power divisions and combinations are as shown. The first division/combination has two 100-Ω lines branching from the 50-Ω feed (two 100-Ω resistances in parallel matches 50 Ω). A section of

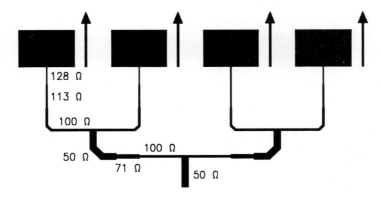

Figure 9-52. A Four-Element Microstrip Array.

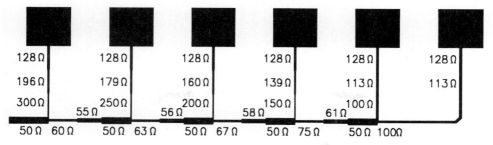

Figure 9-53. An End-Fed Microstrip Array.

transmission line 1/4 wavelength long with an impedance which is the geometric mean of the impedances at its ends forms a transformer which steps the impedance from 100 Ω to 50 Ω. After branching again into 100-Ω lines, there are 1/4 wavelength sections of 113 Ω, which match the 100-Ω lines to the 128-Ω radiating elements. This process can be repeated to form larger arrays. This is a *corporate feed*.

An alternative implementation is an end-fed manifold using microstrip couplers (Fig. 9-53). The manifold shown yields uniform illumination, with one-sixth of the input response going to each patch. The second patch gets one-fifth of that left at its coupler, the third gets one-fourth of that left, and so forth. Note in the first coupler that 300 Ω in parallel with 60 Ω is 50 Ω, and that a 1/4 wave transformer of 55 Ω matches 60 Ω to 50 Ω. Illumination taper is easily handled by using a different coupling ratio for each tap with the appropriate quarter-wave matching transformer between tap and patch. The tap impedances are found from power division relationships in circuit analysis.

Another advantage of microstrip is that circular polarization is easily implemented. The basis of circular polarization methods is to implement an antenna which responds to two orthogonal polarizations (such as horizontal and vertical), phased 90° apart. Figure 9-54 shows a microstrip realization of this basic principle. The patch is fed from two points, one producing horizontal polarization (the feed at the left) and one vertical (at the bottom). The feed trace is offset so that the signal arrives at the left feed 90° before the bottom. The result is a circularly polarized wave radiated normal to the patch (out of the paper). Except at the power divider/combiner, impedance matching is not addressed in this figure. In all likelihood, matching transformers would have to be inserted between the 50/100 Ω divider/combiner and the patch.

An example of a microstrip array is the Raytheon R-20 small boat radar (Fig. 9-55). This antenna, which operates in the X-Band, is about 20 inches wide and serves a transmitter of about 3-kW peak power and 1-W average power. It has a center-fed microstrip power divider, with illumination taper, under the metal "cage" at the bottom of the microstrip board.

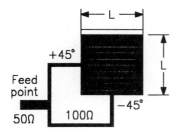

Figure 9-54. Circularly Polarized Microstrip Patch.

9-7. Mechanically Scanned Arrays

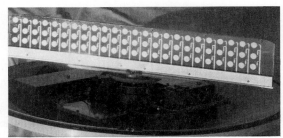

Figure 9-55. Microstrip Array Radar Antenna — Raytheon R-20.

9-8. Electronically Phase-Steered Arrays

Array beams can be steered by controlling the phase of the illumination, as shown in Fig. 9-12. The relationship between phase tilt across the array (phase difference between adjacent elements) is given in Eqs. 9-7 and 9-8. Electronically steered arrays exist in several configurations:

— *Conventional array:* In this array type, the elements are fed from a common source through power dividers and combiners (Fig. 9-56). The power divider/combiners can have many configurations, some of which are shown in Fig. 9-57.

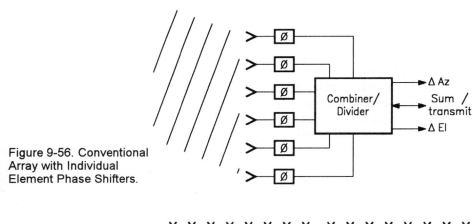

Figure 9-56. Conventional Array with Individual Element Phase Shifters.

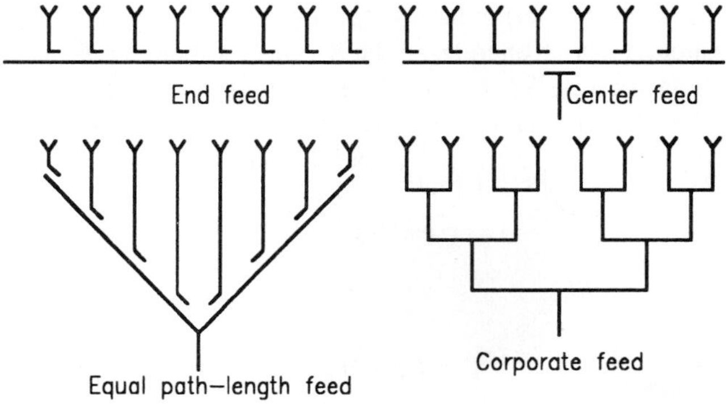

Figure 9-57. Conventional Array Feeds.

450   Chap. 9. Radar Antennas

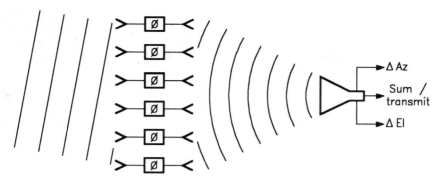

Figure 9-58. Space-Fed Lens Array.

— *Lens array:* In a lens array, the elements are fed from a single source which radiates to one side of the elements (on transmit). These elements receive, process, and re-radiate the signal from the other side. On receive, the process is reversed. This feed method is known as space feed (Fig. 9-58).

— *Reflector array (reflectarray):* This configuration is space-fed, like the lens array, except that the elements receive the feed from one side, processes the signal in the element, and reflect it back out the same side. In this case, the signal passes through each element's processing twice (Fig. 9-59).

Steering in two directions: Phase steering can be accomplished in two directions (e.g., azimuth and elevation) simultaneously by applying the phase tilt described in Eqs. 9-7 and 9-8 independently to each axis. As an example, consider a periodic array operating at 2.50 GHz with element spacings of 0.10 m horizontally and vertically. If it is desired to steer the beam 6.7° to the right of the array's axis and 9.6° up, Eq. 9-8 indicates a phase tilt of 35° horizontally and 50° vertically. *Steering is always away from the leading phase*, so the

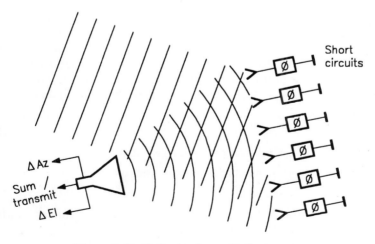

Figure 9-59. Reflector Array (Reflectarray).

9-8. Electronically Phase-Steered Arrays 451

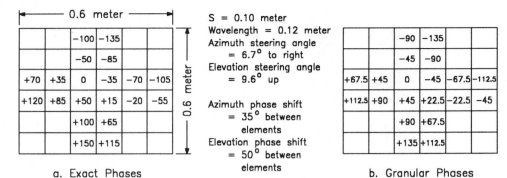

Figure 9-60. Phase Steering Example.

phase of the left elements must lead the right, and the lower elements must lead the upper. The required phases of the elements is shown in Fig. 9-60. Filling in the remainder of the figure is left as an exercise.

Phase granularity: In practice, phase shifters are usually either digital in nature or are digitally controlled. Thus only certain discrete phase shifts are available. For example, if the phase shifter or control were four binary bits, the smallest phase increment would be 22.5° ($360°/2^4$). Therefore the actual phases chosen for the elements in Fig 9-60a can only have 1 of 16 possible values. The phases actually chosen would simply be those nearest the desired value. Figure 9-60b shows the phases in this case.

If the width and height of the array are large (many elements covering many wavelengths), the average phase tilt is usually very close to that desired, and the net beam position is correct. The granularity of the phase shifters, however, causes the phase tilt to be noisy. The result of the phase noise is shown in Fig. 9-9; the array's gain is lowered slightly, but more importantly the sidelobes are raised.

Phase errors in arrays: Phase errors in the elements of an array affect both gain and sidelobes. The gain and sidelobe effects are similar to those of a conventional reflector antenna. Phase errors also result from the granularity of the phase-controlling mechanisms in arrays (phase shifters, in the case of phase-steered arrays). Sidelobes resulting from the granularity of the phase controls, from Eq. 9-12 [11], are shown in Fig. 9-61. The sidelobes from phase granularity *are in addition to* sidelobes produced by illumination taper and aperture blockage.

$$SL_{\Delta\phi} \approx 5/(2^{2P} N_E) \qquad (9\text{-}12)$$

$SL_{\Delta\phi}$ = sidelobes from phase granularity
P = the number of phase-control bits
N_E = the number of elements in the array

Amplitude errors also increase sidelobes. The following equations can be used to estimate the sidelobes caused by both phase and amplitude errors [12].

$$G_S \approx (\delta^2 + \Delta^2 + P_f) G_e \qquad (9\text{-}13)$$
$$G_M \approx \eta G_e N_e \qquad (9\text{-}14)$$

$$SLL = G_S / G_M \approx (\delta^2 + \Delta^2 + P_f) / (\eta N_e) \tag{9-15}$$

G_S = the maximum sidelobe gain caused by element amplitude and phase errors
G_M = the peak main lobe gain
SLL = the ratio of sidelobe gain to main lobe gain
δ = the element rms phase error
Δ = the element rms fractional amplitude error
P_f = the number of radiating elements not functioning
G_e = the element gain
N_e = the number of elements
η = the efficiency

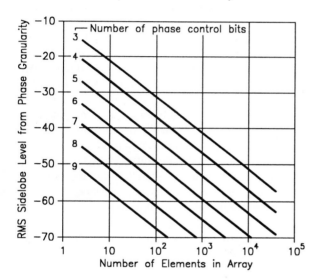

Figure 9-61. Sidelobes from Phase Granularity.

Phase shifters: Many different types of phase shifters (sometimes called phasers) are available for controlling the beam positions in phased arrays. A type commonly used where high RF power is required is the *ferrite latching phase shifter*, shown in Fig. 9-62. In it, a waveguide contains several sections (four, in this case) of ferrite material which acts as the internal transmission medium. The velocity of electromagnetic propagation in the medium is a function of the velocity of propagation in the internal medium, which in turn is a function of the medium's permeability [13].

Ferrites have the property that their permeability is a function of the DC magnetic field through them. By changing the electromagnetic field, the propagation velocity changes, as does the phase shift across the waveguide section. By placing a DC current in a coil of wire around the ferrite, it can be magnetized. When the current is removed, the ferrite retains some permanent magnetism. If the current is applied in a direction opposite the original, the ferrite

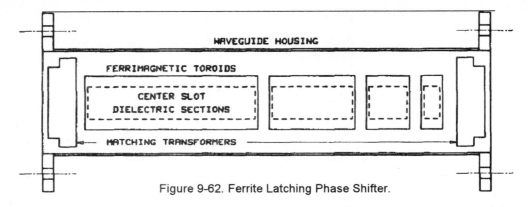

Figure 9-62. Ferrite Latching Phase Shifter.

is magnetized in the opposite polarity, which it retains when the current is removed. The ferrite thus has two residual DC magnetic states, having the same field amplitude and opposite polarity.

The largest section of ferrite is sized such that one state produces a propagation velocity which results in a relative phase shift of 0° and the other state's propagation velocity produces a phase of 180°. The second largest produces 0° or 90°; the third 0° or 45°; and the fourth 0° or 22.5°. Sixteen phase shifts are available, from a relative 0° to 337.5°, in 22.5° increments.

This type of phase shifter is non-reciprocal, meaning that the velocity of propagation in the ferrite is different to signals moving right-to-left than it is for signals moving in the opposite direction. The DC magnetic field in the largest section which produces 0° phase to signal in one direction produces 180° in the other. If this phase shifter is used in a single-antenna radar, each segment of each phase shifter must be reversed between the time the transmitter finishes its pulse and the time the antenna is ready to receive echoes.

Several ferrite-based variants are used. The *Reggia-Spencer* phase shifter is a reciprocal ferrite phase shifter which uses a DC current flowing continuously in a solenoid. Although this phase shifter is not inherently granular in its phase shift, the current drivers for the solenoid usually are, with the same result.

YIG (Yttrium Iron Garnet) *phasers* are again ferrites with permeabilities variable with DC magnetic field.

Varactor phase shifters function because of the fact that a reverse-biased diode is a capacitor (two conducting regions separated by an insulator – the depletion region). Placed in a circuit where this capacitance controls signal phase, the reverse-biased diode (varactor) introduces a phase shift variable with the DC bias voltage. This is a low-power reciprocal phase shifter.

Diode phase shifters, shown in Fig. 9-63, are another reciprocal low power phaser (they are sometimes known as meander line phase shifters). They are implemented in microstrip and works on the principle that time delays can be introduced by simply forcing signals to travel longer paths. Diode switches placed as shown cause the signals to either travel straight across the loops or go around them. In this four-bit shifter, loops are designed for 180°, 90°, 45°, and 22.5°. Once the design is completed, this is an inexpensive circuit to manufacture and is the phase shifter of choice in many phased arrays where high power per element is not required or where phase-shifting is done at low power.

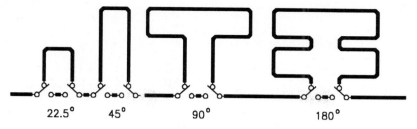

Figure 9-63. Diode (Delay Line) Phase Shifter.

Diode phase shifters have one further advantage. Please refer to Fig. 9-13 and Eq. 9-7. Even though the equation is in terms of phase shift, it is evident from the figure that the parameter which actually determines steering angle is the *time delay between elements*. An equation can be derived relating steering angle to time delay; this equation does not contain frequency. In systems where phase is the controllable parameter, frequency agility requires different phase shifter settings for different frequencies at the same steering angle. Where time delay is the settable parameter, the beam steering angle is independent of frequency.

$$\sin \theta = v_m \, \delta t / S \tag{9-16}$$

θ = the steering angle

v_m = the velocity of propagation in the medium (atmosphere)

δt = the incremental time delay between adjacent elements in a periodic array

S = the periodic element spacing

Phased array examples: Despite their primary disadvantage, cost, the advantages of phased array systems dictate their use in many systems. One outstanding example is the U.S. Navy's AEGIS system and its radars, the AN/SPY-1A, -1B, and -1D. They have four conventional array antennas, each an octagon approximately 12 feet across. Each array has approximately 4000 horn elements. The placement of the system on the *USS Arleigh Burke* (DDG-51) class of destroyers is shown in Fig. 9-64. The forward pair of arrays is visible in the figure. There is another pair aft. Each array covers more than 90° of azimuth and all four are served by one transmitter. In the CG-47 (*USS Ticonderoga*) class cruisers, there are two transmitters, one forward and one aft. The radar searches, tracks, and controls missile illuminators (Mark 99), one of which is the reflector antenna above the array (see also Fig. 9-34). The white dome immediately below the center of the bridge is a radar-aimed gun used for last-ditch defense against sea-skimming anti-ship missiles. This is the Phalanx close-in weapon system (CIWS—"sea-whiz").

The phasers are either 4-bit ferrite devices in earlier models (the AN/SPY-1A), and 7-bit in later versions (AN/SPY-1B and AN/SPY-1D). The increase in phase-setting precision reduces antenna sidelobes (Fig. 9-61).

9-8. Electronically Phase-Steered Arrays

Figure 9-64. USS *Arleigh Burke* with AN/SPY-1D (Two arrays visible on slanted panels below and to the side of the bridge). (Photograph courtesy of Bath Iron Works)

Figure 9-65 is the Hughes AN/TPQ-37 Firefinder radar, which had much success in the Persian Gulf. Its purpose is to find and track incoming artillery and missiles and calculate their point of origin. This S-Band pulse-Doppler radar uses diode phasers.

Another radar which achieved success and became a household name during the Persian Gulf War is the AN/MPQ-53, which is the sensor for the PATRIOT anti-aircraft and anti-missile system. Figures 9-66 shows the antenna. The main radar antenna (the large circle) is a space-fed lens array, with the feed assembly behind the array. Several other antennas are visible in the picture. These antennas have IFF, EW, and missile control functions [14].

The AN/FPS-115 PAVE PAWS early warning radar antenna (Fig. 9-67) is used to search over long distances (to 5000 km or more). Each system has two array faces 72.5 feet in diameter with 2677 element positions, 1792 of which are active (this is a thinned array). The arrays are oriented 120° apart, and each must scan $\pm 60°$, unusually wide angles. The elements are pairs of bent dipoles. They have circular polarization capability and are bent to give them a wide enough beam to accommodate the unusually wide scans. This is an active

Figure 9-65. AN/TPQ-37 Firefinder Weapon Locating Radar.

(Photograph courtesy of Hughes Aircraft Company, Ground Systems Group.)

Figure 9-66. AN/MPQ-53 PATRIOT Radar.

(Photograph courtesy of Raytheon Company)

Figure 9-67. AN/FPS-115 PAVE PAWS Early Warning Radar Array Antenna.

(Photograph courtesy of Raytheon Company)

9-8. Electronically Phase-Steered Arrays

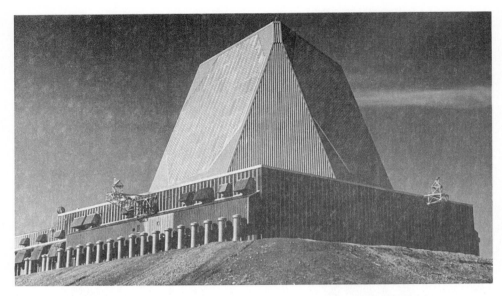

Figure 9-68. Ballistic Missile Early Warning System (BMEWS) Radar.
(Photograph courtesy of Raytheon Company)

array, in that each element has its own transmitter and receiver module (Sec. 8-6). The upgraded BMEWS (Ballistic Missile Early Warning System) radar is similar (Fig. 9-68).

The L-Band AN/FPS-108 COBRA DANE radar (Fig. 9-69) has a single array 29 m (95 ft) in diameter and a 0.92-MW *average power* transmitter. This multi-mission system gathers intellegence on missile test firings within the former Soviet Union, provides space tracking support, and supports ICBM early warning. This radar features very high resolution in range (less than one meter), requiring that the response of various parts of the antenna be synchronized in time. For example, the range difference across the array to a beam steered

Figure 9-69. AN/FPS-108 COBRA DANE Multi-Function Array Radar.

(Photograph courtesy of Raytheon Company)

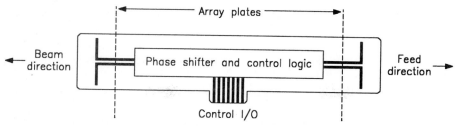

Figure 9-70. MIR Lens Phase Shifter Module.

20° from broadside is almost ten meters and without time synchronization, the range resolution would be smeared to at least 10 m. With appropriate delays across the aperture, the resolving capability of the illumination waveform is preserved [12].

The Raytheon MIR (Multitarget Instrumentation Radar, not shown) array is a twelve-foot diameter space-fed lens with 8,973 elements, each of which consists of two microstrip dipoles, a phase shifter, and phase shifter control logic. One of the phase shift modules is shown schematically in Fig. 9-70. The modules are mounted between two aluminum plates, with just the antenna portions of the circuit boards projecting through slits in the plates. The plate serves as a reflector for the dipoles. Because dipole-over-ground elements have relatively narrow beamwidths, the phase steering is limited to $\pm 30°$.

A newer radar which performs a similar instrumentation function is the General Electric AN/MPS-39 multiple object tracking radar — MOTR (Fig. 9-71). It uses a 12-ft diameter space-fed array with 8359 elements and 3-bit diode phase shifters. Beam steering is limited to $\pm 30°$. The antenna's highest sidelobes are 28.4 dB below the main-lobe response, compared to 20 dB for its predecessor, the AN/FPS-16 reflector-antenna radar. It tracks ten objects simultaneously (either primary radar or using beacons). It can compress a 50 μs illumination wave to 0.25 μs.

Figure 9-71. AN/MPS-39 Multiple Object Tracking Radar — MOTR.

(Photograph courtesy of General Electric)

9-8. Electronically Phase-Steered Arrays

The AN/APQ-164 phased array radar in the U.S. Air Force B-1B is shown in Fig. 9-72. It is shaped and oriented as shown to lower its radar cross-section (energy impinging on the antenna reflects downward rather than back to the radar—Fig. 9-73). The antenna beams are scanned electronically. To reduce the scan angles, the antenna can be physically moved in azimuth to three detented positions.

Figure 9-72. AN/APQ-164 (B1-B) Antenna and Elements.
(Photograph courtesy of Westinghouse Electric Corporation)

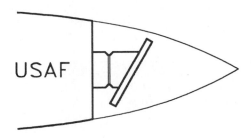

Figure 9-73. AN/APQ-164 Profile.

Lens-fed (bootlace) arrays: The bootlace lens arrays (Fig. 9-74) are a class of antennas which permit multiple independently steerable beams to be generated simultaneously. It consists of a lens, which is simply a specially shaped transmission line, multiple feed points connected to the lens, and multiple ports for radiating elements.

If a single beam is to be transmitted, the signal is fed to a single feed port on the lens, which distributes it such that the array elements will steer a beam in a particular direction. To move the beam, a different feed point is chosen. In the figure, a feed point near the bottom of the lens steers the beam upward. A feed point in the center steers the beam broadside, and a feed at the top deflects the beam downward. Multiple beams are transmitted by simultaneously feeding more than one port.

On receive, energy from a particular direction arrives at the lens from the array elements in a particular phase pattern, which sums coherently at one of the feed ports. The port depends

460   Chap. 9. Radar Antennas

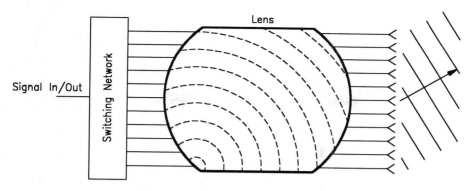

Figure 9-74. Bootlace Lens Array Principle.

on the direction of arrival of the signal. Multiple receive beams are formed by simply attaching multiple receivers to the desired feed ports, one per beam.

Two-directional steering is accomplished by using one elevation lens, with each element port feeding an azimuth lens (or one azimuth feeding multiple elevation).

Lenses of this type are known generically as Bootlace Lenses [15]. The two most common varieties are the Rotman (or Rotman-Turner) lens and the Archer lens. They differ primarily in that the Archer lens is a "sandwich" containing a dielectric material which reduces the velocity of propagation across the lens and thus its physical size. Bootlace lens antennas find wide use in EW systems.

9-9. Frequency-Steered Arrays

An additional method of phase scanning is to change element phases by changing the frequency of a signal traveling through a fixed length transmission line. Consider the waveguide feed system shown in Fig. 9-75. In this example, energy is fed to and removed from the top of the waveguide serpentine, also known as the sinuous feed. It is a length of waveguide folded such that the signal travels much further from node-to-node than the actual external distance between the nodes (S). If the electrical length of each loop is an integer number of wavelengths, the signals at the nodes (which feed array elements) are in-phase with one another. If the frequency increases, the electrical distance between nodes becomes, in phase terms, longer, and lower nodes have phases which lag those of the nodes above. This steers the beam downward. Conversely, lower frequencies cause lower nodes to lead the upper ones, steering beams upward.

Because propagation velocity in TE_{10} waveguides is frequency dependent, the analysis is not terribly simple. After the pertinent relationships are established, a numeric example will be offered to illustrate the calculation of beam steering. The basic beam steering relationship, from Sec. 9-2, is

$$\text{Sin } \theta = \frac{\lambda_0 \Delta\phi}{2\pi S} \quad (\Delta\phi \text{ in radians}) \qquad (9\text{-}17)$$

θ = the steering angle from broadside

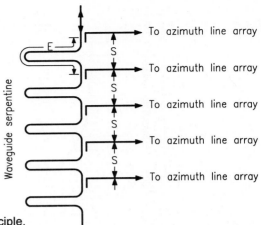

Figure 9-75. Frequency Scanning Principle.

λ_0 = the wavelength in the medium adjacent to the antenna
$\Delta\phi$ = the phase shift between adjacent elements (radians)
S = the element spacing (λ and S must be in the same units)

The length of the waveguide between adjacent nodes (elements) is E. The phase shift between adjacent elements is

$$\Delta\phi = \frac{2\pi E}{\lambda_G} - m\,2\pi \qquad \text{(radians)} \qquad (9\text{-}18)$$

E = the waveguide length between adjacent elements (meters)
λ_G = the guide wavelength (meters)
m = the integer necessary to reduce the expression to an absolute value of less than 2π.

The $m2\pi$ term removes full cycles of phase shift. Substituting Eq. 9-18 into 9-17 gives

$$\sin\theta = \frac{\lambda_0}{S}\frac{E}{\lambda_G} - \frac{m\lambda_0}{S} \qquad (9\text{-}19)$$

But $\lambda_G = \lambda_0 / [1 - (\lambda/\lambda_C)^2]^{1/2}$ for TE_{10} waveguide, where λ_C is twice the waveguide "a" dimension (see Ch. 8). Equation 9-19 becomes

$$\sin\theta = (\lambda_0/S)\{[E/\lambda_0][1-(\lambda_0/\lambda_C)^2]^{1/2} - m\} \qquad (9\text{-}20)$$

In frequency terms, Eq. 9-20 becomes

$$\sin\theta = [c/(fS)]\{[Ef/c][1-(f_C/f)^2]^{1/2} - m\} \qquad (9\text{-}21)$$

c = the propagation velocity in the propagation medium adjacent to the antenna
f = the frequency

In many cases, the sinuous feed nodes attach to horizontal waveguide line arrays, and the frequency changes cause beam steering in elevation. The U.S. Navy's AN/SPS-48 (Fig. 1-1) and AN/SPS-52 (Fig. 9-76) 3-D search radars scan in elevation using this principle. The vertical structure to the left of the AN/SPS-52 main antenna is the sinuous feed. In both of these antennas, the horizontal waveguide runs are traveling wave edgewall slot arrays.

Figure 9-76. AN/SPS-52 Shipboard 3-D Air Search Radar.
(Photograph courtesy of Hughes Aircraft Company, Ground Systems Group)

With the steering relationships established, a numeric example is appropriate, where we calculate beam steering angles from antenna parameters and frequencies. Finding frequencies for given steering angles is more difficult, since it requires an estimation of the factor "m" before attempting to solve the problem.

Example 9-2: An S-Band antenna is composed of horizontal waveguide runs, each being an edgewall slot line array. The vertical spacing between the waveguide runs is 8.0 cm. The length of the serpentine between each node is 138.8 cm. The waveguide itself is WR-284 with a cutoff frequency of 2.080 GHz. The waveguide dielectric is air, and the dielectric propagation velocity is assumed to be exactly 3.00×10^8 m/s (it isn't, so when doing this for real, one may need a more accurate velocity). The entire array is tilted back 10° from vertical. Find the beam steering angles from horizontal at frequencies of 3.00 GHz, 2.98 GHz, and 3.02 GHz.

Solution: At 3.00 GHz, the phase velocity of WR-284 waveguide is 4.1631×10^8 m/s (Eq. 8-10) and the guide wavelength is 13.88 cm (Eq. 8-9). This velocity is greater than the speed of light, but not to worry — that is allowed (actually necessary) for phase velocity in air dielectric waveguides. At this frequency, the electrical distance between nodes is exactly 10 wavelengths. Thus if the top node is 0° phase, the next node down is $-3,600°$ (or zero degrees, after the multiples of 360° are removed), and the next is $-7,200°$ (zero degrees), and so forth. The beam is aimed broadside to the array and is thus elevated 10° above the horizon.

At 2.98 GHz, the phase velocity is 4.1893×10^8 m/s and the guide wavelength is 14.06 cm. The sinuous feed is now 9.8713 wavelengths between nodes. If the top node is 0° phase, the next node down is −3,553.9°, or +46.1°. The free-space wavelength is 10.07 cm. The basic beam steering equation (Eq. 9-16 with 360° substituted for 2π, since the problem is being worked in degrees) gives a steering angle of 9.28° from the plane of the array, away from the leading phase (up). The beam is thus elevated 19.28° from the horizon.

At 3.02 GHz, the phase velocity is 4.1379×10^8 m/s, the guide wavelength is 13.70 cm, and the sinuous feed is 10.1314 wavelengths between nodes. The phase between nodes is −3,647.3°, or −47.3°. The steering angle is 9.52° downward from the plane of the array, for a beam elevation angle of 0.48° above the horizon.

9-10. Antenna Pedestals and Data Pick-Off Devices

Pedestals are the electromechanical devices which position the antenna. They are sometimes called positioners.

Drives: The two main drives used for pedestals are electric and hydraulic. Electric drive has the advantage of simplicity and the disadvantage that the drive motors are relatively large and heavy for their power output and in many cases require active cooling. Hydraulic motors are powerful for their size and are self-cooling (through the hydraulic fluid). Their disadvantage is the complexity of their support system, consisting of pumps, accumulators, heat exchangers, reservoirs, filters, and hydraulic lines.

Bearings: Antenna positioners are typically large and heavy and must rotate with minimum friction. The two classes of bearings used to separate the rotating part of the positioner from the stationary are mechanical bearings (usually roller-type) and hydrostatic bearings. The mechanical bearings are simple, although in many systems they must be actively lubricated, and have moderate friction.

Hydrostatic bearings are simply two large plates with lubricant pumped between them. The rotating plate slides on a film of lubricant over the stationary plate. Once in motion, they are virtually friction-free (very low dynamic friction). If allowed to stop moving, however, the lubricant's molecules align in such a way as to produce what could almost be called a "glue" between the plates, which causes considerable torque to be required to put the positioner in motion (high static friction − sometimes called stiction). Many tracking positioners which use this type bearing superimpose a small high frequency (10 Hz or so) sinusoid on the servo commands. This *dither* keeps the rotating part of the positioner in constant motion and prevents static friction from being a consideration.

Data pick-off devices: It was shown in Ch. 1 that a target's location in azimuth and elevation is found from the position of the pedestal when the detection occurs, or during tracking. A number of different devices are used to sense this position, and these are the subject of this section. Chapter 14 has more information on coordinate systems and conversion.

Potentiometers: Some older tracking systems find pedestal and target position with potentiometers. The range tracker outputs a voltage proportional to target range. This voltage is applied to two potentiometers in the pedestal, one for elevation and one for

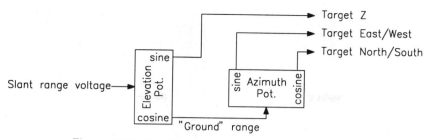

Figure 9-77. Sine/Cosine Potentiometer Connections.

azimuth. Each of these potentiometers, called sine-cosine potentiometers, has its shaft connected to the rotating part of the pedestal. There are two outputs. One is proportional to the input voltage times the sine of the angle from some reference and one proportional to the input times the cosine of the angle. Figure 9-77 shows the connections.

The outputs of the elevation potentiometer are a voltage proportional to the target's Z dimension, its height above the plane of the radar, and a voltage proportional to the target's "ground range," which is its distance from the radar projected onto the horizontal plane of the radar. The azimuth outputs are voltages proportional to the target's East/West and North/South distances from the radar.

Synchros: Synchros are transformers with four windings (Fig. 9-78). Three of the windings, called stator windings, are stationary and are separated from one another by 120°. The fourth, the rotor winding, turns with the input shaft. In synchro transmitters, a standard voltage (120 V 60 Hz, for example) is applied to the rotor winding. The rotor is attached to the rotating part of the pedestal. For each position of the rotor, there is a unique set of couplings to the stator windings and a unique set of stator voltages.

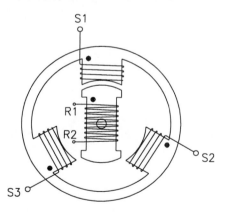

Figure 9-78. Synchro.

Synchros are used in two ways. In the first, the torque-producing system, where the three transmitter stators are connected to the three receiver stators, the receiver rotor is connected to the same supply as the transmitter rotor, and its shaft is free to rotate (usually connected to an indicator dial). Figure 9-79 shows the connections. If the rotors of the two synchros are aligned, their stators put out identical voltages and no current flows in the stator windings. If they are misaligned, the stator voltages are different and current flows. This current produces torque in the receiver, causing it to rotate to the aligned position. In this

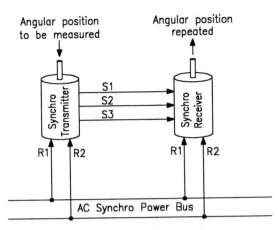

Figure 9-79. Torque-Producing Synchro System.

way, the receiver synchro repeats the angular position of the transmitter synchro. Positions are somewhat inaccurate because near alignment the torque becomes very small and receive rotor friction can prevent perfect alignment.

The second method of using synchros is more accurate at the expense of being more complex. In it (Fig. 9-80), the transmitter synchro stators are connected to the receiver synchro as before. The receiver synchro (which is now called a control transformed – CT) rotor is not connected to the synchro power bus, but to a servo. If the rotors are aligned, the CT rotor has the same voltage as the transmitter rotor. If they are 90° from one another, the CT rotor has no voltage (called a null). The amplified CT rotor is demodulated with the synchro bus wave to produce a DC error, which drives a motor/gear train, which in turn drives the CT rotor to its null position.

In some cases, the CT is electronic and the synchro stator voltages are converted to digital words (in a synchro-to-digital converter).

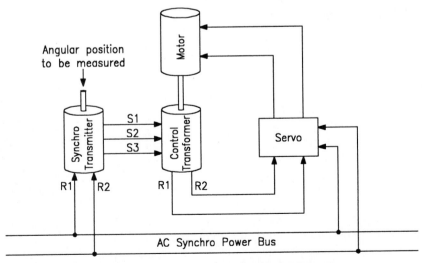

Figure 9-80. Servoed Synchro Positioning System.

Chap. 9. Radar Antennas

Synchros of both of the above types are used many times in pairs, one rotating once for each pedestal rotation (the 1:1 synchro) and one rotating once for each 10° of pedestal rotation (the 36:1 synchro).

In tracking radar manual modes, often the pedestal is slaved to a console synchro. When the pedestal enters automatic track, the handwheel synchro is disconnected. On exit from track, the pedestal would rotate back to the handwheel synchro's position. To prevent this, during track the console handwheel synchro is slaved to the pedestal position. This is called a synchro follow-up. Follow-up is also necessary when the servos are off and the pedestal is drifting in the wind.

Incremental digital encoders: The incremental digital encoder, shown in Fig. 9-81, consists of two devices with shafts connected to the rotating part of the pedestal. One generates a single pulse for each pedestal revolution, often as it passes north, and serves as a position reference. The other produces many pulses per revolution (e.g., some produce 20 pulses per degree of rotation).

Both encoders are connected to a counter (Fig 9-82). The one-pulse-per-revolution encoder resets the counter and the many-pulses-per-revolution encoder serves as its clock. Thus after the first revolution after power-up, the counter contains the rotational position of the pedestal. Incremental encoders are normally used in systems where rotation is continuous and in one direction only. Their primary use in radars is on search radars.

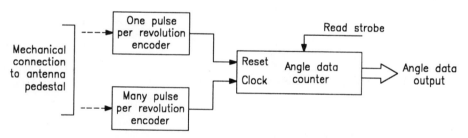

Figure 9-81. Incremental Digital Encoder for Search Radar.

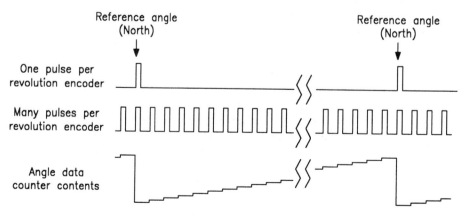

Figure 9-82. Incremental Encoder Timing.

9-10. Antenna Pedestals and Data Pick-Off Devices

Direct digital encoders: Direct encoders sense angular position as a digital word and are unaffected by rate or direction of rotation. They are often glass discs on which are etched concentric tracks, one track per bit in the digital word. Some segments of each track are left clear, perhaps representing binary 0s and others are etched opaque, representing 1s. A light shining through a slit and through the code-wheel is read by a photocell array on the other side of the wheel.

Two primary kinds of codes are used (Fig. 9-83). The binary coded wheel on the left directly generates a binary position word. Its problem is that when it transitions from one specific position to another, there is uncertainty as to the value from each track. Consider the transition from state 3 to state 4. Since each bit changes, each is uncertain and the encoder could put out any value. The only way around this is an elaborate system of lights and photocells which samples the encoder in three different positions. Some radar encoders work this way.

The other way uses a code sequence called a Gray code (Fig 9-83 right). This code is not weighted (as is a binary code), and each bit does not represent a specific value. Note from the figure that only one bit changes at each transition, eliminating the ambiguity of the binary code. Conversion of Gray to binary is quite simple and is illustrated in Fig. 9-84.

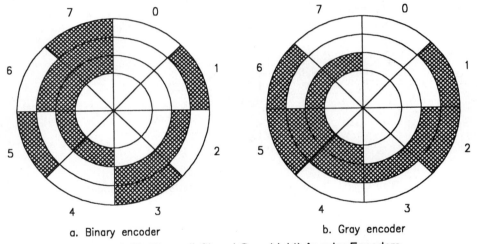

a. Binary encoder b. Gray encoder

Figure 9-83. Binary (left) and Gray (right) Angular Encoders.

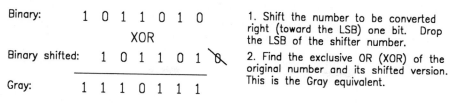

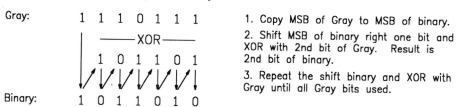

Figure 9-84. Binary / Gray and Gray / Binary Conversions.

9-11. Further Information

General references:

[1] R.C. Johnson and H. Jasik, *Antenna Engineering Handbook*, New York: McGraw-Hill, 1984.

[2] W.L. Stutzman and G.A. Thiele, *Antenna Theory and Design*, New York: John Wiley & Sons, 1981.

[3] H.E Schrank et al. in Ch. 6, and J. Frank in Ch. 7 of M.I. Skolnik (ed.), *Radar Handbook*, 2nd ed., New York: McGraw-Hill, 1990.

[4] *IEEE Standard Test Procedures for Antennas* (IEEE Standard 149), New York: Institute of Electrical and Electronic Engineers, various publishing dates.

Cited references:

[5] J.M. Payne, "Millimeter and Submillimeter Wavelength Radio Astronomy," *Proceedings of the IEEE*, vol. 77, no. 7, July 1989, pp. 993-1017.

[6] M.A. Johnson and D.C. Stoner, "ECM from the Radar Designer's Viewpoint," *IEEE ELECTRO-75 Record*, Part 30, pp. 30-5-1 − 30-5-11, May 1976.

[7] K.S. Kelleher and G. Hyde, Ch. 17 in R.C. Johnson and H. Jasik, *Antenna Engineering Handbook*, New York: McGraw-Hill, 1984.

[8] H.Y. Yee, Ch. 9 in R.C. Johnson and H. Jasik, *Antenna Engineering Handbook*, New York: McGraw-Hill, 1984.

[9] P.N. Richardson and H.Y. Yee, "Design and Analysis of Slotted Waveguide Antenna Arrays," *Microwave Journal*, June 1988.

[10] H.Y. Yee and P.N. Richardson, "Slotted Waveguide Antenna Arrays," *IEEE Antennas and Propagation Society Newsletter*, December 1982.

[11] T.C. Cheston and J. Frank, Ch. 11 of M.l. Skolnik (ed.), *Radar Handbook*, 1st ed., New York: McGraw-Hill, 1970, pp. 11-57.

[12] E. Brookner, Ch. 2 in E. Brookner (ed.), *Aspects of Modern Radar*, Norwood MA: Artech House, 1988, pp. 101-102.

[13] S. Ramo and J.R. Whinnery, *Fields and Waves in Modern Radio*, New York: John Wiley & Sons, 1953.

[14] D.R. Carey and W. Evans, "The Patriot Radar in Tactical Air Defense," *Microwave Journal*, May 1988, pp. 325-332.

[15] G.D.M. Peeler, Ch. 16 in R.C. Johnson and H. Jasik, *Antenna Engineering Handbook*, New York: McGraw-Hill, 1984.

10

RECEIVERS AND DISPLAYS

10-1. Receiver Description and Functions

Much of the introductory material on receivers is in Sec. 2-6, which may need to be reviewed at this point. The functional block diagram of a receiver is given as Fig. 10-1. The four functions of a radar's receiver are listed below.

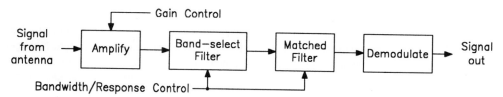

Figure 10-1. Basic Functional Receiver Block Diagram.

- Amplification function: to increase the amplitude of echoes and interference to levels usable in the signal processor and other outputs. Control of this amplification is covered in Ch. 3.
- Channel selection and out-of-channel filtering function: to reject out-of-band interference.
- Matched filtering function: to shape the signals and interference to gain the maximum signal-to-interference ratio. The response of this filter must be variable if diverse waveforms (such as switchable pulse widths) are used.
- Demodulation function: to remove the carrier sinusoid and reduce the signals and interference to their baseband, or information, frequencies.

Receiver types: Several different receivers have been and are used in radars and radar EW, with one type predominating. Early receivers were of the crystal video type, shown in Fig. 10-2. In it, the signal and interference are filtered and possibly amplified and immediately converted to baseband frequency in the crystal (diode) demodulator. The bulk of the amplification is done in the video amplifier. This type of receiver has the advantage of being simple and inexpensive. Its disadvantages from a radar standpoint are, however,

471

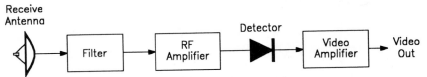

Figure 10-2. Crystal Video Receiver.

overwhelming. Selection of a narrow bandwidth signal (the channel-select filter function) must be done at echo frequencies, making the filter design difficult. Sensitivity is poor. To tune the receiver requires that the RF filter be tuned, making its design even more difficult. Demodulation is of the signal amplitude only, making this receiver is unusable in Doppler systems. Its major use today is in low sensitivity systems such as radar warning receivers (RWRs), where its small size and low cost allow the construction of multi-channel channelized crystal video receivers, each channel covering a particular frequency band.

A second receiver type which finds limited radar use is the homodyne receiver, in which the incoming signal is filtered and (perhaps) amplified, and then immediately converted in a mixer to base band frequencies. See Fig. 10-3. Again, the primary advantage is simplicity. The disadvantages include poor sensitivity and the inability to filter the signal except at the radio frequencies, where precise filtering to suppress interference is difficult. Some filtering can also take place in the video amplifier. Because it demodulates by mixing, this receiver can recover signal phase and is thus useful in Doppler systems. Its main applications are in simple Doppler systems such as laboratory receivers for target measurement, and in fuze radars. Other types finding use in radars and radar electronic warfare systems include Bragg-cell receivers, instantaneous frequency measurement receivers, compressive (microscan) receivers, and the dominant type, the superheterodyne receiver. Hybrid types are also used. See Tsui for more information [1].

Receivers used in radars are almost always of the superheterodyne configuration. In one, the bulk of the amplification and filtering takes place at a frequency lower than that of the received echoes and interference, called the intermediate frequency. The basic block diagram of a superheterodyne receiver is found in Sec. 3-6, with a more detailed description in Fig. 10-4. In the superheterodyne receiver, the bulk of the amplification and filtering functions are carried out in the IF amplifier. Because the frequency shifts are accomplished through mixing, signal phase is preserved.

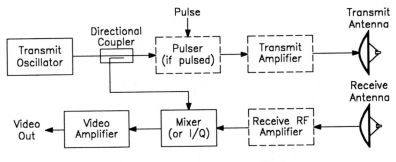

Figure 10-3. Homodyne Receiver.

10-2. Superheterodyne Principles

As shown previously, the superheterodyne receiver does its main signal amplification and filtering at an intermediate frequency, usually lower than the frequency at which the echoes and interference are received. This allows easier and better filter and amplifier design, at the expense of a second frequency band to which the receiver responds (the *image*). See Sec. 2-6 for a discussion of image fundamentals. Another advantage of the superheterodyne receiver is that tuning is accomplished by simply changing the frequency of the local oscillator, assuming the RF filters are wide enough to pass all the desired frequencies. A superheterodyne receiver, showing one method of generating transmit frequencies locked to the receiver, is illustrated in Fig. 10-4.

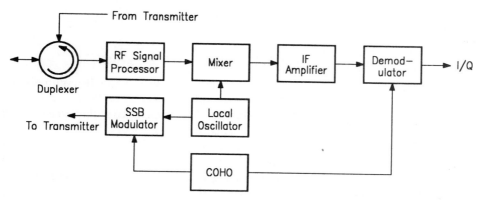

Figure 10-4. Superheterodyne Receiver.

The image response should be suppressed for two reasons. First, it allows a second receiver bandwidth's worth of noise to enter the system, reducing the signal-to-noise ratio by 3 dB. Second, the image is another bandwidth by which ECM can enter the system (so-called image jamming). In monopulse tracking radars, image jamming may cause inverted track errors, driving the radar away from the target rather than toward it.

Much of the complexity of superheterodyne receiver design is in suppressing the image response. Two methods are used: filtering and cancellation. In systems which image suppress by filtering, a bandpass filter is included in the RF processor which allows one of the possible received frequencies through and attenuates the other (Fig. 10-5). Since the image frequency is twice the intermediate frequency removed from the normal signal frequency (Sec. 2-6), it is advantageous from the standpoint of the image-reject filter to have

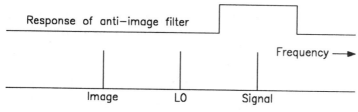

Figure 10-5. Image-Reject Filter.

as high an IF as possible. An IF of ten percent of the radio frequency allows relatively simple RF filters to suppress the image. It is also desirable to make the IF high so that the frequency agility of the receiver does not require that the image-reject filter be tunable. The following example illustrates.

Example 10-1: A radar receiver must be tunable from 9.6 to 10.4 GHz. The filters in the RF processor are to be fixed, with enough bandwidth to accommodate the entire receiver tunable range. It requires at least 300 MHz on each side of its bandpass for rolloff. What is the lowest possible intermediate frequency to meet these conditions?

Solution: By the specifications of the example, no image frequencies may exist within the bandpass 9.3 to 10.7 GHz. Assuming the local oscillator is lower in frequency than the signal, for a signal of 10.4 GHz, the image cannot be greater than 9.3 GHz. This gives an IF of 550 MHz (half the difference between the signal and its image). If the LO is above the signal, a signal frequency of 9.6 GHz must have an image no lower than 10.7 GHz. Again, the lowest IF is 550 MHz.

If, however, the intermediate frequency is too high, the design of the IF amplifiers and filters is complicated. For microwave receivers, it is difficult to have an IF high enough for image rejection and low enough for good filtering. For this reason, multiple-conversion superheterodyne receivers were developed.

Figure 10-6 shows a double-conversion superhet. The first IF is high enough so that images can be suppressed in an RF filter. The second IF is low enough to provide good amplification and filtering. Please note that images can also occur between the first and second IF. This means that the second IF cannot be too much lower than the first (again, a ratio of about 1:10 approaches maximum). Sometimes, then, if the RF is high (say 10 GHz) and the main amplifying and filtering IF is low (say 30 MHz), triple conversion may be required (Fig. 10-7).

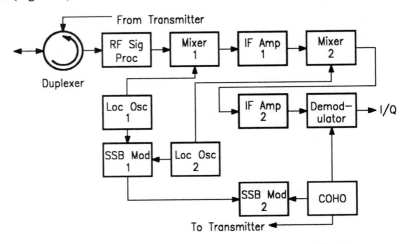

Figure 10-6. Double-Conversion Superheterodyne Receiver.

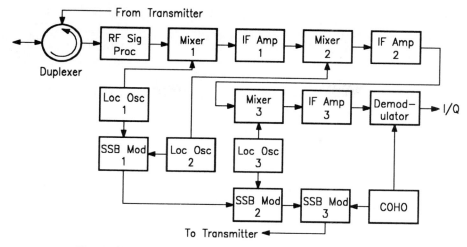

Figure 10-7. Triple-Conversion Superheterodyne Receiver.

Another problem exists in superheterodyne receivers, particularly multiple-conversion superhets, in that harmonics of the various signals and local oscillators plus their mixing products may fall within some of the signal bandpasses. This causes spurious interfering signals known as "spurs" and "birdies." One result is that spurs from local oscillators and a strong interfering signal (clutter or ECM) may interfere with the desired weaker signal. The tool for evaluating this is the so-called spur chart, given in many receiver books [1]. Spurs are avoided by judicious choice of local oscillator and intermediate frequencies.

Proper hardware design and selection can also help avoid spurs. Some mixer types, for example, are less vulnerable than others. Balanced and double-balanced mixers are far less susceptible to even harmonics of signals and local oscillators than are unbalanced "single-ended" types. The result is that few radar receivers use unbalanced mixers in any form (there are other reasons for choosing some balanced mixer type — see Sec. 10-4).

The following sections describe the major blocks in the superhet receiver.

10-3. Radio-Frequency (RF) Processor

The RF processor, block diagramed in Fig. 10-8, is where the signal and interference are amplified, attenuated, and filtered at the signal echo frequency. The radio frequency is a band centered around the transmit frequency, offset by the Doppler shift. In many systems, there is also a signal limiter between the duplexer and the attenuator. It is addressed in the duplexer discussion (Sec. 8-12).

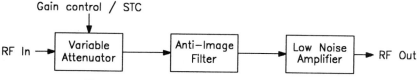

Figure 10-8. Radio-Frequency (RF) Processor.

The primary functions of the RF processor are to filter unwanted signals, especially at the image frequency and to amplify the signal-plus-noise to a level where the noise generated in later stages does not materially contribute to the signal-to-noise ratio. The idea is that the RF processor amplifiers introduce as little noise as possible, but amplify that noise, along with the signal, enough so that it swamps (overwhelms) the noise generated in later stages. The numeric example of Fig. 10-9 illustrates the effectiveness of an RF amplifier.

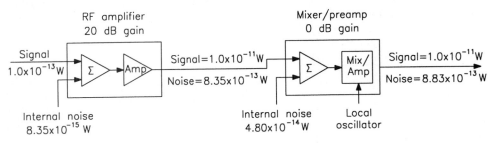

Figure 10-9. Example of RF Amplifier Noise.

Assume a signal input of 1.0×10^{-13} W (-100.0 dBm), a noise bandwidth of 1.2 MHz, an RF amplifier with 20 dB of gain and a noise figure of 2.4 dB, and a mixer/IF preamplifier with a gain of 0 dB and a noise figure of 10 dB. The effective input noise power to the RF amplifier is 8.35×10^{-15} W (-110.8 dBm). The signal power out of the RF amplifier is 1.0×10^{-11} W (100 times the input $-$ 20 dB gain), the noise power out of the RF amplifier is 8.35×10^{-13} W (100 times the input). The signal-to-noise ratio at the output of the RF amplifier is 12.0 (10.8 dB).

The effective noise power at the input to the mixer/preamp, generated by the mixer/preamp, is 4.80×10^{-14} W, which is almost six times the noise of the RF amplifier. The total effective noise into and out of the unity-gain mixer/preamp is 8.83×10^{-13} W, which is the internally generated noise of the mixer/preamp plus the amplified noise out of the RF amplifier. The signal into and out of the mixer/preamp is 1.0×10^{-11} W. The signal-to-noise ratio out of the mixer/preamp is 11.3 (10.5 dB).

Even though the mixer/preamp generated almost six times the noise as the RF amplifier, its effect on the total S/N was almost negligible (0.3 dB). This is because even though the RF amplifier generated a very small amount of noise, this noise was amplified to a level where it effectively swamped the noise of the mixer/preamp.

The noise figure for a multi-stage amplifier is usually determined almost exclusively by its first stage. Figure 10-10 depicts a multi-stage system. The overall noise factor (the noise figure in ratio form) is

$$F_T = F_1 + [(F_2 - 1)/G_1] + [(F_3 - 1)/(G_1 G_2)] + ... \qquad (10\text{-}1)$$

F = the overall noise factor (power ratio)
F_n = the noise factor of the nth stage
G_n = the gain of the nth stage

An important parameter of any low-noise amplifier is the product of its noise factor and gain. The larger this product, the less effect later stages have on the signal-to-noise ratio. It should be made large by the gain, not the noise factor.

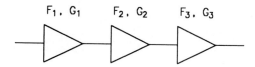

Figure 10-10. Noise Factor in Multi-Stage Amplifiers.

RF attenuators: The RF attenuator prevents strong incoming signals from saturating later parts of the receiver. In tracking radars, it is often mechanically switched, controlled by sensing the signal level (AGC voltage) and comparing it to a threshold. In search radars, the attenuator is usually continuously variable and is controlled with the receiver STC (see Ch. 2). A schematic diagram of one type of variable attenuator, constructed with PIN diodes, is shown in Fig. 10-11.

A PIN diode, so named because it is made in three semiconductor layers — P-type, intrinsic, and N-type — is operated in forward bias. At high forward bias, it is essentially a microwave short circuit. At low forward bias, it is a microwave open circuit. Connected as shown, the signal is split into two equal components by the hybrid. At low DC bias, the diodes have little effect and the signal components combine at the output hybrid. At high bias, the signal reflects at the diode short circuits and combines in the input hybrid to appear at its terminated port. At medium bias, the diodes are mismatches to the transmission lines, but do not reflect completely. Thus part of the signal passes through to the output and part reflects to the termination. The effect at the output is one of attenuation.

RF filters: RF filters limit the response of the signals fed to the receiver amplifiers and to reject image responses. They can be implemented in any of several technologies, including waveguide, coaxial line, and microstrip. The design and evaluation of RF filters is beyond the scope of this book, but Fig. 10-12 gives a typical response curve of a commercially available filter. Their critical specifications include their response curve, insertion loss, and input/output reflections (VSWR or return loss).

Radio frequency amplifiers: These amplifiers, also known as low-noise amplifiers — LNAs — were discussed in the introduction to this section. They amplify the signal and interference while introducing as little noise as possible. Their gain is such that the noise produced is amplified enough to swamp the noise from the rest of the receiver. They also introduce signal distortion and can saturate. Some radar receivers do not have them. Their critical specifications are gain, gain control, noise figure, saturation level, and VSWR.

Several types of amplifiers used in radar receivers are described here, although most modern radar receivers have either no RF amplifier, or a gallium arsenide field-effect transistor (GaAs FET) amplifier.

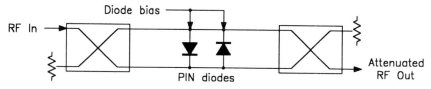

Figure 10-11. PIN Variable Attenuator.

10-3. Radio-Frequency Processor

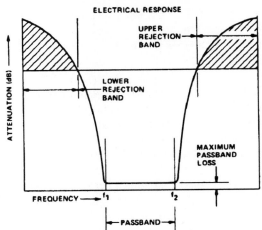

Figure 10-12. Response of RF Filter. (Diagram courtesy of Narda Microwave Division of Loral)

Negative resistance amplifiers: A large group of devices exhibit negative resistance under certain conditions. Tunnel (or Esaki) diodes are one. The tunnel diode is a heavily doped GaAs diode operated in forward bias, with the bias too low to elicit normal diode forward conduction (with depletion region closed). Its response curve is shown in Fig. 10-13.

Placed in a circuit, the tunnel diode terminates a transmission line. If the bias is such that its resistance is negative to signals, the reflection coefficient can be greater than unity. The following equation gives the reflection coefficient at any junction.

$$\Gamma = \frac{V_R}{V_F} = \frac{Z_L - Z_0}{Z_L + Z_0} \tag{10-2}$$

Γ = the voltage reflection coefficient (dimensionless — vector quantity)

V_R = the peak (or rms) value of the voltage reflected at the junction (volts — vector quantity)

V_F = the peak (or rms) value of the forward voltage at the junction (volts — vector quantity)

Z_L = the impedance of the load (Ohms — vector quantity)

Z_0 = the characteristic impedance of the transmission line feeding the load (Ohms — vector quantity)

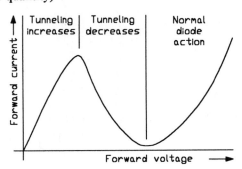

Figure 10-13. Tunnel Diode Characteristic.

478   Chap. 10. Receivers and Displays

Because both the input and output signals use the same port, some way must be found to separate them. The circulator circuit of Fig. 10-14 is commonly used. The input signal is directed to the amplifying device, from which an amplified copy of the signal reflects. This amplified copy is reflected into the same circulator port, where it is directed to the output port. Amplifier stability demands that the loop gain be less than one. Feedback to the amplifying device is by the circulator amplifier port reflecting amplified signal back into the amplifier. To keep loop gain less than unity, the return loss (Eq. 10-3) of the circulator's amplifier port must be less than the gain of the amplifying device.

$$L_R = -20 \, \text{Log}_{10} \, (|\Gamma|) \qquad (10\text{-}3)$$

L_R = the return loss (dB)

$|\Gamma|$ = the amplitude of the reflection coefficient

Example 10-2: A tunnel diode amplifier with a negative resistance of 28 W is connected to a circulator as shown in Fig. 10-14. The system is 50 W characteristic impedance, the circulator insertion loss between adjacent ports is 0.3 dB, and the VSWR of the circulator ports is 1.25. Find the gain of the amplifier. Is the amplifying system stable?

Solution: Equation 10-2 tells us that the reflection coefficient of a 50-Ω line terminated in −28 Ω is −3.55. The power reflection coefficient is 12.6 (Eq. 10-3), giving the amplifier a gain of 11.0 dB. The signal must pass through two insertion losses, so the overall amplifier gain is 10.4 dB. For a VSWR of 1.25, the return loss is 19.1 dB. Since the loss in the reflection at the circulator exceeds the device gain, the amplifier is stable.

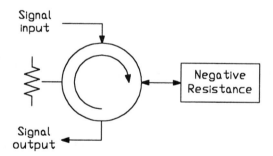

Figure 10-14. Circulator-Type Negative Resistance Amplifier.

Another negative resistance amplifier is the parametric amplifier (*paramp*). In it, one or more varactor diodes, acting as voltage-variable capacitors, behave as negative resistances when biased with a sinusoid of somewhat higher frequency than the signal. This AC bias source is called a pump oscillator. There are many parametric amplifier modes, a complete discussion of which is beyond our purpose. For an easy-to-understand explanation of how they amplify (these are not simple devices), see Roddy [5]. A full treatment of parametric action can be found in any advanced communications text which derives and explains the Manly-Rowe equations.

The greatest advantage of parametric amplifiers is that they produce little noise; they are the quietest commonly available microwave amplifiers. Their greatest disadvantage is that they are complex, expensive, large, and heavy compared to other RF amplifier types.

10-3. Radio-Frequency Processor

Recent advances in GaAs FET technology has resulted in transistor amplifiers which are almost as quiet as paramps, and much smaller, lighter, and cheaper.

Transistor amplifiers: Because of this technology advance, most newly designed receivers use gallium arsenide field-effect transistors (*GaAs FETs*, pronounced "gas'-fet") as their RF amplifiers. Modern GaAs FET amplifiers have noise figures of less than 3.5 dB in the X-Band, 2.0 dB in the C-Band, and less in lower bands.

Unlike negative resistance amplifiers, transistor amplifiers are two-port devices with good isolation between output and input and do not require a circulator. Internally, they are usually designed as balanced amplifiers to minimize input and output VSWR and minimize noise. Figure 10-15 shows a balanced configuration. If the reflections from the two amplifiers are the same, input reflected power goes to the lower-left port, maintaining a low VSWR at the signal input port. Table 10-1 gives some typical performance figures for commercially available amplifiers.

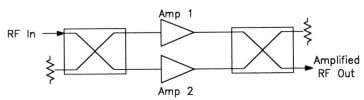

Figure 10-15. Balanced Amplifier.

Table 10-1. Typical GaAs FET Amplifier Performance.
(After Narda Microwave)

Frequency Range (GHz)	Noise Figure (dB)	Gain (dB)	1-dB Compression (dBm)
5.4 - 5.9	3.5	30	+10
9.0 - 9.5	3.0	27	+10
10.0 - 10.5	3.0	27	+10
2.0 - 4.0	3.5	33*	+13
8.0 - 12.0	5.5	34*	+10

* Many gains available.

Useful amplifiers can also be made of bipolar junction transistors (BJTs). They are, however, usually limited to the lower microwave bands and are rapidly being displaced by GaAs FETs.

One disadvantage of all transistor amplifiers is that they are highly sensitive to temperature, particularly in their gain and phase-shift characteristics. Figure 10-16a shows the gain a typical wideband microwave GaAs FET amplifier without internal temperature compensation, with gain variations exceeding 8 dB over its operating temperature range. Figure 10-16b shows the gain of a similar amplifier with temperature compensation, with less than 2 dB variation. Matching the gains of several receiver channels is necessary for the operation of some radar receivers, notably monopulse tracking receivers, and amplifier temperature compensation is important.

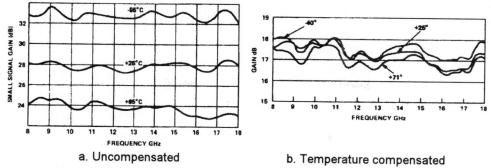

a. Uncompensated   b. Temperature compensated

Figure 10-16. Gain/Temperature Characteristic of a Solid-State Microwave Amplifier. (Diagrams courtesy of Narda Microwave Division of Loral)

Compensation can be accomplished in several ways. Since the gain of solid-state amplifiers usually increases with increased temperature, some compensation networks function by using temperature-sensitive resistors to produce a matching gain decrease with temperature. Other receiver designs place the sensitive components in an oven, so they always operate at the same temperature, regardless of the ambient temperature.

10-4. Mixers

The inputs to the mixer are the signal and interference information on the radio frequency carrier and a sinusoid known as the local oscillator. The output is the signal and interference information on a different carrier wave, known as the intermediate frequency. If the intermediate frequency is much lower that the RF, which is the normal case, the process is known as down-conversion mixing. Mixing is the process of multiplying the RF signal and interference by the local oscillator (the STALO of Sec. 2-1), and then low-pass filtering the product (Fig. 10-17). Mixing is described mathematically in the equations below.

Mixers have three ports (Fig. 10-18): the R port is where the signal at radio frequencies is injected, the L port is where the local oscillator is applied, and the I port is where the signal at intermediate frequency emerges.

The output of the multiplication is the product of two sinusoids, as given in Eqs. 10-4 and 10-5. The signal of Eq. 10-5 is after the high frequency term is removed by a low-pass filter in the mixer. Note that the mixer output is identical to its RF input, including phase, except for a translation in frequency.

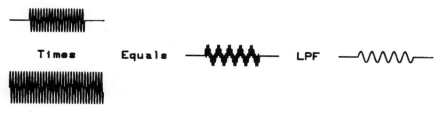

Figure 10-17. Mixing.

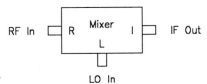

Figure 10-18. Mixer Ports.

$$v_R(t) = V_R \text{Cos}[2\pi f_R t + \phi(t)] \quad (10\text{-}4)$$

$$v_S(t) = 2 \text{Cos}[2\pi f_S t] \quad (10\text{-}5)$$

$$v_{IM}(t) = v_R(t) \, v_S(t) \quad (10\text{-}6)$$

$$v_{IM}(t) = V_R \{\text{Cos}[2\pi(f_R + f_S)t + \phi(t)] + \text{Cos}[2\pi(f_R - f_S)t + \phi(t)]\} \quad (10\text{-}7)$$

$$v_I(t) = V_R \text{Cos}[2\pi(f_C - f_d)t + \phi(t)] \quad (10\text{-}8)$$

$v_R(t)$ = the instantaneous received echo voltage (out of the RF processor)

$v_S(t)$ = the instantaneous local oscillator (STALO) voltage

V_R = the magnitude of the echo voltage

V_S = the magnitude of the local oscillator voltage

f_R = the received echo frequency

f_S = the local oscillator frequency

$v_{IM}(t)$ = the instantaneous mixer output voltage before low pass filtering

$v_I(t)$ = the instantaneous mixer output voltage after low pass filtering

f_C and f_d = the COHO and Doppler frequencies, respectively (see Table 2-1 for the frequency relationships)

$\phi(t)$ = the phase of the IF signal with respect to the COHO

Mixer specifications: The critical specifications and parameters of mixers are

— *Noise figure*, a measure of the noise generated in the mixer
— *Conversion loss*, the ratio of the signal power at the radio frequency into the mixer, to the signal power at the intermediate frequency out of the mixer

$$L_{MC} = P_{RF} / P_{IF} \quad (10\text{-}9)$$

L_{MC} = the mixer conversion loss (ratio)

P_{RF} = the RF signal power into the R port

P_{IF} = the IF signal power out of the I port

— *Isolation* is the attenuation at the various ports to signals injected into other ports. There are four isolations of importance.

$$L_{R-L} = P_{RI} / P_{LO} \quad (10\text{-}10)$$

$$L_{L-R} = P_{LI}/P_{RO} \qquad (10\text{-}11)$$
$$L_{R-I} = P_{RI}/P_{IO} \qquad (10\text{-}12)$$
$$L_{L-I} = P_{LI}/P_{IO} \qquad (10\text{-}13)$$

L_{R-L} = the loss from the R port to the L port
L_{L-R} = the loss from the L port to the R port
L_{R-I} = the loss from the R port to the I port
L_{L-I} = the loss from the L port to the I port

L_{R-I} is important to prevent signal leakage in multi-channel receivers having a common local oscillator, such as monopulse tracking receivers (Sec. 10-8). If too low, signals entering the R port in one channel leak to the others through the local oscillator, causing loss of track accuracy.

L_{L-R} prevents the local oscillator from radiating out the R port and out the antenna (called reradiation). Excessive reradiation clutters the RF spectrum and allows detection of the presence of the radar, even if the transmitter is silent.

L_{R-I} is not important if the RF and IF band are widely separated in frequency. In some modern EW and RCS measurement receivers, the two band may actually overlap, in which case this specification assumes some importance.

L_{L-I} is the least important of the isolations in that the LO is normally separated in frequency from the IF as to be filterable at the IF port.

— *Harmonics and spurs* are caused by non-linearities in the mixer and other receiver components.

— *Port VSWRs* or *return losses* measure the reflections at each of the ports.

Configurations: The most commonly used mixers are shown schematically in the following figure. Figure 10-19a is a balanced mixer. It has two diodes and the hybrid can be either 90° or 180°. If 90°, the internal reflections of the RF signal appear at the LO port, giving a low input VSWR but poor RF to LO port isolation. If 180°, the R port VSWR is higher, but spurious responses caused by even harmonics at one of the ports (R or L) are suppressed. Isolation from LO to RF is good.

The double-balanced mixer, Fig. 10-19b, has four diodes. It has the advantages of the balanced configuration, plus better R to L isolation and lower spurious responses.

The image-reject mixer of Fig. 10-19c treats the two signal frequencies which produce the same IF such that they appear at different output ports. Internally, it has two mixers, two hybrids, and a power divider. In the LO and RF ports, there must be one hybrid and one power divider. It makes no difference which is which (instead of the configuration in the figure, the RF input can be through a hybrid and the LO through a power divider). The overall performance of the mixer depends on the match between the two mixers. This mixer type is normally used in receivers where the image cannot be rejected with RF filters, such as where the IF is a very small percentage of the RF and the signal and image frequencies are close to one another.

Some typical specifications for a commercial mixer are shown in Fig. 10-20. This unit is double balanced and covers the radar X-Band (8.0 to 12.5 GHz).

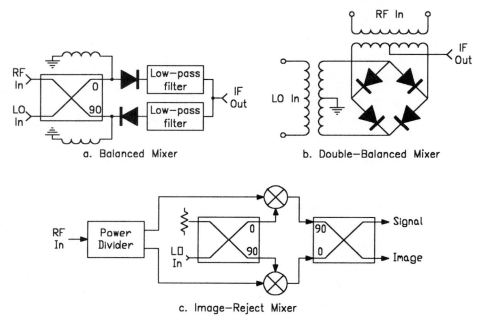

Figure 10-19. Mixer Configurations.

10-5. Local Oscillators

The local oscillator provides the RF power needed by the mixer for frequency conversion. The local oscillator power required depends on the mixer, but usually ranges from about 10 μW to 10 mW. The configuration of the local oscillator depends on the type and age of the radar, but four LO types predominate in microwave receivers: reflex klystron, Gunn oscillator, crystal oscillator and multiplier, and frequency synthesizer. Virtually all coherent receivers use one of the last two types.

Reflex klystron: The reflex klystron, shown in Fig. 10-21, is a single-cavity klystron with feedback from output to input through the electron beam (see Sec. 8-2 for a general discussion of klystron action). Reflex klystrons are also used as pump oscillators in parametric amplifiers.

An electron beam is generated in the electron gun and is accelerated by the anode. As it passes through the cavity, noise in the beam within the cavity's bandpass generates an electromagnetic field within the cavity. The electric field component, acting across the cavity gap, velocity modulates the beam. The beam drifts until it reaches the influence of the negatively charged repeller, where its course is reversed and it drifts through the cavity a second time. Bunching is occurring all this time. During the second pass through the cavity, RF energy developed from the bunching couples back into the cavity. Since the RF energy in the beam during its second pass exceeds that from the first pass, gain exists. The gain is from the cavity back to the cavity. Thus the loop gain exceeds unity and oscillation occurs. A detailed description of the oscillation mechanism and modes can be found in microwave texts such as [6].

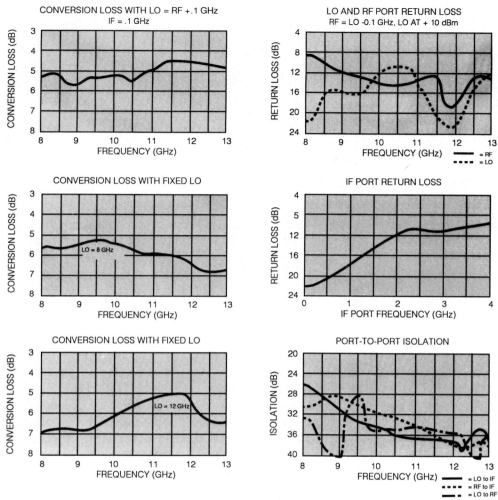

Figure 10-20. Typical Mixer Specifications.
(Plots courtesy of MITEQ)

The frequency on which a klystron oscillates depends on the cavity resonant frequency and the repeller voltage. The power output is maximum when the oscillation frequency is the resonant frequency of the cavity.

Gunn device: Another commonly used local oscillator is the Gunn device [7], or Gunn diode (Fig. 10-22). It is a bulk semiconductor device (usually configured as a diode) which is unstable at very high forward voltage gradients. The oscillation is small compared to the DC voltage and thus the device has low efficiency (<10%).

In the Gunn diode's primitive mode, the frequency of oscillation is controlled by the charge drift time across the device, set by charge mobility and device length. However, it also has modes wherein the oscillation frequency is controlled by an external resonant circuit and these are the useful modes. The resonant circuit can be a cavity, a varactor-tuned transmission line circuit, or any one of many other resonant circuits.

10-5. Local Oscillators

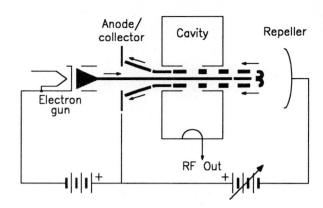

Figure 10-21. Reflex Klystron.

An interesting and extremely useful Gunn diode oscillator uses a YIG-core transformer as the resonant circuit. YIG (*Y*ttrium *I*ron *G*arnet) has the unusual property that its permeability is sensitive to an externally-generated DC magnetic field. The YIG core thus conducts RF magnetic fields across the transformer well at only one frequency, set by the external magnetic field. Since the transformer is in the oscillator feedback circuit, the oscillator frequency is set by its core's resonant frequency, which is set by a magnetic field, which in turn is set by the current in an external solenoid electromagnet. This YIG-tuned Gunn oscillator is probably the most common Gunn oscillator configuration.

Crystal oscillator/multiplier: These local oscillators are commonly used in coherent systems, where accurate frequencies are required. A simple example is shown in Fig. 10-23. The oscillators are usually FET or BJT amplifiers with feedback through a quartz crystal filter. Their oscillation stability is basically determined by the mechanical stability of the crystal. In high accuracy oscillators, the crystal is sometimes placed in an oven to hold its temperature constant. The multipliers are non-linear devices which generate harmonics of their input signals with filters to select the desired harmonics. Varactor diodes are commonly used as the non-linear devices. In Fig. 10-23, a 64.5 MHz crystal oscillator is multiplied by

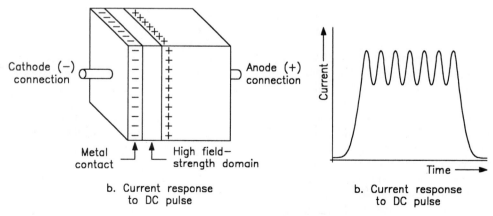

Figure 10-22. Gunn Device and Oscillation.

Chap. 10. Receivers and Displays

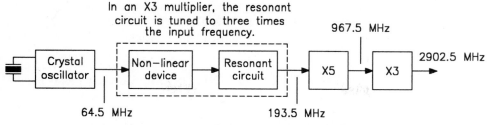

Figure 10-23. Crystal Oscillator / Frequency Multiplier.

45 to yield a local oscillator of 2902.5 MHz. Frequency agility requires that a crystal be supplied for each frequency. For example, to get an output frequency of 2903.0 MHz, the crystal would have to be changed to 64.5111 MHz.

Frequency synthesizers: These are the most accurate and flexible sources. One synthesizer is shown in Fig. 10-24. In it, a standard oscillator (often an atomic clock) is divided in a pre-scaling counter to the frequency of the desired channel spacing. In the figure, the output can be switched in 0.50 MHz increments. The 0.5 MHz sinusoid is fed to a phase detector, whose output controls a voltage-controlled oscillator (VCO). The VCO output is divided by some value, and the result forms the other input to the phase detector. The VCO is servoed to a frequency which produces zero error out of the phase detector. For the servo loop to lock the VCO to the input frequency (the 0.50 MHz), the two inputs to the phase detector must be at the same frequency. Thus the output of the divide by N counter must be at the same frequency as the output of the pre-scale counter.

For lock to occur, the input to the divide-by-N counter (the VCO) therefore must have a frequency of N times that of the pre-scale output. The combination of phase detector, filter, and VCO is known as a phase-lock loop. The output frequency of this type synthesizer is therefore

$$f_{SYN} = (f_{STD} / M) N \qquad (10\text{-}14)$$

f_{SYN} = the synthesized frequency (VCO frequency at lock)
f_{STD} = the standard frequency input
M = the divide ratio of the pre-scale counter
N = the divide ratio of the channel-select counter

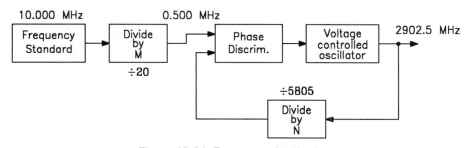

Figure 10-24. Frequency Synthesizer.

Frequency agility is readily achieved by simply changing the divide-by-N counter's ratio. To change the frequency in the example to 2903.0 MHz, the divide-by-N counter is simply changed to 5806. Synthesizers are also used with multipliers to generate higher frequencies than are practical with counters.

Automatic frequency control (AFC) and COHO locking: Coherent radars require that the local oscillator, transmit, and COHO frequencies be locked together (Ch. 2). This is readily achieved with synthesizers because the transmit synthesizer, local oscillator synthesizer, and COHO synthesizer can use the same frequency standard. With magnetron transmitters, however, the local oscillator and COHO must be locked to the magnetron's frequency (or the magnetron locked to the local oscillator).

Automatic frequency control is the method of locking the local oscillator (STALO) to the transmitter. A basic block diagram of AFC is given in Fig. 10-25.

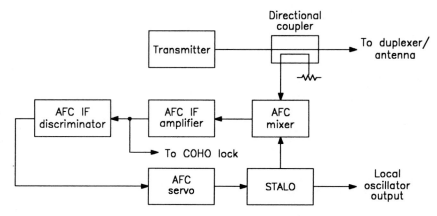

Figure 10-25. Automatic Frequency Control (AFC).

A sample of the transmitter's output is mixed with the local oscillator to produce a copy of the transmit pulse translated to the intermediate frequency (usually the last intermediate frequency in multiple conversion receivers). Its frequency is measured in a discriminator and the frequency error, through the AFC servo, corrects the STALO. This process places the LO and IF at the correct frequencies with relation to the transmitter, but the phase of the transmitter is still random compared to the LO.

The phase of the transmitter is contained on the AFC IF transmitter sample. This wave, at the intermediate frequency (which is also the COHO frequency), locks the COHO oscillator. The COHO then "remembers" the phase of the last transmit pulse until the transmitter fires again, at which time the COHO locks to the phase of the new pulse. The process is shown in Fig. 10-26.

The COHO is an oscillator whose loop gain is greater than one, but not much greater. Thus it does not break into oscillation rapidly. The COHO is shut off (quenched) just before a transmitter pulse. Before it can start oscillating again, the transmitter fires and the COHO is "tickled" by the AFC IF pulse. This tickling action starts the COHO and sets it off at the phase of the AFC IF pulse, which is the phase of the transmitter. The system then relies on the stability of the COHO to oscillate at this phase until it is quenched and re-locked by the next transmit pulse.

To summarize, in a coherent-on-receive system, the frequency of the illumination locks the frequency of the STALO. The phase of the illumination locks the phase of the COHO, producing a system which can, on a pulse-by-pulse basis, measure the phase of the received echoes.

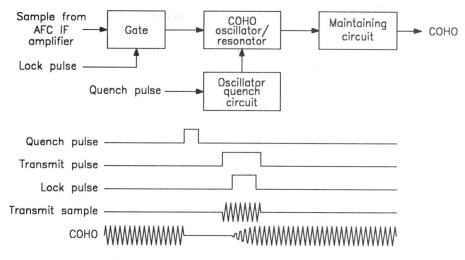

Figure 10-26. COHO Locking.

10-6. Intermediate Frequency Amplifiers

The IF amplifier is where most of the gain and much of the filtering occur. Typical IF amplifiers have gains of up to 120 dB, which is high enough that special care must be taken to stabilize the amplifier, particularly in preventing waveguide feedback modes. This is the reason many IF amplifiers are long and narrow in shape. Waveguide at well below its cutoff frequency acts as an attenuator with about 27 dB of loss for every width — the "a" dimension (Ch. 8) — along its length. Thus an amplifier in a box two inches wide could have no more than 27 dB of gain for every two inches of length.

IF filter design comments: The design of the IF filters involves the matched filter response of the transmit waveform and the maximum anticipated Doppler shift. In CW and pulse Doppler systems, the bandwidth of a search receiver must simply be wide enough to accommodate the maximum Doppler shift (pulse Doppler systems do not try to match the pulse — their narrow bandwidths essentially reject all spectral components except the central line). For example, an X-Band radar designed for Mach 5.5 closing and opening rates should have a bandwidth of about 250 KHz (Mach 5.5 produces about 125 KHz in this band, and the design must accommodate Mach 5.5 closing and opening).

Tracking radars, after target acquisition is complete, can have arbitrarily narrow bandwidths because the Doppler tracker (Ch. 7) pulls the local oscillator to keep the signal intermediate frequency constant. This narrow bandwidth requirement is one of the reasons, incidentally, that the longest range modes in most airborne radars are pulse Doppler modes.

Pulsed systems, whether low or medium PRF, extract range-resolving information from the echo pulses and require sufficient receiver bandwidth to match the illumination waveforms. In addition, they must also accommodate Doppler shift. A pulsed radar with 1.0-μs pulses in the X-Band would require a receiver bandwidth of about 1.45 MHz — 1.2 MHz for the pulse (Sec. 2-4) and 125 KHz on each end for the anticipated Doppler shift.

IF amplifier configurations and specifications: There are several approaches to IF amplifier design. Some amplifiers have fixed gain (few) and some are voltage gain-controlled (VGC). Some have a linear response and some are log. Some IF amplifiers distribute the filtering throughout the amplifier and some use wideband amplifiers and lump the filtering into a single circuit. Critical specifications include the following (Fig. 10-27 gives typical values).

— *Gain, gain control, and gain control linearity:* Most IF amplifiers have high gains and electronic (voltage-controlled) gain control. Some amplifiers require very high gain control linearity (the relationship between gain control voltage and gain). Multi-channel tracking receivers, as found in monopulse radars (Ch. 7), have the same gain control voltage applied to all the receivers, and they must each, over their gain control range (90 dB or more), maintain the same gain within very tight tolerances. Maintaining identical gains in multiple channels over wide gain control excursions is called gain tracking.

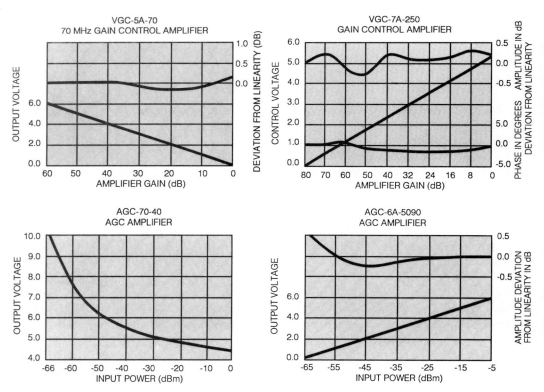

Figure 10-27. IF Amplifier Gain Characteristics.
(Plots courtesy of MITEQ)

— *Phase constancy with gain:* Monopulse tracking receiver channels must not only track in gain, they must also track in phase. This is because the track errors in monopulse systems are developed by comparing the error channel signals in amplitude and phase with the sum channel signal. See Ch. 7.

— *Temperature characteristics:* If the value of gain is important in the receiver's operation, as it is in search and monopulse tracking radars, temperature compensation is required. As noted in Sec. 10-4, the gain of semiconductor devices varies widely with temperature, requiring either feedback stabilization or operating the components at a constant temperature (as in an oven).

Log IF amplifiers: Log amplifiers are used where wide signal dynamic ranges must be handled simultaneously, such as in weather radars and modes, and to counteract some types of ECM. Their output voltage is proportional to the logarithm is input power.

There are two basic types of log amplifiers. The first, called the log detector amplifier, produces a video output — phase is lost (Fig. 10-28). It consists of a number of limiting amplifier stages, each connected to an envelope demodulator (see the next section for information on demodulators). The demodulated video signals are summed at the output.

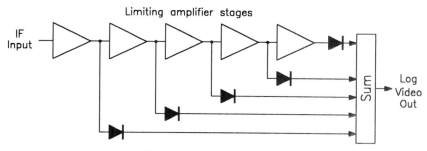

Figure 10-28. Log Detector.

For example, assume that each stage has a gain of 20 dB and the saturated output power is 0 dBm. Assume further that the saturated output produces 1.0 V of video. Table 10-2 shows the output power and voltage for each stage and the total voltage out. The characteristic is approximately logarithmic.

Table 10-2. Log Detector Example.

IF Input	Stage 1 Output		Stage 2 Output		Stage 3 Output		Stage 4 Output		Stage 5 Output		Total Output
Power (dBm)	Power (dBm)	Voltage (V)	Power (dBm)	Voltage (V)	Power (dBm)	Voltage (V)	Power (dBm)	Voltage (V)	Power (dBm)	Voltage (V)	Voltage (V)
−100	−80	10^{-4}	−60	10^{-3}	−40	0.01	−20	0.10	0	1.0	1.11
−80	−60	10^{-3}	−40	0.01	−20	0.10	0	1.0	0	1.0	2.11
−60	−40	0.01	−20	0.10	0	1.0	0	1.0	0	1.0	3.11
−40	−20	0.10	0	1.0	0	1.0	0	1.0	0	1.0	4.1
−20	0	1.0	0	1.0	0	1.0	0	1.0	0	1.0	5.0

10-6. Intermediate Frequency Amplifiers

Another type of log amplifier is the dual-gain amplifier, which functions totally at the intermediate frequency. If desired, a phase-sensitive demodulator can be used with this IF amplifier. As shown in Fig. 10-29, for low input signals the amplifier has a high gain, and for higher inputs it has a lower gain. By setting the gains properly, an approximate log characteristic can be achieved.

Figure 10-30 shows the amplitude response of a commercially available log IF amplifier.

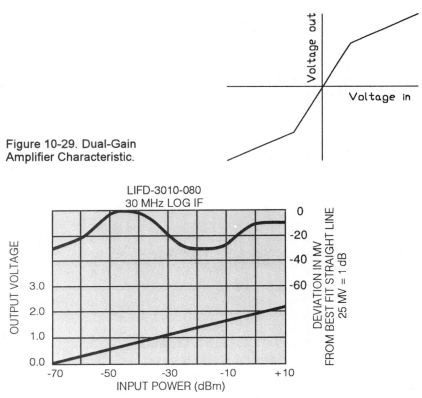

Figure 10-29. Dual-Gain Amplifier Characteristic.

Figure 10-30. Log IF Amplifier Characteristic. (Plot courtesy of MITEQ)

10-7. Demodulators

Demodulation is the process of translating the signal to its information (or baseband) frequencies from the intermediate frequency. Three types of demodulators are commonly used: envelope, synchronous, and I/Q. Only synchronous (rarely) and I/Q are used in coherent and coherent-on-receive radars.

An envelope demodulator, Fig. 10-31, recovers only the amplitude of the signal. Envelope demodulation discards the phase of the signal and thus any Doppler information it may contain. The loss of phase also results in a signal-to-noise ratio loss of 3 dB. Much

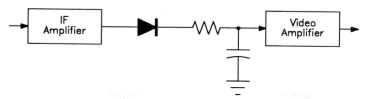

Figure 10-31. Envelope Demodulator.

has been written about envelope demodulators, particularly those using diodes biased to the knee of their current/voltage curve. These are the square-law demodulators. Since modern radars seldom use them, they will not be discussed further here. Further information can be found in most traditional radar and communication texts ([3], [8]).

A synchronous demodulator, Fig. 10-32, compares the signal to a reference oscillator (the COHO) such that the result is the amplitude of the signal times the cosine of the phase between it and the COHO.

$$V_{SD} = V_S \text{Cos}(\phi_S - \phi_C) \tag{10-15}$$

V_{SD} = the peak voltage of the synchronously demodulated signal
V_S = the peak value of the signal voltage prior to demodulation
ϕ_S = the phase of the signal
ϕ_C = the phase of the COHO (normally assumed to be zero)

Synchronous demodulation has three problems. First, it does not recover enough information about the signal to tell whether the Doppler shift is positive or negative. Second, only those components of the signal which are in-phase with the COHO are recovered; those at 90° phase (the quadrature components) are lost. This results in a reduction in signal-to-noise ratio of 3 dB from that which would be available if all components of the signal were recovered. Third, signal of certain phases, known as blind phases, are lost in synchronous demodulation. These are signals whose phase with respect to the COHO are $\pm 90°$, the cosine of which are zero. Some authors use the term "synchronous demodulation" synonymously with "I/Q demodulation."

An I/Q demodulator, Fig. 10-33, recovers all components of the signal, gives sufficient information to discriminate between positive and negative Doppler, and has no blind phases. It is two synchronous demodulators, one comparing the signal to the COHO and the other comparing the signal to the COHO shifted 90°. The signals recovered are given in the equations below. I/Q demodulation is described further in Sec. 6-2.

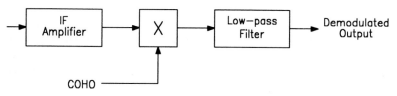

Figure 10-32. Synchronous Demodulator.

10-7. Demodulators

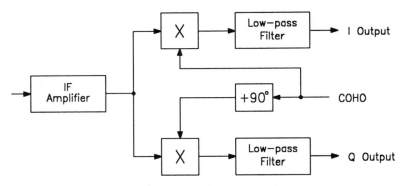

Figure 10-33. I/Q Demodulator.

$$V_I = V_S \cos[\phi(t)] \qquad (10\text{-}16)$$
$$V_Q = V_S \sin[\phi(t)] \qquad (10\text{-}17)$$

V_I = the voltage out of the I demodulator
V_Q = the voltage out of the Q demodulator
V_S = the signal peak voltage
$\phi(t)$ = the signal phase as a function of time, reference the COHO

10-8. Radar Receiver Examples

The following figures give examples of several different radar receivers. Search receiver 1 (Fig. 10-34) is typical of air search radars, such as those found in air traffic control military ground installations and on military ships. It has three IF channels. The normal IF is used for air search in areas where clutter is not present, generally at longer ranges where the clutter is either below the horizon or blocked by intervening terrain. The MTI IF amplifier is used where airborne targets may be masked by clutter, usually at shorter ranges. At some intermediate range, the displays and data processor switch from MTI video to normal video, since MTI does not perform well in noise. The log channel is used for weather displays and to counter certain types of ECM.

Search receiver 2 (Fig. 10-35) is typical of surface search systems where Doppler is not used. Of its two IF channels, the normal channel is used for most operations and the log channel is used for weather displays and in ECM.

Search receiver 3 (Fig. 10-36) is a channelized receiver capable of receiving signals on N frequencies simultaneously. It is used in 3-D air search radars in conjunction with frequency-scanning antennas (see Sec. 9-9). These systems often transmit a complex pulse with subpulses at several frequencies. Each of these frequencies produces a beam at a different elevation angle. Since echoes from all elevation angles occur at the same time, a receiver must be provided for each frequency.

The monopulse tracking receiver shown in Fig. 10-37 is a three-channel system, one for each signal produced by a monopulse comparator (the sum signal, the azimuth error signal, and the elevation error signal). The three receivers must be served by the same local

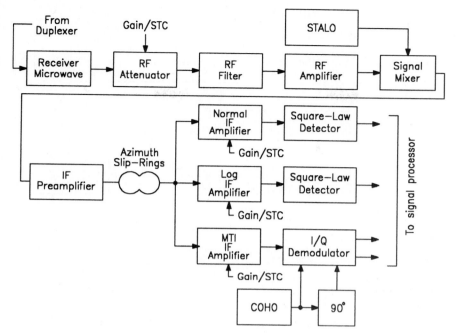

Figure 10-34. Search Receiver 1.

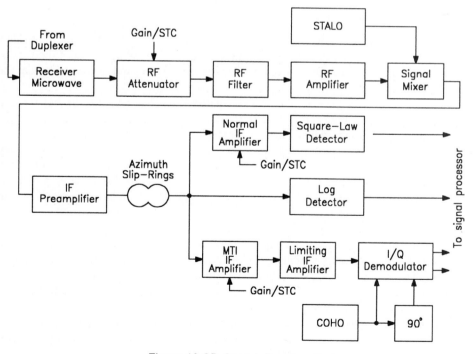

Figure 10-35. Search Receiver 2.

10-8. Radar Receiver Examples

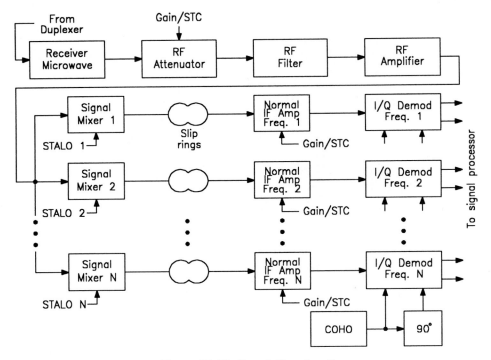

Figure 10-36. Search Receiver 3.

oscillator because the signal phase of one channel compared to the others is important. They must be controlled by the same automatic gain control (AGC) voltage because their gains must match at all times. Chapter 7 explains monopulse tracking and further addresses the receiver requirements.

The monopulse receiver's main outputs are gated sum video, which is the signal being tracked restricted by a range gate. It is used for range tracking and the development of the AGC voltage. In this monopulse receiver there are several auxiliary outputs, generally associated with displays. The ungated sum video is used for normal tracking displays. Gated video is not usable for most acquisition and display purposes — operators and acquisition logic need to see signals other than those in the gates. Non-track video is not AGCd and is useful for seeing small targets while tracking large signals, which turn the receiver gain down. Separations of small targets from large (for example, missile launches) and some types of ECM are handled this way.

A second and more modern monopulse receiver (Fig. 10-38) is identical up to the IF amplifiers. Its three primary outputs are I/Q demodulated, digitized, and processed in a computer. Track errors, AGC, gates, and other signals are generated in the computer.

Another configuration (not shown here — see Fig. 7-23) time-division multiplexes the azimuth and elevation error signals into a single receiver error channel, thus saving the weight, space, and cost of one receiver. With modern RF technology, however, the multiplexing switches may be larger than the third receiver and this configuration is less often used than in the past.

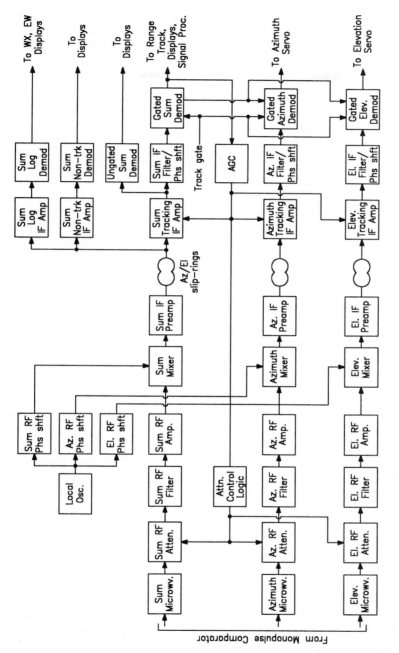

Figure 10-37. Monopulse Tracking Receiver 1.

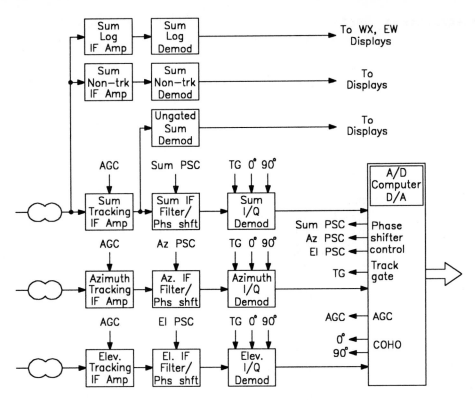

Figure 10-38. Monopulse Tracking Receiver 2.

10-9. Receiver Testing and Monitoring

Radar receivers usually are tested in the field on a regular basis, over and above those procedures which could be called "depot alignments." The latter are primarily filter alignments and will not be discussed here. The former include sensitivity checks and phase alignments for monopulse tracking.

MDS test: The test for minimum discernible signal (MDS) determines the smallest signal which produces a standard receiver output signal-to-noise ratio. Two configurations are commonly used. The injected signal method, shown in Fig. 10-39, applies a test signal from a calibrated signal generator to a port built into the radar. The test signal simulates a target echo in pulse width, pulse shape, and frequency. The attenuator on the test generator is adjusted so that the signal is just visible ("minimally discerned") on either an A-Scope display or a rotating PPI (see Sec. 10-10 — with the rotating PPI method, the signal shows as a ring on the scope). One of two procedures can be used. In the first, a visible signal is placed on the display, and then the test generator attenuation is increased until the signal is barely visible. In the second, the attenuation is increased until the signal vanishes and then reduced until the signal reappears. The method used depends on the manufacturer's

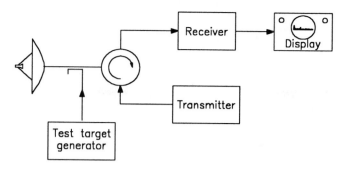

Figure 10-39. Injection Method of MDS Measurement.

recommendations and the standards for the particular installation. The same method should be used each time the test is performed.

The actual minimum discernible signal is found from the test generator's attenuator reading at MDS, corrected for cables and the attenuation of the injection port.

$$MDS = P_{TG} - L_{TC} - L_{CAB} \qquad (10\text{-}17)$$

MDS = the minimum discernible signal (dBm)

P_{TG} = the test generator output power when the MDS criterion used is met

L_{TC} = the ratio (loss) of the test port directional coupler, including cable losses to the port.

L_{CAB} = the cable loss from the test generator to the test port

MDS can be measured with the transmitter off or on. The test is more realistic with the transmitter on, but assurance must be made that the test generator will not be harmed. To a trained operator, MDS occurs at a signal-to-noise ratio of about unity.

Tracking radars usually need a test signal from a known location to align the tracking receivers (Ch. 7) and to assure that the positional data is accurate. Many of them have boresight towers (BSTs) for this purpose (Fig. 10-40). MDS can also be measured from the boresight tower in the same manner as with the injected signal. Comparative measurements using the test generator power output readings only are almost as easy as with injected signals. Deriving actual MDS, however, is a bit more difficult, in that a one-way communication equation is needed to find the space loss (Ch. 3).

$$MDS = \frac{P_{BST} G_{BST} G_{RDR} \lambda^2}{(4\pi)^2 R^2 L_{BST}} \qquad (10\text{-}18)$$

MDS = the minimum discernible signal (milliwatts in this equation)

P_{BST} = the BST test signal power at MDS (read off the test target generator − milliwatts in this equation)

G_{BST} = the gain of the BST antenna

G_{RDR} = the gain of the radar antenna

10-9. Receiver Testing and Monitoring

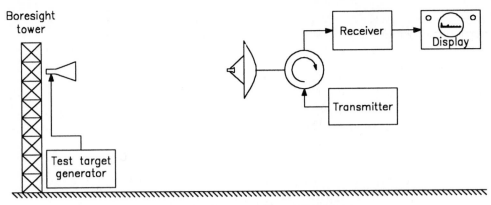

Figure 10-40. Boresight Tower Test Setup.

λ = the wavelength (same units as range)

R = the range from the BST to the radar

L_{BST} = the loss from the test target generator to the BST antenna

Other than the space loss calculation, receiver sensitivity checks with boresight towers are identical to those with injected signals.

Tangential sensitivity (TSS): This is another criterion for receiver sensitivity. The test setup is the same as those above and can use either an injected signal or a boresight tower. TSS must be taken from an A-Scope display and the concept is shown in Fig. 10-41. At TSS, the upper average limit of the noise is tangential to the lower average limit of signal-plus-noise. TSS occurs at a signal-to-noise ratio of about 8 dB. One must be careful taking TSS measurements because such things as the intensity of the display CRT affect the readings.

Although the display shown in Fig. 10-41a appears in many texts and conveys nicely the concept of TSS, this is *not* what a noisy radar signal looks like. In this case, the receiver was not matched to the signal, resulting in the square pulse. In radar receivers, signals and

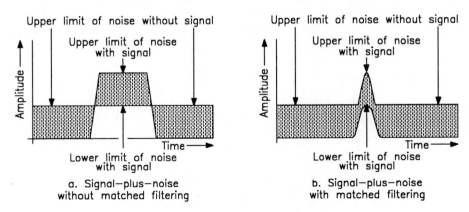

Figure 10-41. Tangential Sensitivity.

500　　　　　　　　　　　　　　　　　　　　　　　Chap. 10. Receivers and Displays

noise look exactly the same, and a noisy signal is simply a roughly Gaussian pulse with amplitude uncertainty. For TSS, the bottom of the signal amplitude uncertainty is made tangent to the average top of the noise. See Fig. 3-21, bottom trace, and Fig. 10-41b for a more realistic display.

Noise figure measurement: Noise figure is a method of describing the noise produced in a receiver compared to the noise which must, by the laws of physics, be produced. See Sec. 2-6 for a description of the noise figure concept. Noise figure is measured with a test setup as shown in Fig. 10-42.

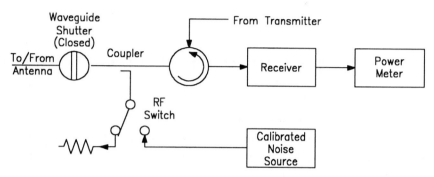

Figure 10-42. Noise Figure Test.

Two measurements are made: one with the calibrated noise source on and the other with it off. With the noise source off, the output of the receiver is

$$P_{N(OFF)} = K T_0 B_N F G_R \qquad (10\text{-}20)$$

$P_{N(OFF)}$ = the noise power out of the receiver with the noise source off.
K = Boltzmann's constant (1.38×10^{-23} J/°K)
T_0 = 290°K
B_N = the receiver's noise bandwidth
F = the noise factor (ratio — noise *figure* when converted to dB)
G_R = the gain of the receiver

One might think that this is the only measurement required to find noise factor (figure). The gain of the receiver, however, is seldom known accurately enough to measure noise factor directly.

With the noise source on, the noise power out of the receiver becomes

$$P_{N(ON)} = [(K T_0 B_N F) + P_{NS}'] G_R \qquad (10\text{-}21)$$

$P_{N(ON)}$ = the noise power out of the receiver with the noise source on.
P_{NS}' = the noise power of the calibrated source injected into the receiver chain (source corrected for cables and couplers)

Equations 10-20 and 10-21 can be divided to form the ratio of $P_{N(ON)}$ to $P_{N(OFF)}$ and solved for noise factor. Receiver gain is eliminated in the ratio.

10-9. Receiver Testing and Monitoring

$$F = \frac{P_{NS}}{K T_0 B_N [(P_{N(ON)} / P_{N(OFF)}) - 1]} \qquad (10\text{-}22)$$

Noise figure measurements are continuous (not pulsed), and interfering signals cannot be time gated. Thus the antenna must be disconnected from the receiver during this test. Because of this, the transmitter cannot be on. Noise figure is not as complete a test of the receiver as is MDS with a boresight tower, but in those systems which have built-in noise figure tests, it is a quick, convenient, and relatively accurate measure of the receiver's sensitivity.

10-10. Radar Displays

Following is a summary of the displays used in radars. Some of these displays are used in modern radars (A-scope, B-scope, F-scope, PPI, RHI). Others are either obsolete or are found only in very specialized applications. Signals displayed on these scopes can be either raw video, processed video, synthetic video (detected video), or computer-generated symbols. This summary is adapted from the IEEE standard definitions [9] and from [10], with additions and modifications by the author.

— **A-Scope:** The A-scope (Fig. 10-43a) is a deflection modulated display where the vertical coordinate is proportional to target amplitude (either linear or log) and the horizontal coordinate is proportional to time since transmit (or since the start of a PRF cycle) or range. If maintenance displays are counted, it is probably the most commonly used radar display.

— **A/R-Scope:** An A/R-scope (Fig. 10-43b), not in *IEEE-686*, is an A-scope with an adjustable segment of the time base expanded. It is common in tracking radars.

— **R-Scope:** The R-scope (Fig. 10-43c), which is not in *IEEE-686*, is an A-scope which displays a limited range segment around an adjustable center, usually a tracking range gate. It is common in tracking radars.

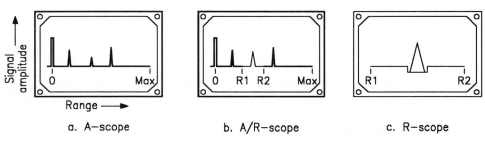

Figure 10-43. Range Displays.

— **B-Scope:** This is an intensity modulated rectangular display with azimuth angle indicated by the horizontal coordinate and range by the vertical coordinate. Target amplitude is shown by the brightness of the spot. This is a commonly used air-to-air combat display because it gives the operator target intercept information. The vehicle carrying the radar is on an intercept course with any target descending vertically down the B-scope (since the entire lower edge of the display is the radar's location). The B-scope shows true range, but distorts the cross-range dimension (see Fig. 10-44a). Two targets separated by a constant cross-range show different separations at different ranges, as illustrated by the two target pairs in the figure. Each pair is separated by the same distance in cross-range. The "bug" at the bottom of the display is the antenna's present azimuth angle.

Flight attitude information is often displayed on the B-Scope, represented by the broken horizontal line (artificial horizon) and the aircraft symbol. This display is "inside-out," showing the horizontal reference from inside the aircraft. A few displays show the attitude from the "outside-in," and show a variable aircraft symbol with a fixed horizon. In this illustration, the aircraft carrying the radar has its nose up and is banked to the right. This display is often projected onto the aircraft's head-up display (HUD).

— **B-Prime-Scope:** The B-Prime display (Fig. 10-44b — not in *IEEE-686*) is identical to the B-Scope except that target closing velocity is displayed rather than range. The center of the display is, in this case, zero relative velocity; a target along the horizon line has no radial velocity to the radar. Targets above this line are closing on the radar and targets below are opening. This is often the same physical display as the B-Scope; whether B- or B-Prime depends on the radar's mode.

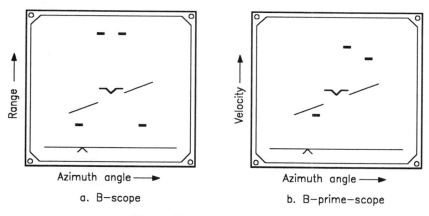

Figure 10-44. B-Type Displays.

— **C-Scope:** The C-scope (Fig. 10-45a) is an intensity modulated rectangular or circular display with azimuth angle indicated by the horizontal coordinate and elevation angle by the vertical. It is often confused with the F-scope, which shows in the same fashion azimuth and elevation errors. Some system displays labeled C-scopes are actually F-scopes. It is seldom used in modern radars.

— **D-Scope:** This is a C-Scope in which the blips extend vertically to give a rough estimate of range. This scope is seldom used.

— **E-Scope:** An intensity modulated rectangular display in which range is indicated by the horizontal coordinate and the vertical coordinate indicates elevation angle. It is by the vertical coordinate. An E-scope is a similar to the B-scope with elevation instead of azimuth, but sees less use than the B-scope. It is a distorted RHI.

10-10. Radar Displays

— **F-Scope:** The F-scope (Fig. 10-45b) is a rectangular display where the center is the axis of the antenna beam and the blip's displacement from center indicates target position relative to the beam axis. Horizontal displacement is azimuth angle error and vertical is elevation. It is basically a tracking error scope and is often confused with the C-scope. Many displays called C-scopes are actually F-scopes. This scope has some use in modern radars.

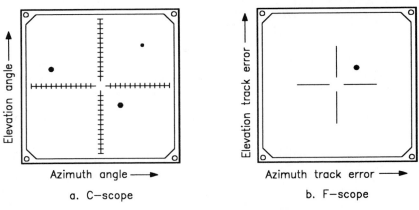

Figure 10-45. Azimuth versus Elevation Displays.

— **G-Scope:** This rectangular display is similar to the F-scope in that the blip's offset from the center indicates a difference between the target's location and beam pointing angle. The G-scope's difference is that targets become horizontally elongated (appear to grow "wings") as range diminishes. It is essentially an F-scope with some range information displayed. It is seldom used in modern radars.

— **H-Scope:** This is similar to the B-scope except that it displays some elevation information. Targets appear as short diagonal lines whose slope is proportional to the sine of the elevation angle. It is seldom used.

— **I-Scope:** In an I-scope, the target appears as a complete circle when the antenna beam is pointing at it, and the radius of the circle is proportional to range. Beam pointing errors change the circle to an arc whose length is inversely proportional to the pointing error, and the position of the gap in the circle is the pointing direction of the antenna beam. It is seldom found in modern systems.

— **J-Scope:** A J-scope is simply an A-scope where the range line is changed to a circle. Signal amplitude is indicated as radial deflection from the center, and range is the arc distance from the start of the sweep. It is a remnant of the days when large cathode ray tubes (for long traces and good range display resolution) were unavailable.

— **K-Scope:** This is a modified A-scope in which a target appears as a pair of vertical deflections differing slightly in range. With the antenna beam axis azimuth pointed at the target, the deflections are of equal height. The height difference is proportional to target azimuth angle displacement from the beam axis. It is seldom used in modern radars.

— **L-Scope:** The L-scope is the same as a K-scope, but rotated 90° to show elevation pointing error, and is seldom used.

- **M-Scope:** This is an A-Scope with an adjustable pedestal. When the target coincides with the pedestal, the control readout give range more accurately than could be obtained directly from the CRT sweep. It is seldom used.

- **N-Scope:** The N-scope is a hybrid between the K-scope and the M- or O-scope. A target causes two deflections, the amplitude difference of which indicates azimuth pointing error. An adjustable pedestal indicates accurate range. This scope is rarely used now.

- **O-Scope:** This is the same as an M-scope except that a notch is used for measuring range instead of a pedestal. It is seldom used.

- **PPI, Plan-Position Indicator, or P-Scope:** All three names are used to describe this display (Fig. 10-46). It is an intensity modulated circular display where echo signals are shown as though on a map (plan position) viewed from above the radar. Range is indicated by radial position from the radar and azimuth angle indicated by the target's angle clockwise from the top of the scope (which is usually north). Offset PPIs (Fig. 10-46b) have the radar in a location other than the center of the display (these are also called sector PPIs). The radar may be completely off the display. It is the most commonly used search radar display.

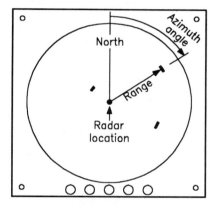
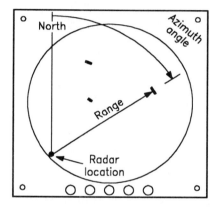

a. Radar–centered PPI   b. Sector (offset) PPI

Figure 10-46. Plan-Position Indicators.

- **RHI or Range-Height Indicator:** This is an intensity modulated display with elevation angle as a counterclockwise angular deflection from horizontal and range as radial distance from the radar (Fig. 10-47). It is basically a sector PPI tipped on its side to show elevation instead of azimuth. This is the primary height-finding display. A PPI and RHI are often paired as a three-dimensional search display. The example shown is part of a precision approach radar (PAR) display. Note the line indicating the desired aircraft glide angle. The other display in the set is a sector PPI.

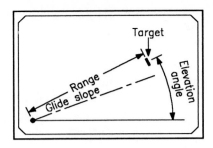

Figure 10-47. Range-Height Indicator (RHI).

10-11. Further Information

General references:

Information on receivers can be found in most communication texts. Microwave texts are also good at providing information on receiver components. The following sources give a good overview of the topic. Although [2] contains some good information, the lack of an index somewhat limits its usefulness.

[1]   J. B. Tsui, *Microwave Receivers with Electronic Warfare Applications*, New York: John Wiley & Sons, 1986.

[2]   S. J. Erst, *Receiving System Design*, Norwood MA: Artech House, 1984.

[3]   J.W. Taylor Jr., Ch. 3 in M.I. Skolnik (ed.), *Radar Handbook*, 2nd ed., New York: McGraw-Hill, 1990.

[4]   *The ARRL Handbook for the Radio Amateur*, Newington CT: American Radio Relay League, published annually.

Cited references:

[5]   D. Roddy, *Microwave Technology*, Englewood Cliffs NJ: Reston Prentice Hall, 1986.

[6]   S.Y. Liao, *Microwave Devices and Circuits*, Englewood Cliffs NJ: Prentice Hall, 1980.

[7]   J.B. Gunn, "Instabilities of Current in III-V Semiconductors," *IBM J. Res. Develop.*, vol. 8, April 1964.

[8]   J.W. Taylor, Jr. and J. Mattern, Ch. 5 in M.I. Skolnik (ed.), *Radar Handbook*, 1st ed., New York: McGraw-Hill, 1970.

[9]   "IEEE Standard Radar Definitions," *IEEE Standard 686-82*, New York: The Institute of Electrical and Electronics Engineers, 1982.

[10] M.I. Skolnik, *Introduction to Radar Systems*, 2nd ed., New York: McGraw-Hill, 1980, Ch. 9.

11

RADAR SIGNAL PROCESSING I:
Introduction and Background

11-1. Introduction

The objectives of signal processing in radar are fourfold:
- Improve both the signal-to-interference ratio and the detection of targets
- Make the radar less vulnerable to targets hiding in clutter
- Make the radar less vulnerable to ECM
- Extract information about the characteristics and behavior of the targets

Signal processing depends on there being some difference between the characteristics of desired targets and those of interfering signals. The first difference is *orderliness versus randomness*. The phases of target echoes, reference the transmit phase, are orderly. Their amplitude may or may not be orderly because of RCS scintillation. Noise and noise jamming are random in both phase and amplitude.

The second difference is the *rate of change of the phases of orderly signals*. Signal components from targets which move radially toward or away from the radar exhibit hit-to-hit phase changes. Those from targets at constant ranges do not. By sorting signals based on phase change, moving targets can thus be discriminated from clutter.

The essence of signal processing is to perform processes which treat desired target echoes and interference from the receiver such that the echoes are enhanced and the interference suppressed. Three basic processes are used.

- *Signal integration:* Signal integration sums composite signals within the same range bin for several hits. The components which are orderly from hit-to-hit produce a larger sum than those which are random.
- *Correlation:* Correlation is the process of measuring the similarity between two functions. Signal-plus-interference is compared to a function matching what a desired target signal would look like, if present. The degree of match gives the likelihood that the composite signal contains a target echo (or sophisticated ECM). Pulse compression is the process of correlation in the time domain.
- *Filtering and spectrum analysis:* In this frequency domain process, composites of target echoes and interference are correlated with a number of complex sinusoids,

separating the signals into their frequency (Doppler) components. The likelihood that a signal component is a desired target depends on its spectrum. The components which concentrate into one or a few bins are target echoes (or some types of ECM). Of these, the ones which concentrate to non-zero Doppler bins are from targets which are moving radially to or from the radar. Those which concentrate in the zero Doppler bin are from targets which are at a constant range from the radar, such as clutter. The contents of those bins likely to contain desired targets are retained and the rest are discarded. Noise or noiselike interference spreads equally among the bins.

In addition to those given above, two other processes are used to assist in performing the basic functions and in suppressing processing artifacts.

- *Windowing:* For processing, signals must be limited in time and the process itself must be finite. The result is processing errors that cause signals which should occupy one output bin to spread into other bins. This anomaly, called spectral leakage, degrades the output and can allow strong interfering signals to mask weaker target echoes. Windowing is used in correlations and spectrum analyses to reduce leakage errors. Windowing, however, modifies the signal itself and information is lost, but the overall result is more useful than if the window were not applied.

- *Convolution:* Convolution in one domain (time or frequency) has the same effect as multiplication in the other domain. In spectrum analysis, for example, the window is applied by multiplying in the time domain. It can, however, be applied instead in the frequency domain by convolving. Thus the process of convolution provides flexibility in certain signal processes and is also useful as an analysis tool for understanding and developing signal processes.

Block diagram: The block diagrams of two typical signal processors are shown in Figs. 11-1 and 11-2. The first is a totally digital processor and the second is a hybrid. Not all signal processors have all the functions shown, nor are they always in the order shown, nor all implemented digitally. The processors shown however, are representative.

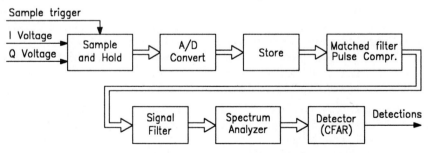

Figure 11-1. Typical Signal Processor, Digital Pulse Compression.

- The *sample-and-hold* function samples the signal voltage at the appropriate time, usually once per range bin. Sampling usually occurs at approximately the center of the bin and is done in both I and Q. The result is a complex voltage $V_I + jV_Q$, from which the magnitude and phase of the signal can be derived.

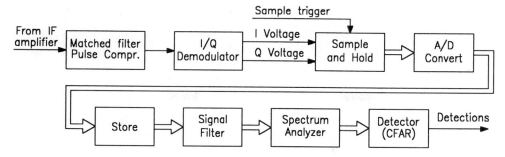

Figure 11-2. Typical Signal Processor, Analog Pulse Compression.

- The *A/D converter* transforms the analog I and Q signals to digital words, and is, of course, present only in those systems having digital signal processing. Any analog process, such as matched-filter pulse compression, must occur before the sample-and-hold and A-to-D conversion.
- The *signal storage* function provides a place to keep digitized signals temporarily while all the signals for a particular process are gathered.
- The *matched filter* function correlates the echo wave with a delayed copy of the transmitted signal. If pulse compression is present, it is performed here. If pulse compression is absent, this block is absent.
- *Signal filtering* removes a portion of the Doppler spectrum, usually that containing clutter. It is often used alone, without the spectrum analyzer, especially in low-PRF MTI radars. In some CW and pulse Doppler radars, it is absent. If the signal filter is used with a spectrum analyzer, it serves as a pre-filter to remove as much clutter as possible so that the clutter does not take up an inordinate fraction of the spectrum analyzer's dynamic range.
- *Spectrum analysis* segregates signal components by Doppler shift. In doing so, it concentrates targets and clutter into their respective Doppler bins, and distributes random signals throughout all the bins (Fig. 1-45). This process allows moving targets and clutter to be segregated. It also improves the signal-to-noise or signal-to-noise-jamming ratio in each bin, because the bin containing the target has most of the target's energy but only a small fraction of the random interference's energy.
- The *CFAR* (*Constant False-Alarm Rate*) function sets and applies the detection threshold, thereby applying the statistics discussed in Ch. 6. This is the function where the decision is made as to whether or not a target is present. Some authors place the CFAR function in the radar's data processor, rather than the signal processor. We shall consider it a signal processing function, although it makes little difference.

In radars, the principles of analog and digital signal processing are applied to achieve the desired result. General information on signal processing can be found in a number of sources. Four useful resources are [1] through [4].

11-2. System Fundamentals and Definitions as Applied to Radars

A system in the signal processing context is an entity which takes an input signal, modifies it in accordance with the system's parameters, and outputs the result. Signal processing systems can be analog or digital, depending on whether or not the signal is sampled prior to processing. The following definitions provide the major distinctions between various types of systems.

- *Linearity:* A linear system is one where an input composed of a sum of signals causes an output which is the sum of the outputs which would be caused by the signals acting separately. If $x_i(t)$ at the input causes $y_i(t)$ at the output and an input signal $x(t)$ composed of the sum $x_1(t) + x_2(t) + x_3(t) + ...$ causes an output which is $y_1(t) + y_2(t) + y_3(t) + ...$, the system is linear. A linear system is restricted to additions and subtractions of signal components. Multiplications can occur, but only by constants. A non-linear system is one where the output can contain signal components arising from the *product* of input signal components.

- *Time invariance:* A time invariant system is one where the actions performed on signals do not change with time. If an input $x(t)$ produces an output $y(t)$, then that same input shifted in time $x(t - \tau)$ will produce the same output shifted by the same amount of time, $y(t - \tau)$. Radar systems with STC, which varies the receiver gain with time, appear to be time variant. However, normal signal processing takes place using data from consecutive hits within the same range bin. Since an individual bin (for example, bin 17) has the same gain for all its occurrences, the system is actually time invariant.

- *Causality:* A system is causal if an input is required to produce an output, *and* if the input must occur in time before the output. A causal system is one which cannot predict the input. Note that in a causal system, the output may continue after the input has become zero, so long as the input is what produced the output.

- *Recursiveness:* A system is non-recursive if its present output is determined by its parameters and present and past inputs, but not by past outputs. A system is recursive if its present output is determined by its parameters, present and past inputs, and past outputs. A recursive system has feedback.

- *System impulse response:* A system has a finite impulse response (FIR) if at some time $nT > NT$ (N finite), the contribution to the output of the input $x(mT)$ ($m < n$) becomes and remains zero. Non-recursive systems have FIR responses. A system has an infinite impulse response (IIR) if the contribution to the output at time $nT > NT$ of the input $x(mT)$ ($m < n$) does not remain zero for any finite N. Most recursive systems have IIR responses.

11-3. Signal Integration

Signal integration is the process of summing the contents of several samples of the same range bin. It increases the signal-to-interference ratio well if the interference is random. Integration can be either *coherent*, incorporating both the signal's amplitude and phase, or *non-coherent*, using just the amplitude. Older terminology calls coherent integration *pre-detection*, or *pre-demodulation* integration, since this type integration takes place with the signal modulated onto the IF carrier. Non-coherent integration was formerly called *post-detection* integration and occurred after the IF carrier was removed.

Most modern radars use some form of I/Q demodulation, which preserves phase, and coherent integration takes place after I/Q demodulation. Modern radars with Doppler processing do not generally integrate signals separately from the Doppler process (Sec. 11-5 and Ch. 12).

Coherent integration suppresses noise, clutter, and some forms of ECM. Noise is suppressed by its randomness, and clutter by the fact that it occupies a different Doppler bin than the target echo. Non-coherent integration suppresses only noise.

Analysis: Signal-to-noise ratio is improved by integration through the orderliness of target echoes and the randomness of noise. The amount of improvement depends on the number of samples summed in the process, known as the *integration number* (N_L). If the integration number is eight, eight consecutive samples of range bin 1 will be summed together to form an integrated signal for range bin 1, eight samples of range bin 2 will be summed, and so on (Sec. 3-5).

If interference is random, the signal-to-interference ratio is increased in proportion to the effective integration number, which is the integration number adjusted for imperfect integration.

$$N_{eff} = N_L / L_i \qquad (11\text{-}1)$$

N_{eff} = the effective integration number, which is the number by which the single-hit signal-to-interference ratio is multiplied to find the integrated signal-to-interference ratio

N_L = the actual number of pulses in the look

L_i = the integration loss (Ch. 3).

Integration improves the signal-to-random-interference ratio because orderly and random entities sum differently. Orderly signal components – target echoes and clutter – sum as voltages, while random components – noise and noise-like ECM – sum as power. Conceptually, the summation of orderly echo signals is

$$V_{OS} = V_{IS1} + V_{IS2} + \ldots + V_{ISN} \qquad (11\text{-}2)$$

V_{OS} = the output voltage after summation, ignoring the integration loss

V_{ISn} = the input voltage for the *n*th sample of an orderly signal

If all V_{IS}s are equal (no target scintillation) and the integration loss is again ignored, the summation becomes

$$V_{OS} = N_L V_{IS} \qquad (11\text{-}3)$$

Random values sum as power.

$$P_{OR} = P_{IR1} + P_{IR2} + \ldots + P_{IRN} \qquad (11\text{-}4)$$

P_{OR} = the output power after summation

P_{IRn} = the input power for an average single sample of a random signal

If the average (rms) level of the random values does not change

$$P_{OR} = N_L P_{IR} \tag{11-5}$$

Since voltage is proportional to the square root of power, the summation of random signal becomes

$$V_{OR}^{1/2} = V_{IR1}^{1/2} + V_{IR2}^{1/2} + \ldots + V_{IRN}^{1/2} \tag{11-6}$$

V_{OR} = the output voltage after summation

V_{IRn} = the input rms voltage for one sample of a random signal

If interfering power is a constant

$$V_{OR} = N_L^{1/2} V_{IRn} \tag{11-7}$$

The voltage signal-to-random-interference ratio becomes

$$\frac{V_{OS}}{V_{OR}} = \frac{V_{IS}}{V_{IR}} N_L^{1/2} \tag{11-8}$$

Power signal-to-random-interference is the square of voltage

$$S/N_O' = \frac{P_{OS}}{P_{OR}} = \frac{P_{IS}}{P_{IR}} N_L = N_L S/I_I \tag{11-9}$$

S/I_O' = the output signal-to-interference ratio for random interference after integration with no integration loss

S/I_I = the input signal-to-interference ratio for one target hit

Applying the integration loss completes the general discussion (Ch. 5 has information on the calculation of the integration loss).

$$S/N_O = S/N_I (N_L / L_i) \tag{11-10}$$

S/N_O = the integrated signal-to-interference ratio for random interference including the integration loss

If target scintillation occurs, the integrated signal-to-noise ratio becomes the sum of the signal-to-noise ratios of the individual samples [10].

$$S/N_O = (1/L_i) \sum_{k=1}^{N_L-1} S/N_k \tag{11-11}$$

S/N_k = the signal-to-noise ratio of the kth sample

Non-coherent integration: Non-coherent integration is a simple summation of signal amplitudes, as shown in Eq. 11-12.

$$\Xi_{N(NC)} = (1/L_i) \sum_{k=0}^{N_L-1} |\Xi_i(k)| \tag{11-12}$$

$\Xi_{N(NC)}$ = the signal after non-coherent integration of N hits

$|\Xi_i(k)|$ = the amplitude of the signal from the kth hit

The effective integration number for non-coherent integration depends on many factors (see later in this section), a major one of which is the number of pulses integrated. For small values, the effective integration number approaches the actual integration number (the integration loss approaches unity).

$$\lim_{N_L \to 0} N_{eff} = N_L \qquad (11\text{-}13)$$

As the integration number approaches a large value, the loss increases and the effective integration number approaches the square root of N_L. The integration loss is covered in Ch. 5 and in [10].

$$\lim_{N_L \to \infty} N_{eff} = \sqrt{N_L} \qquad (11\text{-}14)$$

Figure 11-3 shows a non-coherent integration of a moving target echo with interfering noise. The signal sum is greater than the noise, but not as much greater as it would be if the integration were coherent. Note that with non-coherent integration, noise can never sum to zero, as it can with coherent integration.

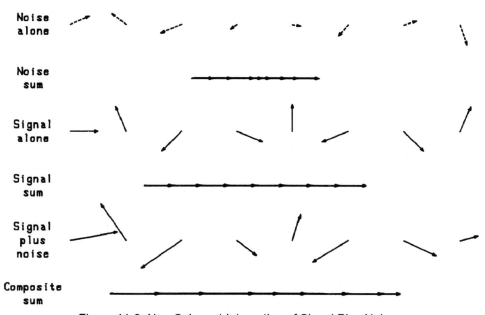

Figure 11-3. Non-Coherent Integration of Signal Plus Noise.

Unlike coherent integration, non-coherent integration does not improve signal-to-clutter ratios. Clutter is a signal; without Doppler phase shifts to discriminate between clutter and moving targets, they integrate identically. Figure 11-4 shows an example of non-coherent integration of signal-plus-clutter. Its primary use is in non-coherent radars, where it is one of the few processes available for improving signal-to-noise ratios.

11-3. Signal Integration   513

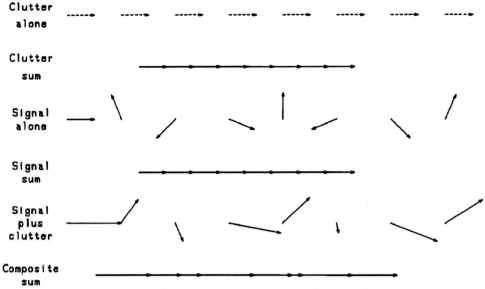

Figure 11-4. Non-Coherent Integration of Signal Plus Clutter.

Coherent integration: Non-coherent integration is a simple summation of signal amplitudes. Coherent integration differs in that the sample-to-sample phase change of the signal is accounted for. The problem is that in all but the simplest analyses, the sample-to-sample phase of signals is unknown prior to the process, and yet must be known for the process to succeed. The two signal samples in Fig. 11-5 have unity amplitude and differ in phase by 180°. If the two are simply summed, the result is zero. If one is rotated 180°, the result has a value of two.

Figure 11-5. Summation of Two Signals.

The process of coherent integration is summation after phase compensation for target motion. In summation notation, coherent signal integration is

$$\Xi_{N(C)} = (1/L_i) \sum_{k=0}^{N_L - 1} \Xi_i(k)\, e^{-j\phi(k)} \tag{11-15}$$

$\Xi_{N(C)}$ = signal after coherent integration of N hits
$\Xi_i(k)$ = the signal from the kth hit
$\phi(k)$ = the phase compensation for the kth hit

Figure 11-6 shows coherent integration of a stationary target. In this figure and those following, the top row shows eight consecutive samples of the signal from a single range bin.

The first phasor is $\Xi_i(0)$, the second is $\Xi_i(1)$, and so forth. The left column of phasors represents the phase compensation. The center column is the summation after the signal phasors are rotated by the angle of the phase compensation. The right column shows the final sum. This same data would of course be produced by a moving target whose phase changed 360°, or 720°, and so forth between hits. Therefore, this target could be stationary, or could be a *blind Doppler shift*, produced by a target moving at a *blind velocity*. Without further information this target's Doppler shift is ambiguous.

With the stationary (or blind) target, the bin-0 phase compensation matches the signal and its sum is large. In all other bins the compensation is mismatched to the signal and the sum is zero. This phase-compensate-and-sum process is a *matched filter* (see Ch. 2) to the stationary signal in Doppler bin 0. All other phase compensations implement filters which are mismatched to this signal.

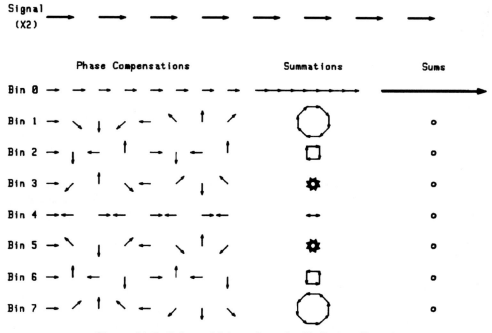

Figure 11-6. Coherent Integration of a Stationary Target.

Figure 11-7 shows the same process applied to a target which matches the bin-1 compensation. Here, a matched filter is implemented in bin 1; a mismatch occurs in all other bins. Figure 11-7 shows eight consecutive hits on a target which is approaching the radar at such a velocity that the phase of the echo advances 45° between hits. If the target's Doppler shift is Nyquist sampled, the phases shown represent the actual target motion. The same data would be produced, however, if the target's phase were advancing 405° per hit, or 765°, or 1125°, and so forth. The target could also be moving away from the radar with 335° of phase *delay* per hit. With the information given, the target is ambiguous. The rate at which the signal phasor rotates is the *apparent Doppler shift* (Ch. 6) of the target. The bin-1 filter matches multiple target motions, which is characteristic of sampled systems.

11-3. Signal Integration

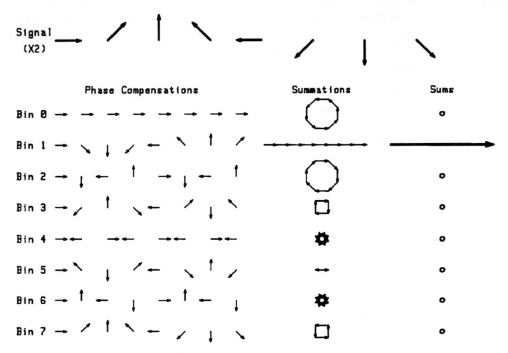

Figure 11-7. Coherent Integration of a Bin-1 Moving Target.

Figure 11-8 shows the integration of a bin-5 target. Note that its per-bin phase rotation is the same as that of Fig. 11-7 but is in the opposite direction. Given Nyquist sampling, Fig. 11-7 represents an inbound target and the target in Fig. 11-8 is outbound. If the magnitude and phase of the signal are fully known, the coherent integration process places them in separate Doppler bins. As with all sampled signals, this target's Doppler shift is ambiguous without further information about it.

Figure 11-9 represents coherent integration of a target which is apparently advancing 67.5° per hit. It would be matched to a filter with phase rotations of −67.5° per sample. However, in this example no compensation of this value exists and there is no filter matched to this signal. The result is twofold: The signal is divided between bins 1 (45° per hit) and 2 (90° per hit), and it *leaks* into the other bins. This is a major problem in signal processing because it is a mechanism by which a large interfering signal can mask lower-level desired targets. This problem and its solution are discussed in Sec 11-8.

Figure 11-10 represents the coherent integration of two scatterers in the same range bin, such as would be found in pulse Doppler radars and modes and with targets exhibiting jet engine modulations (JEM). They are in Doppler bins 1 and 6, with the bin-1 target having four times the RCS of the target in bin 6 (2:1 in voltage). Filters are implemented which match both scatterers and there is no leakage into bins containing no signal. Coherent integration separates the two scatterers and preserves their relative amplitudes.

Figure 11-11 shows two scatterers in the same range bin which would match Doppler bins 1.2 and 6, again with a 4:1 RCS ratio. Note again the leakage caused by the mismatch from the target centered at frequency 1.2.

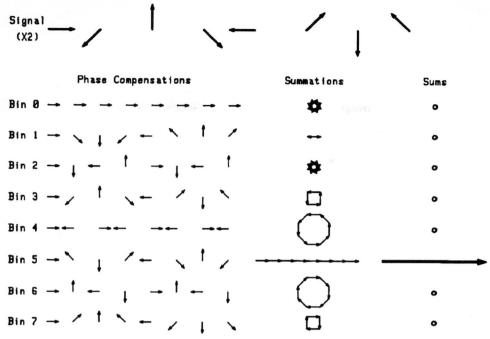

Figure 11-8. Coherent Integration of a Bin-5 Moving Target.

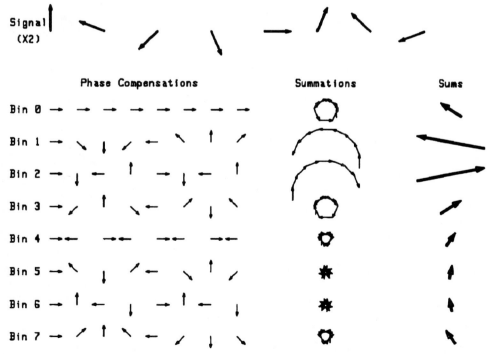

Figure 11-9. Coherent Integration of a Bin-1.5 Moving Target.

11-3. Signal Integration

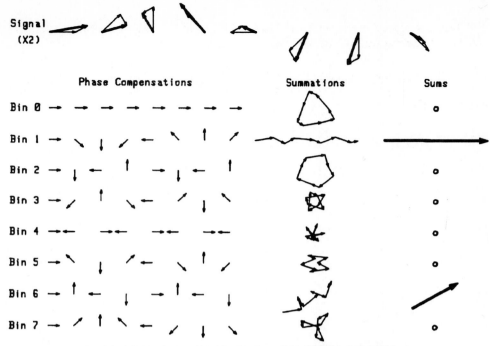

Figure 11-10. Coherent Integration of Two Moving Scatterers.

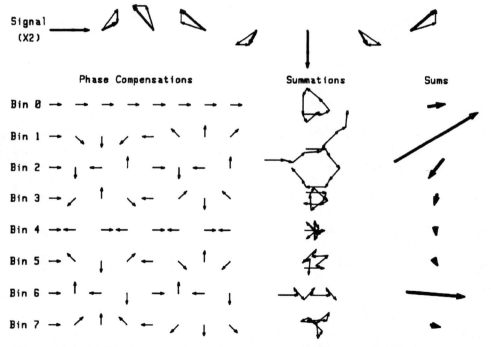

Figure 11-11. Coherent Integration of Two Moving Scatterers (one matched, one not).

Chap. 11. Radar Signal Processing I

Noise is shown coherently integrated in Fig. 11-12. Its randomness results in relatively equal energy among the bins and a much smaller summation in each bin than would result from the same amplitude of coherent signal. Because direction (phase) is used in the summation, the result is smaller than would be achieved with non-coherent integration (Fig. 11-3). If a larger number of samples were used, the distribution of noise among the Doppler bins would be expected to be more uniform.

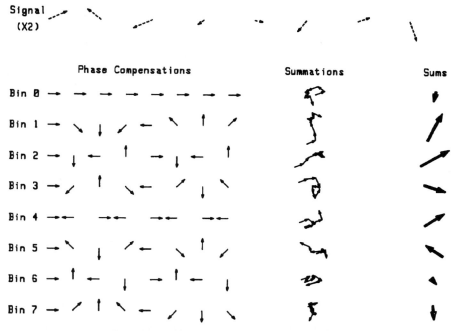

Figure 11-12. Coherent Integration of Noise.

Noise plus a moving target are integrated in Fig. 11-13. The noise spreads roughly equally among the output bins and the signal from the moving target concentrates into one bin. This target is matched to bin 1. As in previous examples, if the signal were not matched to one bin leakage would occur.

Figure 11-14 shows the coherent integration of clutter alone and Fig. 11-15 shows integration of signal plus clutter. Clutter concentrates into bin 0. The phase of the clutter does not have to be zero, but does not change from sample-to-sample. If the clutter did not match bin 0, but were in bin 0.3, leakage from the clutter could mask the moving target. This non-zero-motion clutter is common in moving radars and those which must observe sea clutter. See Ch. 4 for a description of moving clutter and Ch. 12 for some methods of compensating for moving radars and clutter. Separation of moving target echoes from clutter is the basis of moving target indicator (MTI) and moving target detector (MTD) radars. See Ch. 12.

Compensation for any motion: The examples above applied several phase compensation patterns to each signal set. If one of the anticipated motions was correct, a large sum resulted. If the motion anticipated did not match the target's actual motion, the sum was

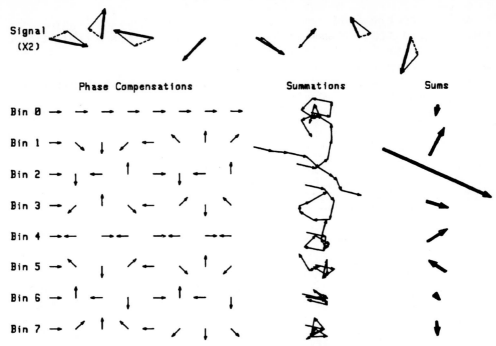

Figure 11-13. Coherent Integration of Signal Plus Noise.

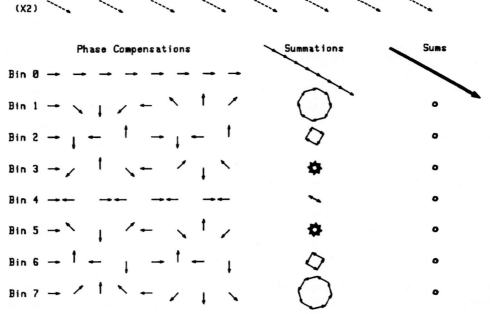

Figure 11-14. Coherent Integration of Clutter.

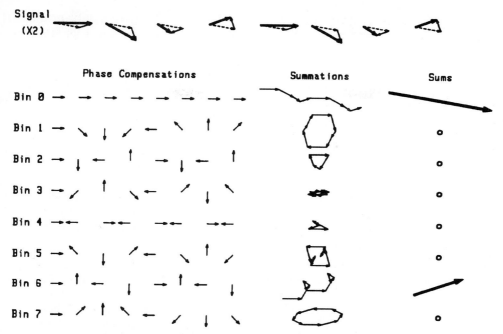

Figure 11-15. Coherent Integration of Signal Plus Clutter.

smaller and leakage occurred. The process shown in Figs. 11-6 through 11-15 is implemented in radars as the discrete Fourier transform (DFT). It is, of course, not possible to anticipate all target motions prior to processing, and therefore the DFT must use a selected phase-compensation set. The more points used in the DFT, the more likely it is that a phase compensation will come close to matching the signal. Thus leakage from large targets (such as clutter and ECM) masking smaller targets with similar phase patterns is less with large-order DFTs than with small. See Secs. 11-7 and 11-8 for further details.

Integration loss: The integration loss is a function of a number of factors:

- The type of integration (coherent or non-coherent)
- The number of pulses integrated
- The required detection and false alarm probabilities
- The target fluctuation statistics
- The processing window used (Sec. 11-8)

Coherent integration has a smaller integration loss than non-coherent, but has the disadvantage that moving targets must have motion compensation applied before summation can take place (see the spectrum analysis section of this chapter for details). Coherent integration is virtually universally used in modern radars having amplifying transmitters and digital signal processors. The coherent integration loss is determined by only two factors: the processing window used (Sec. 11-8) and target fluctuation statistics. The window loss for most windows is less than 3 dB.

As noted in Ch. 5, a detailed analysis of signal detection using coherently integrated signals, like that for non-coherent integration, is beyond the scope of this book. Using a reference such as Meyer and Mayer [10], detection after coherent integration is treated as follows:

1. Calculate the effective integration number from Eq. 11-1, using the integration loss found from Table 11-1 in Sec. 11-8;
2. Calculate the integrated signal-to-noise ratio using the radar equation;
3. On the appropriate Meyer chart for the desired Swerling case target and false alarm probability, find the detection probability using the integrated S/N and an integration number of one.

Example 11-1: A radar coherently integrates 128 pulses using a Dolph-Chebyshev window with 70-dB sidelobes (a=3.5). On a particular Swerling case 2 target, the single-hit signal-to-noise ratio is -3.0 dB. Find the detection probability if the required false alarm probability is 1.0×10^{-5}.

Solution: Table 11-1 gives the integration loss of the Dolph-Chebyshev (a=3.5) window as 1.62, or 2.10 dB. The effective integration number is 79 (19.0 dB $-$ Eq. 11-1). The integrated signal-to-noise ratio is therefore 16.0 dB (Eqs. 3-11 and 3-12). Figure 5-11 is the Meyer plot for the conditions given (Swerling case 2, false alarm probability of 10^{-5}). For an integration number of one and a S/N of 16 dB, the detection probability is 0.75.

For non-coherent integration, the calculation of integration loss is considerably more complicated. The reader is referred to a reference such as Meyer and Mayer [10].

11-4. Correlation

Correlation is the process of matching two waveforms, usually in the time domain, to determine their degree of "fit" and to determine the time at which the maximum correlation coefficient, or "best fit," occurs. Correlation can occur in either the continuous or discrete realms.

Continuous correlation: This correlation form involves signals which are continuous and aperiodic. A mathematical description of continuous correlation is

$$z(t) = \int_{-\infty}^{+\infty} x(\tau) \, h(t+\tau) \, d\tau \qquad (11\text{-}16)$$

$z(t)$ = the correlation function of displacement time t.

$x(\tau)$ = one function (of integration time τ)

$h(t+\tau)$ = the other function (of both integration and displacement times)

In the process, one signal $[x(\tau)]$ is held stationary in time and the other $[h(t+\tau)]$ is displaced in time and "slides" across it. At each point in the displacement, or "sliding," process, the product of x and h is taken and the area under the product found. This area is the correlation of x and h at time t. The variable tau (τ) is time for purposes of finding the area under the product. The parameter t represents the amount of displacement of h from its normal time position (where t = 0). Note that on the right of the equation, t is a parameter; on the left, it is a variable. Thus the process is truly a *transform* in that it changes the independent variable. The correlation process is shown in Fig. 11-16.

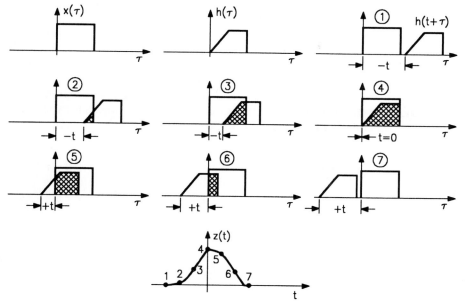

Figure 11-16. Correlation.

Two types of correlation exist. If the two waves being matched are different, $x(\tau) \neq h(\tau)$, a *cross-correlation* is performed. If they are the same, $x(\tau) = h(\tau)$, it is an *autocorrelation*. Autocorrelations are used to evaluate the suitability of waveforms to certain radar tasks and cross-correlations are used primarily in pulse compression. Both applications are described in Ch. 13.

Discrete correlation: Discrete correlation, the corresponding process performed on sampled signals in digital computers, is described below. Equation 11-17 describes the process mathematically.

$$z(kT) = \sum_{i=0}^{N-1} x(iT)\, h[(k+i)T] \qquad (11\text{-}17)$$

$z(kT)$ = the discrete correlation of x and h

N = the total number of samples in one period of the signal (including any zero padding present)

k = the sample number of displacement time (corresponds to t in the continuous realm)

11-4. Correlation

i	= the sample number of the time used to find the area under the product (corresponds to τ in the continuous)
T	= the time between samples of the discrete signals and the time granularity of the displacement of h
$x(iT)$	= the function fixed in time
$h[(k+i)T]$	= the function displaced in time

Discrete correlation can only be performed on periodic functions. If the functions are not periodic, they must be assumed (and made) so before this process can be applied. Generally, this entails repeating the functions with appropriate numbers of zeros appended to each period. Any good signal processing reference describes this clearly [6].

An application of discrete correlation is found in digital pulse compression. Figure 11-17 shows a data stream from an I/Q demodulator. It contains noise and two embedded targets. The correlation function clearly identifies the targets.

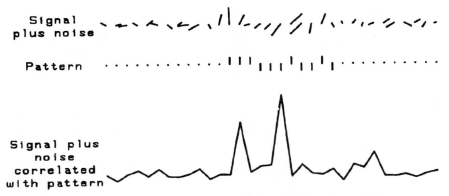

Figure 11-17. Correlation Example Using Pulse Compression Waveform.

Process gain in correlation: Correlation is basically a coherent integration of the signal in time with a process gain. The correlated signal component concentrates into a short time span, while the weakly correlated components spread over many time bins. If the composite signal is target echo plus noise, the echo concentrates and the noise spreads. In this case, the process gain is the ratio of the echo width into the correlator to that of its correlation function.

$$G_{P(COR)} = \tau_{IN} / \tau_{COR} \qquad (11\text{-}18)$$

$G_{P(COR)}$	= the process gain from correlation
τ_{IN}	= the input signal width (seconds)
τ_{COR}	= the correlated signal width (seconds)

If the correlation results in pulse compression and the interference is totally uncorrelated to the transmitted pattern, the process gain is given below. (See Sec. 3-7 for more information.)

$$G_{P(PC)} = \tau_E / \tau_C \qquad (11\text{-}19)$$

$G_{P(PC)}$ = the process gain for pulse compression
τ_E = the expanded (transmitted) pulse width
τ_C = the compressed pulse width

11-5. Convolution

Convolution is a process by which multiplications are transferred from one domain to the other. The relationship between multiplication and convolution is given below and in Fig. 11-18.

$$FT[f(t) \cdot w(t)] = FT[f(t)] * FT[w(t)] = F(f) * W(f) \qquad (11\text{-}20)$$

$f(t)$ = the first signal as a function of time
$w(t)$ = the second signal as a function of time
$F(f)$ = the first signal as a function of frequency
$W(f)$ = the second signal as a function of frequency
$FT[x(t)]$ = the Fourier transform of $x(t)$ and is $X(f)$

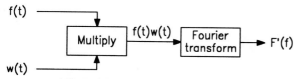

a. Window application in time domain

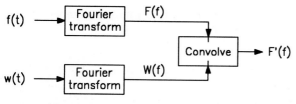

b. Window application in frequency domain

Figure 11-18. Relationship between Multiplication and Convolution.

Continuous convolution: The process of convolution is almost identical to that of correlation. The only difference is that one of the signals (it makes no difference which one) is reversed, or folded, in time before the displace-multiply-integrate operations. Convolution of continuous functions is given below.

$$y(t) = \int_{-\infty}^{+\infty} x(\tau) \, h(t-\tau) \, d\tau \qquad (11\text{-}21)$$

$y(t)$ = the convolution of x and h as a function of displacement time t.
$x(\tau)$ = one signal as a function of integration time τ

$h(-\tau)$ = the second signal reversed in integration time τ

$h(t-\tau) = h(\tau)$ reversed and displaced

Graphically, convolution is shown in Fig. 11-19. Note its similarity to correlation (Fig. 11-16); the only difference is the time-reversal of h.

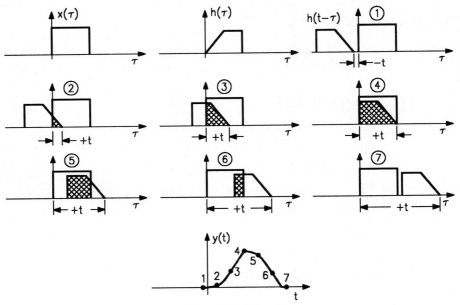

Figure 11-19. Convolution.

Discrete convolution: Discrete convolution is applied to sampled signals and is described below. Again, note its similarity to discrete correlation.

$$y(kT) = \sum_{i=0}^{N-1} x(iT)\, h[(k-i)T] \qquad (11\text{-}22)$$

$y(kT)$ = the discrete convolution of x and h

N = the total number of samples in one period of the signal (including any zero padding present)

k = the sample number of displacement time (corresponds to t in the continuous realm)

i = the sample number of the time used to find the area under the product (corresponds to τ in the continuous)

T = the time between samples of the discrete signals and the time granularity of the displacement of h

$x(iT)$ = the first function fixed in time

$h[(k-i)T]$ = the second function folded and displaced in time

Chap. 11. Radar Signal Processing I

Like discrete correlation, discrete convolution can be performed on periodic functions only. If the functions are not periodic, they must be made so before this process can be applied. For more information, see one of the signal processing references in Sec. 11-13.

Convolution with impulses: Many radar convolution applications involve impulses. An impulse in the continuous world is a rectangular pulse, having a width of zero, infinite amplitude, and an area of one. Continuous convolution with impulses is quite simple. The function being convolved with the impulses is copied at the location of each impulse.

Figure 11-20 illustrates. In it, a gated CW pulse is formed by multiplying in the time domain a cosine wave with a rectangle. Since multiplication in one domain is convolution in the other, the spectrum of the gated CW wave is the convolution of the spectrum of the cosine wave and that of the rectangular pulse. The spectrum of a cosine is two real impulses, one at the frequency of the cosine and one at the negative of its frequency (Sec. 6-2). The spectrum of the time rectangle is a sinc function centered at zero frequency. Therefore the spectrum of the gated CW wave is two sinc functions, one centered at the cosine's frequency and one at its negative frequency. In other words, the spectrum of the product of the pulse and the cosine is the spectrum of the pulse copied at the location of each of the cosine's spectrum's impulses.

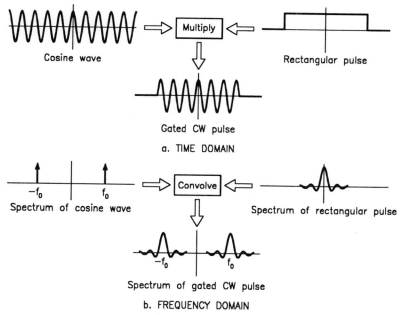

Figure 11-20. Spectrum of Gated CW Wave from Convolution.

11-6. Spectrum Analysis

Spectrum analysis is the process of dividing functions into their frequency components. Radars, as stated before, use spectrum analysis to separate moving targets from clutter and other types of interference.

The basic tool of spectrum analysis is the Fourier transform (FT), which transforms functions of time to functions of frequency. The inverse Fourier transform (IFT) converts functions of frequency to functions of time.

$$G(f) = FT[g(t)] \tag{11-23}$$

$G(f)$ = a function of frequency
$g(t)$ = the corresponding function of time
$FT[\ ]$ = the Fourier transform of a function

$$g(t) = IFT[G(f)] \tag{11-24}$$

$IFT[\ ]$ = the inverse Fourier transform

There are three varieties of the Fourier transform.

— The *continuous Fourier transform* (CFT) describes the frequency components of a signal which is continuous and aperiodic in time. The resulting spectrum is continuous and aperiodic in frequency.

— The *Fourier series* gives the spectrum of a function which is continuous and periodic in time. The spectrum is continuous, but has non-zero values at only certain discrete frequencies. These frequencies are harmonically related to the sample frequency. The spectrum is also aperiodic.

— The *discrete Fourier transform* computes the spectrum of a function which is discrete (sampled) in time. Whether or not the time function is periodic, its spectrum is discrete and periodic and is the spectrum of a periodic time function.

Continuous Fourier transform: The continuous Fourier transform, or CFT, performs spectral analysis on signals which are continuous and aperiodic in the time domain. The resulting spectrum is continuous and aperiodic. The CFT, being continuous, is performed with integration.

$$G(f) = \int_{-\infty}^{\infty} g(t)e^{-j2\pi ft}\,dt \tag{11-25}$$

$G(f)$ = the spectrum of $g(t)$
$g(t)$ = the function in the time domain
f = frequency
t = time

The inverse CFT (ICFT) finds the function in the time domain from its spectrum. The ICFT is

$$g(t) = \int_{-\infty}^{\infty} G(f)e^{+j2\pi ft}\,df \tag{11-26}$$

The CFT of a rectangular pulse in the time domain is a *sinc function* (Sinc(x) = Sin(πx) / πx). The peak value of the spectrum is the area under the pulse. The spectral nulls occur at n/L, where L is the width of the pulse and n is any non-zero integer. The absolute value of the minor lobes in the spectrum (spectral sidelobes) is the pulse area times $2/(k\pi)$, where k is any integer except -1, 0, and $+1$. Figure 11-21 shows this spectrum using the CFT.

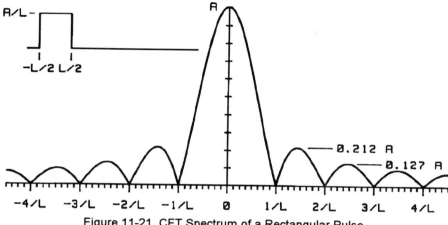

Figure 11-21. CFT Spectrum of a Rectangular Pulse.

Properties of the Fourier transform: The Fourier transform is *linear*, meaning that signals which are sums of components in the time domain yield spectra which are the sums of the spectra of the individual signals. It also means that the real and imaginary components of complex signals ($a_i + jb_i$) can be processed as separate entities.

$$G(f) + H(f) = \text{FT}[g(t)] + \text{FT}[h(t)] = \text{FT}[g(t) + h(t)] \qquad (11\text{-}27)$$

$G(f)$ and $H(f)$ = the spectra of $g(t)$ and $h(t)$, respectively

The transformation from one domain to the other has an *area-amplitude* relationship. The peak amplitude of the spectrum is a linear function of the area under the time envelope. Likewise, the area of the spectrum is a linear function of the time domain peak amplitude. This leads to the Fourier transform scaling properties, for which the reader is referred to signal processing texts, such as [1], [2], [3], and [4].

Fourier series: The Fourier series describes continuous periodic functions. Periodicity in time causes the formation of a line spectrum, whose components are frequency impulses. A frequency impulse represents a complex sinusoid. The spectrum of a periodic time function is a summation of sinusoids.

Periodicity in frequency causes the time domain signal to be a sequence of impulses, and vice versa. Since a time-sampled wave is a series of impulses, this explains the phenomenon of aliasing of sampled waves — Ch. 6.

$$y(t) = \sum_{n=-\infty}^{\infty} c(n) \exp(j2\pi n f_0 t) \tag{11-28}$$

$y(t)$ = a wave composed of an infinite series of complex sinusoids.

$c(n)$ = the coefficients and are complex (see below)

f_0 = the fundamental frequency of the wave

n = any integer

The Fourier series actually functions in one domain only, with the other inferred from the sequence of sinusoids necessary to describe the wave. In the above equation, each spectral line is one term of the summation. The spectrum itself is composed of a sequence of frequency-domain impulses. The ith impulse is at frequency nf_0 and has amplitude $c(n)$.

It is clear from the equation that the spectrum contains discrete frequencies at integer multiples of the wave's fundamental frequency. The coefficients $c(i)$ contain the time domain information and are evaluated as

$$c(n) = 1/P \int_{-P/2}^{P/2} y(t) \exp(-j2\pi n f_0 t)\, dt \tag{11-29}$$

P = the period of the wave

The Fourier series is often expressed in trigonometric form as

$$y(t) = a(0)/2 + \sum_{n=-\infty}^{\infty} [a(m) \cos(2\pi m f_0 t) + b(m) \sin(2\pi m f_0 t)] \tag{11-30}$$

The trigonometric coefficients are evaluated as

$$a(0) = 2/P \int_{-P/2}^{P/2} y(t)\, dt \tag{11-31}$$

$$a(m) = 2/P \int_{-P/2}^{P/2} y(t) \cos(2\pi m f_0 t)\, dt \tag{11-32}$$

$$b(m) = 2/P \int_{-P/2}^{P/2} y(t) \sin(2\pi m f_0 t)\, dt \tag{11-33}$$

m = any integer greater than zero

The spectrum of an infinite periodic train of continuous DC pulses is shown in Fig. 11-22. The spectrum of a periodic train of gated CW waves is identical to this spectrum except that its center is at the frequency of the gated CW.

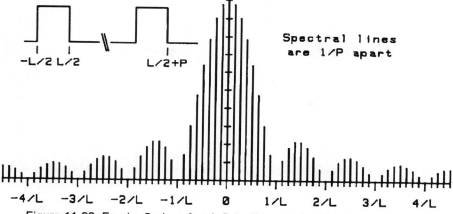
Figure 11-22. Fourier Series of an Infinite Train of Rectangular DC Pulses.

Discrete Fourier transform: The discrete Fourier transform (DFT) changes time to frequency and *vice versa* for sampled functions. The DFT is

$$G(n/NT) = 1/N \sum_{k=0}^{N-1} g(kT) \exp(-j2\pi nk/N) \qquad (11\text{-}34)$$

$G(n/NT)$ = the spectrum of the function $g(kT)$ at frequency n (see below)

n = the frequency sample number

n/NT = the frequency of sample n

N = the total number of time samples

T = the time between samples (reciprocal of sample frequency)

k = the time sample number

kT = the time since the start of the time function

nk/N = frequency times time

In this primitive spectrum analysis form, the process can be thought of as follows. For each desired spectral output, a complex sinusoid at that frequency is created. This complex sinusoid is then essentially correlated with the function being tested to see how well they match. The degree of match is the value of the spectral component at that frequency. The frequency is then changed and the process repeated until the desired spectral components have been evaluated. Figure 11-23 is the DFT of the sequence of time samples shown in the upper part of the figure. The time signal is analogous to a rectangular pulse.

The interpretation of a DFT follows. All frequencies are interpreted as though they were Nyquist sampled, whether they actually were or not, explained further later in this section. Positive signal frequencies land in bins 0 through $N/2-1$, with DC in bin 0 and increasing bin numbers corresponding to increasing frequency. Bins $N-1$ through $N/2+1$ contain the negative frequencies, with the lowest negative frequency in bin $N-1$ and *decreasing* bin

11-6. Spectrum Analysis

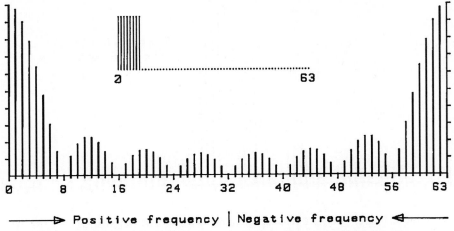

Figure 11-23. Discrete Fourier Transform of a Discrete Pulse Waveform.

number corresponding to increasing negative frequency. Bin N, if it existed, would be at the sample frequency, which in sampled systems is the same as zero frequency.

Bin $N/2$, at one-half the sample frequency, is not Nyquist sampled and some information in signals containing this frequency is not recoverable. The primary information lost is the sense of up and down (positive and negative frequency). It would be unknown whether a target in this bin was inbound or outbound.

The ends of the horizontal axis of Fig. 11-23 are actually the same point. It is useful to think of the horizontal axis as a circle instead of a straight line.

Figure 11-23 presents the spectrum in a somewhat distorted fashion, in that the frequency scale abruptly changes from maximum positive to maximum negative. Before presenting the spectrum, many signal processors perform a routine which moves the frequencies to their more natural positions (sometimes called SWAP).

The result after using SWAP is in Fig. 11-24. Zero frequency (and the sample frequency, and twice the sample frequency, and so forth) is now in the center and the maximum positive and negative frequencies are at the ends. Again, bin $N/2$ (32 in this case) is not Nyquist sampled. As before, the horizontal axis is actually circular.

The inverse discrete Fourier transform (IDFT) converts frequency to time.

$$g(kT) = \sum_{k=0}^{N-1} G(n/NT) \exp(j2\pi nk/N) \tag{11-35}$$

Granularity, frequencies, and span: With the DFT, the total number of frequency samples calculated is unrestricted (≥ 1). The frequency granularity of a spectrum calculated with the DFT can also have value, since n does not have to be an integer. In the cases where n is an integer, the granularity is $1/NT$, which is the same as the total dwell time T_d.

$$\Delta f_G = 1/NT \tag{11-36}$$

$$\Delta f_G = 1/T_d \tag{11-37}$$

$$T = 1/f_S \tag{11-38}$$

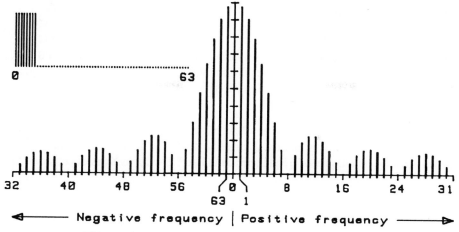

Figure 11-24. Spectrum from Figure 11-23 after SWAP.

Δf_G = the frequency granularity of the DFT if n is an integer
N = the number of time samples
T = the time between samples
T_d = the data gathering time, defined in Ch. 1
f_S = the sample frequency (the PRF in a pulsed radar)

When the Fast Fourier Transform (FFT) algorithm is used to calculate the DFT (Sec. 11-7), several restrictions apply. The number of frequency samples calculated must equal the number of time samples, values of n can only be integers, and the frequency granularity is as described in Eqs. 11-36 and 11-37.

The interpretation of each bin in the DFT output is a matter of calculating the signal frequency which centers in that bin. The argument of the left (frequency) side of the DFT is n / NT, and is the key to finding frequencies.

$$n = [F(n) \text{ MOD } f_S] \, N / f_S = [F(n) \text{ MOD } f_S] \, NT \qquad (11\text{-}39)$$

n = the bin containing the frequency $F(n)$
$F(n)$ = the frequency associated with the nth Doppler bin

As stated above, the output frequency is interpreted to be from a component which is Nyquist sampled, whether or not it actually is.

$$F(n) = n f_S / N \qquad (11\text{-}40a)$$

or

$$F(n) = (n f_S / N) - f_S \qquad (11\text{-}40b)$$

whichever has the smaller absolute value.

11-6. Spectrum Analysis

Example 11-2: A 2400-Hz signal is sampled 1000 times at 3200 samples per second. In what bin will the result appear and what is the interpreted frequency?

Solution: The signal is undersampled. It will appear in bin 750 (Eq. 11-39). Equations 11-40a and 11-40b give the interpreted frequency as 2400 and -800 Hz, respectively. Since Eq. 11-37b gives the smallest absolute value, the interpreted frequency is -800 Hz. Note that -800 Hz is Nyquist sampled at 3200 samples per second; 2400 Hz is not.

The maximum frequency span of any form of the DFT is the sample rate, which is also the bandwidth of the transformation. Using the FFT, the total span is the frequency granularity times the number of frequencies calculated. Owing to Nyquist considerations, this bandwidth is from $-f_S/2$ to $f_S/2$.

$$B_{DFT} = \Delta f_G \, N \tag{11-41}$$

$$B_{DFT} = [1 / NT] \, N \tag{11-42}$$

$$B_{DFT} = 1 / T \tag{11-43}$$

$$B_{DFT} = f_S \tag{11-44}$$

B_{DFT} = the maximum possible bandwidth of the DFT and the bandwidth of the FFT

Example 11-3: A pulse Doppler radar has a PRF of 278,528 pps and performs 2048 point FFTs for its spectrum analysis. Find the transform bandwidth and frequency granularity, and interpret the results. (What Doppler frequency corresponds to each Doppler bin?)

Solution: The sample time T is 3.5903+ μs (Eq. 11-43). There are 2048 samples, for a total data time of 7.3529+ ms. Equation 11-37 gives the frequency granularity as 136.00 Hz. The total span (or transform bandwidth) is, by Eq. 11-41 or 11-43, 278,528 Hz.

The zero-th bin is centered at 0 Hz (and 278,528 Hz, and 557,056 Hz, and 835,584 Hz, and so forth). A signal centered in bin 1 is 136 Hz. Bin 2 contains 272 Hz, and so forth. Bin 1023 ($N/2-1$) contains the highest positive frequency analyzed, 139,128 Hz (Eq. 11-40). Bin 1024 is not Nyquist sampled and contains $\pm 139,264$ Hz; it is unknown whether the frequency is positive or negative. Bin 1025 contains $-139,128$ Hz (Eq. 11-40), bin 1026 contains $-138,992$ Hz, and so forth. A signal of -272 Hz is centered in bin 2046, and bin 2047 ($N-1$) has as its center -136 Hz.

Aliasing and the DFT: Sampling theory tells us that spectral periodicity must occur as a result of sampling in the time domain. This can be seen from Fig. 11-24 and its discussion, and from the bottom row of Fig. 11-30 in a later discussion. The interpretation of frequency of each bin in the DFT output, given in Eq. 11-40, is of a signal that was Nyquist sampled, whether or not it actually was. It must be remembered, however, that undersampled signals also fall into one or more of the calculated bins. For example, consider a "square" DFT calculating eight frequency bins from eight time samples, with each frequency bin 100 Hz wide (Fig. 11-25), and a sample rate of 800 samples per second. A 200 Hz signal analyzed

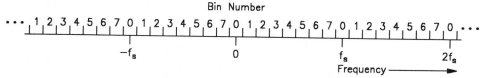

Figure 11-25. Aliasing and the DFT.

by this system lands in frequency bin 2. By previous discussions (Ch. 6 and above), however, a 1000 Hz signal also lands in bin 2, as does an 1800 Hz signal, a 2400 Hz signal, and a -600 Hz signal. Finding the actual frequency of a signal in a particular bin depends on having more information than is available from a single DFT. See Ch. 6 for methods of securing this information.

DFT gain: The discrete Fourier transform inherently has a gain of N, which is removed by the factor $1/N$, as seen in Eq. 11-34. The IDFT has unity gain. In many processes, the absolute gain is immaterial, since either the processed signal amplitude is not used or is calibrated out. In these cases, a small amount of processing time can be saved by eliminating the $1/N$ calculation and accepting that a DFT's output (including that from an FFT algorithm) is a factor of N larger than natural.

In processes which use DFT/IDFT pairs on the same data, the gain must be removed in either the forward or inverse transform; otherwise the signal grows by N for each transform/inverse pair. Since in most of these processes the IDFT is much less common than the DFT, applying the $1/N$ to the IDFT saves a small amount of processing time.

Fourier transforms in two dimensions: Some signal processes require that Fourier transforms be performed in more than one dimension. Some synthetic aperture radar (SAR) processes are like this (Ch. 13). If the data are in rectangular coordinates (X and Y, for example), the computation is an extension of the normal FFT. Figure 11-26 shows a 2-D FFT in rectangular coordinates.

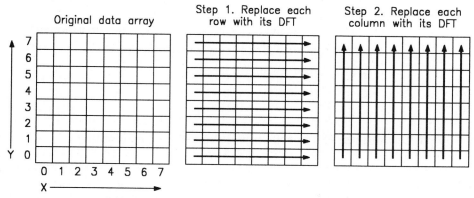

Figure 11-26. Two-Dimensional Rectangular DFT.

The process is as follows:
- Arrange the data in a two-dimensional array as shown in the figure.
- Treating each row independently, find the DFT of each row and replace the data in that row by its DFT.
- Once all rows are complete, find the DFT of each column and replace the data in the column with its DFT.

When all columns are complete, the array contains the 2-D transform of the original data in the array. A mathematical representation of the 2-D Fourier transform in rectangular coordinates is

$$F(u,v) = \int_{-\infty}^{\infty} \int_{-\infty}^{\infty} f(x,y) \exp[-j2\pi(ux + vy)] \, dx \, dy \qquad (11\text{-}45)$$

u,v = the independent variables after transformation
x,y = the independent variables before transformation
$F(u,v)$ = the transformed function
$f(x,y)$ = the function to be transformed

Fourier transforms in other coordinate systems: The Fourier transforms shown so far function in Cartesian coordinates. Some processes require Fourier transforms in other coordinate systems. For example, radar imaging using some SAR techniques require that a Fourier transform be solved in polar coordinates. The polar DFT, defined in the following equation, has no FFT equivalent (Sec. 11-7) and the computation requires so much time that real-time processing is impractical. Fortunately a method has been derived for artificially resampling the polar coordinate data into rectangular coordinates, thereby allowing the FFT to be used [11] [12]. These resampling techniques themselves involve interpolation using Fourier transforms and are discussed in Sec. 11-10 and Ch. 13.

$$F(Q,\Phi) = \int_0^{2\pi} \int_0^{\infty} f(r,\theta) \exp\{-j2\pi[Q \, r \, \cos(\theta-\Phi)]\} \, r \, dr \, d\theta \qquad (11\text{-}46)$$

Q and Φ = the radius and angle, respectively, after transformation
r and θ = the radius and angle before transformation
$F(Q,\Phi)$ = the transformed function
$f(r,\theta)$ = the function to be transformed

11-7. Fast Algorithms: FFT, Fast Convolution, Fast Correlation

The discrete Fourier transform, if implemented as Eq. 11-34, can require vast amount of computation if the number of samples is large. For example, if there are 1024 time samples, each summation for a single frequency requires 1024 complex multiplications and 1023 complex additions. If a like number (1024) of frequency samples are found, which is the most common situation, the total computation requires 1024^2, or 1,048,576 complex multiplications and 1,047,552 complex additions. This assumes the exponentials are found in a table and not computed. If they are calculated, the processing load increases markedly. Implemented as Eq. 11-34, the minimum calculation load for a DFT is

$$N_{CMUL} = N^2 \qquad (11\text{-}47)$$

N_{CMUL} = the number of complex multiplications
N = the number of time data points and the number of frequency samples

$$N_{CADD} = N^2 - 1 \qquad (11\text{-}48)$$

N_{CADD} = the number of complex additions in the transform

Recognizing that there are four real multiplications and two real additions in one complex multiplication, and two real additions per complex addition, one can see that the computation load becomes unwieldy rapidly.

In the 1950s and early 1960s, several people recognized that there is considerable redundancy in the DFT of Eq. 11-31, in that many calculations are identical. In 1965, J.W. Cooley and J.W. Tukey published a famous paper [13] which identified and removed these redundancies.

The derivation of Cooley and Tukey's Fast Fourier Transform (FFT) is beyond the scope of this book, but the principle is quite simple. It is possible to compute an N point square transform (same number of time and frequency samples) as two N/2 point transforms plus some overhead. Thus a 1024 point transform, instead of having 1024^2 or 1,048,576 complex multiplication, has 2×512^2 or 524,388 complex multiplications. (For this argument, we shall ignore the additions, which are also reduced in the same manner.) Since a 512 point transform can be calculated as two 256 point transforms, the computation load is further reduced to $2 \times 2 \times 256^2$ or 262,144 complex multiplications. If one carries this to the limit, one finds that the same 1024 point transform can be calculated as 512 two-point transforms, plus the overhead. Although the computation reduction is not as great as predicted by this simplistic approach, the FFT requires only $(N \log_2 N)/2$ complex multiplications, which for 1024 points is 5,120. This is a factor of 200 smaller than the direct approach. The development of the FFT made possible the use of the DFT in radar spectrum analysis for real-time clutter suppression and target velocity identification.

This two-point transform, which forms the basis of the FFT, is called a *butterfly*, because of the form of its signal flow diagram (Fig. 11-27). Not all FFT algorithms are based on the two-point butterfly. Some compute four points directly and some use more.

There are many good references which derive the FFT. Three of the best, in this author's opinion, are found in [5], [6], and [7].

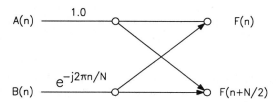

Figure 11-27. DFT Butterfly

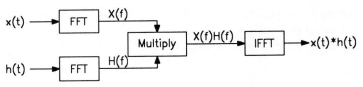

Figure 11-28. Fast Convolution.

Fast convolution and correlation: No algorithms exist, per se, for rapidly calculating the convolution and correlation in the same manner as the FFT. Taking advantage of the convolution theorem (Eq. 11-20 and Fig 11-18) it is possible to quickly calculate the convolution using the FFT.

Figure 11-28 shows the principle. Since the convolution of two functions in one domain is equivalent to multiplying their transforms in the other domain, one simply transforms the two functions, multiplies their transforms, and inverse transforms the product back to the original domain.

A comparison of the convolution and correlation processes shows that the only difference is that one of the functions in correlation is reversed in time compared to convolution (it is actually reversed in convolution). Reversing time in the time domain is identical to taking the complex conjugate of the spectrum in the frequency domain.

$$FT[h(-t)] = H^*(f) \qquad (11\text{-}49)$$

$H(f)$ = the transform of $h(t)$

$h(-t)$ = $h(t)$ reversed in time

$H^*(f)$ = the complex conjugate of $H(f)$

FT = the Fourier transform

This fact suggests a method for computing the correlation using FFTs similar to the fast correlation. The method is shown in Fig. 11-29. In practice, many correlations in radars are performed using dedicated optimized hardware, which allows more rapid calculation of the correlation function than even the method shown here.

In these two uses of the DFT/IDFT (as FFT/IFFT algorithms), either the DFT or the IDFT must have the $1/N$ factor so that the gain across the convolution or correlation is unity. As implemented in the above block diagrams, a small amount of processing time can sometimes be saved by applying the correction to the IDFT, which is used half as often as the DFT.

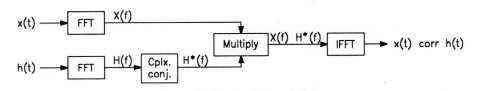

Figure 11-29. Fast Correlation.

11-8. Processing Errors and Windows

Calculation of the discrete Fourier transform introduces processing errors, caused primarily by the fact that processing resources are limited. These errors are illustrated beautifully in a figure originated by E.O. Brigham [5], reproduced here as Fig. 11-30.

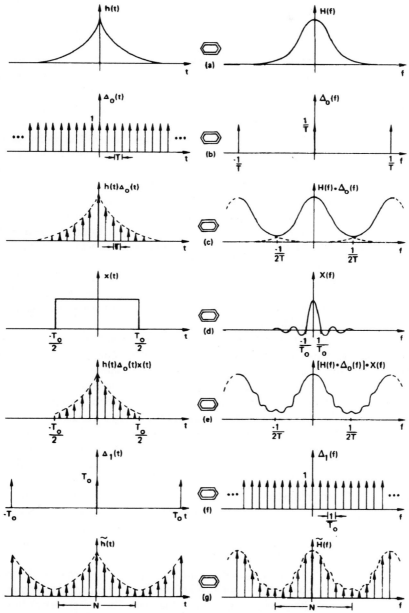

Figure 11-30. DFT Development Showing Errors Introduced.
(Reprinted from E.O. Brigham [5]. © Prentice Hall; used with permission)

In the figure, the left-hand column is the time domain and the right-hand column is frequency. Each row shows the same function, time on the left and frequency on the right. The first row, Fig. 11-30a, shows the original signal and its spectrum. This spectrum, in the upper right, is the goal of the DFT process.

Before computation of the spectrum, the signal must be sampled by the waveform shown in Fig. 11-30b. Time sampling is accomplished by multiplying the time domain signal by a periodic train of time impulses. The spectrum of the sampling wave is a train of frequency impulses spaced the sample frequency apart (frequency impulse spacing is the reciprocal of time impulse spacing).

The sampled wave and its spectrum are given in Fig. 11-30c. The sampled wave in time is the product of the original wave and the sampling waveform. The spectrum of the sampled wave is the convolution of the original wave's spectrum with that of the sampling waveform. This convolution is simply a matter of copying the original spectrum at the location of each of the impulses in the sampling spectrum.

Next (Fig. 11-30d), the sampled signal is truncated, where the signal is given a beginning and an end in time, and only a limited number of samples are retained. Truncation is necessary for three reasons. First, resources in signal processing computers are limited, and storage and processing are simply not available for observing targets for very long periods of time. Second, radars must gather and process signal from a segment of their search volume quickly and move on to the next if they are to cover the volume in a reasonable time. Third, targets move and maneuver over time, and the radar's look times must be short enough so that the targets' behavior is evident. The long look times necessary to minimize truncation effects tend to average, or "smear," the motion of targets.

Truncation is done by multiplying (in time) the sampled signal with a rectangle of zero value outside the processing window and unity within it. In the frequency domain, a convolution takes place between the spectrum of the sampled wave and the spectrum of the window (in this case, a sinc function).

The result of this truncation is spectral leakage, where a signal having a single frequency shows up in more than one spectral bin. The spectral leakage concept and its remedies are illustrated later in this section. In our example, the result is shown in Fig. 11-30e, in that the spectrum is now "rippled" as a result of spectral leakage of strong components of the original signal caused by the window. See Fig. 11-34 for a view of a single-line spectrum which leaks into many other frequencies because of the signal's truncation.

The spectrum of the signal is computed in Figs. 11-30f and 11-30g, by *sampling* the signal's frequency content (by computing the DFT for selected values on n). The process is identical to multiplication by impulses $1/T_0$ apart in the frequency domain, which causes convolution with impulses T_0 apart in time.

Thus the spectrum computed in the DFT process is the desired spectrum with aliases (from the time sampling), with leakage (from truncation or windowing in time), and at discrete frequency samples only (from using the DFT). We started with the time signal in Fig. 11-30a left, and were trying to find its spectrum (Fig. 11-30a right). We found instead the spectrum of Fig. 11-30g right, which is the spectrum of Fig. 11-30g left.

The three errors introduced were *aliasing*, *leakage*, and *non-continuity* in the spectrum. Aliasing errors are dealt with by Nyquist sampling the time signal. Non-continuity is alleviated by placing the frequency samples close together. The remainder of this section deals with reduction of leakage errors.

Windowing: Spectral leakage is suppressed by windowing the signal with a shape which has itself as narrow a spectrum as possible. The ideal spectrum of a window is a single impulse at zero frequency. This would produce no change in transitioning from Fig. 11-30c to 11-30d. Unfortunately, the time signal which produces this spectrum is a constant (DC) starting at the beginning of time (the big bang) and ending at the end of time (the big crunch), and is equivalent to no window at all. *Windowing cannot be avoided, since it is absolutely necessary to make the spectral computation finite.*

The process of windowing is multiplication in time, as seen in Fig. 11-30d. The Doppler signals viewed by radars, being sinusoids, are easy to visualize windowed. See Figs. 11-31 and 11-32. Figure 11-31 shows a sinusoid windowed with the rectangle (also called a *uniform* or *Dirichlet* window). Figure 11-32 shows a sinusoid modified by a cosine-squared function, called a Hann window.

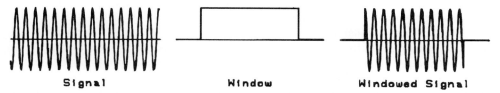

Figure 11-31. Sinusoidal Signal Windowed with a Rectangular Window.

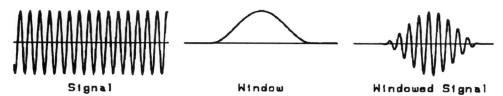

Figure 11-32. Sinusoidal Signal Windowed with a Hann Window.

Many windows are available, and several which are useful in radar Doppler processing, pulse compression, and SAR processing are discussed below. The reader is referred to a classic paper by Harris [8] for details of window design and definitions and performance data on many more windows.

The following series of figures shows a sampling of windows. Each figure gives the result of treating a cosine function with a different window. There are three parts to each figure. The first shows the time data (64 points) after application of the window. The second is the spectrum obtained from the time samples. In the third, the time samples were padded with 192 zeros and the spectrum taken. This padding operation effectively reduces the frequency granularity by a factor of 4, so that details can be seen. Section 11-10 has more information on padding and interpolation.

Figures 11-33 and 11-34 illustrate the rectangular window and the leakage expected with it. In Fig. 11-33, 64 samples were gathered of a signal having a single Doppler shift of exactly ten cycles in the data-gathering time, for a frequency of 10/NT. This signal was I/Q demodulated and is complex; therefore its spectrum is one-sided and retains the target's flight direction (in this case, inbound). The signal is truncated after 64 samples and processed in an FFT without any further modification. The resulting unpadded spectrum occupies only

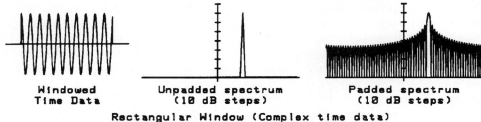

Figure 11-33. Rectangular Window, Bin-Centered Signal.

a single Doppler bin. Signals where an exact integer number of cycles are processed are known as being *bin-centered*, corresponding to the example of Fig. 11-7.

It is to be expected that the spectrum of a sinusoid after rectangular windowing would be a sinc function, with its associated leakage. The padded spectrum shows that it is. The frequency samples in the unpadded spectrum happen to land in the *spectral nulls*, and the leakage is present but hidden. The padded spectrum shows the classic sinc shape.

Contrast this with Fig. 11-34, which is the same rectangular window and sinusoidal signal, except in this case there are 10.5 cycles of Doppler shift in the 64 samples (the signal is not bin-centered). The spectrum is now a sinc function where the unpadded frequency samples are taken at the *peaks* of the sinc lobes. The spectral leakage is now obvious. The padded spectrum shows exactly the same result as a bin-centered shifted 1/2 unpadded bin (2 padded bins).

The rectangular windows used above simply turn the signal on and shut it off. Within the window, the signal's original amplitude is retained. This results in high spectral leakage. As a rule of thumb, the more abruptly the time signal's amplitude changes, the higher the spectral leakage. The rectangular window is described mathematically in Eq. 11-50. All the following window descriptions are from Harris [8].

$$w(kT) = K \qquad (11\text{-}50)$$

$w(kT)$ = the window as a function of time

k = the time counter, usually ranging from 0 to $N-1$

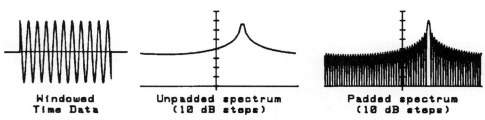

Figure 11-34. Rectangular Window, Non-Centered Signal.

T = the time between window (and signal) samples

K = any constant, usually unity

Windows are almost always complex. Unless otherwise specified, the real and imaginary parts of the window are equal.

In spectrum analysis, the window is applied to the signal by multiplying each time sample of the signal by the corresponding time sample of the window. The real and imaginary components are treated separately. The I portion of the signal is multiplied by the real part of the window and the Q signal is multiplied by the imaginary window. Window application *is not* a complex multiplication.

$$g_I'(kT) = g_I(kT)\, w_I(kT) \tag{11-51}$$

$$g_Q'(kT) = g_Q(kT)\, w_Q(kT) \tag{11-52}$$

$g_I'(kT)$ and $g_Q'(kT)$ = the real and imaginary parts of the windowed signal, respectively

$g_I(kT)$ and $g_Q(kT)$ = the real and imaginary parts of the unwindowed signal

$w_I(kT)$ and $w_Q(kT)$ = the real and imaginary parts of the window

The discrete Fourier transform with a window applied is

$$G'(n/NT) = 1/N \sum_{k=0}^{N-1} g(kT)\, w(kT)\, \exp(-j2\pi nk/N) \tag{11-53}$$

$G'(n/NT)$ = the spectrum of the function $g(kT)$ at frequency n with a window applied to $g(kT)$

The summation limits can be any value, as long as they encompass one complete period of the window. For efficient computation, the summation limits should match those of the window.

In the following paragraphs, a few major windows used in radars are defined, with figures showing their form and the result of spectrum analyzing a single complex sinusoid with each. There are many more windows than described here. The best reference for further information is Harris [8].

Cosine family of windows: The cosine family is made up of windows of the form $\text{Cos}^\alpha(kT)$, α being an integer and determining the window type. Their mathematical descriptions are given below. Note that for odd values of α, the form of the final window is actually a sine.

$$w(k) = \sin^\alpha[\pi k/N] \quad \text{(general)} \tag{11-54}$$

$w(k)$ = the window value (real and imaginary) for that k

k = a counter ranging from 0 to $N-1$

α = an integer determining the window type

N = the number of points in the signal/window

11-8. Processing Errors and Windows

The special case of $\alpha=2$ is known as the *Hann window*. Some call it the *Hanning window*, but since there is a different window known as *Hamming* (see below), the term Hann is best to avoid confusion.

$$w(k) = \sin^2[\pi k/N] \quad (\alpha = 2 \text{ or Hann window}) \tag{11-55}$$

Applying the appropriate trigonometric identity shows this window to be one full cycle of a cosine raised by its peak value.

$$w(k) = 0.5\,[\,1 - \cos(2\pi k/N)] \quad \text{(Hann window)} \tag{11-56}$$

Figures 11-35 and 11-36 show the Hann window and its result. Figure 11-35 shows a bin-centered signal, and 11-36 shows one which is not centered.

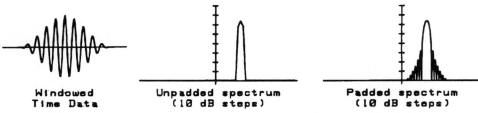

Windowed Time Data   Unpadded spectrum (10 dB steps)   Padded spectrum (10 dB steps)
Cosine Squared (Hann or Hanning) Window (Complex time data)
Frequency: 10   Amplitude: 1   N: 64
Peak Spectrum Magnitude: Unpadded -6.0 dB, Padded -6.0 dB

Figure 11-35. Hann Window — Bin-Centered Signal.

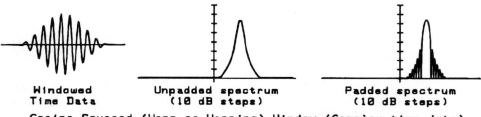

Windowed Time Data   Unpadded spectrum (10 dB steps)   Padded spectrum (10 dB steps)
Cosine Squared (Hann or Hanning) Window (Complex time data)
Frequency: 10.5   Amplitude: 1   N: 64
Peak Spectrum Magnitude: Unpadded -7.4 dB, Padded -6.0 dB

Figure 11-36. Hann Window — Non-Centered Signal.

The Hamming window is of the cosine family and is similar to the Hann, except that the ends do not go to zero. Since this window has data discontinuities at its ends, one would expect that its sidelobes would be higher than Hann. In fact, the close-in sidelobes are 10 dB lower. The sidelobes do not fall off as rapidly as Hann, however, which makes this a less than satisfactory signal processing window (see Sec. 11-9). The Hamming window is shown in Fig. 11-37.

$$w(k) = 0.54 - 0.46\,\cos(2\pi k/N) \quad \text{(Hamming window)} \tag{11-57}$$

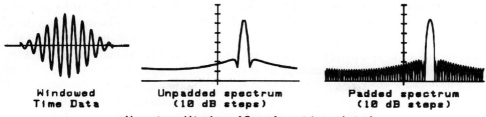

Hamming Window (Complex time data)
Frequency: 10.5   Amplitude: 1   N: 64
Peak Spectrum Magnitude: Unpadded -7.1 dB, Padded -5.4 dB

Figure 11-37. Hamming Window — Non-Centered Signal.

Kaiser-Bessel windows: Members of this family of windows approach the ideal windows for signal processing, in that they have very low sidelobes (leakage) and their sidelobes roll off rapidly with frequency from the center of their transformed output. They resolve well in the presence of widely separated signal amplitudes (Sec. 11-9). This property is especially important where a large interfering signal is close in Doppler shift to a small desired signal. Kaiser-Bessel windows are described mathematically as

$$w(k) = \frac{I_0\{\pi a\,[1.0 - (2k/N)^2]^{1/2}\}}{I_0[\pi a]} \qquad (11\text{-}58)$$

I_0 = the zero-order modified Bessel function of the first kind

$$I_0(A) = \sum_{i=0}^{\infty} [(A/2)^i / i!]^2 \qquad (11\text{-}59)$$

$I_0(A)$ = the zero order modified Bessel function of the first kind with argument A

Figure 11-38 shows the Kaiser-Bessel window with $\alpha = 3.0$ for a non-centered signal. With these more effective windows, centering has little effect on the spectrum. The effect of centering is discussed further in the scalloping description later in this section.

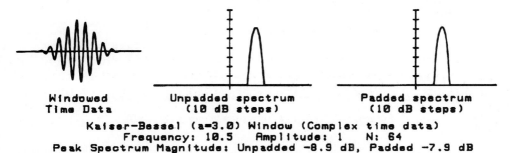

Kaiser-Bessel (a=3.0) Window (Complex time data)
Frequency: 10.5   Amplitude: 1   N: 64
Peak Spectrum Magnitude: Unpadded -8.9 dB, Padded -7.9 dB

Figure 11-38. Kaiser-Bessel (α=3.0) Window — Non-Centered Signal.

Dolph-Chebyshev windows: The Dolph-Chebyshev windows are some of the most effective and widely used in radar signal processing. They are unique for two reasons. First, they are defined in the domain requiring convolution rather than the multiplication domain. Second, although their sidelobes are low, they do not roll off with distance from the center.

They have three highly desirable properties for radar signal processing. First, of all window families, they concentrate the most energy into the main spectral lobe for a given sidelobe level. Second, they present one of the best compromises available for good resolution of very close signals of approximately the same amplitude (narrow main lobe) and low leakage (sidelobes). Third, they resolve signals with widely separated amplitudes well (Sec. 11-9).

Dolph-Chebyshev windows are defined by Harris [8] as follows.

$$w(k) = \text{IFT}\left\{(-1)^n \frac{\cos[N \cos^{-1}[\beta \cos(\pi n/N)]]}{\cosh[N \cosh-1(\beta)]}\right\} \quad (11\text{-}60)$$

$\quad$ IFT = the inverse Fourier transform
$\quad k$ = the time sample number
$\quad n$ = the frequency sample number
$\quad N$ = the number of time/frequency points
$\quad \beta$ = defined below

$$\beta = \cosh[(1/N) \cosh^{-1}(10\alpha)] \quad (11\text{-}61)$$

$\quad \alpha$ = the window parameter — 2.5 and 3.0 are used in the figures

Note that complex functions must be used to compute these windows because the argument of the inverse cosine can exceed unity. A Dolph-Chebyshev windows for $\alpha=3.0$ is shown in Fig. 11-39. The non-sinusoidal appearance of the windowed time data is because of phase reversals in the window.

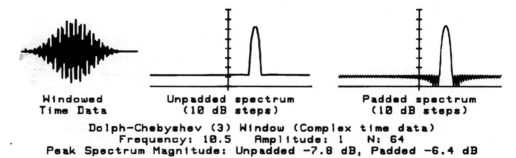

Windowed Time Data $\quad$ Unpadded spectrum (10 dB steps) $\quad$ Padded spectrum (10 dB steps)
Dolph-Chebyshev (3) Window (Complex time data)
Frequency: 10.5 $\quad$ Amplitude: 1 $\quad$ N: 64
Peak Spectrum Magnitude: Unpadded -7.8 dB, Padded -6.4 dB

Figure 11-39. Dolph-Chebyshev Window ($\alpha = 3.0$).

Taylor windows: This family of windows is used in some analog pulse compression systems. Their performance approximates that of the Dolph-Chebyshev (D-C) family and they are physically realizable in circumstances where D-C is not [14]. Their major property is, like D-C, narrow main lobes for a given sidelobe level. In regions near the center frequency, sidelobes fall off little; for points farther removed, rolloff is 6 dB per octave (or

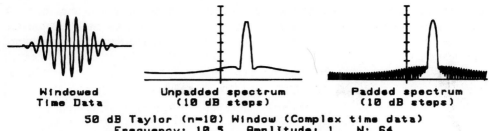

Windowed　　　　Unpadded spectrum　　　　Padded spectrum
Time Data　　　　(10 dB steps)　　　　　　(10 dB steps)
50 dB Taylor (n=10) Window (Complex time data)
Frequency: 10.5　　Amplitude: 1　　N: 64
Peak Spectrum Magnitude: Unpadded -7.4 dB, Padded -5.8 dB

Figure 11-40. 50-dB Taylor Window — Non-Centered Signal.

time doubling, if used in pulse compression). They are defined below in the time domain, for consistency with previous discussions. For use in pulse compression, windows are defined in frequency and transformed to time. These windows are not covered by Harris [8]. The reader is referred to Farnett and Stevens [14] for additional data. Figure 11-40 gives the performance of a 50-dB Taylor (n = 10) window.

$$w(k) = 1 + 2 \sum_{m=1}^{n-1} F_m \cos [2\pi m(k/N)] \qquad (11\text{-}62)$$

$w(k)$ = the window as a function of k

n = a parameter similar to α in D-C windows. A value of 6 gives 40-dB sidelobes

F_m = Taylor coefficients (see [14])

m = a counter

N = the total number of points in the window

Clipped windows: To recover low RCS moving targets in high RCS clutter requires that the signal process have as little leakage as possible. If limiting occurs in the signal processor after windowing, the resulting clipped window will cause much leakage. See Fig. 11-41 for an example. This is a Kaiser-Bessel (α=3.0) window which has been clipped at the 50-percent level. As can be seen by comparing Figs. 11-38 and 11-41, clipping must be avoided.

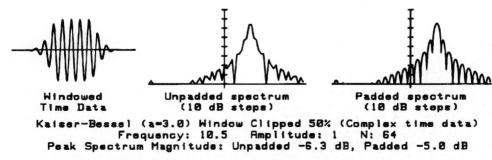

Windowed　　　　Unpadded spectrum　　　　Padded spectrum
Time Data　　　　(10 dB steps)　　　　　　(10 dB steps)
Kaiser-Bessel (a=3.0) Window Clipped 50% (Complex time data)
Frequency: 10.5　　Amplitude: 1　　N: 64
Peak Spectrum Magnitude: Unpadded -6.3 dB, Padded -5.0 dB

Figure 11-41. Kaiser-Bessel Window Clipped at 50-percent Level.

Window parameters: Several parameters indicate the suitability of windows for radar signal processing. Some of these parameters for the windows discussed in this section are summarized in Table 11-1. See the references by Harris [8] and Farnett [14] for more information.

Table 11-1. Windows and Their Parameters.
(From Harris [8]. U.S. Government work not protected by U.S. copyright.)

Window	Highest side-lobe level (dB)	Side-lobe roll-off (dB/oct.)	Coherent Gain	Equiv. noise BW (bins)	3 dB BW (bins)	Scallop loss (dB)	Worse-case process loss (dB)
Rectangle	−13.4	−6	1.00	1.00	0.89	3.92	3.92
Triangle	−26.8	−12	0.50	1.33	1.28	1.82	3.07
Cosine	−23.	−12	0.64	1.23	1.20	2.10	3.01
Hann	−32.	−18	0.50	1.50	1.44	1.42	3.18
Hamming	−43.	−6	0.54	1.36	1.30	1.78	3.10
K-B 2.5	−57.	−6	0.44	1.65	1.57	1.20	3.38
K-B 3.0	−69.	−6	0.40	1.80	1.71	1.02	3.56
D-C 2.5	−50.	0	0.53	1.39	1.33	1.70	3.12
D-C 3.0	−60	0	0.48	1.51	1.44	1.44	3.23

— *Window name:* This is the name as given by Harris.
— *Highest sidelobe level:* This is the highest response to a signal within bins other than those containing the fundamental signal itself. It is expressed in decibels below the peak response to the signal.
— *Sidelobe roll-off:* This is a measure of how rapidly the sidelobes are attenuated as frequency from the bin(s) containing the signal increases. It is expressed in decibels per frequency doubling (octave) from the center frequency.
— *Coherent gain:* Coherent gain is related to the equivalent noise bandwidth (below) and is a measure of the loss in signal-to-noise ratio caused by the window. It is defined as the ratio of the output S/N to the input S/N within one bin. The power gain of the window is the square of coherent gain.
— *Equivalent noise bandwidth (ENBW):* This is the noise bandwidth of the window in the sense of the amount of noise passing through each bin. It is the B_N in $K T_S B_N$ (see Ch. 2). *The integration loss in coherent integration is the equivalent noise bandwidth.*
— *3-dB bandwidth:* This factor is simply the bandwidth of each bin between the points where the response is within 3 dB (1/2 power or 0.707 voltage) of the peak response.
— *Scalloping loss:* The scalloping loss is the amplitude uncertainty caused by the window. It is the ratio of the gain of the window to a signal located with respect to the output bins such as to give the maximum response to the gain at the location of the minimum response. It is also known as the picket fence effect. Figure 11-42 shows

the principle, and Fig. 11-43 shows the response of two windows to signals at various locations with respect to bin-centering. Note that scalloping causes wide fluctuations in output amplitude for the rectangular window, but that output amplitude with the Kaiser-Bessel ($\alpha = 3.0$) window is essentially independent of frequency placement. Note also that the ratio of the largest peak amplitude to the smallest is the scalloping loss listed in Table 11-1 for that window.

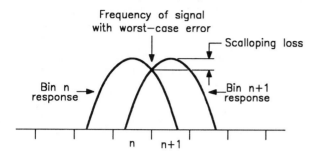

Figure 11-42. Scalloping Loss Effect.

— *Worst-case process loss:* This factor is the sum (dB) of the processing loss (negative of the coherent gain in dB) and the scalloping loss. It represents the maximum possible degradation in signal-to-noise ratio cause by the window (versus the no-loss condition) and the worst possible frequency location with respect to bin centers.

11-9. Windows and Resolution

The parameters shown in Table 11-1 give much valuable information as to the suitability of a particular window in radar spectrum analysis and correlation. In the final analysis, however, one requirement as to whether a window functions well in radar signal processing stands out above all the rest. That requirement is that the window handle large interfering signals in such a way that their leakage does not mask small targets. In spectrum analysis terms, this means that a large signal situated such that it produces maximum leakage (usually, but not always halfway between two output bins) causes minimum interference with a small signal located a few bins from it.

Harris, in both papers cited, [7] and [8], suggests an experimental way to evaluate windows from this standpoint. In the experiment, two bin-centered signals are generated, each in different bins of the DFT. The signals have widely divergent amplitudes. Harris used signals located in bins 10 and 16 with 40 dB of amplitude separation (1:100 in voltage, 1:10,000 in power) and this arrangement is used here. The following series of figures give the results. In each figure the left half shows the signals bin-centered and the right half shows the larger signal moved toward the smaller by 1/2 Doppler bin.

Figure 11-44 shows the result from using a rectangular window. With bin-centering (left figure), there is no leakage and the signals are perfectly resolved. In the right figure, the large signal is moved to bin 10.5, producing the maximum leakage. It now completely hides the small signal. This figure alone illustrates the futility of using rectangular windows in radar signal processing, even though it discards no signal and yields the best signal-to-noise ratio of any window.

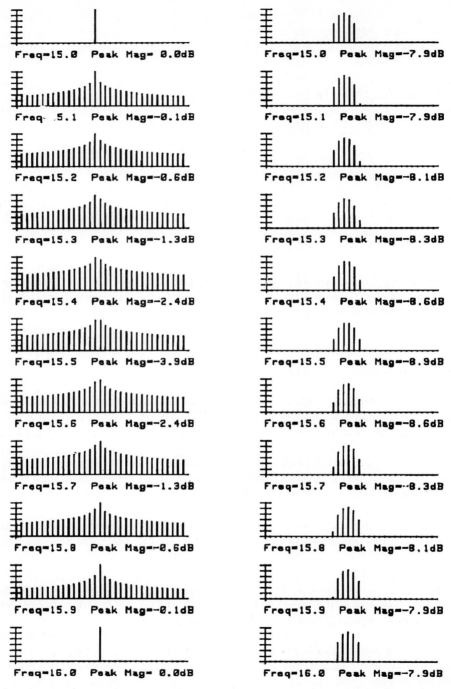

Figure 11-43. Scalloping. Left column with rectangular window; right column with Kaiser-Bessel (α = 3.0) window.

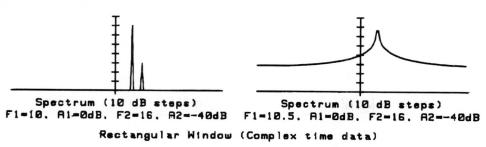

Figure 11-44. Resolution with the Rectangular Window.

Figures 11-45 through 11-48 show the same data processed with several different windows. In all cases, the worst frequency positioning is shown on the right. Some windows which are commonly used in signal processing do not perform well in this test, notably the Hann and Hamming (not shown) windows. The Kaiser-Bessel and Dolph-Chebyshev windows, on the other hand, do very well with this selection criterion, leading to their widespread use in situations requiring resolution of widely divergent signal amplitudes.

Signal limiting spoils the effect of windows which in themselves resolve well. Figure 11-49 gives the effect of limiting on the Kaiser-Bessel ($\alpha = 3.0$) window, which normally does a good job of resolving signals with widely divergent amplitudes.

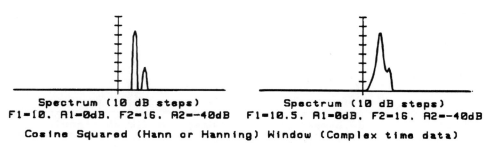

Figure 11-45. Resolution with the Hann Window.

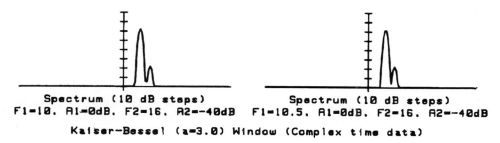

Figure 11-46. Resolution with the Kaiser-Bessel ($\alpha = 3.0$) Window.

11-9. Windows and Resolution

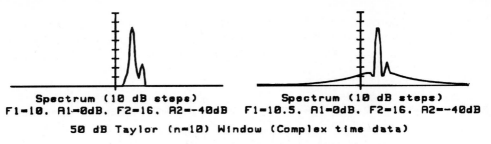

Figure 11-47. Resolution with the 50-dB Taylor Window.

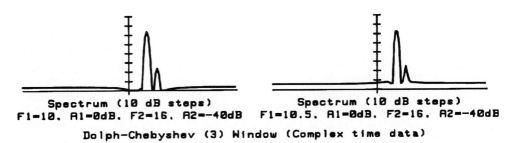

Figure 11-48. Resolution with the Dolph-Chebyshev (α = 3.0) Window.

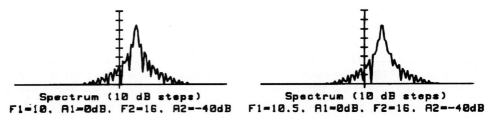

Figure 11-49. Resolution with a 50-percent Clipped Kaiser-Bessel (3.0) Window.

11-10. Recovery from Samples — Interpolation

Recovery of signals from samples, resampling for synthetic aperture focusing (Ch. 13), data presentation as smooth curves, and some other signal processes require a method of interpolating data points which do not in fact exist. For methods of interpolation and data recovery, consider Fig. 11-50.

The simplest way to recover the information represented by samples is to "connect the dots" between samples, as shown in Fig. 11-50b. The result is obviously not the original function.

The easiest way to interpolate points between samples is through linear interpolation, also shown in Fig. 11-50b. The result is identical to the "connect the dots" and is equally inaccurate. In a more rigorous sense, this linear interpolation scheme has added information which simply does not exist. The spectrum of the original samples plus the linearly interpolated points is quite different from the spectrum of the original samples alone. An examination of the figure shows that higher frequencies are necessary to produce the curve bounded by the original plus interpolated samples than by the original.

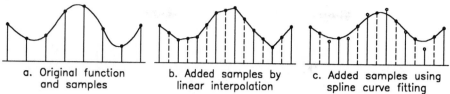

Figure 11-50. Methods of Recovery of Information from Samples.

There are many commonly used interpolation methods which, while producing smooth curves with approximately the same bandwidth as the original, do not yield mathematically exact results and distort the information in the samples. These methods are therefore not suitable for any task requiring the preservation of the information in the samples. The spline algorithms (cubic spline, for example) produce a smooth and apparently properly interpolated continua from samples, but in fact the smoothed data produced by these algorithms does not even necessarily land on the original data points. Figure 11-50c shows a spline curve fitted to the original samples (this one by the spline algorithm in AutoCAD®). Note that although the new curve is smooth and pleasing to the eye, it again does not represent the information from which the samples were taken.

A totally different approach to curve-fitting and information recovery from samples uses the discrete Fourier transform. Thirty-two samples of an original function (different from the previous figure) are shown in Fig 11-51a. These samples meet the Nyquist criterion. Figure 11-51b shows a "connect the dots" interpretation of the samples, which, as discussed above, is the simplest way to recover the information. This form of the recovered function doesn't look too much like the original and it is obvious that information has been lost.

The Fourier transform of the samples is shown in Fig. 11-51c, as though the function were in time and the transform in frequency (the function can actually be in any domain). The sample after the highest spectral line is the sample rate.

In Fig. 11-51d, 96 zeros were appended to the function's spectrum. This action, since it extended the spectrum by a factor of four, in effect quadrupled the sample frequency (the new "spectral line" immediately after the highest one in the spectrum being the new sample rate). Even though the action increased the effective sample rate, it added no new information. The new spectrum is exactly the same over the original bandwidth as the original. If the original function was Nyquist sampled, there is no information greater than at the sample rate (counting negative frequencies) and these new zero points, had they existed in the spectrum of the original unpadded signal, would have been zero. Note that the zeros must be appended to the center of the spectrum if they are not to distort the frequency relationships, since the positive frequencies of the original spectrum are in the first 16 samples and the negatives in the last 16.

Taking the inverse DFT of the padded "spectrum" puts us back into the original domain (time in this example) with four times the original number of samples, shown in Fig. 11-51e. The original samples lie on the original function, as do the interpolated points.

11-10. Recovery from Samples

Thus, from 32 samples, 128 points on the original function were recovered. If the 32 points in the original spectrum had been padded with 224 zeroes and an IDFT performed, the original function would have been shown with eight times the original number of samples within the same total time.

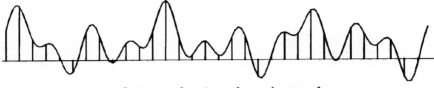

a. Original signal and samples

b. "Connect the dots" recovery from samples

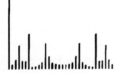

c. Spectrum of samples

d. Padded spectrum of samples

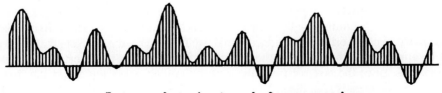

e. Interpolated signal from samples

Figure 11-51. Interpolation of Sampled Signals.

11-11. Synthesis of Complex Data from Magnitude-Only

An interesting process is sometimes possible which makes I/Q (complex) data available where only direct sampling was used. A digital spectrum analysis (DFT) of the magnitude-only data yields a two-sided spectrum. *If it is known* that only positive (or negative) frequencies are present, the negative (or positive) side of the spectrum can be set to zero. Transformation to the time domain *via* a discrete inverse Fourier transform returns complex time data where none existed before. The energy in the derived complex signal, and hence its signal-to-noise ratio, is one-half that of the original, since half of the information present was discarded. These normally discarded samples can actually be folded and summed with those retained, preserving the total signal energy. The process, shown in Fig 11-52, is useful where one side of the spectrum is known to be absent and is applicable to some types of radar processing [11].

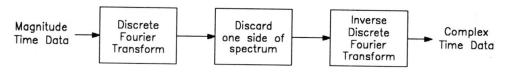

Figure 11-52. Magnitude-Only to Complex Transformation.

11-12. Digital Filter Fundamentals

Clutter is suppressed in pulsed radars with a periodic filter whose response is as shown in Fig. 6-29. It is a digital filter, operating on a sampled input. The response must be periodic because aliasing places copies of the clutter at multiples of the sample rate (the PRF). The filter used is high-pass over its Nyquist sampling interval (Fig. 11-53a). Digital filters, implemented as discrete sampled entities, have aliased responses and are thus periodic (Fig. 11-53b).

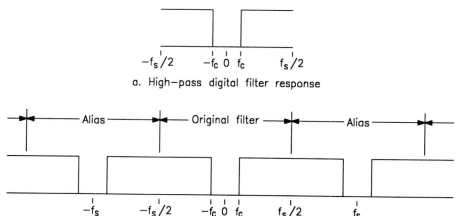

Figure 11-53. Clutter Filter Synthesis and Response.

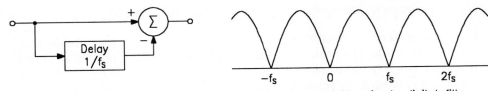

a. Simple digital filter   b. Response of the simple digital filter

Figure 11-54. A Simple Digital Filter and its Amplitude Response.

A simple filter is shown in Fig. 11-54a. It is composed of a delay line and a summing junction, with all gains equal to unity. The magnitude of its response is shown in Fig. 11-54b. This filter and other digital filters are analyzed and synthesized using the *Z-transform*, reviewed below. The filter in Fig. 11-54 is analyzed in Ex. 11-4.

Z-Transform: There are several transforms which describe the relationship between time and frequency. The most familiar, the continuous Fourier transform, deals with continuous signals in the steady state; the Fourier transform does not analyze system transients. To analyze transients and the steady state in the continuous world requires the use of the Laplace transform. The continuous Fourier transform is thus a special case of the Laplace transform.

Likewise, in the discrete realm the discrete Fourier transform treats the steady-state case and ignores transients. The total solution, including transients, is found from the Z-transform. The discrete Fourier transform is a special case of the Z-transform. The four transforms are

$$X(s) = \int_{-\infty}^{\infty} x(t)e^{-st}\, dt \qquad \text{Laplace transform} \qquad (11\text{-}63)$$

$$G(f) = \int_{-\infty}^{\infty} g(t)e^{-j2\pi ft}\, dt \qquad \text{Fourier transform} \qquad (11\text{-}64)$$

$X(s)$ = the Laplace transform of $x(t)$
$G(f)$ = the Fourier transform of $g(t)$
s = a complex variable of the form $\alpha + j2\pi f$
α = a real value
f = frequency
t = time

$$X(z) = \sum_{k=0}^{\infty} x(k)z^{-k} \qquad \text{Z-transform} \qquad (11\text{-}65)$$

$$G(n/NT) = 1/N \sum_{k=0}^{N-1} g(kT)\exp(-j2\pi nk/N) \qquad \text{DFT} \qquad (11\text{-}66)$$

$X(z)$ = the Z-transform of $x(k)$
z = a complex variable

k	= the time sample number
$G(n/NT)$	= the DFT of the time function $g(kT)$
n	= the frequency sample number
n/NT	= the frequency of sample n
N	= the total number of *time* samples
T	= the time between samples (reciprocal of sample frequency)
kT	= the time since the start of the time function
nk/N	= frequency times time

The Z-transform is used to develop the periodic high-pass filters needed for signal filtering. The discrete Fourier transform, as noted above, is the Z-transform special case which solves for the steady-state, ignoring transients. The DFT is the Z-transform with z replaced by $e^{j2\pi f/T}$.

A filter transfer function described using the Z-transform contains factors of z^{-k}, where k is an integer. It can be shown (in [2], for example) that z^{-k} is a time delay of k samples.

Example 11-4: Find the magnitude of the steady-state response of the simple digital filter of Fig. 11-54a, for a delay time of $T_0 = 1/f_S$ (f_S is the sample rate – the PRF in an MTI signal filter).

Solution: By inspection of the diagram, the time transfer function is

$$v_O/v_i = 1 - z^{-1}$$

v_O	= the output voltage
v_i	= the input voltage
z^{-1}	= a time delay of one sample.
v_O/v_i	= the transfer function in time

Substitute the transfer function into the Z-transform definition (Eq. 11-65) with $x(0) = 1$ and $x(-1) = -1$. All other $x(nT) = 0$. $x(0)$ is the present signal sample (in I and Q) and $x(-1)$ is the previous sample. Only these two samples exist in this filter.

$$X(z) = x(0) + x(-1) z^{-1}$$

$$X(z) = 1 - z^{-1}$$

To find the steady-state response (the DFT) from the Z-transform of the time function, substitute $z = e^{j2\pi fT} = e^{j2\pi f/PRF}$.

$$X(f) = 1 - e^{-j2\pi f/PRF}$$

The magnitude of any function $X(f)$ is given by

$$|X(f)| = [X(f) X^*(f)]^{1/2} \tag{11-67}$$

$X^*(f)$ = the complex conjugate of $X(f)$

Substituting the trigonometric form of the exponential (Euler's identity) gives

$$X(f) = 1 - \cos(2\pi f/PRF) + j\sin(2\pi f/PRF)$$

$$X^*(f) = 1 - \cos(2\pi f/PRF) - j\sin(2\pi f/PRF)$$

Multiplying and gathering terms (no simple feat) yields

$$|X(f)| = 2\sqrt{[1 - \cos(2\pi f/PRF)]/2}$$

Applying one of the half-angle trigonometric identities and taking the magnitude of the right side of the equation gives the magnitude of the filter's transfer function in frequency, which is plotted in Fig. 11-52b.

$$|X(f)| = 2|\sin(\pi f/PRF)|$$

Figure 11-55 shows the response of this simple filter using the decibel scale. This response is of the correct form for rejecting clutter, but the notches are much too narrow (see Secs. 4-11, 4-13, and 6-6).

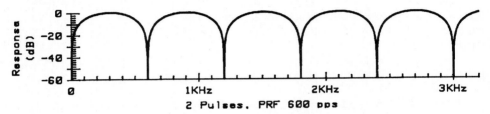

Figure 11-55. Non-Recursive (FIR) Filter Response -- One Filtering Stage.

If two sections of this filter are cascaded (Fig. 11-56), the response (Fig. 11-57) is better, but far from ideal. The response of a cascades of three stages is shown in Fig. 11-58.

The responses of higher orders of the simple non-recursive filter show better cancelling characteristics in that the notches are wider, but as stages are added the characteristic between notches becomes increasingly peaked. The response is thus diverging from the ideal of Fig. 11-53. The response of our non-recursive digital filter can be shaped by changing the gain constants in the summations (the simple filter uses + or − unity gains). The methods of synthesizing non-recursive digital filters are beyond the scope of this book, but can be found in a complete signal processing text [1].

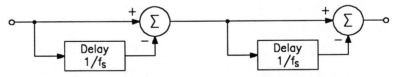

Figure 11-56. Non-Recursive (FIR) Filter — Two Filtering Stages.

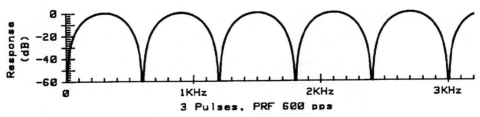

Figure 11-57. Non-Recursive (FIR) Filter Response — Two Filtering Stages.

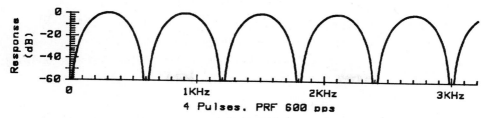

Figure 11-58. Non-Recursive (FIR) Filter Response — Three Filtering Stages.

Recursive filter synthesis: The synthesis of recursive filters, on the other hand, is straightforward and is discussed here. The advantage of using recursive filters is that (1) they are relatively easy to synthesize, and (2) less hardware and/or computation is needed to realize a suitable filter than with non-recursive (many radar signal filters are realized in digital computers). The process, known as the *bi-linear transform* and offered without proof, is outlined below (after [1]).

1. *Choose* the desired filter model transfer function in s ($s = \alpha + j\omega$) from compilations of standard filters, or design the desired response. Tables of standard analog filters, normalized to a critical frequency of one radian per second, can be found in signal processing texts and other sources. A listing of Butterworth (maximally flat) filters is given in Table 11-2.

Table 11-2. Butterworth High-Pass Filter Models.

Filter order	Model response
1	$\dfrac{s}{s+1}$
2	$\dfrac{s^2}{s^2 + 1.414s + 1}$
3	$\dfrac{s^3}{s^3 + 2s^2 + 2s + 1}$
4	$\dfrac{s^4}{s^4 + 2.6131s^3 + 3.4142s^2 + 2.6131s + 1}$

2. *Pre-warp* the critical frequency ω_C to find the filter's equivalent analog frequency ω_A.

$$\omega_A = \text{Tan}(\omega_C T / 2) \quad (11\text{-}68)$$
$$T = 1 / PRF \quad (11\text{-}69)$$

ω_A = the equivalent analog frequency
ω_C = the critical frequency of the model filter
T = the time between samples
PRF = the sample frequency

11-12. Digital Filter Fundamentals

3. *Frequency scale* the model transfer function by substituting s/ω_A for s.
$$H'(s) = H(s/\omega_A) \qquad (11\text{-}70)$$
$H'(s)$ = the scaled transfer function of the model

4. *Change* the model transfer function from the s domain to the z domain by the following substitution
$$H(z) = H'[(z-1)/(z+1)] \qquad (11\text{-}71)$$
$H(z)$ = the transfer function in z

5. *Transform* the transfer function in z to time using the following equivalence
$$H(z) = \frac{a_0 + a_1 z^{-1} + a_2 z^{-2} + \ldots + a_k z^{-k} + \ldots}{1 + b_1 z^{-1} + b_2 z^{-2} + \ldots + b_k z^{-k} + \ldots} \qquad (11\text{-}72)$$
a_k = the coefficient of the kth numerator term
b_k = the coefficient of the kth denominator term

$$y(n) = a_0 x(n) + a_1 x(n-1) + a_2 x(n-2) + \ldots + a_k x(n-k)$$
$$- b_1 y(n-1) - b_2 y(n-2) - \ldots - b_k y(n-k) \qquad (11\text{-}73)$$
$y(n)$ = the output at sample time n
$x(n)$ = the input at sample time n

6. *Implement* the filter by realizing Eq. 11-73 as a signal block diagram.

Example 11-5: Synthesize a third-order Butterworth high-pass filter to reject clutter (3 dB down) of up to ± 10 Kt for a PRF of 1500 pps and a transmit frequency of 10.0 GHz. Show a possible physical realization of the filter and plot its steady-state response to at least twice the sample rate. Estimate the parameters of a similar filter where the 10-Kt clutter is attenuated by 30 dB.

Solution: The Doppler shift of 10-Kt motion at 10.0 GHz is 343 Hz. The critical frequency is 2155 radians per second (2π times 343 Hz). From Table 11-2, the third-order Butterworth high-pass model is
$$H(s) = \frac{s^3}{s^3 + 2s^2 + 2s + 1}$$
The equivalent analog frequency, (Eqs. 11-68 and 11-69), is 0.8742. The scaled transfer function (Eq. 11-70) is
$$H'(s) = \frac{1.498\, s^3}{1.498\, s^3 + 2.618\, s^2 + 2.288\, s + 1}$$
After factoring 1.498 from numerator and denominator, the transfer function in the z domain is (Eq. 11-71)
$$H(z) = \frac{[(z-1)/(z+1)]^3}{[(z-1)/(z+1)]^3 + 1.748\,[(z-1)/(z+1)]^2 + 1.528\,(z-1)/(z+1) + 0.668}$$
Factoring $1/(z+1)^3$ from numerator and denominator, clearing parentheses, and gathering terms gives

$$H(z) = \frac{z^3 - 3z^2 + 3z - 1}{4.943\,z^3 - 1.218\,z^2 + 1.727\,z - 0.112}$$

An examination of the above equation shows a serious problem. z^{-k} is a time delay of k samples. Unfortunately, z^{+k} looks into the *future* k samples. Crystal balls (and modified DeLoreans) have limited use in filter design; thus something must be done. Factoring z^3 from the numerator and denominator puts us comfortably back in the past.

$$H(z) = \frac{1 - 3z^{-1} + 3z^{-2} - z^{-3}}{4.944 - 1.218\,z^{-1} + 1.727\,z^{-2} - 0.112z^{-3}}$$

Finally, 4.944 is factored from the numerator and denominator to put this latest equation in the form necessary to transform to the time domain (Eq. 11-72).

$$H(z) = \frac{0.202 - 0.607\,z^{-1} + 0.607\,z^{-2} - 0.202z^{-3}}{1 - 0.246\,z^{-1} + 0.349\,z^{-2} - 0.0227z^{-3}}$$

The time domain function of this example is given in the following difference equation (Eq. 11-73) [1]. Figure 11-59 shows a possible realization of the filter.

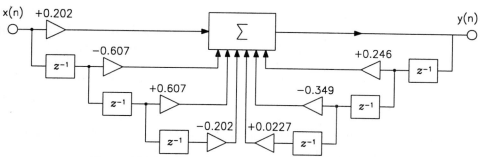

Figure 11-59. A Possible Realization of the Third-Order Butterworth High-Pass Filter of Example 11-5.

$$y(n) = 0.202\,x(n) - 0.607\,x(n-1) + 0.607\,x(n-2) - 0.202\,x(n-3)$$
$$+ 0.246\,y(n-1) - 0.349\,y(n-2) + 0.0227\,y(n-3)$$

The steady-state frequency response of the filter is found by solving the Z-transform of its transfer function for $z = e^{j2\pi f/PRF}$.

$$H(f) = \frac{0.202 - 0.607\,e^{-j2\pi f/PRF} + 0.607\,e^{-j4\pi f/PRF} - 0.202\,e^{-j6\pi f/PRF}}{1 - 0.246\,e^{-j2\pi f/PRF} + 0.349\,e^{-j4\pi f/PRF} - 0.0227\,e^{-j6\pi f/PRF}}$$

After considerable manipulation of the above equation, the steady-state frequency response can be calculated and plotted (Fig. 11-60).

The author knows of no exact method for finding the filter parameters for 30 dB of attenuation at a particular frequency. However, once the design of the basic filter is automated, it is easy to try different 3-dB cutoff frequencies until the response is down 30 dB at the desired frequency of 343 Hz. An examination of Fig. 11-60 shows that a good place to start might be with a critical frequency of about 540 Hz.

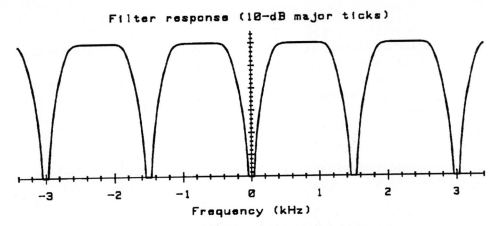

Figure 11-60. Response of the Filter of Example 11-5.

11-13. Further Information

Many good signal processing texts exist. Useful examples include [1] through [4]. Ahmed and Natarajan's book is a good general signal processing text, and is particularly useful in the filter design examples. Oppenheim and Willsky is good and has many appropriate examples. Oppenheim and Schafer is widely used and is considered a classic in this field.

[1]   N. Ahmed and T. Natarajan, *Discrete-Time Signals and Systems*, Englewood Cliffs NJ: Reston Publishing, 1983.

[2]   A.V. Oppenheim and R.W. Schafer, *Digital Signal Processing*, Englewood Cliffs NJ: Prentice Hall, 1975.

[3]   A.V. Oppenheim and A.S. Willsky, *Signals and Systems*, Englewood Cliffs NJ: Prentice Hall, 1983.

[4]   L.R. Rabiner and B. Gold, *Theory and Application of Digital Signal Processing*, Englewood Cliffs NJ: Prentice Hall, 1975.

In the field of Fourier methods and windowing, two books by E.O. Brigham and two papers by F.J. Harris stand out. Brigham's books give highly readable developments of Fourier methods. The latter book [6] includes most of the material in the first [5], plus some interesting material on DFT applications.

Harris's first paper [7] develops the discrete Fourier transform and the FFT algorithm in an extremely understandable way. The paper on windows [8] defines a large variety of popular window functions and has a good section on criteria by which signal processing windows can be evaluated.

[5]   E.O. Brigham, *The Fast Fourier Transform*, Englewood Cliffs NJ: Prentice Hall, 1974.

[6]   E.O. Brigham, *The Fast Fourier Transform and its Applications*, Englewood Cliffs NJ: Prentice Hall, 1988.

[7] F.J. Harris, The *Trigonometric Transform*, Technical Publication DSP-005, San Diego: Spectral Dynamics Inc. (a division of Scientific Atlanta), 1977.

[8] F.J. Harris, "On the Use of Windows for Harmonic Analysis with the Discrete Fourier Transform," *Proceedings of the IEEE*, vol. 66, no. 1, January, 1978, pp. 51-83.

To do any signal processing analysis and synthesis requires a good reference to mathematical functions such as trigonometric identities, series expansions of Bessel functions, integrals, and the like. One of the least expensive and best of these references is [9].

[9] M.R. Spiegel, *Mathematical Handbook*, Schaum's Outline Series, New York: McGraw-Hill, 1968.

Signal integration is presented in exquisite detail in reference [10]. Non-coherent integration is stressed, but coherent integration is covered.

[10] D.P. Meyer and H.A. Mayer, *Radar Target Detection: Handbook of Theory and Practice*, New York: Academic Press, 1973.

Signal processing for high resolution radar is covered very nicely by D.L. Mensa [11]. In the book, Mensa explains many signal processing techniques which are of use to radar specialists and to those in several other fields.

[11] D.L. Mensa, *High Resolution Radar Cross-Section Imaging*, 2nd ed., Norwood MA: Artech House, 1991

[12] D.L. Mensa and K. Vaccaro, "Two-Dimensional RCS Image Focusing," *Proceedings of the Antenna Measurement Techniques Association*, 1987.

[13] J.W. Cooley and J.W. Tukey, "An Algorithm for Machine Calculation of Complex Fourier Series," *Math. Computation*, vol. 19, April 1965, pp. 297-301.

The Taylor window is covered in the following reference.

[14] E.C. Farnett and G.H. Stevens, Ch. 10 in M.I. Skolnik (ed.), *Radar Handbook*, 2nd ed., New York: McGraw-Hill, 1990.

12

RADAR SIGNAL PROCESSING II:
Moving Target Indicators and Doppler Processing

12-1. Doppler and Moving Target Indicator (MTI) Fundamentals

The principles of the Doppler shift were introduced in Ch. 1 and analyzed in Ch. 6. Following is a review of the fundamental relationships. The exact Doppler shift is derived in Ch. 6.

$$f_d = f_T \left[\frac{1 + v_R/c}{1 - v_R/c} - 1 \right] \qquad (12\text{-}1)$$

f_T = the transmit frequency
v_R = the target-to-radar radial velocity (closing is positive)
c = the velocity of propagation

Unless the target-to-radar radial velocity approaches relativistic proportions, the approximate Doppler shift used throughout this text (and most others) produces negligible errors. It is, by assuming $v_R \ll c$,

$$f_d \approx 2 f_T \frac{v_R}{c} \qquad (12\text{-}2)$$

The primary goal of Doppler signal processing, particularly MTI processing, is to resolve moving targets and stationary (or moving) clutter occurring in the same time bin. Figure 12-1 shows the effect of moving target indication.

In CW systems, the Doppler shift can be recovered directly by comparing the echo frequency to that of the illumination. In pulsed radars, the Doppler shift is seldom sufficient to produce a measurable frequency difference between the transmitted and received signals for a single pulse. In these systems, the Doppler shift is found by observing the change in signal phase from consecutive echoes from the target (Fig. 12-2). If the Doppler shift is Nyquist sampled, the samples contain all the information present in the signal. If not, information about target motion is lost. In any case, the Doppler shift is recovered by multiplying the received signal by a copy of the transmitted signal, delayed in time to the range bin of the echo.

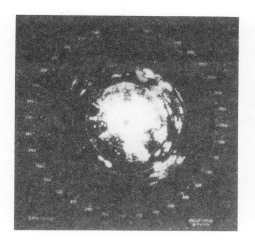 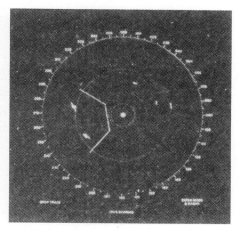

Figure 12-1. Normal (left) and MTI Video of the Same Terrain.
(Photographs courtesy of Hughes Aircraft, Ground Systems Group)

Four conditions in Doppler sampling occur; sampling for a coherent radar is illustrated in Fig. 12-2. Figure 12-2a shows a target which is stationary with respect to the radar. The phases shown are representative; the phases do not have to have the value shown but in this case are equal to one another.

A Nyquist-sampled moving target is shown in Fig. 12-2b. Nyquist sampling requires that the phase between samples change less than 180°.

A target moving at a blind velocity is shown in Fig. 12-2c. Blind velocities cause the target's phase to change in integer multiples of 360° between hits.

The target in Fig. 12-2d has too much Doppler shift to be Nyquist sampled by the PRF used. In this case, the target phase changes $-315°$ per pulse, which after sampling appears as $+45°$ per pulse. The fact that the target is moving is apparent, but its velocity cannot ordinarily be recovered because of undersampling. This is the normal case in MTI and MTD.

The apparent Doppler shift is the frequency resulting from aliasing of undersampled signals, and is found from whichever of the following two equations produces the smaller absolute value. See Sec. 6-4.

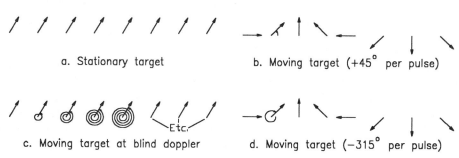

Figure 12-2. Doppler Sampling Conditions.

12-1. Doppler and MTI Fundamentals

$$f_A = f_d \text{ MOD } PRF \qquad (12\text{-}3\text{a})$$

or

$$f_A = f_d \text{ MOD } PRF - PRF \qquad (12\text{-}3\text{b})$$

f_A = the aliased, or apparent, Doppler shift

f_d = the actual Doppler shift

It was shown in Ch. 6 that Doppler recovery with pulse-to-pulse frequency agility usually, at the present state of the art, requires a more accurate knowledge of target range than is available.

MTI improvement factor (MTI-I, or I): The improvement factor is the process gain to clutter. It is defined as

$$MTI\text{-}I = I \equiv \frac{S/C_o}{S/C_i} \qquad (12\text{-}4)$$

$MTI\text{-}I$ = the MTI improvement factor, equivalent to I, the improvement factor

S/C_o = the signal-to-clutter ratio out of the MTI process

S/C_i = the signal-to-clutter ratio into the MTI process

Another way of measuring Doppler clutter rejection is with the sub-clutter visibility, defined as the clutter-to-signal ratio into the MTI process for a "visible" signal at the output.

$$SCV = C/S_i \,|_{\text{"Visible" output signal}} \qquad (12\text{-}5)$$

SCV = the sub-clutter visibility

C/S_i = the clutter-to-signal ratio at the input to the MTI process

12-2. MTI Principles and Methods

Moving target indication is achieved by comparing the received echoes to the transmitted wave for several samples. If the radar is stationary, those signals which change from pulse to pulse are assumed to be from moving targets. Moving radar MTI is discussed later in this chapter. Figure 12-3 is a simplified block diagram, with intermediate frequencies, of a basic MTI system. The received signal is I/Q demodulated with the transmitted wave and filtered to see if the signal changes from pulse to pulse compared to the transmitter.

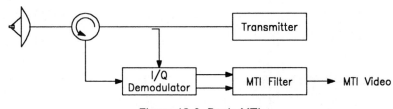

Figure 12-3. Basic MTI.

a. Stationary target

b. Clear moving target

c. Composite target

Figure 12-4. Target Types.

The outputs of the I/Q demodulator, called bipolar video, represent the magnitude and phase of the signal compared to the transmit wave (Sec. 6-2). For a given range bin, the signal can be noise, clutter, moving target, ECM, EMI, spillover, or any combination of them. Filtering is done on the composite signal, three types of which are represented in Fig. 12-4.

One channel of bipolar video (I or Q) is shown in Fig. 12-5. The figure shows many range bins, and several range sweeps from several transmit pulses are overlaid. The stationary targets do not change from pulse to pulse and thus show the same I/Q output for each range sweep. The clear moving target varies through 360° of phase and thus has outputs of both polarities. The composite of clutter and a moving target does not vary through 360° of phase and shows its own distinctive pattern. The analog signals out of the I/Q demodulator are rounded, and a clear target's bipolar video is sometimes called an "onion" because of its appearance. The squared signal shown is after the sample and hold circuit (the boxcar generator), triggered at the range bin rate, in preparation for A/D conversion.

MTI processing is done on multiple samples from the same range bin. Information in range bin 7 from pulse 3 is compared to that from range bin 7 from pulses 2 and 1, and so forth. Information from bin 7 is not compared to that from any other range bin. The MTI look (dwell) times are usually short enough that movement of a target from one range bin to another (range bin walking) over the time of the look is unlikely. To cover the instrumented range of the radar, separate processes are required for each range bin. Figure 12-6 illustrates.

MTI filter types: Two fundamental methods are used to extract moving targets from clutter: digital notch filters (often referred to as cancellers) and filter banks, usually implemented with discrete Fourier transforms. Figure 12-7 shows the processes and the difference between them.

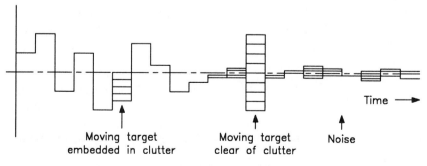

Figure 12-5. Bipolar Video.

12-2. MTI Principles and Methods

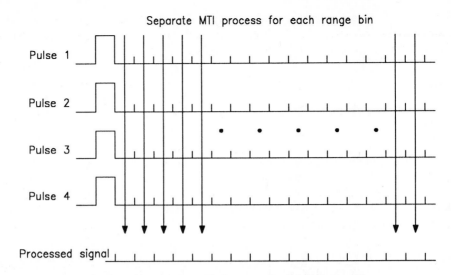

Figure 12-6. Separate MTI Process for Each Range Bin.

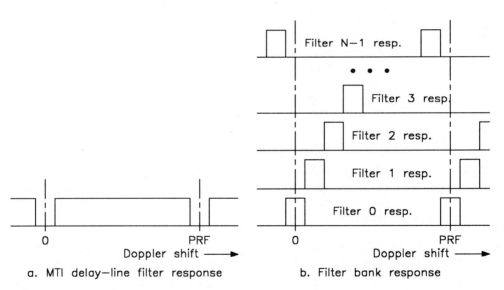

a. MTI delay-line filter response   b. Filter bank response

Figure 12-7. MTI Filter Types.

Cancellers reject frequency bands centered at zero frequency and at multiples of the sample rate. The width of the notch, discussed in Sec. 6-4, must be sufficient to remove moving components of clutter (Sec. 4-11) without occupying so much of the spectrum that a large percentage of the moving targets are also filtered out by them. Figure 12-8 shows the trade-off.

Cancellers exploit pulse-to-pulse signal changes in moving targets. These changes can occur in amplitude and phase (Fig. 12-9). Moving targets which are clear of clutter vary in phase only, unless they scintillate rapidly, as do Swerling case 2 and 4 targets. Signals which are composites of stationary target echoes and those from moving targets vary in both phase and amplitude. Canceller filter design is introduced in Sec. 12-12.

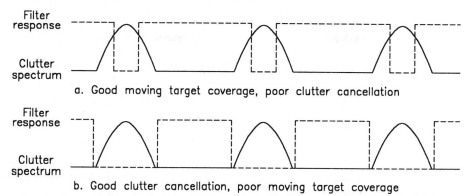

Figure 12-8. Canceller Trade-off: Clutter Cancellation versus Moving Target Coverage.

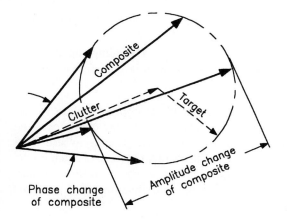

Figure 12-9. Clutter and Moving Target Composite.

MTI types: Three distinct types of canceller-type MTIs exist, based on what signal changes they detect, shown in Fig. 12-10. Their characteristics are listed below.

— *Phase MTI* exploits phase differences between echoes. Phase MTIs have limiters in their IF amplifiers to remove target amplitude. The limiters discard part of the signal and thus reduce the signal-to-noise ratio. In addition, the extreme non-linearity of the limiters severely restricts the improvement factor in the presence of a scanning antenna. Please see Sec. 12-6 for MTI-I limitations. Phase MTI is simple to implement and is found most often in older radars.

— *Amplitude-only MTIs* exploit the amplitude differences between samples of the clutter/moving target composite (Fig. 12-10). Their most common use is as an add-on to a non-coherent radar. This ability to do MTI without any vestige of coherence is its primary advantage. It has two significant disadvantages. First, improvement factors are generally

12-2. MTI Principles and Methods

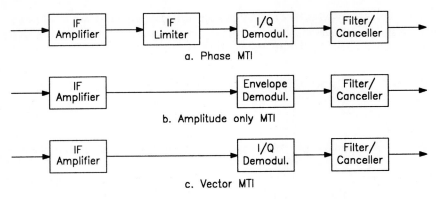

Figure 12-10. Canceller Types.

poor. Second, moving targets which do not scintillate over a cancellation dwell are often rejected along with clutter.

— *Vector MTIs* process both the phase and amplitude changes of the signal, and are capable of higher improvement factors than either of the other types. Their primary disadvantage is the complexity of implementation, but modern technology makes that a minor problem. Most MTIs currently produced are of this type.

Filter bank MTI, also known as moving target detection (MTD), is implemented with discrete Fourier transforms (Sec. 11-6). Their improvement factor is a function of the size of the transform and the window used (Sec. 11-8).

MTI as a function of range: MTI processing is usually done only at short ranges (Fig. 12-11). One reason is that the earth's curvature places long-range clutter over the horizon. This is particularly true of radars which are used over oceans, where local mountains do not interfere. The distance from the radar to the radar horizon depends on the radar's height above the clutter surface and the height of the targets above the same surface. The distance to the radar horizon is given in the following approximation, which accounts for refraction by assigning the earth's radius of 4/3 of its actual value (Sec. 6-7).

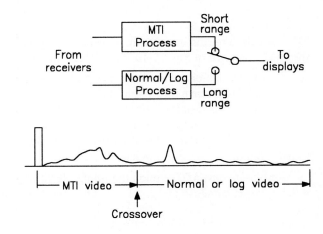

Figure 12-11. MTI and Normal/Log Video Use as a Function of Range.

$$R_{RH} \approx 4.15 \sqrt{h_R} \qquad (R_{RH} \text{ in kilometers}, h_R \text{ in meters}) \qquad (12\text{-}6a)$$

$$R_{RH} \approx 1.2 \sqrt{h_R} \qquad (R_{RH} \text{ in nautical miles}, h_R \text{ in feet}) \qquad (12\text{-}6b)$$

R_{RH} = the distance to the radar horizon

h_R = the height of the radar above the surface

Another reason to use MTI only at short ranges is that limiting and amplitude-only MTI filters discard part of the signal and degrade the signal-to-noise ratio. At ranges where noise is the limiting interference, these types of MTI are detrimental to target detection. These systems usually allow selection of the range at which the crossover between MTI and normal video occurs.

Some systems also allow selection based on azimuth sector, so that local clutter conditions are accommodated (Fig. 12-12).

MTIs implemented with filter banks concentrate coherent targets into one or a few Doppler bins and spread noise equally among all the bins. These MTI types enhance signal-to-noise ratios and can be used at all ranges.

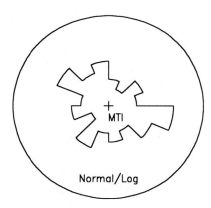

Figure 12-12. MTI and Normal or Log Video as a Function of Range and Azimuth.

12-3. Blind Doppler Shifts and PRF Stagger

Doppler undersampling, the rule in low PRF MTI and MTD, can cause moving targets to appear stationary (Fig. 12-2c). Blind Dopplers occur when the frequency shift is an integer multiple of the sample rate (PRF). A blind velocity is a target radial velocity which causes a blind Doppler shift.

$$f_B = n\, PRF \qquad (12\text{-}7)$$

$$v_{RB} = \frac{c\, n\, PRF}{2 f_T} \qquad (12\text{-}8)$$

f_B = a blind Doppler shift

n = an integer ≥ 0

v_{RB} = a blind radial velocity

f_T = the transmit frequency

PRF = the sample rate

Blind Doppler shifts occur because the Doppler phase is sampled at the same point in each cycle. The problem is alleviated by sampling in different positions, using PRF stagger. PRF stagger can be pulse-to-pulse, look-to-look (dwell-to-dwell), or scan-to-scan. Each has its advantages and disadvantages.

Figure 12-13 shows pulse-to-pulse PRF stagger in a stationary radar, viewing stationary clutter and a moving target (Sec. 12-5 treats moving clutter). In Fig. 12-13a, the normal position of the kth range bin at a constant PRF is shown. Figure 12-13b shows the phase of the clutter and the sampling points for range bin k. Figure 12-13c shows a moving target's Doppler phase and the bin k sampling points. Since the phase is the same for each sample, the target appears to have zero Doppler shift.

In Fig. 12-13d, samples shifted in time (staggered), are shown. The stationary clutter produces the same phase with stagger as without (Fig. 12-13e). The moving target, however, is sampled at different phases and no longer appears stationary; the radar is no longer blind to it.

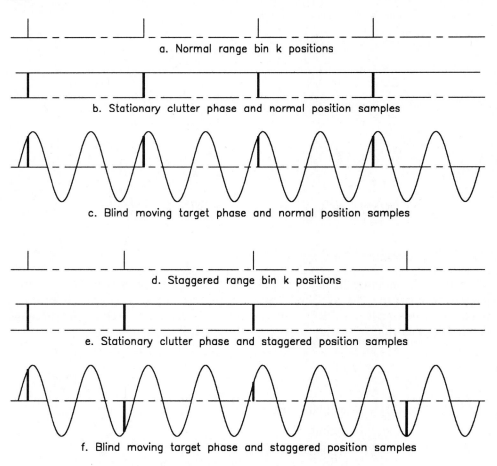

Figure 12-13. Pulse-to-Pulse PRF Stagger.

Figure 12-14 shows the same target and clutter with look-to-look and scan-to-scan PRF stagger. Pulse-to-pulse stagger has the advantage of allowing the fastest antenna scan with good moving target recovery, since only one process is needed for each antenna beamwidth. It is widely used in canceller MTIs, but is poorly suited to filter bank MTI, as discrete Fourier transforms require constant sample spacings. It is also vulnerable to clutter in the second and greater range zones in coherent-on-receive systems, since the COHO is re-phased with each transmitted pulse.

Non-recursive filters process pulse-to-pulse stagger easily, since the delays can be switched to "track" varying pulse spacings as several pulses pass through the filter. To use recursive filters, the pulses must be de-staggered before being applied to the filter, since each delay line simultaneously contains information from several consecutive pulses. De-stagger is difficult to implement in analog cancellers, but is easy in digital circuits, since the delay lines are generally just random access memories with highly flexible write and read times.

Look-to-look stagger is used in filter bank MTD. Its primary disadvantage is that if blind Doppler shifts are to be eliminated within each scan, there must be several pulse groups at

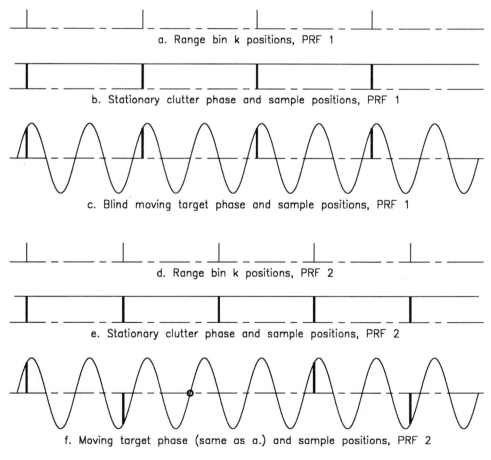

Figure 12-14. Look-to-Look and Scan-to-Scan PRF Stagger.

12-3. Blind Doppler Shifts and PRF Stagger

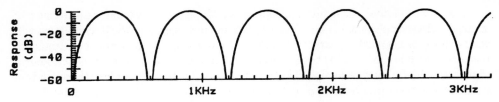
Figure 12-15. Three-Delay Non-Recursive Filter Response — 600 pps.

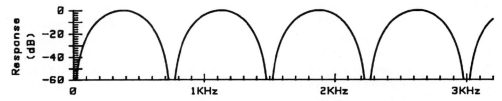
Figure 12-16. Three-Delay Non-Recursive Filter Response — 750 pps.

different PRFs within each antenna beam. This slows the antenna scan rate. It is also somewhat vulnerable to multiple-time-around (range zone > 1) clutter.

Scan-to-scan stagger loses moving targets for whole scans if the PRF for that scan causes a blind Doppler. Its primary advantage is that with a coherent radar, multiple-time-around clutter is suppressed.

PRF selections: The goal of PRF designs is to choose PRFs so the radar is never totally blind to any moving target within the its instrumented parameters. Consider the filter response of Fig. 12-15. It is a triple delay line non-recursive filter, described in Sec. 11-12, at a PRF of 600 pulses per second. The notch widths are small, but finite. At this PRF, the radar is blind to Doppler shifts of 600 Hz, 1200 Hz, 1800 Hz, 2400 Hz, 3000 Hz and so forth.

Figure 12-16 shows the same filter modified (delays shortened) to accommodate a PRF of 750 pps. This filter is blind to 750, 1500, 2250, 3000 and so forth.

The use of both PRFs and filters over a long enough time to allow both PRFs to totally charge its filter produces a result which is the sum of those of the individual PRF/filter combinations (Fig. 12-17). This combination of two PRFs and filters produces a first blind Doppler at 3000 Hz (a second at 6000 Hz and so forth), which is the least common multiple of 600 and 750 Hz.

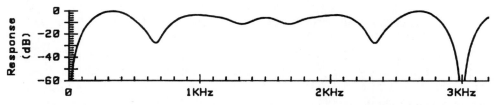
Figure 12-17. Same Filters as in Figs. 12-15 and 12-16 — 600- and 750-pps Stagger.

Note that, although the first blind Doppler is 3000 Hz, the filter does not respond well to target velocities producing Doppler shifts of about 650 and 2350 Hz. The total response can be smoothed by adding a third PRF of 1000 Hz. Its individual response is shown in Fig. 12-18 and the response of the three-PRF stagger (600, 750, and 1000 pps) is shown in Fig. 12-19. This combination has the same first blind Doppler, 3000 Hz, but the overall response shows less moving target attenuation between 0 and 3000 Hz. The response could be smoothed further by adding more PRFs.

The process of multiple PRFs described above applies to pulse-to-pulse and look-to-look stagger, but not scan-to-scan.

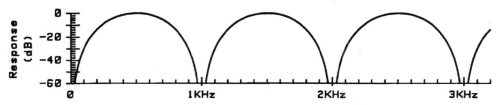

Figure 12-18. Three-Delay Non-Recursive Filter Response — 1000 pps.

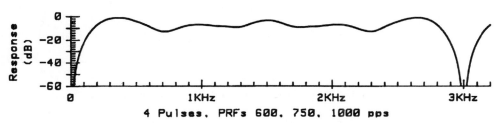

Figure 12-19. Same Filters as in Figs. 12-15, 12-16, and 12-18
— 600-, 750-, and 1000-pps Stagger.

Equivalent PRF: The Doppler equivalent PRF of a stagger, for purposes of evaluating blind Doppler shifts, is simply the least common multiple (LCM) of the individual PRFs. Finding the least common multiple is a two-step process.

1. Factor the individual PRFs into their primes. For example, 600 is 2·2·2·3·5·5. 750 is 2·3·5·5·5. 1000 is 2·2·2·5·5·5.
2. Form a product of the prime factors, using each factor as many times as the largest number of times it appears in the individual PRFs. For instance, three twos must be used (from 600 and 1000), one three (from 600 and 750) and three fives (from 750 and 1000). The product is the LCM, 2·2·2·3·5·5·5, or 3000.

It is apparent that the fewer prime factors which are shared by the individual PRFs and the larger the prime factors are, the larger will be their least common multiple, the Doppler equivalent PRF, and first blind Doppler. Figure 12-20 shows a one-delay non-recursive canceller using PRFs of 2145 (3·5·11·13) and 3003 pps (3·7·11·13), which have an LCM of 15015 (3·5·7·11·13). The filter response *appears* highly suitable to MTI.

12-3. Blind Doppler Shifts and PRF Stagger

Now consider the same PRFs using three-delay non-recursive filters (Fig. 12-21). Whereas the one-delay filters had very narrow notches, the three-delay filter notches are somewhat wider, and even though there is no common null near 7500 Hz, the total response at this frequency is very low.

This problem is even worse when the individual PRFs are closer together and have several common prime factors, as in Fig. 12-22. The PRFs shown here, 1309 (7·11·17) and 1463 (7·11·19) have an LCM of 24871, but virtual nulls occur at about 11800 and 13100 Hz.

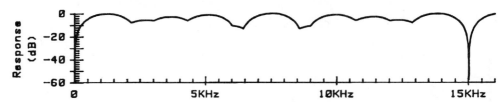

Figure 12-20. One-Delay Non-Recursive Filter — 2145- and 3003-pps Stagger.

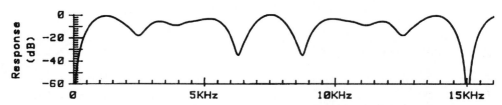

Figure 12-21. Three-Delay Non-Recursive Filter — 2145- and 3003-pps Stagger.

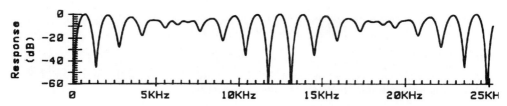

Figure 12-22. Three-Delay Non-Recursive Filter — 1309- and 1463-pps Stagger.

This phenomenon is the result not of notches coinciding, but overlapping (Fig. 12-23). The wider the notches, which is good for rejecting clutter, the worse the overlap. This notch width problem makes the design of PRF staggers less analytical and more empirical. The LCM is a starting point, but computer simulations are usually needed to evaluate PRF designs.

Finally, Fig. 12-24 gives the response of a three-delay non-recursive filter with a stagger of 1200, 1173, 1120, 1050, 950, and 713 pps. This is the PRF stagger, but not the filter, for the ASR-7 radar, built in the 1970s by Texas Instruments and operated at approximately 3 GHz. The LCM is 203,632,800 Hz. The first blind velocity, from the LCM and Eq. 12-1 solved for v_R, is about 9,847,000 m/s (19,143,000 Kt), or 3.28% of the speed of light. The actual first blind velocity, from notches somewhat wider than those shown, is 2200 Kt, and occurs at this speed because of clutter notch overlaps.

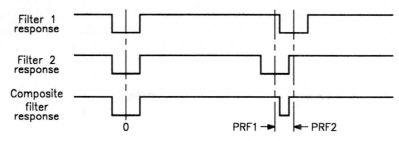

Figure 12-23. Clutter Notch Overlap.

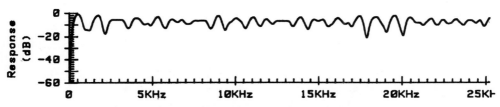

Figure 12-24. ASR-7 MTI Approximate Response.

12-4. De-Staggering and Processing

Non-recursive filters can process pulse-to-pulse stagger at low PRF if the delays are switchable to all the PRIs used in the stagger and if the delays are properly programmed. Recursive filters, which have more than one pulse in each delay simultaneously, cannot filter staggered samples. Before they can be used with pulse-to-pulse stagger, the signals must be de-staggered. This process, illustrated in Fig. 12-25, is done by inserting appropriate switched delays into the echo signals to make them appear to be at the rate. Small switchable delays must be used in analog processing. In digital processing, it is simply a matter of correctly addressing the data stored in digital memories. De-stagger works only with low PRF; medium and high PRF do not use pulse-to-pulse stagger. De-stagger also generally is not used in MTD, since pulse-to-pulse stagger is not often used.

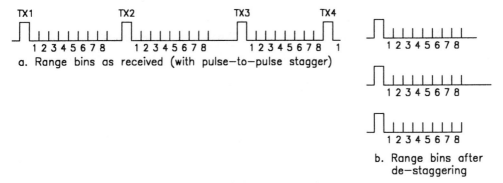

Figure 12-25. De-Stagger Principle.

The signals in a staggered MTI are shown in Fig. 12-26. In it, the signal phasors from stationary and moving targets are followed through the stagger and de-stagger processes. In de-stagger, delays are introduced into each range sweep to align it with the sweep producing the most delay from the normal (un-staggered) pulse positions. This is necessary to prevent the need to anticipate.

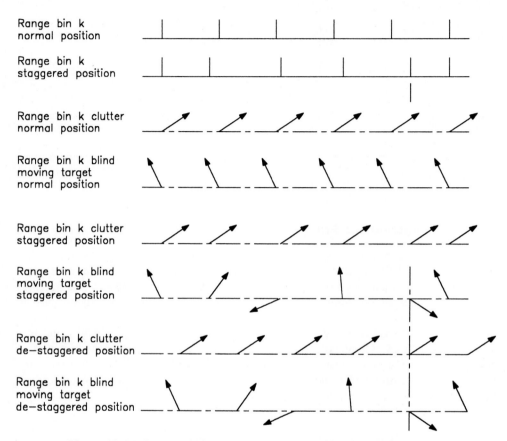

Figure 12-26. De-Stagger Principles — Signals from One Range Bin.

12-5. MTI and MTD with Moving Radars and Moving Clutter

The previous discussions assumed, unless otherwise stated, that the radar was stationary and the clutter had no net velocity. Both of course can move. Moving radars are those on aircraft, ships, and ground vehicles. Sea and weather clutter move, as does chaff. Several different conditions exist for the motion of the radar and the clutter, illustrated in Fig. 12-27 and discussed below.

Before starting the discussion, it needs to be stated that in canceller MTI, it is very difficult to implement the clutter notch at any frequency other than zero *with respect to the radar platform*, and some of the conditions of the figure cannot be implemented directly. It will be shown shortly, however, that it is easy to shift the clutter Doppler so that it is always centered at zero. In other words, even if the radar is in an aircraft, the ground and sea clutter over which the aircraft flies can be offset to zero Doppler shift. See the discussions of Doppler offset and clutter locking, later in this section.

In Fig. 12-27a, the radar and clutter are stationary. In this case, the clutter notch is centered at zero Doppler shift and its width is set by the spectral width of the clutter and the radar's PRF. Note that with low PRF and high bandwidth clutter it may not be possible to have notches wide enough to attenuate the clutter sufficiently to prevent it from interfering with targets.

In the next condition, the radar is stationary and the clutter moving. This condition is found in land-based radars observing sea clutter. In Fig. 12-27b, there is no offset applied to either the clutter or notch center and the clutter may appear as a moving target. Two methods can be used to compensate for the clutter motion. First, the clutter or the clutter notch can be shifted in frequency (Fig. 12-27c) so that the notch lies on the clutter spectrum (the clutter is almost always the one shifted). Second (Fig. 12-27d), the notch can be centered at zero, but made wide enough to filter out the clutter. This may not be practical for low PRF and high frequency systems.

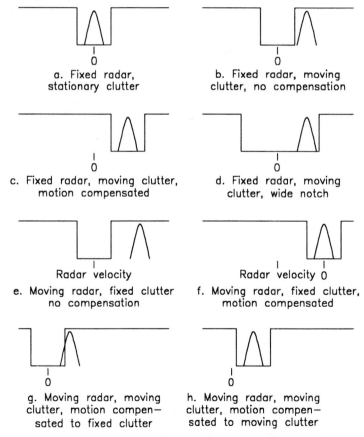

Figure 12-27. Moving Radar and Clutter Conditions.

Example 12-1: Can an S-Band radar with transmit frequency of 3.0 GHz and PRF of 300 pps use the method of Fig. 12-27d to reject sea clutter with 10 knots net radial motion closing and a spectral width of ± 5 Kt?

Solution: The worst case is the component of the 5-Kt spectral width moving toward the radar. This, added to the 10-Kt net motion requires a notch at least 15 Kt from zero Doppler to reject the clutter. At 3.0 GHz, 15 Kt produces 154 Hz of Doppler. The notch centered at zero extends from -154 to $+154$ Hz and its first alias, centered at 300 Hz, extends from $+146$ to $+454$ Hz. The notches overlap, there is no frequency at which the filter responds, and no moving target can pass through the filter. The answer is no.

Next, the radar can be in motion and the clutter stationary. This is the common situation with airborne and shipboard radars viewing land clutter. In Fig. 12-27e, the clutter notch is centered at the radar's velocity (zero absolute Doppler shift) and the clutter appears as a moving target. In Fig. 12-27f, the notch is shifted to the clutter frequency (actually, it is more likely that the clutter is shifted to the notch frequency) and the clutter is suppressed.

Finally, both the radar and clutter can move. Now it is not sufficient to match the notch to the radar's motion (Fig. 12-27g). Knowledge of the clutter's velocity is necessary to shift the clutter to the center of the notch. Clutter-locking circuits and computations are commonly used at low PRF in this situation. An alternative is to center the clutter in the notch as though the clutter were stationary and widen the notch to accommodate the moving clutter. This is the commonly used technique in airborne radars operating at medium and high PRF, but because of the notch width required is not practical at low PRF.

Clutter Doppler offset: As stated earlier, it is impractical in most cases to move the center frequency of the clutter notch, although its width can be changed easily by changing the filter constants (Sec. 11-12). The commonly used technique is to offset the receiver's local oscillator by the amount of the calculated fixed-clutter Doppler shift, pulling the clutter to zero absolute Doppler. The frequency relationships are

$$f_{IF} = f_T + f_d - f_{LO} \quad \text{(with no LO offset)} \tag{12-9}$$

But

$$f_T = f_C + f_{LO} \tag{12-10}$$

Hence

$$f_{IF} = f_C + f_d \quad \text{(with no LO offset)} \tag{12-11}$$

$$f_{IF(O)} = f_T + f_d - f_{LO(O)} \quad \text{(with LO offset)} \tag{12-12}$$

But

$$f_{LO(O)} = f_{LO} - f_d \tag{12-13}$$

Hence

$$f_{IF(O)} = f_C \quad \text{(with LO offset)} \tag{12-14}$$

f_{IF} = the actual IF signal frequency with Doppler shift and without LO offset

f_T = the transmit frequency

f_d = the target Doppler shift

f_{LO} = the local oscillator frequency without LO offset

f_C = the COHO frequency, also the nominal IF frequency

$f_{IF(O)}$ = the actual IF signal frequency with LO offset *if the local oscillator is at a lower frequency than the signal*

$f_{LO(O)}$ = the local oscillator frequency offset to match the stationary clutter Doppler shift

These equations assume that the local oscillator is at a frequency lower than the signal. If the LO is above the signal, the normal intermediate frequency is $f_C - f_d$ (Eq. 12-11), and the offset LO frequency is $f_{LO} + f_d$ (from Eq. 12-13).

Clutter Doppler offset for stationary clutter: The amount of local oscillator offset needed to center stationary clutter in the notch depends on the velocity of the radar over the clutter surface (the radar-carrying vehicle's track), the direction in which the radar's antenna beam is pointing with respect to the vehicle's track and the local horizontal, and the transmit frequency. Knowing the vehicle's airspeed and heading (airborne radars), is not sufficient. Wind drift may make the over-ground (or -sea) speed and direction different from these parameters, as shown in Fig. 12-28. If the vehicle contains an inertial navigation system (INS), the INS provides track information. If not, one of the clutter-locking techniques, introduced later in this section, must be used. Once the track is known, antenna position encoder readouts and the present transmit frequency are used to calculate the clutter Doppler shift. This offset is applied as an offset to the receiver local oscillator, using Eq. 12-13 or 12-13 modified, depending on the LO placement.

$$v_{CL} = v_{VG} \cos \theta_{AzG} \cos \theta_{DepH} \qquad (12\text{-}15)$$

v_{CL} = the clutter velocity

v_{VG} = the vehicle's ground velocity

θ_{AzG} = the antenna's azimuth angle from the vehicle's track

θ_{DepH} = the depression angle from local horizontal

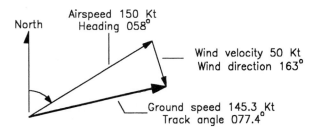

Figure 12-28. Flight-Path Geometry.

The INS method works fine for correcting the velocity of the radar over a stationary reference and is suitable for rejecting land clutter. Sea clutter, however, moves with respect to the reference coordinate system, and there is no mechanism for removing the average velocity of the clutter. This is often not a serious problem in airborne radars because in Doppler modes they operate at either medium or high PRF and the clutter notches are wide enough to accommodate moving clutter.

12-5. MTI and MTD With Moving Radars and Moving Clutter

For systems whose clutter notches are narrow (including almost all low PRF radars and modes), the INS estimation of clutter Doppler is unsatisfactory. Consider an S-Band radar with PRF of 600 pps and a clutter notch 150 Hz wide, with clutter whose spectrum is ± 3 Kt wide. The Doppler spectrum of the clutter is about 62 Hz wide (± 31 Hz). The net clutter Doppler must be less than 44 Hz (75 − 31) if the clutter is to remain in the notch. This is a clutter velocity of only 4.3 knots (steady-state sea state 1 − Table 4-5).

For the low PRF case viewing sea clutter, a clutter locking system is often used (Fig. 12-29). In it, a circuit or algorithm senses prominent clutter, usually by time gating of signals from close to the radar. The Doppler shift of this prominent clutter is determined and the local oscillator offset by that amount. This method compensates for clutter motion, radar-carrying platform motion, and antenna angle in one step. It is often used in shipboard radars.

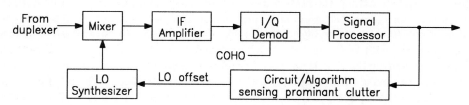

Figure 12-29. Clutter Locking.

12-6. Limitations on the Improvement Factor

Several factors work to limit the signal-to-clutter ratio improvement. With a good canceller or filter bank design, cancellation can be essentially perfect if

— The antenna is stationary (not scanning)
— The clutter is totally stationary, with a zero width spectrum
— Enough range sweeps are gathered to totally charge the canceller, or in the case of a filter bank, the number of points processed is large
— The system is totally linear
— Pulse-to-pulse stagger is not necessary to avoid blind Doppler shifts

In practice, of course, antennas scan, clutter has bandwidth, a limited number of points are processed, systems are always somewhat non-linear, and pulse-to-pulse stagger is needed if scan times are to be kept within limits.

Many MTI systems are specified and tested with the antenna stationary. Scanning is, of all the factors listed, the most important in limiting the improvement of MTI and MTD. The reason is simple, as shown in Fig. 12-30. Without scanning or with step-scanning (where the beam is stationary during a dwell and then steps to the next position), the same antenna gain is pointed at the clutter throughout the dwell and the echo from non-moving clutter is constant (Fig. 12-30a). With scanning, Fig. 12-30b, the antenna gain pointed at the clutter during the dwell changes, producing amplitude modulation on the clutter. This amplitude change is sometimes called scan modulation. If the antenna pedestal wobbles, there may also be phase modulation.

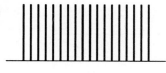 

a. Step scan　　　　　b. Continuous scan;　　　c. Continuous scan;
　　　　　　　　　　　　many hits per beam　　　few hits per beam

Figure 12-30. Antenna Scan Effects on Clutter.

Figure 12-30b and 12-30c show the effect of many versus few pulses received during the antenna's scan across a piece of clutter. The differences between adjacent echoes are smaller with more hits per scan and the improvement factor can be larger than with few hits per scan.

Barton [11] developed models of the improvement factor limits imposed by clutter spectrum, given in the equations below for one, two, and three delay-line non-recursive filters. They assume linear systems and Gaussian clutter spectra.

$$I_1 \approx 2 \, [PRF / (2\pi \delta f_C)]^2 \tag{12-16}$$

$$I_2 \approx 2 \, [PRF / (2\pi \delta f_C)]^4 \tag{12-17}$$

$$I_3 \approx 1.333 \, [PRF / (2\pi \delta f_C)]^6 \tag{12-18}$$

$I_1, I_2,$ and I_3 = the improvement factor limits imposed on one, two, and three delay cancellers, respectively

δf_C = the rms power bandwidth of the clutter spectrum

The rms power bandwidth of the clutter, caused by scanning of a Gaussian beam antenna is shown by Barton to be

$$\delta f_C = [\sqrt{Ln\,2} / \pi] \, [PRF / N_{HB}] \tag{12-19a}$$

$$\delta f_C = 0.265 \, [PRF / N_{HB}] \text{ Hz} \tag{12-19b}$$

N_{HB} = the number of hits received as the antenna scans one beamwidth

Substituting the scan-induced bandwidth from Eq. 12-19 into Eqs. 12-16 through 12-18 produces a model of the improvement factor limitation from scanning, given the assumptions stated above. Equations 12-20 through 12-22 give the model for 1, 2, and 3 delay-line non-recursive cancellers. Figure 12-31 shows Barton's antenna scan improvement factor limits in graphic form. The improvement factor limits are power ratios and convert to decibels as $10 \, Log_{10}(I)$.

$$I_1 \approx 0.721 \, N_{HB}^2 \tag{12-20}$$

$$I_2 \approx 0.260 \, N_{HB}^4 \tag{12-21}$$

$$I_3 \approx 0.0626 \, N_{HB}^6 \tag{12-22}$$

Step-scanning is highly desirable for good MTI/MTD and eliminates scan modulation and the above limitation, but is presently available only with electronically steered antennas (phased arrays − Ch. 9). Step-scanning also is limited to feed-forward (FIR) filter realizations because of the limited number of pulses and the long time necessary to load feedback (IIR) filters. With step-scanning, the filter output is not available until the $(N+1)$th pulse where N is the number of delay lines in the filter.

12-6. Limitations on the Improvement Factor

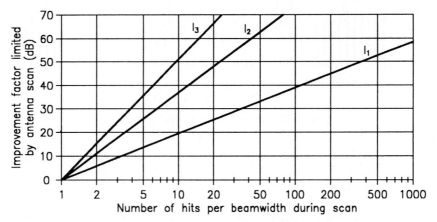

Figure 12-31. MTI Improvement Limited by Antenna Scan Modulation.

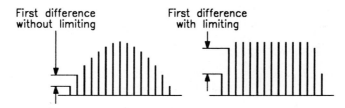

Figure 12-32. Scan Modulation without and with Signal Limiting.

A further limitation imposed by scan modulation occurs if the system saturates. Figure 12-32 shows that if the signal is limited, the scan modulation differences at the edges of the beam increase, increasing the part of the clutter which appears to be a moving target. This is an inherent limitation in the performance of phase MTI, which purposely limits the signal.

The spectrum of the clutter reduces the improvement factor by allowing clutter components outside the notches (Sec. 4-11). Figure 12-33 shows the effect of some specific clutter types and velocities on the improvement factor for a particular three-delay non-recursive filter. The improvement values given assume that the net velocity of the clutter has been compensated and the system treats it as zero. The figure is derived from Eq. 12-26, which is Eq. 12-18 with δv substituted for δf_C per Eq. 12-23 [4]. Values for other filters can be found from Eqs. 12-24 and 12-25, which are Eqs. 12-16 and 12-17 with the substitution from Eq. 12-23. The parameters in Fig. 12-33 are first blind velocities v_{B1} (Eq. 12-27).

$$\delta f_C \approx 2 f_T (\delta v / c) \tag{12-23}$$

$$I_1 \approx 2 [(c\,PRF) / (4\pi f_T \delta v)]^2 \tag{12-24}$$

$$I_2 \approx 2 [(c\,PRF) / (4\pi f_T \delta v)]^4 \tag{12-25}$$

$$I_3 \approx 1.333 [(c\,PRF) / (4\pi f_T \delta v)]^6 \tag{12-26}$$

$$v_{B1} = (c\,PRF) / (2 f_T) \tag{12-27}$$

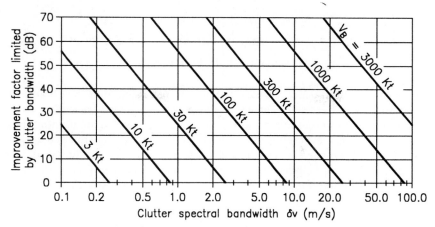

Figure 12-33. MTI Improvement Limit from Clutter Spectra.
Net clutter velocity is assumed to be zero.

δv = the spectral bandwidth of the clutter (m/s rms)
f_T = the transmit frequency
c = the propagation velocity
v_{B1} = the first blind velocity (Ch. 6)

Pulse-to-pulse PRF stagger also reduces the effectiveness of MTI. The amount of improvement limiting is a function of the breadth of the stagger, defined by the factor γ, below. An approximate model for improvement limitation caused by PRF stagger is in Eq. 12-28 [4]. Figure 12-34 gives the improvement factor limit because of pulse-to-pulse stagger.

$$I_{STAG} \approx [(2.5 / N_{HB}) (\gamma - 1)]^2 \quad (12\text{-}28)$$
$$\gamma = PRF_{max} / PRF_{min} \quad (12\text{-}29)$$

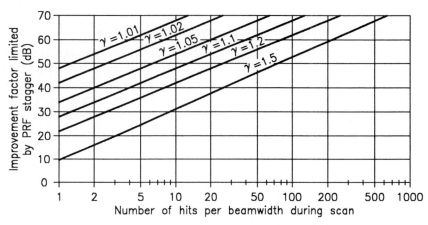

Figure 12-34. Improvement Limitation from PRF Stagger.

12-6. Limitations on the Improvement Factor

I_{STAG} = the improvement factor limited by stagger
PRF_{max} = the maximum PRF in the stagger
PRF_{min} = the minimum PRF in the stagger

Clutter maps: Clutter residue in older systems either caused high false alarm rates or forced inordinately high detection thresholds and poor moving target detection. In modern systems, the detection threshold adapts to the clutter residue in each azimuth/range bin. This is often accomplished by averaging the signal in each azimuth/range bin and establishing a detection threshold for each bin based on its average content. This technique of storing a clutter map and using it to help set the detection threshold reduces the false alarm rate while maintaining the optimum threshold for moving target detection. See Ch. 5.

12-7. CW, High PRF, and Medium PRF Doppler Processing

Almost without exception, Doppler processing in modern CW, high PRF, and medium PRF radars and modes is accomplished with filter banks using discrete Fourier transforms. With CW and high PRF, sampling is Nyquist and there is no Doppler need for stagger, although ECCM and ranging requirements for jitter may continue to exist. The primary Doppler design decisions in CW and high PRF are given below.

— In the case of CW, an anti-alias filter is selected, if used. Fig. 12-35 shows a block diagram of a CW processor (compare it with the signal processing block diagrams of Sec. 12-1). The purpose of the low-pass anti-alias filter is to remove signal components which would be undersampled at the rate used. The filter, if present, must be analog since it is applied before sampling.

Figure 12-36 shows simplified responses of low-pass filters with a variety of poles in their transfer functions. The selection of the sample rate depends on the desired attenuation of signals which would be undersampled Ex. 12-2).

Pulsed radars do not use anti-alias filters, since sampling by the illumination pulse is completed before there is any opportunity to apply a filter. It may be best to consider not using these filters in some CW systems also, on the premise that it is better to accept aliased targets than to completely miss the presence of an unexpectedly fast target.

— The sample rate is selected, based on (1) the maximum expected radar/target closing rate, (2) the transmit frequency, and (3) the anti-alias filter response, if such a filter is present. In pulsed systems, hardware limitations may also influence the sample rate selection. The design closing velocity is the sum of the maximum speed of the radar-carrying platform and that of expected targets. Generally, systems of this type oversample by about 20 to 30 percent.

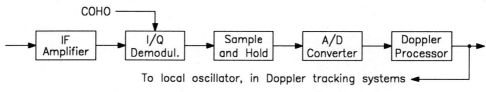

Figure 12-35. CW Processor with Anti-Alias Filter.

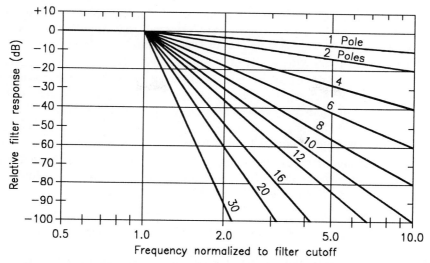

Figure 12-36. Response of Low-Pass Filter versus Its Number of Poles.

Example 12-2: A CW radar operating at 9.98 GHz is designed to work with targets at maximum closing rates of Mach 5.5. If the response to undersampled signals is to be down at least 72 dB and the anti-alias filter has 20 poles, find the required sample rate.

Solution: The Doppler shift of a Mach 5.5 target (1860 m/s) at 9.98 GHz is 123,700 Hz, which is set as the cutoff frequency of the anti-alias filter. Figure 12-36 shows that to be down 72 dB with a 20-pole filter requires a frequency 1.32 times the cutoff frequency. If this frequency is to be Nyquist sampled, the sample rate must be 2.64 times the cutoff frequency (so there is no aliasing for signals whose response is less than 72 dB down). The sample rate, or PRF, thus must exceed 326,600 samples per second.

— The number of points to be used in the discrete Fourier transform is then selected. More points yield smaller Doppler bins. Lower sample frequencies, because they have more total data time for a given number of samples, also reduce the Doppler bin width.

$$\delta f_d = PRF / N_L \qquad (12\text{-}30)$$
$$\Delta f_d = \delta f_d B_W \qquad (12\text{-}31)$$

δf_d = the granularity of the Doppler bins
Δf_d = the Doppler resolution
PRF = the sample rate
N_L = the number of time samples in the discrete Fourier transform
B_W = the equivalent bandwidth of the window used in the DFT (see Table 11-1)

12-7. CW, High PRF, and Medium PRF

Example 12-3: Assume that the radar in Ex. 12-2 is pulsed at the calculated sample rate, and must resolve a 10-Kt ground moving target from clutter with the radar moving at 400 Kt. What number of samples must be processed?

Solution: The 10-Kt difference is 342 Hz of Doppler resolution. Using a Kaiser-Bessel 3.0 window, the granularity must be a factor of 1.44 less than the resolution (Table 11-1, 3-dB bandwidth). Thus the Doppler granularity must be no greater than 235 Hz. Equation 12-30 indicates that the number of points in the transform must exceed 1374. FFT algorithms work best if the number of points is an integer power of 2. The number of samples thus must be 2048 (2^{11}).

Note that FFT algorithms exist for numbers other than integer powers of two. They execute, however, much slower than those with 2^K points, where K is an integer, and are seldom used in real-time embedded signal processing.

Figure 12-37 shows a hypothetical situation involving the airborne radar described in the previous two examples operating at high PRF. The radar is moving over the ground at 400 Kt and is trying to detect a ground moving vehicle with Doppler. The vehicle is moving 15 Kt radial to the radar and has a radar cross-section 10 dB below the clutter. The vehicle is resolved, but it is very difficult to see on the plot. The plot shows clearly that the signals are resolved. The responses below 80 dB are artifacts of the process and window. The window used (Kaiser Bessel 3.0) produces extraordinarily low processing artifacts.

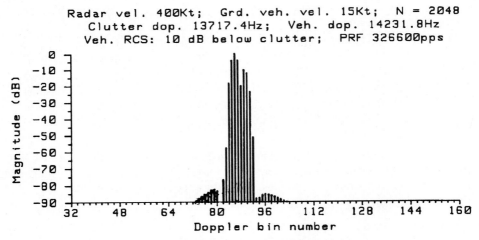

Figure 12-37. Ground Moving Target Example 1.

Figure 12-38 shows the same situation at a medium PRF of 20,000 pps with 512 samples, giving a Doppler bin granularity of 40 Hz. The ground moving target's velocity is reduced to 7.5 Kt and its RCS is reduced to 40 dB below the clutter. The two signals are easily resolved. The primary reason the targets of Fig. 12-38 are resolved better than Fig. 12-37 is that its dwell time is much longer (25.6 ms in Fig. 12-38 versus 6.29 ms in Fig. 12-37).

Note that Figs. 12-37 and 12-38 are an exercise in spectrum analysis and do not represent a totally realistic radar situation. The clutter itself is assumed to have zero bandwidth and spectral spreading because of the antenna's beamwidth and sidelobes are ignored (Ch. 4).

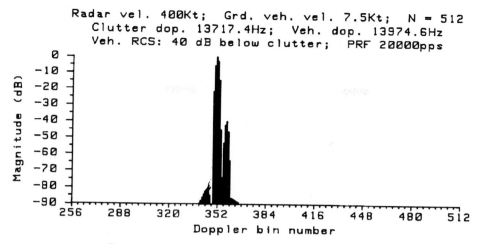

Figure 12-38. Ground Moving Target Example 2.

— The processing window is selected, based primarily on the trade-off between the loss of energy (and signal-to-noise ratio) and the gain in ability to resolve in Doppler targets which are near one another in frequency, with one having much higher RCS than the other (Secs. 12-8 and 12-9).

— Finally, the FFT algorithm is chosen and the process implemented.

Pure CW systems seldom resolve in range, although the modulated CW can find target range and can resolve using modulation techniques discussed in Sec. 6-5. High PRF (pulse Doppler) in many cases also does not resolve in range, with the time between transmit pulses constituting one range bin. Medium PRF, on the other hand, almost always resolves in both range and Doppler. In this case, a Doppler analysis must be done for each range bin (as shown in Fig. 12-6). From the standpoint of range gates (bins) and Doppler filters (bins), CW, pulse Doppler, and medium PRF Doppler processing are shown in Fig. 12-39.

12-8. MTI Testing

MTI testing is primarily a matter of measuring the MTI improvement factor. It is done by comparing the signal-to-clutter ratio with the MTI on, versus the same ratio with it off. Unlike normal MTI operation, the signal and clutter do not have to be measured at the same time, or even in the same range bin. If moving target signal and clutter come from different range bins, however, STC must be disabled or accounted for. From measurements of the system response to stationary and moving target, the improvement factor is

$$MTI\text{-}I = \frac{\text{(Response to moving target / Response to fixed) with MTI on}}{\text{(Response to moving target / Response to fixed with MTI off}} \qquad (12\text{-}32)$$

$MTI\text{-}I = (I) =$ the improvement factor

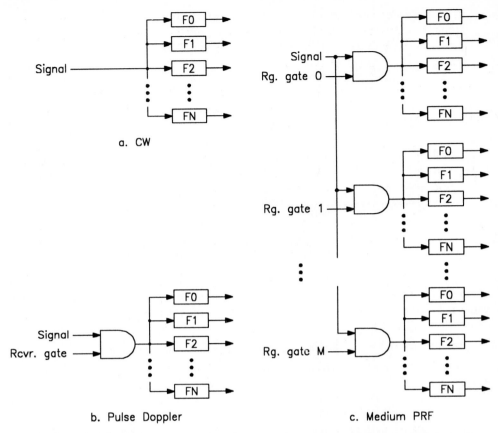

Figure 12-39. Doppler DFT Processing in CW, High PRF, and Medium PRF Radars.

To perform the test requires that stationary and moving targets of known radar cross-section be found, placed, or generated.

Stationary calibration target: The stationary calibration target must have zero Doppler shift, placing it in the center of the nulls in the MTI/MTD filter. It can be one of the following:

- A stationary target of opportunity, such as a water tower or other large, fixed, unchanging structure. The structure should be clear of other clutter. Its distance from the radar is surveyed and the RCS measured by the radar. The radar antenna and the target must be in each other's far field.
- A corner reflector, Luneburg lens, or flat-plate reflector placed in an area clear of other clutter. The range from the radar is surveyed, and the RCS is calculated, specified by the target manufacturer, or measured. The far-field criterion must be maintained.
- An active augmentor (sometimes called an active reflector) in a fixed far-field location and returning an unmodified signal to the radar. It should also be in a location where it is clear of other clutter. The distance is surveyed and the equivalent RCS calculated or measured.

- A signal generator, *phase locked to the radar*, whose signal is injected into the radar through a calibrated coupling port or from a boresight tower. In either case, the signal power injected into the radar and the target's equivalent radar cross-section must be calculated.

Moving calibration target: The moving calibration target should give a response in the center of the MTI filter, which is equivalent to a Doppler shift of half the PRF. Some moving calibration targets are:

- A signal generator *not locked* to the radar. The signal is injected into a (calibrated) port in the radar or from a boresight tower. Most signal generators drift enough to assure that for at least part of the time their signal is in the center of the MTI response. Because of the uncertain placement of its Doppler shift, this type of moving target gives a less accurate measurement than some others.
- A signal generator *phase locked to the radar*, whose output phase changes 180° from pulse to pulse. This is the equivalent of a Doppler shift of one-half the PRF and guarantees placement of the signal in the center of the MTI/MTD filter.
- An active augmentor whose phase shift changes 180° from pulse-to-pulse. One circuit to perform this function is shown in Fig. 12-40.

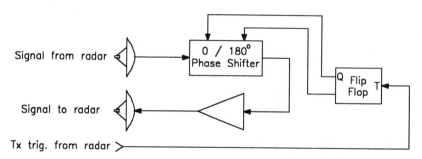

Figure 12-40. Active Augmentor with Selectable Phase Shift.

Antenna scanning: The stationary and moving targets external to the radar (water tower, corner reflector, active augmentor, signal generator and boresight tower) present the opportunity to make the MTI-I measurement with the antenna scanning. The question becomes whether or not it should do so. The system, when in operation, is of course scanning and it can be argued that this is the only legitimate measurement. On the other hand, scanning introduces improvement factor limitations which are not the "fault" of the MTI system. It can thus be argued that to test the MTI/MTD without interference from other factors, the antenna should not scan.

Regardless of whether the measurement is with the antenna scanning or not, consistence in how this measurement is made is most important, because the scanning/non-scanning results will be very different. An excellent discussion of the pros and cons of scanning can be found in W.W. Shrader's article [12]. It is, in this author's opinion, important reading for those who must make MTI measurements.

12-9. Further Information

The purpose of this chapter is to describe clutter suppression techniques and to give system information without providing the details necessary to design the filters. Information at the detailed design depth is available in several sources, although no one of them may provide all the necessary data. Many of the classic papers on Doppler radar (MTI and pulse Doppler) are found in the first two references.

[1] D. Curtis Schleher (ed.), *MTI Radar*, Norwood MA: Artech House, 1978.

[2] D.K. Barton, *CW and Pulse Doppler Radar*, Norwood MA: Artech House Radar series, 1978.

The two editions of the *Radar Handbook* contain much useful information on Doppler radar.

[3] W.K. Saunders, "CW and FM Radar," Ch. 14 in M.I. Skolnik (ed.), *Radar Handbook* 2nd ed., New York: McGraw-Hill, 1990.

[4] W.W. Shrader and V. Gregers-Hansen, "MTI Radar," Ch. 15 in M.I. Skolnik (ed.), *Radar Handbook*, 2nd ed., New York: McGraw-Hill, 1990.

[5] F. M. Staudaher, "Airborne MTI," Ch. 16 in M.I. Skolnik (ed.), *Radar Handbook* 2nd ed., New York: McGraw-Hill, 1990.

[6] W.H. Long, D.H. Mooney, and W.A. Skillman, "Pulse Doppler Radar," Ch. 17 in M.I. Skolnik (ed.), *Radar Handbook*, 2nd ed., New York: McGraw-Hill, 1990.

[7] W.K. Saunders, "CW and FM Radar," Ch. 16 in M.I. Skolnik (ed.), *Radar Handbook*, 1st ed., New York: McGraw-Hill, 1970.

[8] W.W. Shrader, "MTI Radar," Ch. 17 in M.I. Skolnik (ed.), *Radar Handbook*, 1st ed., New York: McGraw-Hill, 1970.

[9] F. M. Staudaher, "Airborne MTI," Ch. 18 in M.I. Skolnik (ed.), *Radar Handbook*, 1st ed., New York: McGraw-Hill, 1970.

[10] D.H. Mooney, and W.A. Skillman, "Pulse-Doppler Radar," Ch. 19 in M.I. Skolnik (ed.), *Radar Handbook*, 1st ed., New York: McGraw-Hill, 1970.

[11] D.K. Barton, *Radar Systems Analysis*, Englewood Cliffs NJ: Prentice Hall, 1964.

[12] W.W. Shrader, "Radar Technology Applied to Air Traffic Control," *IEEE Transactions on Communications,* vol. COM-21, May 1973, pp. 591-605.

13

HIGH RESOLUTION RADAR

13-1. Resolution Review

When a low resolution radar views a target such as an aircraft, it sees a single point. For this reason, most targets to these radars are point targets. When a high resolution radar views the same target, it sees a group of scattering centers which may offer additional information about the target, possibly leading to its identification. Figure 13-1 shows the difference between low and high resolution views of the same target.

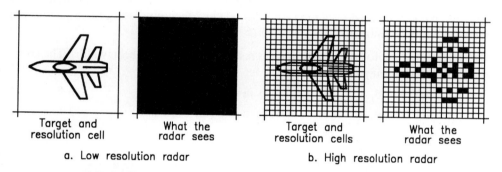

a. Low resolution radar   b. High resolution radar

Figure 13-1. Comparison of Low and High Resolution Radars.

Targets can be resolved (if the radar is so equipped) in four dimensions: range, azimuth cross-range, elevation cross-range, and Doppler. Range resolution is the resolution of multiple scatterers differing in their distance from the radar. Cross-range resolution is the separation of scatterers differing in the dimension normal to range. Horizontal scatterer separation requires azimuth cross-range resolution, vertical requires elevation cross-range resolution. Resolution of scatterers differing in Doppler shift requires Doppler resolution.

Range resolution: Range resolution is a function of the transmitted waveform. In Ch. 1, it was shown that two related factors determine a wave's resolving capability: compressed pulse width and waveform bandwidth. In this section, we will see that these factors are the result of a third waveform property. This property is a wave's autocorrelation function and is the actual determinant of its range resolving capability. Wave evaluation is accomplished using the ambiguity function. (Note that this function is not directly related to the ambiguity

593

in range and Doppler discussed in Ch. 6. Unfortunately, the term *ambiguity* is now ambiguous. The ambiguity function is covered in Sec. 13-2.)

Range resolution can thus be defined in three ways: by compressed pulse width, by signal bandwidth, and by signal autocorrelation function. The three relationships are given below.

$$\Delta R \approx c\tau/2 \tag{13-1}$$
$$\Delta R \approx c\tau_C/2 \tag{13-2}$$
$$\Delta R \approx c/(2B) \tag{13-3}$$
$$\Delta R \approx c\tau_A/2 \tag{13-4}$$

ΔR = the range resolution (meters)
c = the velocity of propagation (meters/second)
τ = the pulse width if there is no compression
τ_C = the compressed pulse width
B = the echo waveform's matched bandwidth (Hertz — Ch. 2)
τ_A = the width of the echo signal's autocorrelation function (seconds — Sec. 13-2)

Enhanced range resolution is achieved through pulse compression, which eliminates the need to trade target detection capability (requiring wide pulses for high energy) for range resolution (requiring narrow pulses). It is the process of transmitting pulses which are too wide to give the required resolution (in Eq. 13-1) and then processing the echoes to the required narrower pulse widths.

Cross-range resolution: Cross-range is resolved with antenna beamwidths. The cross-range resolution of any radar, from Ch. 1, is

$$\Delta X \approx R\theta_3 \quad \text{(if } \theta_3 \text{ is in radians)} \tag{13-5}$$
$$\Delta X \approx R\theta_3(\pi/180) \quad \text{(if } \theta_3 \text{ is in degrees)} \tag{13-6}$$

ΔX = the cross-range resolution (meters)
R = the range from radar to target (meters)
θ_3 = the antenna's 3-dB beamwidth in the direction of the resolution (radians in Eq. 13-5, degrees in Eq. 13-6)

Enhanced cross-range resolution is achieved through large antennas (with narrow beams). If implementing a physical antenna size is impractical, a small antenna can be moved to a number of locations to simulate a large antenna, called a synthetic antenna, or synthetic aperture.

Doppler resolution: Doppler is resolved with spectrum analysis. The Doppler resolution of a radar or mode is a function of how much time is spent gathering signal for the analysis (Ch. 1 and 12).

$$\Delta f_d \approx 1/T_D \tag{13-7}$$

Δf_d = the Doppler resolution (Hertz)
T_D = the time spent gathering data for the analysis (seconds)

In sampled systems, the data gathering time (look time or dwell time) is the sample period times the number of pulses gathered. Pulsed radars sample targets at the PRF rate.

$$T_D = N_L / f_S \tag{13-8}$$

N_L = the number of samples in a look

f_S = the sample rate, which in a pulsed radar equals the PRF

Example 13-1: A radar transmits a 3.5 μs pulse at 5.70 GHz with a bandwidth of 4.0 MHz. The PRF is 550 pps and 64 pulses are processed together. Its antenna beamwidth is 1.2°. Find how far targets must be spaced from one another at a range of 20 km in range, azimuth cross-range, and Doppler.

Solution: The radar uses pulse compression and the bandwidth is much greater than the reciprocal of the pulse width. The range resolution, from Eq. 13-3, is 38 m. The cross-range resolution (Eq. 13-6) at 20 km is 880 m. The Doppler resolution, from Eq. 13-7, is 8.6 Hz. Thus two scatterers at 20 km range must be separated by at least 38 m in range *or* 880 meters in azimuth cross-range *or* 8.6 Hz in Doppler to be resolved.

13-2. Pulse Compression Fundamentals

Pulse compression is the process of transmitting a wide pulse (for energy and detection) and processing it to a narrow pulse (for range resolution). The transmitted pulse is called the expanded pulse, and the processed pulse is called the compressed pulse. The compression ratio is

$$CR \equiv \tau_E / \tau_C \tag{13-9}$$

CR = the compression ratio

τ_E = the expanded (transmitted) pulse width

τ_C = the compressed (processed) pulse width

Two primary classes of pulse compression are used in radars: analog, where the transmit wave contains frequency modulation across the pulse, and digital, where the transmit pulse is phase coded. Characteristic of all pulse compression waveforms is that the bandwidth is much greater than the reciprocal of the expanded pulse width, and is approximately the reciprocal of the compressed pulse width.

$$B \gg 1 / \tau_E \tag{13-10}$$

$$\tau_C \approx 1 / B \tag{13-11}$$

B = the matched bandwidth

The received echoes are compressed either in a filter matched to the transmit wave or by the process of correlation with a delayed copy of the transmit wave. Generally, analog pulse compression is done with matched filters and digital is done by correlation. Figure 13-2 shows the basic outline of a matched filter pulse compression system. Figure 13-3 shows approximate waveforms at various points in the system.

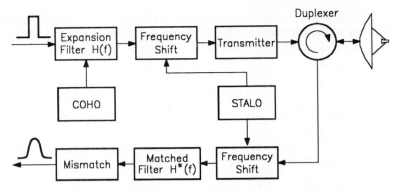

Figure 13-2. Matched Filter Pulse Compression System.

The COHO and a short pulse are fed to $H(f)$, the expansion filter, where the transmit waveform is generated. The pulse is slightly shorter than the desired compressed pulse width. The output of $H(f)$ has a pulse width equal to the expanded pulse width. The expanded pulse is frequency shifted and transmitted. The echoes and interference are received, frequency shifted back to the COHO frequency (plus the Doppler shift) and fed to $H^*(f)$, the compression filter. The response of the compression filter is the complex conjugate of that of the expansion filter.

A property of the complex conjugate in the frequency domain is that in the time domain, time is reversed.

If $H(f)$ = the Fourier transform of $h(t)$,

Then $H^*(f)$ = the Fourier transform of $h(-t)$ (13-12)

Thus any phase change introduced into the signal by $H(f)$ (such as a frequency sweep across the pulse) is undone in $H^*(f)$. If a short burst of the COHO is expanded in $H(f)$, approximately, but not exactly, the same short burst of COHO is recovered in $H^*(f)$.

Because of the finite nature of the signals used, the compressed wave is not exactly the same as the COHO pulse which was expanded. This introduces the undesired characteristic that the compressed pulse "leaks" into times other than that occupied by the echo (Fig. 13-3,

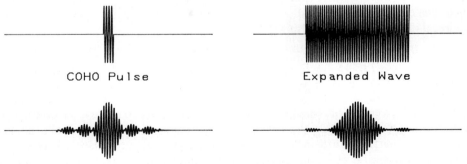

Figure 13-3. Waveforms at Various Points in Pulse Compression System.

lower left). This range leakage allows large interfering signals to hide small desired echoes. For this reason, a mismatch must be introduced, the result of which is Fig. 13-3 (lower right). The mismatch reduces the leakage with the penalty of increasing the compressed pulse width. This mismatch is the same as the window functions introduced in Ch. 11. Table 11-1, although described for spectrum analysis, applies to pulse compression. The compressed pulse is widened by the factor shown in the "3-dB bandwidth" column of the table.

The time-reversal property exhibited in Eq. 13-12 allows an alternative implementation of the matched filter, shown in Fig. 13-4 (left), which allows the same filter to be used for expansion and compression. The time inversion and $H(f)$ in Fig. 13-4 have the same transfer function as the conjugate filter $H^*(f)$ in Fig. 13-3. The problem with this scheme is realizing the time inversion. It is easy to accomplish digitally (because of cheap, flexible data handling and memory), but is difficult to achieve in analog circuits.

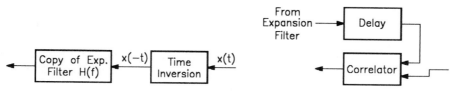

Figure 13-4. Alternate Implementations of the Matched Filter.

The use of matched filtering as a system realization is best adapted to analog pulse compression, as will be shown in Sec. 13-3. The mathematical relationship of the matched filter is shown in Sec. 3-6.

A third implementation of pulse compression correlates the received signal to a delayed copy of that which was transmitted, as shown in Fig. 13-4 (right). The correlation is a cross-correlation, because the echo is different from the transmit waveform (delayed, attenuated, Doppler-shifted, and contaminated with interference). The processing function is given below. It is best adapted to digital pulse compression, as will be shown in Sec. 13-4.

$$z(t) = \int_{-\infty}^{\infty} x(\tau)\, h(t+\tau)\, d\tau \tag{13-13}$$

$z(t)$ = the correlation function of delay time t (see Ch. 11)

$x(\tau)$ = the received waveform

$h(t+\tau)$ = the transmitted waveform, delayed by time t

The combination of the delay and correlator in Fig. 13-4 (right) is functionally identical to the matched filters of the other block diagrams.

Figure 13-5 shows the process of correlation for a 5-segment digital code. The envelope of the signal is shown in the figure, and the plus signs indicate the wave phase is 0° and the minuses indicate 180°. In correlation, one of the waves is held constant [$x(\tau)$ in Eq. 13-13], in this case the delayed version of the transmit wave), and the other [$h(t+\tau)$] is "slid" across it, the received wave, with the displacement indicated by t. The two are multiplied together and the area under the product is found. This area, plotted on the same time scale as Fig 13-5, is shown in Fig. 13-6. It is the correlation function and the compressed wave. It will be shown

13-2. Pulse Compression Fundamentals

in Sec. 13-4 that multiple targets within the expanded pulse will be resolved, as long as they are at least one code segment apart in time (range).

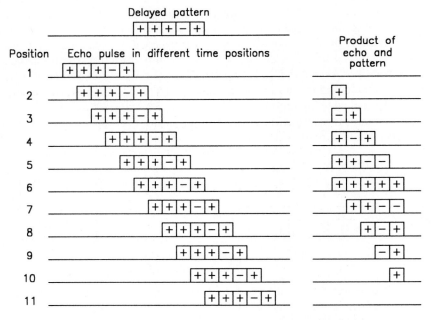

Figure 13-5. Correlation Pulse Compression Process.

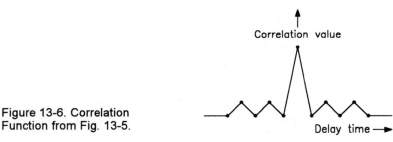

Figure 13-6. Correlation Function from Fig. 13-5.

13-3. Evaluation of Waveforms for Range and Doppler Resolution

The suitability of different transmit waves for range and Doppler resolution is determined by the so-called ambiguity function, which is the 2-D autocorrelation of the wave in time (range) and frequency (Doppler). The plots which display this function are called Rihaczek plots. The 2-D autocorrelation functions can be described as both time and frequency integrals [1].

$$\chi(t,f) = \int_{-\infty}^{\infty} \mu(\tau)\, \mu^*(\tau-t)\, e^{j2\pi f\tau}\, d\tau \qquad (13\text{-}14)$$

$$\chi(t,f) = \int_{-\infty}^{\infty} M^*(\nu) \, M(\nu-f) \, e^{j2\pi\nu t} \, d\nu \qquad (13\text{-}15)$$

$\chi(t,f)$ = the 2-D autocorrelation function (also known as the ambiguity function) in time and frequency

$\mu(\tau)$ = the function described in the time domain

$M(\nu)$ = the function described in the frequency domain

t = the time displacement (the output independent variable in time)

τ = the time variable of integration

f = the frequency displacement (the output independent variable in frequency)

ν = the frequency variable of integration

The ambiguity function of a single pulse of a gated CW wave with rectangular envelope is given (in Rihaczek plot form) as Fig. 13-7. The horizontal axis is time displacement (t) in the autocorrelation and its limits are $\pm$ the pulse width. The slant axis is frequency displacement (f), with limits of 25 times the reciprocal of the pulse width. The vertical axis is the correlation value (χ), plotted in voltage terms on a linear scale.

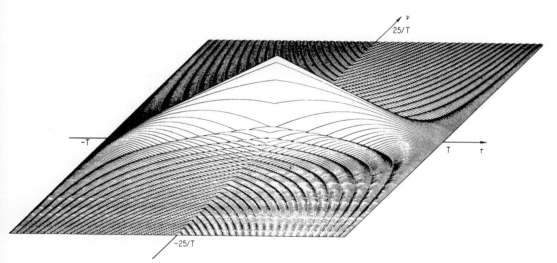

Figure 13-7. Ambiguity Function of a Single Gated CW Pulse.
(Reprinted from Rihaczek [1]; ©1977 Peninsula Press; used with permission)

Profiles of the ambiguity function, taken through its peak value, are shown in Fig. 13-8. The time profile is a triangle (the autocorrelation of a rectangle is a triangle) with a half-voltage width of the pulse width. The frequency profile is a sinc function whose main lobe width is twice the reciprocal of the pulse width.

13-3. Evaluation of Waveforms

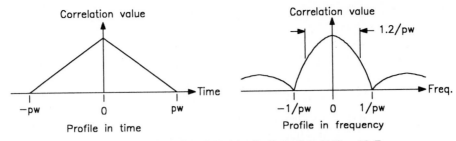

Figure 13-8. Profiles of Ambiguity Function of Fig. 13-7 (not to same scale as Fig13-7).

Example 13-2: A single rectangular gated CW pulse 4.0 μs wide at a frequency of 10.0 GHz illuminates targets. Find the range and Doppler resolving capability of this wave.

Solution: In time, the wave resolves targets separated by more than the width of its autocorrelation function. As seen in Fig. 13-8, the width of the function depends on where the width is measured; in other words, the placement of the detection threshold. In Ch. 2, it was shown that the information from a signal depends on the signal-to-noise ratio, and thus the signal-to-noise ratio plays a role in resolution. In theory, it is possible to resolve to much less than the half-voltage width of the autocorrelation function. In practice, with noisy signals and in particular with targets and interference which may have very different amplitudes, the relationships of Eqs. 13-4 and 13-7 hold. Thus, a 4.0-μs pulse width translates to about 600 m range resolution.

In frequency, resolution is given by Eq. 13-7, with the same arguments as the previous paragraph regarding signal-to-noise ratio. For the single pulse, T_D is the pulse width and Δf_d is 250 KHz. Since this is the Doppler shift of a target moving at about Mach 11, this example shows why Doppler analysis is seldom attempted on a single pulse.

Uncertainty: Another way of evaluating range and Doppler resolving capabilities of waveforms is through a tool called the uncertainty function, which plots to an uncertainty diagram.

Uncertainty is a way of expressing the fact that time can be measured with essentially infinitesimal error, and frequency can be measured with essentially infinitesimal error, but they cannot *both be measured simultaneously* to essentially infinitesimal error (this is a variation of Heisenberg's uncertainty principle). The uncertainty function, as it relates to time and frequency, states that the product of uncertainty in time and uncertainty in frequency cannot be reduced to a value appreciably less than unity.

$$\Delta t \, \Delta f \approx 1 \tag{13-16}$$

Δt = a target's uncertainty in time (range)

Δf = the same target's uncertainty in frequency (Doppler) *for the same measurement*

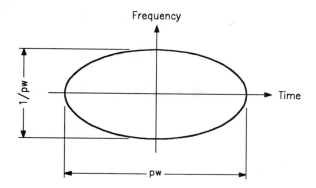

Figure 13-9. Uncertainty Ellipse for Gated CW.

The uncertainty diagram is the intersection of a plane parallel to the time/frequency plane of the ambiguity function and the ambiguity function itself (in other words, a threshold). The resultant figure is an ellipse, which for gated CW is shown in Fig. 13-9. The values shown are for an intersecting plane at one-half the peak autocorrelation value. A more general form of the ellipse, known as Helstom's uncertainty ellipse, is given later.

If more than one target is present, there is one uncertainty ellipse for each target. Resolution occurs if the threshold is placed such that the multiple ellipses do not overlap. Figure 13-10 shows resolved and unresolved targets with respect to their uncertainty ellipses.

Intuitively it would seem that resolution could be increased without limit by simply placing the threshold plane closer to the peak of the ambiguity function, as shown in Fig. 13-11. In practice this does not work because of a lack of knowledge as to the peak correlation value, the presence of noise, and targets with differing amplitudes. The issue is especially complicated if the scatterers to be resolved are not of the same amplitude. For a complete treatment, the reader is referred to Rihaczek [1] and Harris [7]

The above discussion and example show resolution in time and frequency for a single pulse of gated CW. It has been stated previously that radars seldom do much of anything with a single echo pulse. One, but not the only, reason is that for values normally encountered in radar range and Doppler resolution, it is not possible to resolve well enough to accomplish the mission of systems. For this reason, whatever the waveform transmitted, it is usually transmitted as a pulse train. Figure 13-12 shows the ambiguity function of a train of gated CW pulses (medium PRF) and Fig. 13-13 shows its uncertainty.

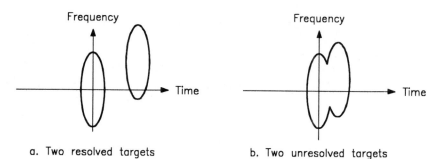

a. Two resolved targets   b. Two unresolved targets

Figure 13-10. Uncertainty with Two Targets (gated CW).

13-3. Evaluation of Waveforms

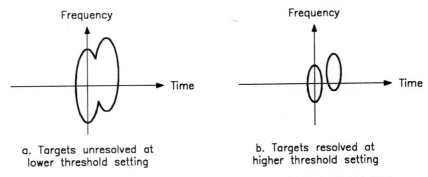

a. Targets unresolved at lower threshold setting

b. Targets resolved at higher threshold setting

Figure 13-11. Effect of Threshold Level on Uncertainty (Gated CW).

With pulse trains, the range resolution is normally determined by the characteristics of a *single pulse*, while Doppler resolution is a function of *the length of the train* used for a look. Range and Doppler resolution are as stated in Eqs. 13-4 and 13-7.

Pulse trains improve in particular Doppler resolution for normal waveforms. The price paid is ambiguity in range and Doppler (Ch. 6). Figure 13-14 shows uncertainty for high PRF. Note the extreme ambiguity in range (actually worse than the figure shows). Figure 13-15 shows low PRF, with its extreme ambiguity in Doppler.

Medium PRF is moderate ambiguous in both range and Doppler. These ambiguities can often be worked out by viewing the target volume with multiple PRFs, as discussed in Ch. 6. Figure 13-16 is uncertainty for a single target with two PRFs. Note that there is only one range and Doppler (at the centers of the diagrams) where the target exists at both PRFs. This is assumed to be the correct range and Doppler.

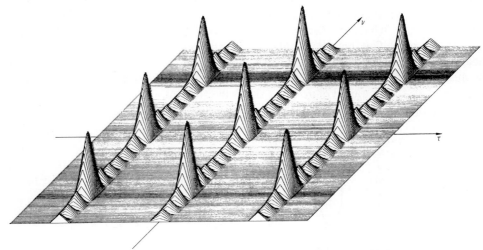

Figure 13-12. Ambiguity Function of Gated CW Pulse Train.
(Reprinted from Rihaczek [1]; ©1977 Peninsula Press; used with permission)

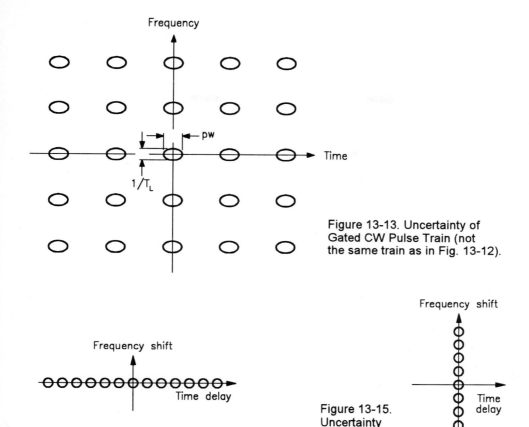

Figure 13-13. Uncertainty of Gated CW Pulse Train (not the same train as in Fig. 13-12).

Figure 13-14. Uncertainty for High PRF.

Figure 13-15. Uncertainty for Low PRF.

Example 13-3: A gated CW wave is 2.5 μs long. Find its range and Doppler resolving capabilities for the following conditions:

a. Single pulse

b. Low PRF - 450 pps, a look is 8 pulses

c. High PRF - 120,000 pps, a look is 1024 pulses

d. Medium PRF - 13,000 pps, a look is 128 pulses

Solution: a. For a single pulse, the range resolution (from Fig. 13-9 and Eq. 13-1) is 375 m. The Doppler resolution (Fig. 13-9 and Eq. 13-7) is 400 KHz (targets must be separated by 400,000 Hz in Doppler to be resolved — not good).

b. The range resolution is 375 m. Low PRF does not normally function well in Doppler, but the theoretical resolution, from Eq. 13-7 with a look time of 17.8 ms. (Eq. 13-8), is 56 Hz.

c. Theoretical range resolution is still 375 m, even though high PRF does not function too well in range. Look time is 8.53 ms and Doppler resolution is 117 Hz.

d. Range resolution is 375 m, look time is 9.85 ms, and Doppler resolution is 102 Hz.

13-3. Evaluation of Waveforms

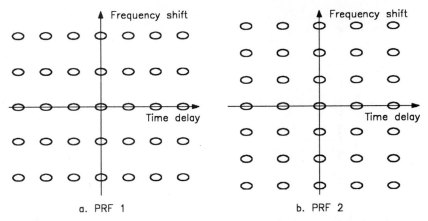

Figure 13-16. Medium PRF Uncertainty — One Target at Two PRFs.

Pulse compression waveforms: Pulse compression waveforms also have ambiguity functions, and they are characterized by being much narrower in time than gated CW for a given pulse width. Figure 13-17 is the ambiguity of a single rectangular pulse of linear FM chirp waveform whose time-bandwidth product is 25. Figure 13-18 is its uncertainty.

Profiles of the ambiguity function in time and frequency are shown in Fig. 13-19. Note that both are sinc functions, indicating both spectral leakage, and range leakage. Most pulse compression waveforms must be processed with windows either prior to or after compression. The discussions of Ch. 12 regarding windows apply to pulse compression.

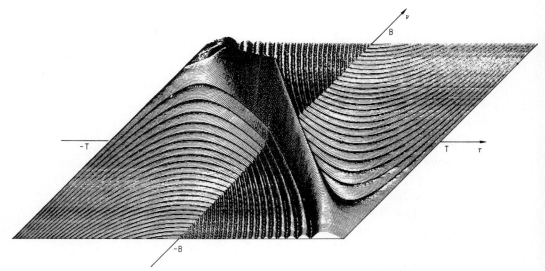

Figure 13-17. Ambiguity Function of a Single Linear FM Pulse (pw B = 25). (Reprinted from Rihaczek [1]; ©1977 Peninsula Press; used with permission)

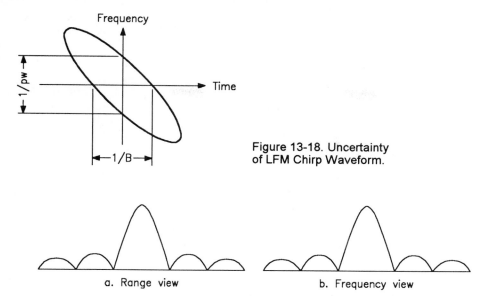

Figure 13-18. Uncertainty of LFM Chirp Waveform.

a. Range view

b. Frequency view

Figure 13-19. Profiles of Ambiguity Function of Fig. 13-17, not to same scale as Fig. 13-17.

Digital pulse compression waveforms likewise are evaluated with ambiguity and uncertainty. Figure 13-20 shows the uncertainty of a single pulse phase-coded with a 13-bit Barker sequence and Fig. 13-21 shows its ambiguity function. The unique feature of Barker sequences is that all minor *range* responses of their autocorrelation functions have the same value, which is 1/N times the value of the peak, N being the number of bits in the sequence. Barker codes are discussed in Sec. 13-4.

The relationship between single pulses of modulated waves and pulse trains of these waveforms is the same as for gated CW. The individual pulse determines the range resolution and the length of the train (the look time) determines Doppler resolution. The uncertainty of a train of LFM chirp pulses is depicted in Fig. 13-22.

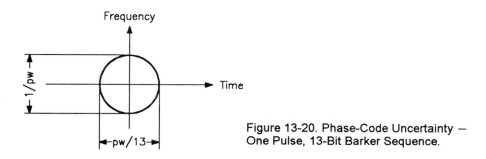

Figure 13-20. Phase-Code Uncertainty — One Pulse, 13-Bit Barker Sequence.

13-3. Evaluation of Waveforms

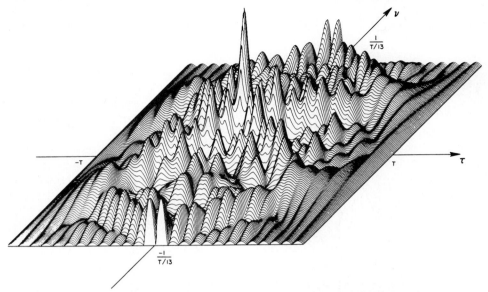

Figure 13-21. Ambiguity Diagram of a Single Phase-Coded Pulse, 13-Bit Barker. (Reprinted from Rihaczek [1]; ©1977 Peninsula Press; used with permission)

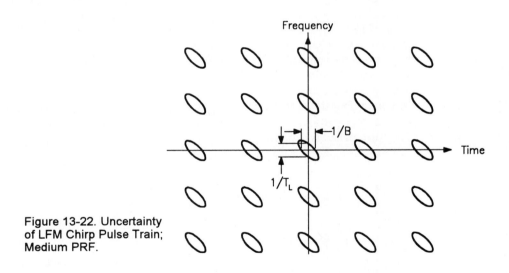

Figure 13-22. Uncertainty of LFM Chirp Pulse Train; Medium PRF.

606 Chap. 13. High Resolution Radar

13-4. Analog Pulse Compression

This type of pulse compression is done with waveforms which contain analog modulation within the pulses, most commonly frequency modulation (FM). In most cases, the modulation is a frequency sweep across the pulse, called chirp.

The FM wave is formed in an expansion filter and the echo is compressed in a compression filter. In Fig. 13-2, $H(f)$ is the expansion filter and $H^*(f)$ is the compression filter. For FM, the filters must be dispersive; that is, different frequencies propagate at different velocities. Waveguide is dispersive and was used in early radars. Most modern analog pulse compression systems use surface acoustic wave (SAW) devices.

A SAW device is a thin slab of a piezoelectric material with transducers at the ends. Piezoelectric material have the property that if an electric potential is placed across them, a mechanical stress results. Likewise, a mechanical stress causes an electric potential. The induced stress propagates as a mechanical (acoustic) wave on the surface of the material at the material's natural velocity of propagation. Figure 13-23, showing a SAW delay line, gives the principle.

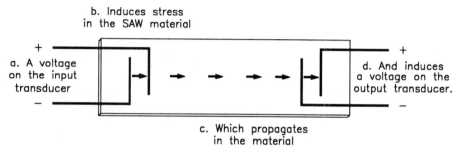

Figure 13-23. SAW Device Principle.

The transducer is conducting material placed on the piezoelectric SAW material using printed circuit or integrated circuit technology. The frequency which couples from transducer to SAW material to transducer is that where the spacing of the transducer elements is one-half wavelength in the SAW material.

$$\lambda_{AC} = v_{AC}/f \qquad (13\text{-}17)$$

λ_{AC} = the acoustic wavelength on the SAW material

v_{AC} = the acoustic wave propagation velocity on the material

f = the frequency

The velocity of propagation depends on the material. Table 13-1 gives some of the properties of selected SAW materials.

FM pulse compression systems consist of a transmit (expansion) section and a processing (compression) section. One type of expansion section is shown in Fig. 13-24.

The system phase reference (COHO) is fed to a time gate which passes a short burst whose width is slightly less than the final compressed pulse (to account for the pulse widening of the compression processing window). The spectrum of the COHO pulse is a sinc

Table 13-1. Surface Acoustic Wave Parameters of Selected Piezoelectric Materials [8]

Material	Cut	Prop. Dir.	Prop. Vel. m/s	Atten. (1GHz) dB/m	Delay Temp. Coeff. ppm/°C
$LiNbO_3$	Y	Z	3485	1.6	85
Quartz	Y	X	3173	4.1	−24
Quartz	ST	X	3158	−	<3
$Bi_{12}GeO_{20}$	(100)	(011)	1681	1.5	−122
PZT	Poled Dir.	Any	2200	6.0	−
$LiTaO_3$	Z	Y	3329	−	67
AlN/Al_2O_3	X	Z	6120	1.7	44

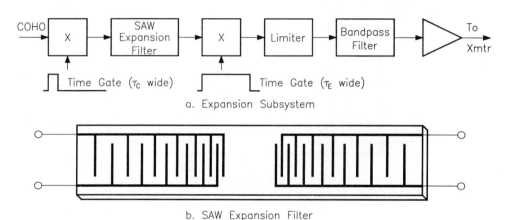

Figure 13-24. Expansion Filter.

function and is shown in Ch. 2. It has a relatively wide spectrum because of its narrow width ($B \approx 1/\tau$). The COHO pulse is then expanded in the expansion filter (Fig. 13-24b).

The COHO burst is applied to all areas of the input transducer essentially simultaneously, since the signal's velocity of propagation on the transducer approaches the speed of light. Velocity on the transducer is much higher than the acoustic propagation velocity in the SAW material. Coupling from transducer to SAW is most effective where the transducer elements are separated by about one-half an acoustic wavelength. Thus the low frequency components of the COHO burst couple near the left-hand side of the input transducer and the high frequency components couple to the right. The signal propagates across the SAW material at its natural propagation velocity.

At the output transducer, the high frequencies couple to the left and the lows to the right. Thus the high frequency components of the COHO burst arrive at the output before the low frequencies, resulting in a down-chirped signal.

There are frequencies in the COHO burst which are not desired in the output (the spectral sidelobes) and all components do not have the same amplitude. The time gate following the expansion filter removes the undesired frequencies and the limiter causes all component amplitudes to be the same. After filtering to remove unwanted transient components, the expanded wave is sent to the transmitter, where it is frequency shifted and amplified. The frequency/time characteristic is set by the spacing of the elements of the transducers. Several curves were shown and described in Sec. 2-4.

The compression subsystem is shown in Fig. 13-25. The echo pulse, which has the same frequency/time characteristic as that transmitted, is amplified and applied to a SAW compression filter, shown in Fig. 13-25b. In this filter, the low frequencies propagate across the filter quicker than the highs. The wave thus collapses on itself, recovering an approximation of the COHO burst. If the filter is designed properly, the output pulse is at the constant frequency of the COHO. Phase is preserved in this process, and the compressed pulse can be I/Q demodulated to recover Doppler.

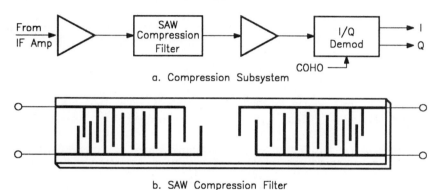

Figure 13-25. Compression Filter.

SAW coupling is related also to the length of the transducer elements. Note that the elements in the compression filter are of different lengths. Because of this, the compression filter couples the middle frequencies more effectively than the highs and lows. This is windowing, used to suppress time leakage. Figure 13-26 shows compression without and with windowing. Window performance in pulse compression is identical to that in spectrum analysis (Ch.11)

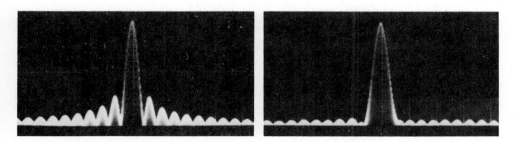

Figure 13-26. Envelope of Compressed Pulse without Windowing and with Dolph-Chebyshev Window. (Reprinted from [9]; © McGraw-Hill; used with permission)

13-4. Analog Pulse Compression

This SAW implementation is not the only way of producing and compressing FM waveforms. Another expansion method uses a voltage controlled oscillator (VCO) to generate the transmit waveform. The VCO is fed a voltage which produces the desired frequency/time characteristic, either linear or non-linear. Compression filters include waveguide (upchirp only) and folded tape meander lines. See Farnett and Stevens [4] for details of these other implementations.

Some characteristics of commercial pulse compression units are given in Tables 13-2 and 13-3 [10]. Table 13-2 is for a linear FM unit and Table 13-3 shows a non-linear FM unit. Figures 13-27 and 13-28 show the waveforms and spectra of the modules.

Table 13-2. A Typical Linear FM Pulse Compression Subsystem (RACAL-MESL, Ltd. Model 510 [10])

Transmitter section:		Receiver section:	
Frequency (f_0)	60 MHz	3-dB compressed pulse width	0.1 μs
Bandwidth (B)	14 MHz	Phase code law	Linear upchirp
Time dispersion (τ_E)	4 μs	Amplitude weighting	45 dB Taylor
Phase code law	Lin. downchirp	Impedance/VSWR	50Ω/<1.5
Center frequency delay	5 μs	Range sidelobes	< -38 dB
Output level	0 dBm	Dynamic range	>60 dB
Expanded pulse S/N	>40 dB	Output level	1V$_{p-p}$

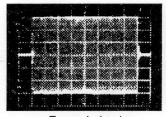 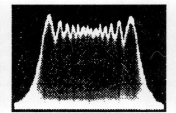 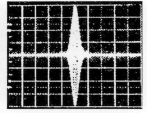

a. Expanded pulse   b. Expanded spectrum   c. Compressed pulse

Figure 13-27. Linear FM Pulse Compression System Waveforms (Model 510). (Diagrams courtesy of RACAL-MESL, Ltd.)

Example 13-4: Using Y-cut quartz, find the transducer element spacing for the two ends of an expansion filter if one end is to couple 58.500 MHz and the other is to couple 61.500 MHz.

Solution: Equation 13-17 and Table 13-1 give the acoustic wavelength of 58.500 MHz in this material as 54.2 μm and the wavelength of 61.500 MHz as 51.6 μm. The spacings are half-wavelength. Therefore they are 27.1 μm and 25.8 μm.

Table 13-3. A Typical Non-Linear FM Pulse Compression Subsystem
(RACAL-MESL, Ltd. Model 511 W3295)

Transmitter section:		Receiver section:	
Frequency (f_0)	60 MHz	3 dB compressed pulse width	0.2 µs
Bandwidth (B)	7.5 MHz	Phase code law	Non-lin. up.
Time dispersion (T)	12.45 µs	Ampl. weighting	45 dB Taylor
Phase code law	Non-lin. down	Range sidelobes	<−42 dB
Output level	0 dBm	Dynamic range	>60 dB
Expanded pulse S/N	>40 dB	Output level	$1V_{p-p}$

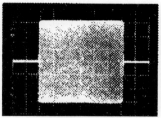  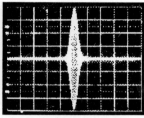

   a. Expanded pulse   b. Expanded spectrum   c. Compressed pulse

Figure 13-28. Non-Linear FM Pulse Compression System Waveforms (Model 511).
(Diagrams courtesy of RACAL-MESL, Ltd.)

13-5. Digital Pulse Compression

Phase-coded waveforms are well adapted to digital pulse compression. Whereas analog systems compress the echo signals before I/Q demodulation, digital systems work with post-demodulated signals. Digital implementations tend to be less complex than analog.

Digital waveforms are usually bi-phase modulated sinusoids, with the two possible phases being 0° and 180°. Bi-phase modulation is used because it yields the widest bandwidth for a given code sequence. The waveform consists of N subpulses of width τ_S, forming a sequence totaling the expanded pulse. The code is applied to the subpulses, one bit in the code sequence per subpulse. The subpulses are coded by their phase with respect to the COHO, with + or 1 being 0° and − or 0 being 180°. Details of the waveforms are given in Sec. 2-4.

Barker codes: Several types of sequences suitable for pulse compression exist. Barker sequences are unique in that their autocorrelation functions can contain only three possible absolute values: 0, 1, and N, where N is the number of bits in the sequence. This property allows them to be processed without windows and they thus yield the best resolution for a given transmitted bandwidth. The disadvantage to using Barker sequences is that there are only nine known examples and the largest has 13 bits. Since there are so few codes and

Table 13-4. Barker Sequences.	
Number of bits	Known sequences
2	11, 10
3	110
4	1101, 1110
5	11101
7	1110010
11	11100010010
13	1111100110101

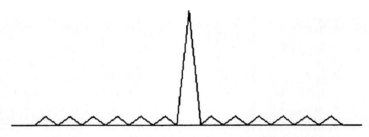

Figure 13-29. Autocorrelation of 13-Bit Barker Sequence.

everyone knows them, signal security is limited. Since the longest is relatively short, the compression ratios of radars using them are small. Despite these disadvantages, many radars use phase coding with Barker sequences. Table 13-4 gives all known sequences of this type (of course, each can be reversed and bit-inverted). The amplitude of the autocorrelation function of a 13-bit Barker sequence is shown in Fig. 13-29. Note that range leakage is small.

Barker codes can be combined into longer sequences, but lose most of their advantages regarding leakage. A 35-bit combination sequence can be constructed, for example, of five sequences, phasing the individual sequences as 7-bit Barker, and setting the phase of each group according to the 5-bit Barker code. The resulting sequence 1110010,1110010,1110010,0001101,1110010 has the autocorrelation function shown in Fig. 13-30.

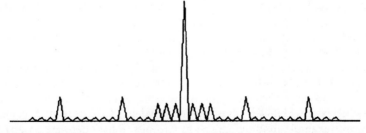

Figure 13-30. Autocorrelation of 35-Bit Combination (5 x 7 bit) Barker Sequence.

Pseudorandom codes: The other class of digital waveforms are phase-coded with pseudorandom sequences. Their advantage is that they can be long, giving high compression ratios, and that there are many of them, allowing secure encyphering of waveforms, if desired. Their disadvantage is that their autocorrelation functions contain partial sums with values greater than one so that windowing must be used to reduce range leakage. The autocorrelation function of a particular 13-bit pseudorandom sequence is shown in Fig. 13-31.

Figure 13-31. Autocorrelation of a 13-Bit Pseudorandom Sequence.

The pseudorandom sequences with minimum sidelobes (leakage) for their length are known as optimal sequences (Barker codes are optimal sequences). For example, there are 222 35-bit sequences which have maximum autocorrelation sidelobes of three. There are none with maximum sidelobes of two or one. A listing of the number of optimal codes for each length up to 40 is given in Table 13-5. The autocorrelation function for the 35-bit sequence given in the table is shown as Fig. 12-32. Compare it to the 35-bit combination Barker sequence of Fig. 13-30.

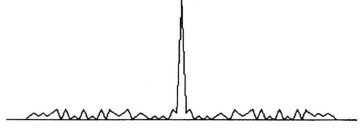

Figure 13-32. Autocorrelation of One of the 35-Bit Optimal Sequences.

A group known as maximal length sequences are of interest because they are easy to generate and can be of virtually any length. They are particularly attractive in cases where signal encryption is required. Figure 13-33 shows shift register generation of maximal length sequences. The code is determined by the placement of the taps for the modulo 2 adder (exclusive OR). The placement shown is only an example.

Another class of codes sometimes considered are the complementary or sidelobe cancelling codes [11]. These are pairs of sequences having the property that the sum of their autocorrelations is, at the peak, twice that of the autocorrelation of a single code. Elsewhere, the pairs sum to zero. Thus taken together, the sidelobes cancel. The Welti codes are general sidelobe cancelling codes. The Golay codes [12] are binary Welti codes, and are thus usable

Table 13-5. Optimal Sequences.
(After Farnett and Stevens [4])

Length of code	Amplitude of maximum sidelobe	Number of codes	Code[b] (Octal)	Length of code	Amplitude of maximum sidelobe	Number of codes	Code[b] (Octal)
2	1[a]	2	2, 3	21	2	6	5204154
3	1[a]	1	6	22	3	756	11273014
4	1[a]	2	15, 16	23	3	1021	32511437
5	1[a]	1	35	24	3	1716	44650367
6	2	8	64	25	2	2	163402511
7	1[a]	1	162	26	3	484	262704136
8	2	16	261	27	3	774	624213647
9	2	20	654	28	2	4	1111240347
10	2	10	1632	29	3	561	3061240333
11	1[a]	1	3422	30	3	172	6162500266
12	2	32	6443	31	3	502	16665201630
13	1[a]	1	17465	32	3	844	37233244307
14	2	18	36324	33	3	278	55524037163
15	2	26	74665	34	3	102	144771604524
16	2	20	141335	35	3	222	223352204341
17	2	8	265014	36	3	322	526311337707
18	2	4	467412	37	3	110	1232767305704
19	2	2	1610445	38	3	34	2251232160063
20	2	6	3731261	39	3	60	4516642774561
				40	3	114	14727057244044

[a] Barker.
[b] If more than one code exists, the number given is an example.

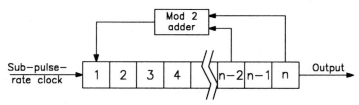

Figure 13-33. Maximal Length Code Generation.

for pulse compression. Despite their theoretical attractiveness, they do not perform well in noisy environments. This is because target views with the two segments must occur at different times, different frequencies, or different polarizations, and the noise components of the two views do not match.

Phase-coded signals can be generated in many ways, with an example shown in Fig. 13-34. In it, the COHO is split into two phases, 0° and 180°. The gates controlled by the code sequence select which is used, and the resulting coded sinusoid is frequency shifted, amplified, and transmitted.

Digital pulse compression is implemented as shown in Fig. 13-35. Both the I and Q compression channels are necessary to eliminate blind phases (echoes at $\pm$ 90° from the COHO) and to preserve the complex data for Doppler analysis.

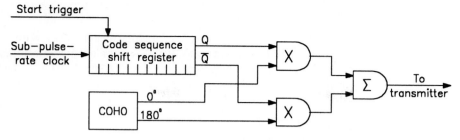

Figure 13-34. Generation of Phase-Coded Signals.

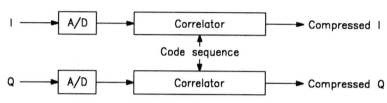

Figure 13-35. Digital Pulse Compressor.

The correlator is shown in Fig 13-36. It consists of a shifting memory, a multiplier, and an adder. Data (I or Q depending on the channel) are shifted through the memory at the subpulse clock rate. At each displacement, the data are multiplied by the code (and by the window, if used), and the products are summed. This implements the discrete correlation discussed in Ch. 11.

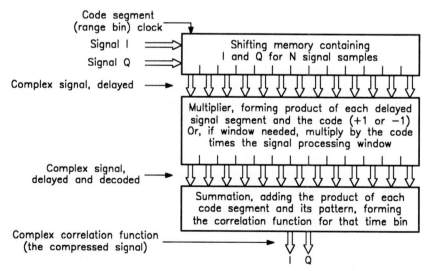

Figure 13-36. Digital Correlation Pulse Compressor — 13-Bit Sequence.

13-5. Digital Pulse Compression

13-6. High Cross-Range Resolution Introduction

Cross-range resolution is enhanced by narrowing the antenna beamwidth. With real antennas, this requires enlarging the physical antenna, which is often not possible because of physical constraints. The same effect can be accomplished with a small real antenna by viewing a target volume from several different locations. See Sec. 2-5 for an overview. The fundamental relationships are given in Eqs. 13-18 and 13-19.

$$\Delta X_R \approx R \lambda / D_{eff} \quad \text{(radians)} \tag{13-18}$$

$$\Delta X_S \approx R \lambda / 2L_{eff} \quad \text{(radians)} \tag{13-19}$$

ΔX_R = the cross-range resolution using a real antenna (meters)

ΔX_S = the cross-range resolution using a synthetic antenna

R = the range to the targets being resolved (meters)

D_{eff} = the effective length of the real antenna (meters)

L_{eff} = the effective length of the synthetic antenna (meters)

λ = the wavelength (meters)

There are three fundamentally different ways of enhancing cross-range resolution: Doppler beam-sharpening (DBS), sidelooking synthetic aperture radar (SAR), and inverse synthetic aperture radar (ISAR) using target rotation.

If the radar is moved to enhance cross-range resolution, the technique is a form of synthetic aperture radar (SAR). If the radar is stationary (or moving) and the target's motion caused the cross-range enhancement, the technique is known as inverse synthetic aperture radar (ISAR).

— **Doppler beam-sharpening:** This method enhances cross-range resolution by exploiting the small Doppler shift differences of scatterers in one range bin and within the azimuth cell of a moving radar. Figure 13-37 shows a moving radar looking at the ground ahead and to one side of the radar's track over the ground.

Note that scatterers having shallower angles from the flight path of the radar have higher velocities radial to the radar than those with steeper angles. Since radial velocity is what causes Doppler shift, the scatterers have slightly different echo Doppler frequencies. If Doppler resolution is sufficient, the scatterers will be resolved.

— **Sidelooking synthetic aperture radar:** This method synthesizes a large antenna by sequentially examining the volume of interest with a small real antenna moved to different locations (Fig. 13-38). In a real antenna, the incremental elements are formed simultaneously, and the beam is formed by summing the outputs of each of the elements (Fig. 13-39a). In a synthetic antenna, the incremental elements are formed sequentially and their outputs must be stored until the full array has been formed. Then, the outputs of the elements are summed and the beam formed (Fig. 13-39b). The effective length of the synthetic array is the distance the real antenna moved. This technique is commonly used for high resolution ground mapping from aircraft.

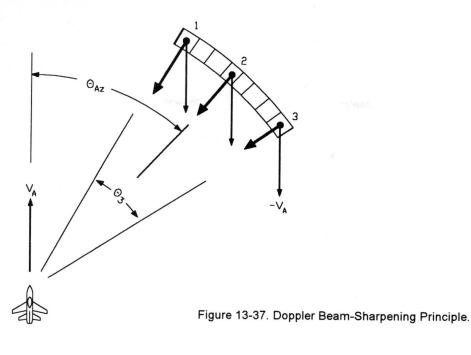

Figure 13-37. Doppler Beam-Sharpening Principle.

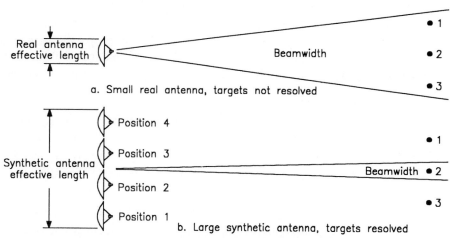

Figure 13-38. Synthetic Aperture Radar Concept.

- **Inverse synthetic aperture radar (ISAR) with rotating targets:** This method of enhancing cross-range resolution takes advantage of differential Doppler shifts from objects at different positions within the antenna's angular resolution cell (Fig. 13-40).

Scatterers are present in four of the range bins. In two of them, multiple scatterers exist. All are within the beam of the real antenna. The cross-range position of the scatterers is found by observing their Doppler phase as the object rotates. In the range bin containing scatterers

13-6. High Cross-Range Resolution Introduction

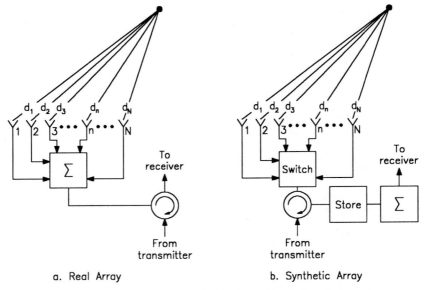

Figure 13-39. Real and Synthetic Array Beam Development.

1, 2, 3, 4, and 5, there will be a signal component with no phase change. The object producing this signal component is on the radar's axis and has a cross-range position of zero (scatterer 1). Another component has a phase pattern with a small down-Doppler shift and is assigned a position near the axis to the radar's left of center (scatterer 2). A third component has a large down-Doppler and if from a scatterer to the radar's left and far from the axis (scatterer 4). Two others have up-Doppler shifts, identifying them as being to the radar's right of the axis (scatterers 3 and 5). Similarly, the cross-range positions of the scatterers in the remaining range bins are found.

This technique is commonly used for imaging tactical targets from airborne, shipboard, and land-based radars. It is also used to analyze the locations of scatterers on targets (diagnostic radars).

13-7. Doppler Beam-Sharpening (DBS)

A simple form of SAR is a technique which uses the radar's Doppler filters to enhance angular resolution. It is generally used to improve azimuth resolution, and is most effective where all scatterers to be resolved are moving in the same direction and at the same velocity with respect to the radar. This process is especially effective for improving the resolution of ground maps taken from aircraft. It is also used in land-based, shipboard, and airborne radars to resolve formations of aircraft, in a process known as raid discrimination.

To understand the principle of Doppler beam-sharpening, an example is instructive. Please refer to Fig. 13-37, showing three targets simultaneously within the beam of the antenna and in the same range bin. Under ordinary circumstances, these would appear to the radar as a single unresolved target. Assume that the antenna's azimuth beamwidth is $2.0°$

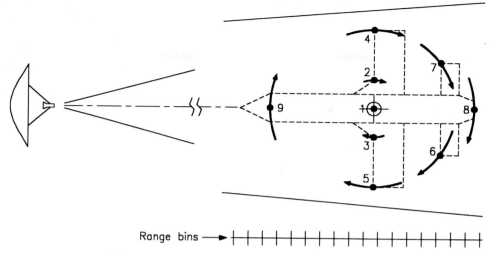

Figure 13-40. Concept of ISAR with Rotating Targets.

and that the beam is pointing 15° to the right of the aircraft's ground track (the line along the ground over which the aircraft flies). Further assume that the ground area of interest is 10.0 km from the radar, that the aircraft (radar) is moving at 400 Kt (205.78 m/s), and that the transmit frequency is 10.00 GHz.

At this range, ordinary resolution techniques require the targets to be about 350 m apart in cross-range for resolution (Eq 13-41). Because of the geometry of this situation, however, note that the angles from the radar's ground track to the three scatterers is slightly different. Because of this, their radial velocities to the radar are also slightly different, as are their Doppler shifts (depression angle is ignored in this example). Table 13-6 summarizes the known target quantities for this example.

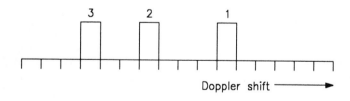

Figure 13-41. Doppler Spectrum for DBS Example.

Table 13-6. DBS Example Data.

Target	Angle from Ground Track	Radial Velocity	Doppler Shift
1	14.125°	199.59 m/s	13,303.9 Hz
2	14.875°	198.88 m/s	13,258.9 Hz
3	15.875°	197.93 m/s	13,195.4 Hz

13-7. Doppler Beam-Sharpening (DBS)

The Doppler shift difference between targets is 45 and 64 Hz, respectively. If there is sufficient Doppler resolution, spectrum analysis will resolve the targets. In this example, the 2.0° beam was divided into eight parts, each 0.25° wide. In this example, adjacent cells are 0.25° apart, which is about 0.22 m/s, or 15 Hz of Doppler shift, resulting in the spectrum of Fig. 13-41. This resolved Doppler spectrum can then be superimposed on the range/azimuth cell of Fig. 13-37, resulting in the resolution situation shown in Fig. 13-42. To obtain this Doppler (azimuth) resolution requires that the look time for the example be approximately 67 ms for 15 Hz of resolution (Eq. 13-7).

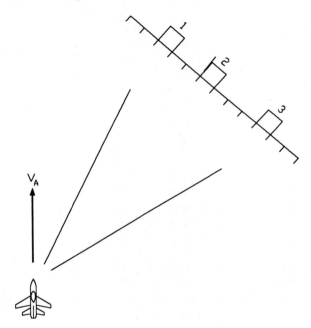

Figure 13-42. Relationship between Doppler Spectrum and Azimuth Target Location in Doppler Beam-Sharpening.

Doppler beam-sharpening interprets Doppler shift as angular offset. If the ground illuminated by the DBS radar includes moving targets, such as vehicles on roads, the vehicles' Doppler shifts will result in their being displayed with a cross-range offset. The vehicles will show on the display in locations different from their actual positions (perhaps off the road), with their offset proportional to their velocity. Given sufficient resolution, the velocity of the vehicles could be obtained from their lateral offset and from the illumination geometry. An example of the result of Doppler beam-sharpening is shown in Fig. 13-43.

DBS processing and data space: The equations for analyzing DBS are developed below. Note that the required Doppler resolution varies with azimuth angle from the radar's ground track, and that as the look angle approaches zero, the look time becomes infinite. Also note that at angles approaching 90° from the track, the azimuth resolution becomes a very non-linear function of look angle. Doppler beam sharpening is practical for look angles from ground track of about 5° to 60°. The following symbols are used for DBS analysis.

θ_0 = the antenna pointing angle, the azimuth angle between the radar's ground track and the center of the antenna beam

Figure 13-43. Doppler Beam-Sharpened Image of a Runway Complex. (Photograph courtesy of Hughes Aircraft Company, Radar Systems Group)

θ_3 = the antenna's 3-dB beamwidth
n = the number of the particular DBS resolution cell in the beam ($n = 0$ is closest to the ground track)
N_X = the total number of DBS resolution cells in the beamwidth
θ_n = the angle to the nth DBS cell in the beam
$\Delta\theta$ = the angular width of each DBS cell
V_A = the aircraft's velocity over the ground
R = the range to the range bin being analyzed
f_T = the transmit frequency
θ_H = the antenna's depression angle from horizontal
v_{Rn} = the radial velocity of the a target in the nth cell
c = the velocity of propagation of the EM waves
f_{dn} = the Doppler shift for a target in the nth cell
Δf_X = the Doppler resolution required for N cells in the θ_3 beamwidth
T_D = the minimum necessary look time
R_{TD} = the distance the aircraft travels in the direction of the beam during T_D
N_P = the number of pulses transmitted during T_D

DBS analysis is performed as follows. The angular width of DBS cells is

$$\Delta\theta = \theta_3 / N_X \qquad (13\text{-}20)$$

The angular position of the nth cell from the edge of the beam closest to the flight path is

$$\theta_n = \theta_0 - \theta_3/2 + n\Delta\theta + \Delta\theta/2 \qquad (13\text{-}21)$$

The radial velocity of the target in the nth cell is

$$v_{Rn} = V \cos\theta_n \cos\theta_H \qquad (13\text{-}22)$$

The Doppler shift of the target in the nth cell is

$$f_{dn} = 2 f_T v_{Rn} / c \qquad (13\text{-}23)$$

The required Doppler resolution to separate targets in adjacent cells is

13-7. Doppler Beam-Sharpening (DBS)

$$\Delta f_X = f_{d1} - f_{d2} \quad \text{(Worst case)} \tag{13-24}$$

The required look time to achieve this Doppler resolution is

$$T_D \approx 1 / \Delta f_X \text{ (maximum)} \tag{13-25}$$

The cross-range resolution after DBS processing is

$$\Delta X_{DBS} \approx \Delta\theta \, (\pi/180) \, R \; (\Delta\theta \text{ in degrees}) \tag{13-26}$$

The distance the radar travels in the look time is

$$R_{Td} \approx V_A T_D \tag{13-27}$$

The number of pulses which are received during the look time is given below, but in no case can be less than N_X. If FFT algorithms are used to extract Doppler information, N_p must usually be an integer power of two. Rather than restrict the process to certain PRFs, the dwell time may be extended to a value greater than specified in Eq. 13-26, resulting in resolutions no worse than those desired.

$$N_p = PRF \, T_D \tag{13-28}$$

In Doppler beam-sharpening, the surface to be imaged is scanned by the antenna as the radar flies past it. Figure 13-44 shows the geometry of one pulse in a DBS process. As with all digital radars, the beam's footprint is divided up into range bins. Unlike many other modes, these bins may have to be stabilized to the target surface rather than to the transmit pulses (Fig. 13-45).

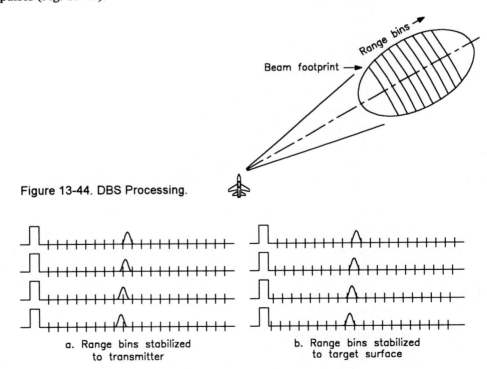

Figure 13-44. DBS Processing.

a. Range bins stabilized to transmitter

b. Range bins stabilized to target surface

Figure 13-45. Range Bins' Stabilization.

After several pulses have been gathered, a spectrum analysis is performed on the data for each range bin. Figure 13-46 shows four echoes from one range/azimuth bin. The result of the spectrum analysis is that the real beam of the antenna is resolved into four approximately equal parts. The displayed image is built from DBS processes on several real beams, as shown in Fig. 13-47.

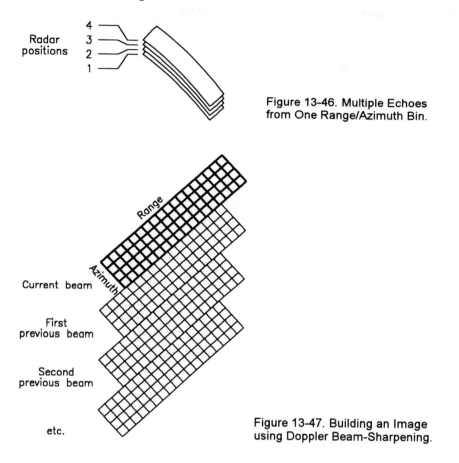

Figure 13-46. Multiple Echoes from One Range/Azimuth Bin.

Figure 13-47. Building an Image using Doppler Beam-Sharpening.

PRF: Doppler beam-sharpening is usually accomplished at medium PRF. One criterion for PRF is that the synthesized antenna elements be close enough to one another to prevent antenna grating lobes from forming. In Ch. 9, it was determined that elements in a real array must be spaced less than one-half wavelength apart to prevent grating. In synthetic arrays, this reduces to one-fourth wavelength. Because the elements of the synthetic array (the real antenna) are highly directional, the array multiplication principle makes some grating tolerable.

If the values of the above example were used, the transmitter must fire each time the aircraft moves 0.0075 m, which is one-fourth the wavelength. At 205.78 meters per second, the aircraft moves 0.0075 m in 36.4 μs, giving a minimum PRF for the grating criterion of 27,440 pps. Because the beamwidth of the real antenna (the "element" in the array multiplication principle — Ch. 9) is relatively narrow, grating on the order resulting from two wavelength element spacings is usually tolerable. This gives a modified minimum PRF of 3,430 pps.

Although medium PRF is ambiguous in range and Doppler, deghosting may not be necessary in DBS modes, particularly when used for ground mapping. Because only ground targets are of interest and because the real antenna's beam footprint on the ground restricts the range and Doppler from which echoes can be received (ignoring sidelobes), it is often possible to restrict echoes to a single range and Doppler zone, provided the PRF is matched to the radar's velocity and the illumination geometry. See Figs. 13-48 and 13-49; Fig. 13-49 shows the Doppler spectrum of the example which introduced this section.

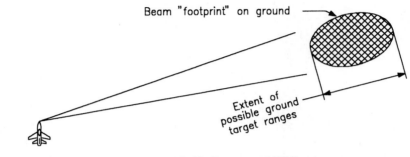

Figure 13-48. Range and DBS.

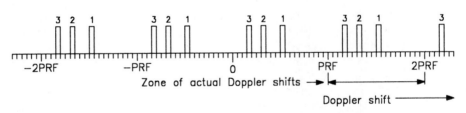

Figure 13-49. Doppler Spectrum of Example — Medium PRF.

If the PRF in the introductory example were 3,500 pps, 233 pulse returns will be needed (67 ms at the PRF). In 67 ms, the aircraft moves (at 400 Kt) about 13.7 m. If the range resolution is less than about 15 m, the targets have a high probability of moving from one bin to another over the data gathering period. This phenomenon is called range bin walking and defocuses the process (reduce its resolution) unless compensated for (Fig. 13-45). Range resolution is usually set to approximately match cross-range resolution.

The discussions of PRF for synthetic aperture radar (Sec. 13-7) also apply to Doppler beam-sharpening.

Example 13-5: A radar traveling at 380 Kt 1000 m above a flat surface is required to make a high resolution map of a roughly square ground sector 24° wide. The center of the segment is 30° from the ground track of the radar and the nominal slant range to the center of the sector is 10.0 km. The real antenna beamwidth is 3.0° wide and is to be divided into 16 segments using Doppler beam-sharpening. The final result is to be an array of 128 x 128 square pixels. The radar's frequency is 10.0 GHz. Everything is to be designed around the *center* pixel in the sector. Find:

a. The number of beams which must be processed
b. The nominal cross-range resolution at the center of the array
c. The range resolution and signal bandwidth
d. The number of range bins
e. The look time for each beamwidth
f. The total scan time and the antenna scan rate
g. The number of points in each Fourier transform and the minimum PRF
h. The minimum PRF and FFT size if the array grating criterion is met
i. The total number of Fourier transforms

Solution: a. The number of beams in the sector is simply the total sector width divided by the beamwidth, or 8.

b. At a range of 10,000 m, the 3° beam footprint is, from basic geometry, about 525 m wide (in the azimuth direction). Since each beam is to be divided into 16 parts, the segments are about 33 meters across, and this is the cross-range resolution.

c. The pixels are given in the problem to be square. Therefore the range resolution is also 33 m. By Eq. 13-3, the signal bandwidth required is about 4.6 MHz.

d. The sector is 8 beams wide (part a), each beam covering 525 m of cross-range (part b), for a total cross-range of about 4,200 m. Since the sector is to be roughly square, 4,200 m of range must also be covered. Since each range bin is 33 m, there are 128 range bins.

e. The segments nearest the flight path are the hardest to resolve (less velocity difference). The angular width of each segment ($\Delta\theta$) is 0.1875° (3° / 16). The first (0th) segment is at an angle of 28.594°. The next (1st) segment is at 28.781°. The depression angle is 5.74° (arcsine of the altitude divided by the range). The radial velocity and Doppler shift in the 0th segment is 170.785 m/s and 11,385.67 Hz, respectively. The first segment's values are 170.480 m/s and 11,365.37 Hz. The Doppler difference (and therefore the Doppler resolution) is 20.3 Hz. The look time (Eq. 13-7) is therefore about 50 ms. Remember, however, that the signal processing window will degrade this by a factor of about 1.5, giving a real look time of about 75 ms.

f. The total time for the scan is the time per beam (75 ms from part e) multiplied by the number of beams in the scan (8 from part a), or 600 ms. The *maximum possible* antenna scan rate is 24° in 600 ms, or 40°/s.

g. Each beam is divided into 16 parts, which dictates a minimum of 16 points in each transform. However, to accommodate the degradation caused by the window, the number of points needs to be about 24. Since FFTs calculate much faster if $N = 2^m$, where m is an integer, the minimum number of points in each transform is 32. Thirty-two pulses in 75 ms gives a minimum PRF of 427 pps.

h. The grating criterion specifies that the radar move one-fourth wavelength or less between samples. The wavelength is 0.03 m and at 380 Kt, the aircraft moves 0.0075 m in 38.4 μs, giving a PRF of 26,080 pps. The look must still be 75 ms, during which time a minimum of 1954 samples are taken. The FFT size must thus be 2048. If the grating criterion is relaxed so there are two wavelengths between samples, the minimum PRF is 3,258 pps and the FFT size is 256.

i. There are 8 beams in the scan (part a), each having 128 range bins, and requiring at least one 32-point FFT per range bin (range bins are processed independently). This is 128, 32-point FFTs per beam times 8 beams, or 1,024, 32-point transforms.

Note: The above results are only approximate because many of the relationships used are approximations and because of the non-linearity of the Doppler shift across the 24° sector. A perfectly flat target surface was also assumed. Reality complicates this considerably.

Flight path perturbations: The above discussions assume that the radar's flight path is a straight line. If the path differs materially from the line over the look time, the signal phases must be corrected before the FFT is taken. The distance error is found from the aircraft's inertial navigation system (INS). The phase correction to be applied to each signal sample is given as

$$\delta\phi_i = 2\pi\,\delta R_i / \lambda \text{ radians} \tag{13-29}$$

$\delta\phi_i$ = the phase correction to be applied to signal sample i

δR_i = the range error from straight-line flight for signal sample i

λ = the wavelength

Inverse Doppler beam-sharpening (IDBS): It is also possible, using DBS principles, to enhance the resolution of moving targets from a fixed radar. This process, shown in Fig. 13-50, is particularly useful in resolving aircraft flying in formation. The radial velocity differences resolve the targets, with the absolute velocity of the formation contaminating the measurement, just as with DBS and moving targets. The average azimuth can, however, be obtained from tracking. As with DBS, IDBS works only when the velocity of the formation is in a direction different from the antenna's axis. This technique is available to help with the so-called raid discrimination problem.

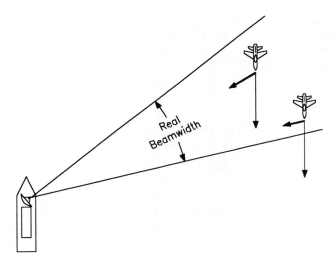

Figure 13-50. Inverse Doppler Beam-Sharpening (IDBS).

Example 13-6: Two targets flying at 350 Kt in formation 120 m apart in cross-range must be resolved. Both targets are at the same range of 35 km. Their flight path is at a 15° angle to the axis of a shipboard tracking radar at a frequency of 3.00 GHz. Find the minimum time the radar must track the targets to resolve them. How far do the targets travel in this time? Is range bin walking a problem?

Solution: From basic geometry, the angular difference between the two targets is 0.196°. Assume one target is at 15.000° and the other is at 15.196° from the radar's axis. Their radial velocities are 173.906 and 173.746 m/s and their Doppler shifts are 3,478.13 and 3,474.92 Hz, respectively. The Doppler shift difference is 3.21 Hz. The time required to resolve them is theoretically 312 m. (Eq. 13-7). Because of processing windows (increasing the resolution required by about 1.5) and the need to separate the targets by at least one empty Doppler bin if they are to be seen as two targets on a display (effectively doubling the resolution requirement), the signal gathering time increases to about one second. During this time, the targets move about 175 m. Since this is a tracking radar, the range gate(s) follows the target and walking does not occur.

13-8. Sidelooking Synthetic Aperture Radar

Sidelooking SAR is an airborne technique which, as its name implies, uses antennas which look perpendicular to the flight path of the radar [13]. Figure 13-38 shows the principle. Its primary use is ground mapping. Military uses are typically for reconnaissance. It has been used on earth satellites to obtain radar maps of the earth's surface, and on other spacecraft to map the surface of planets (for example, the Magellan spacecraft mapping Venus).

Sidelooking SAR ground mapping from aircraft generally illuminates the surface at a low angle, resulting in shadows being cast by higher objects (Fig. 13-51). This gives a dramatic 3-D effect which yields considerable terrain height information.

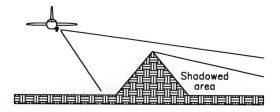

Figure 13-51. Shadowing in Sidelooking SAR.

Consider Figure 13-39. In the real array, signals received by the array elements are summed to form the beam. The signal received at each element is the sum of echoes from the illumination from all elements. The array output, ignoring the space loss (range to targets) and element pattern, is described as

$$v_{RE} = \{ \sum_{n=1}^{N} \exp[-j\,(2\pi/\lambda)\,d_n] \}^2 \tag{13-30}$$

v_{RE} = the voltage out of the summation of the real array

A_n = the amplitude (taper) response of the nth element
d_n = the distance from the nth array element to the target
N = the total number of elements in the array

The antenna beam for sidelooking SAR is formed in much the same way as that of the real array. The primary differences are (a) the array is formed sequentially instead of simultaneously, necessitating storing the elemental signals until all array elements have been formed, and (b) the echo signal received by each element is caused by the illumination from that element only. The output, again ignoring space loss and element pattern, is

$$v_{SY} = \sum_{n=1}^{N} \{ \exp[-j\,(2\pi/\lambda)\,d_n] \}^2 \qquad (13\text{-}31)$$

v_{SY} = the voltage out of the summation of the synthetic array

The resulting patterns, real and synthetic, differ somewhat, as shown in Fig. 13-52, which gives two-way patterns for both. The synthetic array beam is approximately half the width of the real beam from an array of the same extent. The sidelobes of the synthetic array are higher than those from a like-sized and like-tapered real array. The approximate beamwidths for real and synthetic arrays are

$$\theta_{3R} = \lambda / D_{\text{eff}} \qquad (13\text{-}32)$$

θ_{3R} = the 3-dB beamwidth of a real array (radians)
λ = the wavelength
D_{eff} = the effective width of the real array

$$\theta_{3S} = \lambda / 2L_{\text{eff}} \qquad (13\text{-}33)$$

θ_{3S} = the 3-dB beamwidth of a synthetic array (radians)
λ = the wavelength
L_{eff} = the effective length of the synthetic array

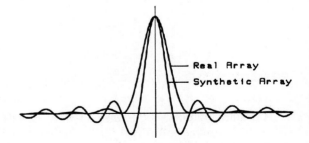

Figure 13-52. Real and Synthetic Beams.

Sidelooking SAR resolution: The cross-range resolution of a sidelooking SAR system is described in terms of the geometry of Fig. 13-53. The effective length of the synthetic antenna (L_{eff}) is the distance the radar moves while a scatterer remains in the beam. It has the same value as the cross-range resolution of the real antenna. For small real beamwidths, this value is

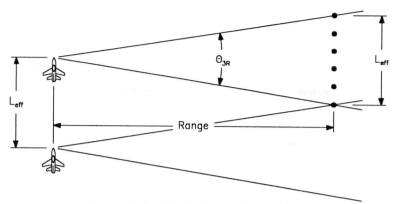

Figure 13-53. Sidelooking SAR Geometry.

$$\Delta X_R = L_{\text{eff}} = R\,\theta_{3R} = R\lambda/D_{\text{eff}} \qquad (13\text{-}34)$$

ΔX_R = the real antenna cross-range resolution at range R

R = the range to the scatterer in question

The cross-range resolution of the synthetic antenna at range R is, by basic resolution relationships

$$\Delta X_S \approx \theta_{3S} R \qquad (13\text{-}35)$$

ΔX_S = the synthetic antenna's cross-range resolution

Substituting Eq. 13-33 into Eq. 13-35 gives

$$\Delta X_S \approx \frac{\lambda R}{2 L_{\text{eff}}} \qquad (13\text{-}36)$$

Substituting Eq. 13-34 into Eq. 13-36 gives

$$\Delta X_S \approx \frac{\lambda R}{2} \frac{D_{\text{eff}}}{R \lambda} \qquad (13\text{-}37)$$

Algebra reduces this equation to

$$\Delta X_S = D_{\text{eff}}/2 \qquad (13\text{-}38)$$

Equation 13-38 is rather amazing in several aspects (and when initially developed was not believed). First, the cross-range resolution is independent of range, implying a non-diverging antenna beam. This is explained by realizing that the synthetic antenna's length is a linear function of range; the synthetic antenna is larger to a long range target than to one at a short range (Fig. 13-54).

Second, cross-range resolution and the synthetic antenna's "beamwidth," is not a function of wavelength. Although the synthetic beamwidth (Eq. 13-32) broadens with longer wavelengths, so does the real beamwidth (Eq. 13-33). Thus, the synthetic antenna is larger at long wavelengths than short, cancelling the broadening of the synthetic beam.

Finally, the synthetic resolution is better with *smaller* real antennas, exactly opposite of the relationship with real antennas. The explanation for this effect is easy if one refers

13-8. Sidelooking Synthetic Aperture Radar

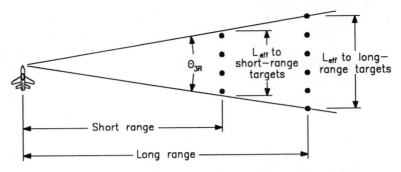

Figure 13-54. Effect of Target Range on Sidelooking SAR.

to Fig. 13-55. The smaller the real antenna, the wider its beamwidth and thus the greater the length of the synthetic antenna. There is, of course, a limit as to how small the real antenna can be, since it needs enough gain and aperture to assure an adequate signal-to-noise ratio.

Example 13-7: Find the length of the synthetic antenna necessary to produce cross-range resolution of 1.0 m at a range of 5.0 km. Find also the maximum length of the real antenna and the distance to the far field of the synthetic antenna. The radar's frequency is 10.0 GHz.

Solution: The synthetic antenna's length, from Eq. 13-36, is 75 m. The maximum length of the real antenna (Eq. 13-38) is 2.0 m. The distance to the far field of the synthetic antenna is $2D^2 / \lambda = 94$ km.

Comment: Obviously, the target space at 5-km range is in the near field of the synthetic antenna. This is significant in that the processing will have to focus the antenna at the 5-km range (actually, the antenna may have to be focused to the range of each pixel in the image). This is the meaning of focused processing in SAR — see below.

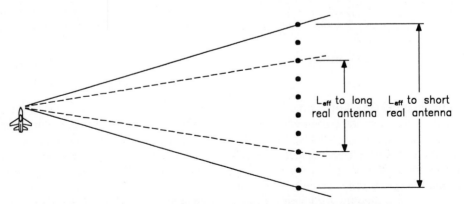

Figure 13-55. Effect of Real Antenna Size on Sidelooking SAR.

Unfocused SAR resolution: The above relationships are realizable only if the data are corrected to simulate a circular flight path around each scatterer. This process, called focusing, is discussed later in this section. If the processing is unfocused, the scatterers must always be in the far field of the synthetic antenna, limiting its length to the following value (see Fig. 13-56).

$$L_{eff}^2 < \lambda R / 2 \tag{13-39}$$

Substituting this maximum value for effective synthetic array length into the SAR resolution relationship (Eq. 13-36) gives the best resolution for an unfocused SAR system.

$$\Delta X_{S(U)} = (R \lambda / 2)^{1/2} \tag{13-40}$$

$\Delta X_{S(U)}$ = the cross-range resolution for unfocused SAR

Actual resolution, whether focused or not, is marginally coarser than that given in Eqs. 13-38 and 13-40 because of the windows used in processing.

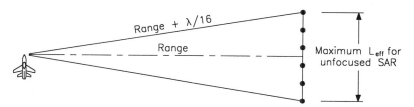

Figure 13-56. Limits of Unfocused Sidelooking SAR.

Sidelooking SAR at angles other than normal to the ground track: Sidelooking SAR techniques can be used in situations where the antenna beam does not point at a right-angle to the ground track. In this case, the resolution is coarser than predicted by Eqs. 13-38 and 13-40, and corrections must be made to the data to remove translational motion. The effective length of the synthetic array is now shorter than the flight path with the target in the beam. The approximate flight path length for a given effective length is calculated from the following equation.

$$L_{FP} \approx L_{eff} / \sin \theta_{AZ} \tag{13-41}$$

L_{FP} = the flight path (ground track) length corresponding to the effective SAR length L_{eff}

θ_{AZ} = the azimuth angle between the ground track and the antenna beam axis

Note that as the azimuth angle approaches zero, the flight path length approaches infinity for a finite synthetic array effective length, and SAR is not possible. *SAR is only possible at azimuth angles away from the ground track.*

When the azimuth angle is less than 90°, each consecutive signal sample is taken with the scatterer of interest at a different range from the radar. The phase of the signal is simply corrected for this translational motion and the processing is similar to sidelooking SAR. The phase correction, from the geometry of Fig. 13-57 and ignoring depression angle (which is not a constant and must be calculated), is given in the following equation.

13-8. Sidelooking Synthetic Aperture Radar

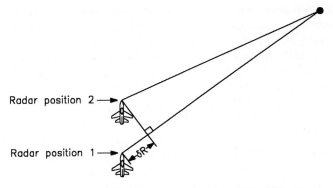

Figure 13-57. SAR at Angles other than 90°.

$$\delta R = S \sin \theta_{AZ} \tag{13-42}$$

$$V_{Si}' = V_{Si} \exp[-j\,2\,\pi\,\delta R / \lambda] \tag{13-43}$$

S = the flight-path distance between samples

δR = the range difference between samples

θ_{AZ} = the azimuth angle

λ = the wavelength

V_{Si}' = the voltage received from the target after correction for non-normal azimuth angle

V_{Si} = the voltage received from the target before correction for non-normal azimuth angle

Processing for unfocused SAR: Unfocused SAR processing is simply a matter of collecting the signals over the required flight path distance, correcting for non-normal azimuth angle and summing the results. It is illustrated in Fig. 13-58 (after [6]).

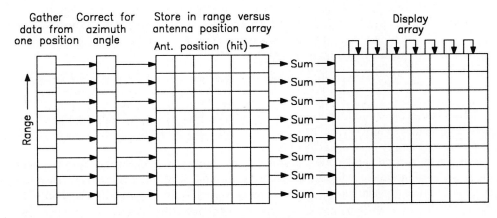

Figure 13-58. Unfocused SAR Processing.

The number of hits to needed to provide the required resolution are first gathered, then corrected for non-normal azimuth angle, and stored. Each hit is divided into range bins. After the necessary number of hits is gathered, the signals from the different antenna positions are coherently summed, one sum per range bin. The result is placed in a display array. To yield a continuously moving display, the display array is shifted for each summing process. The process is best illustrated with a numeric example.

Example 13-8: An unfocused sidelooking SAR radar is moving 480 Kt over the ground. The azimuth angle is 90° and the azimuth and elevation beamwidths are 4.0°. The transmit frequency is 9.75 GHz and the PRF is 4,000 pps. The average range to the footprint is 20,000 m and the radar altitude is 4,500 m above the target surface. The resolution is the maximum possible for the situation. Find the resolution and describe the processing.

Solution: The wavelength is 0.030 m. Antenna theory says that in order for the beam footprint to be in the far field of the synthetic array, the length of the synthetic array cannot exceed 17.3 m ($R > 2D^2/\lambda$). From Eq. 13-40, the resolution is also 17.3 m (about 60 ft).

Flying 17.3 m at 480 Kt (246.9 m/s) requires 70.06 ms. At 4,000 pps, 280 pulses are fired in this distance. The depression angle of the beam center is, from basic trigonometry, the arcsine of 4,500 m divided by 20,000 m, or 13.0°. The depression angles to the edges of the footprint are from 11.0° (the beam-center depression angle minus half the elevation beamwidth) to 15.0° (the beam-center depression angle plus half the elevation beamwidth). The footprint on the ground extends from a slant range of 17,390 m (4,500 m divided by the sine of 15°) to 23,580 m (4,500 / Sin 11°), for a slant length of 6,190 m. If the range resolution is set to match that in cross-range, each range bin is 17.3 meters wide and there are about 350 of them.

Thus the process is (1) pulse the transmitter and gather 350 range bins of I/Q data; (2) at 4,000 pps, repeat this gathering 280 times; (3) sum the 280 samples for each range bin; (4) load this sum into a display column vector approximately 6,190 m (350 bins) in range by 17.3 m in cross-range; (5) repeat steps 1 through 3; (6) shift the display vector from the previous process one column to the right and load the new 6,190 m (range) by 17.3 m (cross-range) vector into the left-most display column; (7) repeat the above steps indefinitely, sliding the display column 17.3 m to the right for each 17.3 m the aircraft flies.

The end result is a display of 17.3-m by 17.3-m pixels which moves as the aircraft moves. Usually, about the same number of cross-range and range columns are maintained. Thus the display is approximately 350 by 350 pixels, or a total of 122,500 pixels. Relative RCS can be calculated and displayed in each pixel with a gray scale or with pseudo-colors (one color for each RCS range).

Processing for focused SAR: Focusing of the synthetic array involves some process which corrects the elemental signal phases such that all elements of the resulting array are at the same distance from the target, after correction for azimuth angles other than 90°. Figure 13-59 shows the data from one sample set as gathered (Fig. 13-59a) and as corrected for range differences (Fig. 13-59b).

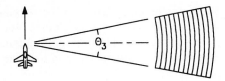 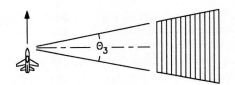

a. Location of data on target surface before focusing correction

b. Location of data on target surface after focusing correction

Figure 13-59. Data before and after Focusing.

Correcting the phase for a single azimuth/range cell is a matter of applying the principle shown in Fig. 13-60, which is to simulate a flight path which is a circle around the cell being processed. The phase correction for the nth array element for the image cell shown is given in the figure and in the equations below. A correction must be applied for each cell in the image if complete focusing is to be done.

By the Pythagorean theorem,

$$(\delta R_n + R)^2 = R^2 + (n\,S)^2 \qquad (13\text{-}44)$$

R = the range from the broadside SAR element to the scatterer position being corrected

δR_n = the range difference between the broadside SAR element and the nth element

n = the number of the element being corrected

S = the flight-path spacing between elements

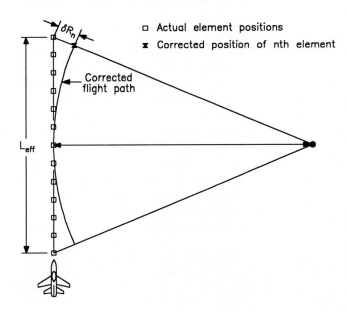

Figure 13-60. Focusing Principle.

Solving the above equation for δR_n assuming $(dR_n / 2R) \ll 1$ gives

$$\delta R_n = \frac{n^2 S^2}{2 R} \qquad (13\text{-}45)$$

The phase error associated with the range error is:

$$\delta\phi_n = \frac{2\pi (2 \delta R_n)}{\lambda} = \frac{2\pi n^2 S^2}{R \lambda} \qquad (13\text{-}46)$$

$\delta\phi_n$ = the phase error associated with the focusing range error

The 2 in $2 \delta R_n$ accounts for the round trip

The process of recovering the image is as shown in Fig. 13-61. The data array consists of Is and Qs from each range bin (columns) for each array element (rows). After correcting for azimuth angle, the data array is "masked" as shown in the figure. The mask represents the real antenna beam, and the data within the mask is the data for one SAR process. Phase corrections are applied to the entire data array within the mask, transforming the data from the representation of Fig. 13-59a to that of Fig. 13-59b. The results are summed across the corrected range bins. This sum is the image pixel at the range and cross-range being processed. The data array is then stepped one image pixel distance further along the flight path and the process repeated. If the processor is fast enough, an image strip appears on the display in essentially real-time. If not, a group of array element data points is gathered, stored, and post-processed to give an image.

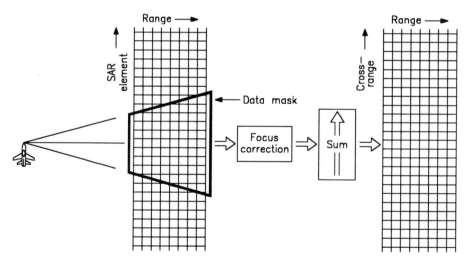

Figure 13-61. Focusing Process (one of many).

Another way to focus is to, for each pixel in the image, calculate what the data array would contain if a scatterer of unity RCS were present in that pixel, and then 2-D correlate this hypothetical data with the actual data. The result is the RCS present in that pixel. This method, although exact, is highly compute-intensive and may not be practical with present-day tactical processors. For example, it was shown in Ch. 11 that a 1-dimensional correlation

13-8. Sidelooking Synthetic Aperture Radar

required three Fourier transforms (two forward and one inverse) plus some overhead. A 2-D $N \times N$ correlation requires N 1-D N-point correlations. Thus to construct a 512 x 512 pixel image requires 512^2 or 262,144 (256K) 2-D correlations. Each 512-point 2-D correlation requires 512 1-D correlations, or 1536 512-point Fourier transforms. The total Fourier transform count for the 512 x 512 image is 402,653,184. Processing shortcuts exist, including unfocused processing of a small group of pixels. Further exploration is beyond the scope of this text.

Yet another method, taking advantage of the apparent rotation of the scatterers as the radar flies by, is presented in Sec. 13-9.

Sampling and PRF for sidelooking SAR: As in Doppler beam-sharpening, three criteria determine the PRFs possible in sidelooking synthetic aperture radar: (a) grating lobe considerations, (b) range ambiguities, and (c) Doppler ambiguities. See Fig. 13-62 and assume the net radial motion of the target space is removed.

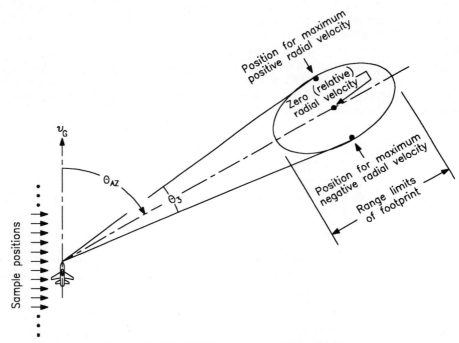

Figure 13-62. SAR Geometry for PRF Selection.

Grating lobes are generated in synthetic arrays if the distance between elements exceeds one-fourth wavelength. The minimum PRF to avoid grating in sidelooking SAR where the beam angle is normal to the ground track is given below. As with DBS, the grating criterion is often not met, with consequences which depend on the real antenna's pattern and the system's tolerance to sidelobe echoes.

$$PRF_{\min(G)} = 4 v_G / \lambda \tag{13-47}$$

$PRF_{min(G)}$ = the minimum PRF to avoid grating
v_G = the radar's ground velocity
λ = the wavelength

The second criterion, range ambiguity, is satisfied if the entire beam footprint is spread over no more than one range zone's extent. If spitting the footprint can be tolerated (Fig. 13-63), this criterion is met if the PRF is less than the time equivalent of the footprint's length along the ground in the direction of the antenna's axis.

$$PRF_{max(R1)} = c / 2 \; \Delta R_F \qquad (13\text{-}48)$$

$PRF_{max(R1)}$ = the maximum PRF so that the footprint length is within the extent of one range zone, but not necessarily within a single zone

c = the electromagnetic velocity of propagation

ΔR_F = the length of the beam footprint along the ground in the direction of the antenna's axis.

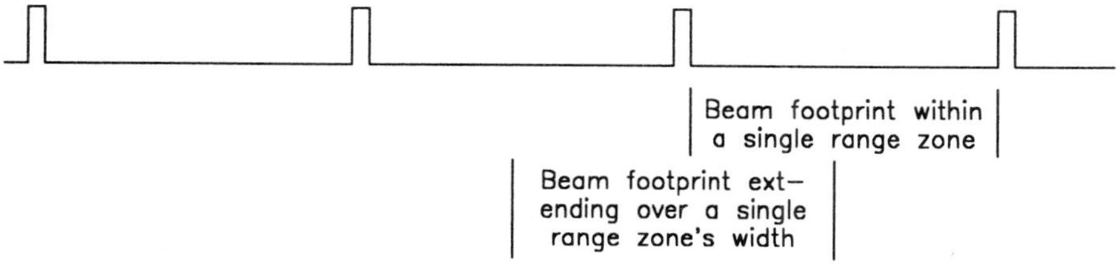

Figure 13-63. Beam Footprint and Range Zones.

If splitting the footprint into two range zones is not tolerable, the PRF must satisfy the following relationships.

$$n_{RL}' = R_{FL} \; \text{DIV} \; (c / 2PRF_{max(R1)}) + 1 \qquad (13\text{-}49)$$

n_{RL}' = the range zone within which the beam footprint's leading edge (closest to the radar) lies for the PRF calculated in Eq. 13-48 (zone numbers are named as in Ch. 2)

R_{FL} = the range to the leading edge of the beam footprint

DIV = the integer division operation. $x \; \text{DIV} \; y = \text{INT}(x/y)$, where INT is "the integer value of"

$$PRF_{(R2)} = [c / 2R_{FL}] \; [n_{RL}' - 1] \qquad (13\text{-}50)$$

$PRF_{(R2)}$ = a sample rate which will place the leading edge of the footprint at the beginning of a range zone (at zero apparent range) and whose range zone is long enough to accommodate the entire footprint. There are other rates which satisfy the range condition.

The third criterion for PRF selection is that the Doppler shifts across the beam footprint lie within the extent of one Doppler zone. Again, if the spectrum can be split between two zones while keeping an extent of one zone or less, the calculation is simpler than if the spectrum must be contained within a single zone. See Fig. 13-64.

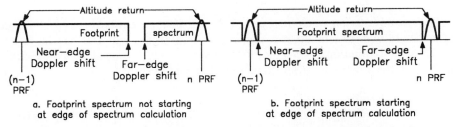

Figure 13-64. Two Possible Footprint Spectra without Doppler Ambiguity.

The radial velocity difference across the footprint is the maximum positive radial velocity possible minus the maximum negative (or minimum positive, depending on the geometry and the azimuth angle) value possible, as shown in Fig. 13-62. The velocity difference is

$$\delta v_r = v_G \cos(\theta_{EL}) [\cos(\theta_{AZ} - \theta_3/2) - \cos(\theta_{AZ} + \theta_3/2)] \quad (13\text{-}51)$$

δv_r = the maximum radial velocity difference across the beam footprint

v_G = the ground velocity of the radar

θ_{EL} = the beam depression angle from horizontal

θ_{AZ} = the beam azimuth angle from the radar's ground track

θ_3 is the azimuth beamwidth

If the footprint spectrum is to be unambiguous but can be spread over two Doppler zones, the required minimum PRF is the Doppler equivalent of the velocity difference.

$$PRF_{\min(D1)} \approx 2 f_T \delta v_r / c \quad (13\text{-}52)$$

$PRF_{\min(D1)}$ = the minimum PRF required for the footprint spectrum to be unambiguous, but where it can cross between two Doppler zones

f_T = the transmit frequency

Use of this PRF may result in part of the footprint spectrum being lost to the altitude return, as shown in Fig. 13-64a.

The set of PRFs which places the lowest footprint spectral component at the lowest usable frequency in the spectrum, and at the same time prevents aliasing of the total footprint spectrum is given below. *The value chosen must be greater than that calculated from Eq. 13-52, plus the bandwidth of the altitude return.*

$$PRF_{(D2)} = [1/n_D] [f_{d(FF)} - \delta f_{d(AR)}/2] \quad (13\text{-}53)$$

$PRF_{(D2)}$ must exceed $PRF_{\min(D1)} + \delta f_{d(AR)}$

$$f_{d(FF)} = 2 f_T v_{r(FF)} / c \quad (13\text{-}54)$$

$$v_{r(FF)} = v_G \cos(\theta_{EL}) \cos(\theta_{AZ} + \theta_3/2) \quad (13\text{-}55)$$

$PRF_{(D2)}$ = the PRF which places the lowest Doppler shift in the footprint at the lowest usable part of the calculated spectrum. *It must be greater than* $PRF_{min(D1)}$, *from Eq. 13-52.*

n_D = an integer which makes $PRF_{(D2)}$ exceed $PRF_{min(D1)}$ + $\delta f_{d(AR)}$

$f_{d(FF)}$ = the Doppler shift at the footprint's far edge (the maximum negative value in Fig. 13-63)

$\delta f_{d(AR)}$ = the spectral bandwidth of the altitude return

$v_{r(FF)}$ = the radial velocity of the footprint's far edge (Fig. 13-62)

Example 13-9: A 9.60 GHz sidelooking radar has a real antenna with a pencil beamwidth of 3.5°. The azimuth angle is 33° from the ground track. The aircraft carrying the radar is flying at 450 Kt, 2000 m above a flat, horizontal ground surface. The center of the beam intersects the ground at a distance of 15 km slant range from the radar. The altitude return has a bandwidth of 700 Hz. Find the possible PRFs to avoid (a) grating, (b) range ambiguity, and (c) Doppler ambiguity.

Solution: (a) The wavelength is 0.03125 m. At 450 Kt (231.48 m/s), the radar travels 7407 wavelengths per second. If the half-wavelength spacing criterion is used, the minimum PRF is 14,814 pps. If a two-wavelength spacing is sufficient, the minimum PRF becomes 3704 pps.

(b) The depression angle is the arcsine of the altitude divided by the slant range, or 7.662°. The leading edge of the footprint touches the horizontal surface at a range of 12,230 m (altitude divided by the sine of the depression angle plus one-half the beamwidth). The footprint trailing edge touches the horizontal surface at a range of 19,420 m (altitude divided by the sine of the depression angle minus one-half the beamwidth). The range span of the footprint is therefore 19,420 − 12,230, or 7190 m. From Eq. 13-48, the maximum PRF if zone-splitting is tolerable is 20,862 pps. If the footprint must be within one zone, Eq. 13-50 gives the PRF as 12,264 pps.

(c) The velocity difference between the near and far edges of the footprint is 7.631 m/s (Eq. 13-51). The Doppler difference corresponding to this velocity difference is 488.4 Hz. If the footprint spectrum does not have to exclude the altitude return, the minimum PRF (Eq. 13-52) is 489 pps. If the spectrum must exclude the altitude return, the PRF (from Eqs. 13-53, 13-54, and 13-55) is 11,314 Hz, divided by an integer such that the quotient exceeds 1189 Hz ($PRF_{min(D1)} + \delta f_{d(AR)}$). One such value of PRF is 1257 pps ($n_D = 9$). Another is 1414 pps ($n_D = 8$), and so forth.

Flight-path perturbation: As with Doppler beam-sharpening, flight-paths are assumed in the processing to be straight lines. If not, a correction must be applied. See Sec. 13-6 for the details.

13-9. Synthetic Aperture With Rotating Objects

In sidelooking synthetic aperture radar (Sec. 13-7), the apparent rotation of targets as the radar flies past them contributes to image defocusing. In this section, methods are discussed to take advantage of the rotation to enhance cross-range resolution. The most fundamental application, inverse synthetic aperture radar (ISAR) with rotating objects is discussed first. In Fig. 13-40, the radar is stationary and the target rotates about an axis. This technique is often used in the evaluation of targets by systems known as diagnostic radars. See Mensa [2] for more information.

As with sidelooking SAR, the processing can be unfocused or focused. In unfocused processing, scatterers are resolved in range using pulse compression. Cross-range information comes from the Doppler spectrum of each range bin. Data is gathered for each range bin as a function of the target rotation angle (aspect − horizontal, and tilt − vertical). Spectrum analysis is performed on the multiple samples within each range bin and the resulting spectrum is displayed as cross-range.

The image is formed as follows (Fig. 13-65). The object to be imaged is rotated to a particular angle (aspect or tilt). The radar views the target with an antenna whose beam is wide enough to encompass the entire target and records a high resolution "range profile" (Fig. 13-65a). The target is then rotated by an amount discussed later in this section and another range profile is taken (Fig. 13-65b). Note that in the second sample, scatterer 3 is a small distance closer to the radar and scatterer 5 is a larger distance closer. Note also that scatterers 2 and 4 are farther from the radar, with 4 having moved farther than 2. This process is repeated until data are gathered for the total rotation angle necessary to give the desired resolution.

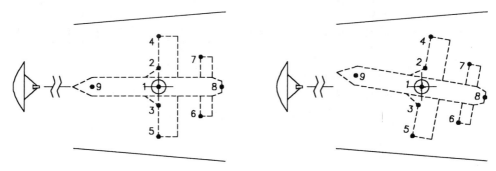

a. Target aspect for first sample.   b. Target aspect for second sample.

Figure 13-65. ISAR with Rotating Targets.

The result is a data array containing I and Q information for each angle/range cell (Fig. 13-66). The figure is somewhat misleading in that arcs are shown as straight lines in the aspect direction. The focusing discussion in this section addresses this concern.

The image is formed by finding the spectrum of the different-aspect samples of each range bin. The process is shown in Fig. 13-67, and a simulated result for the above example is shown in Fig. 13-68. Because the range bin sets are actually curved while a Cartesian coordinate (straight-line) Fourier transform is used, the solution is inexact and hence

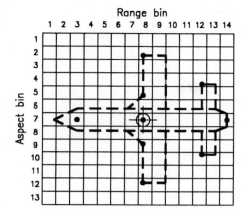

Figure 13-66. Data for Unfocused Processing of Rotating Targets.

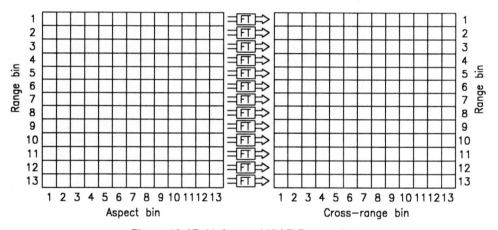

Figure 13-67. Unfocused ISAR Processing.

unfocused. Focusing errors are evident in Fig. 13-68 because of the smearing of some of the point scatterers. Note that scatterer 1, on the axis of rotation, is narrower than the others and is in fact focused. Later discussions of focusing will give methods of effectively placing each scatterer at the axis of rotation. The resultant image is simulated in Fig. 13-69. The image level in each pixel is proportional to the radar cross-section of the scatterer(s) in that pixel.

13-9. Synthetic Aperture With Rotating Objects

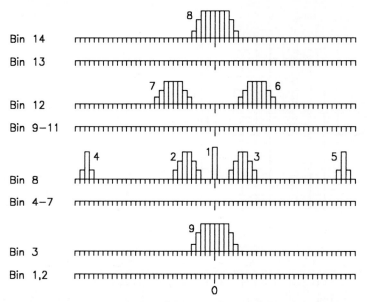

Figure 13-68. Representation of the Spectra of Scatterers in Fig. 13-67.

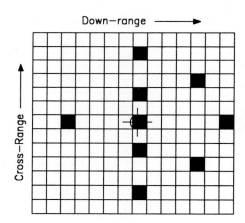

Figure 13-69. Simulated Image of the Aircraft in Fig. 13-67.

Resolution of rotating objects: The cross-range resolution of synthetic aperture processing of rotating objects results from Doppler fundamentals. Objects are resolved in cross-range if their Doppler phase goes through at least one cycle during the total rotation angle. The resolution derivation is assisted by Fig. 13-70.

The differential distance each resolution cell travels toward or away from the radar as the target rotates through the angle $\Theta_T / 2$ is

$$\delta R = \Delta X \sin(\Theta_T / 2) \tag{13-56}$$

δR = the incremental distance (in the down-range direction) traveled by scatterers in adjacent cross-range resolution bins for the rotation angle $\Theta_T / 2$

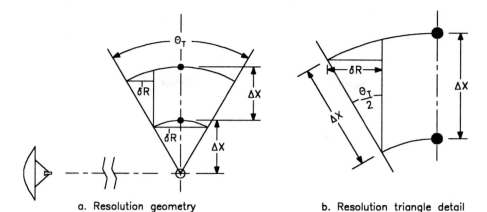

a. Resolution geometry   b. Resolution triangle detail

Figure 13-70. Rotating ISAR Resolution in Cross-Range.

ΔX = the resulting cross-range resolution

Θ_T = the total rotation angle to achieve ΔX

In phase terms, the range differential becomes

$$\delta\phi = 4\pi \Delta X \sin(\Theta_T/2) / \lambda \qquad (13\text{-}57)$$

$\delta\phi$ = the differential phase shift between adjacent cross-range resolution cells for a rotation of $\Theta_T/2$. (The 4 is for the round trip.)

λ = the wavelength

Objects are resolved in cross-range if the phase differential for a rotation of Θ_T ($2\delta\phi$) is greater than 2π radians (or one cycle). Doubling both sides of Eq. 13-57 (to account for rotation of Θ_T) and satisfying the resolution condition gives

$$2\pi = \frac{8\pi \Delta X \sin(\Theta_T/2)}{\lambda} \qquad (13\text{-}58)$$

Solving for ΔX gives the relationship linking total rotation angle, wavelength, and cross-range resolution.

$$\Delta X = \frac{\lambda}{4 \sin(\Theta_T/2)} \qquad (13\text{-}59)$$

For small rotation angles (where the angle and its sine are very nearly equal), the equation for resolution simplifies to

$$\Delta X \approx \lambda / (2\Theta_T) \qquad (13\text{-}60)$$

Sampling criterion for rotating targets: The sample rate at which the radar must view rotating objects is set, as always, by the Nyquist criterion. Each signal must be sampled greater than twice per cycle. The fastest moving objects in this form of SAR are those on an axis at right angles to the axis of the radar and at the maximum distance from the center of rotation (Fig. 13-71). Between samples, these objects must move less than the round-trip phase equivalent of π radians.

13-9. Synthetic Aperture With Rotating Objects

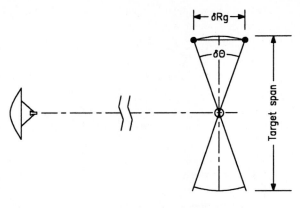

Figure 13-71. Sampling Criterion for Rotating ISAR.

$$\delta Rg / 2 = (S_{Tgt} / 2) \sin(\delta\Theta / 2) \qquad (13\text{-}61)$$

δRg = the maximum range change between samples for scatterers at the limits of the target span

S_{Tgt} = the maximum extent (span) of the target

$\delta\Theta$ = the rotation angle increment between samples

The round-trip phase increment between samples is

$$\delta\phi = \frac{4\pi S_{Tgt} \sin(\delta\Theta / 2)}{\lambda} \qquad (13\text{-}62)$$

$\delta\phi$ = the round-trip incremental phase

For small angle increments (the normal condition), the phase increment becomes

$$\delta\phi \approx \frac{2\pi S_{Tgt} \delta\Theta}{\lambda} \qquad (13\text{-}63)$$

To satisfy the Nyquist criterion, the incremental round-trip phase must be less than π radians. Applying this criterion and solving for $\delta\Theta$ gives

$$\delta\Theta < \frac{\lambda}{2 S_{Tgt}} \qquad (13\text{-}64)$$

Windowing and ISAR processing: To prevent leakage from one cross-range bin to another, windows are applied to the signals prior to spectrum analysis. The criteria for selecting the windows are the same as with Doppler and pulse compression processing, and the reader is referred to those discussions. Don't forget that the resolution is degraded by the window and the values calculated in Eqs. 13-59 and 13-60 must be adjusted (Table 11-1).

Example 13-10: An image of a target having a 6-m maximum span is to have a resolution of 10 cm in range and cross-range. The processing is to be focused and a Dolph-Chebyshev window with 60-dB sidelobes is to be used to suppress leakage in both dimensions. The center frequency is 10.0 GHz. Find the rotation angle required and the sampling and signal processing parameters.

Solution: The Dolph-Chebyshev 3.0 window (60 dB) degrades resolution by a factor of 1.44 (Table 11-1). Without degradation, range resolution of 10 cm requires a waveform with a bandwidth of 1.50 GHz. Accounting for the window requires a waveform bandwidth of 2.2 GHz.

The rotation angle required for 10-cm resolution at a wavelength of 0.03 m is (Eq. 13-60) 0.15 rad, or 8.6°. Accounting for the window increases the angle to 12.4°. Angular sampling must take place at least every 0.0025 rad or 0.14° (Eq. 13-64).

Because of the window, the pixels needed to give 10-cm resolution can be no larger than 6.9 cm (10 cm / 1.44 for the window). Dividing a 6-m span into 6.9-cm pixels requires at least 87 pixels in each direction. Since most of the processing is with FFTs, the number of pixels should be increased to 128. The extra pixels can be used either to increase the resolution or increase the size of the image map.

If the extra samples are used to increase the image size, the range bins remain 6.9 cm long and the total image span in range is 8.8 m. The angular span remains at 12.4° and each pixel is still 6.9 cm wide. Thus the image is also 8.8 m wide in cross-range. If a swept frequency waveform is used, each sample is 17.19 MHz apart (2.2 GHz / 128).

If the extra samples are used to increase resolution, the swept bandwidth increases to 3.2 GHz (2.2 GHz x 128 / 87) and the range resolution, with the window applied, drops to 6.8 cm. The rotation increases to 18.24° (12.4° * 128 / 87) and the cross-range resolution, with the window, is reduced also to 6.8 cm.

Unfocused ISAR results: The process described for finding the spectrum of the signal from each range bin (or range slice) results in an unfocused image of the target. There are two reasons for this. First, as stated above, the scatterer paths are curved, and their radial velocity is not constant over the rotation necessary to produce the required resolution. For small targets, small rotations, and coarse resolution, this effect is minimal. For large targets and large rotations, this phenomenon is limiting on the total resolution achievable.

The second effect is caused, again for large rotation angles, by scatterers walking from one range bin to another during the measurement. Figures 13-72 and 13-73 show the result of both errors. In these images, four equal scatterers are shown. The scatterer at the center is focused and the others are not, because of both their curved paths and range bin walking. Note that defocusing is most severe for scatterers with large cross-range displacements from the center of rotation. In Fig. 13-72, a Hann window was used for the spectrum analysis. Figure 13-73 shows the same four scatterers with a rectangular window; the defocusing effect is even more pronounced.

Focused processing of rotating objects: Focusing of ISAR images is similar to the other focusing processes discussed in previous sections. The image pixel lying on the center of rotation is in focus. The process of focusing is then to correct the data so that the pixel being processed is at the center of rotation (Fig. 13-74).

In the figure, the target was held stationary and the antenna rotated about it, which is exactly the same as holding the antenna stationary and rotating the target. Phase corrections are applied to each signal sample so that it is apparently at the center of rotation of one of the image pixels. After processing for that pixel, the process is repeated for each pixel in the image. The details of the process can be found in several advanced texts, including D.L. Mensa's *High Resolution Radar Imaging* [2].

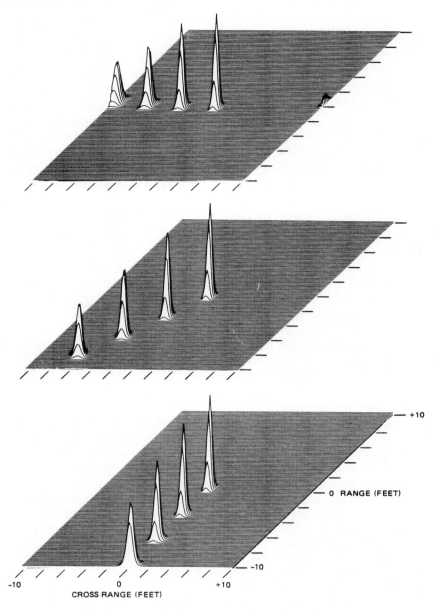

Figure 13-72. Unfocused Processing of Four Equal-RCS Scatterers at Various Aspects Using a Hann Window. (Image courtesy of D.L. Mensa; U.S. Government work not protected by U.S. copyright)

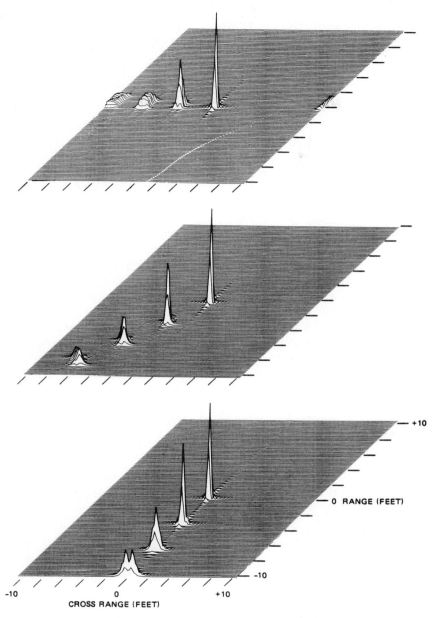

Figure 13-73. Unfocused Processing of Four Equal-RCS Scatterers at Various Aspects Using a Rectangular Window. (Image courtesy of D.L. Mensa; U.S. Government work not protected by U.S. copyright)

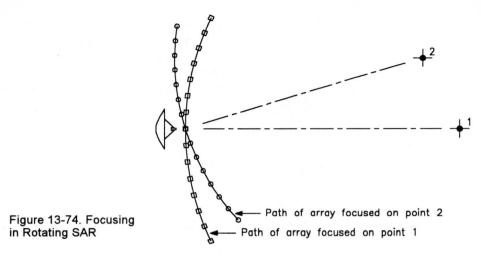

Figure 13-74. Focusing in Rotating SAR

— Path of array focused on point 2
— Path of array focused on point 1

This process is functionally equivalent to hypothesizing a scatterer of unity RCS in one pixel, calculating the data array which would result, and correlating this hypothesized data array to the actual data. As with sidelooking SAR, this process requires extensive computation and may not be practical.

Mensa and Vaccarro [14] have described an algorithm which reduces the computational load significantly, which is described below. Gupta et al. have another method of focusing with reduced computation [15].

Mensa found that if the data from rotating objects are taken as frequency profiles instead of range profiles for each aspect, a 2-D Fourier transform on the frequency/aspect data yields an exact image in range and cross-range (Fig. 13-75). If data are available in range/aspect form, they can be converted to frequency/aspect by replacing each aspect row of N range bins with its Fourier transform.

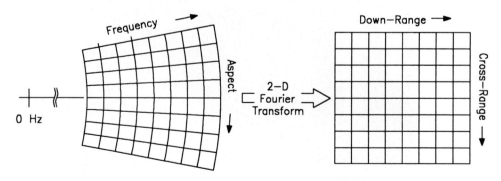

Figure 13-75. Focused Image Processing for Rotating SAR.

The problem lies in the computation. The frequency/aspect data is in polar coordinates. It was shown in Ch. 11 that there is no FFT equivalent to the polar-format Fourier transform, and the processing requires excessive computation. If the polar-format data could be re-sampled in rectangular coordinates, a 2-D FFT would yield an exact, focused image.

The method used by Mensa and Vaccaro [14] is a three-step interpolation of the polar data into Cartesian sample locations. Figure 13-76 shows the desired transformation. They used a four-times interpolation and picked the re-sampled data point closest to that desired. Interpolations are done using the Fourier transform method shown in Sec. 11-10. Figure 13-77 shows the frequency/aspect data of a light aircraft. Figure 13-78 shows the image developed taken using this method.

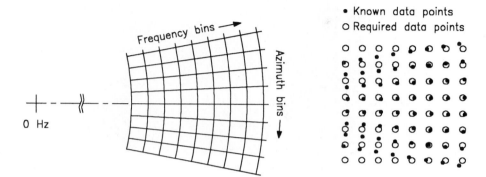

Figure 13-76. Transformation from Polar to Rectangular Data.

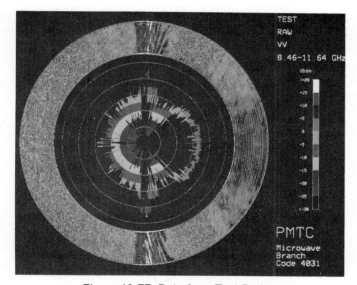

Figure 13-77. Data from Test Body.
Low resolution RCS versus aspect in center of data.
(Photographs courtesy of D.L. Mensa;
U.S. Government work not protected by copyright)

13-9. Synthetic Aperture With Rotating Objects

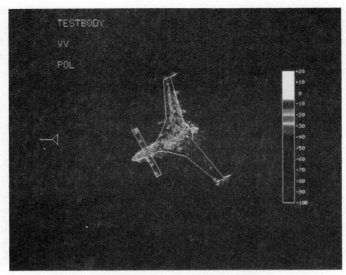

Figure 13-78. Image of Test Body.
(Photograph courtesy of D.L. Mensa;
U.S. Government work not protected by copyright)

Rotational imaging of tactical targets: The imaging method described above can be applied to tactical radars flying in aircraft and to fixed radars using the rotational motion of their targets. This rotational motion is achieved by spotlighting the target, which is pointing the radar at the desired target space and following that space with the radar. Figure 13-79 shows that the targets viewed apparently rotate with respect to the radar.

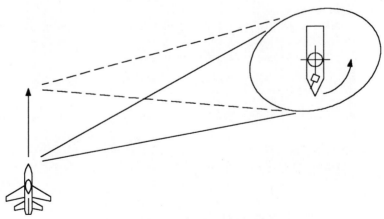

Figure 13-79. Spotlight, or Patch, SAR.

The translational motion found with a moving radar spotlighting a target on the ground or sea must be removed by techniques discussed earlier, involving "correcting" the phase of the received signal. Translation introduces non-linearity, in that the effective rotation rate is not constant for a radar moving in a straight line. This can be compensated by varying sample rate. Perturbations in the aircraft flight path must also be corrected.

This method of producing high resolution in cross-range (as before, pulse compression is used to produce high range resolution) is sometimes called spotlight SAR, or patch SAR

Figure 13-80 shows results obtained from this method with a stationary radar spotlighting a moving target. The image is of a Lockheed L-1011 airliner departing Philadelphia International Airport by a radar located at Valley Forge.

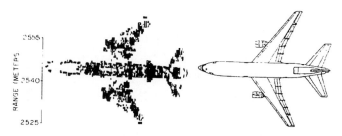

Figure 13-80. Radar Image of L-1011 In Flight.
(Reprinted from Sternberg, *et. al.* [16]. ©1989 IEEE; used with permission)

13-10. Further Information

General references: Two books and two associations are useful for persons wanting detailed information about high resolution radar. The first book presents the theory of high resolution in great detail and is a definitive work in this area. The second reference is particularly useful in its treatment of synthetic aperture radar. It shows techniques for obtaining both unfocused and focused target images of rotating objects. Although emphasizing laboratory techniques, the principles presented extend very nicely to some types of radars used for mapping from aircraft and spacecraft, and for imaging of tactical targets.

[1]   A.W. Rihaczek, *Principles of High Resolution Radar*, Palo Alto CA: Peninsula Press, 1977.

[2]   D.L. Mensa, *High Resolution Radar Imaging* 2nd ed., Norwood MA: Artech House, 1990.

The Institute of Electrical and Electronic Engineers (IEEE) has several journals in which high resolution articles appear from time to time. Of particular usefulness are the Proceedings of the IEEE, the IEEE Transactions on Aerospace and Electronic Systems, and the IEEE Transactions on Antennas and Propagation.

A trade association, the Antenna Measurement Techniques Association (AMTA), holds annual meetings and publishes Proceedings which contain many useful articles on the practical aspects of implementing high resolution radar. Although emphasis is placed on laboratory techniques, the principles extend into other fields.

The following general references provide useful information on the theory and techniques of high resolution radar.

[3] D.R. Wehner, *High Resolution Radar*, Norwood MA: Artech House, 1987.

[4] E.C. Farnett and G.H. Stevens, "Pulse Compression Radar," Ch. 10 in M.I. Skolnik (ed.), *Radar Handbook*, 2nd ed., New York: McGraw-Hill, 1990.

[5] L.J. Cutrona, "Synthetic Aperture Radar," Ch. 21 in M.I. Skolnik (ed.), *Radar Handbook*, 2nd ed., New York: McGraw-Hill, 1990.

[6] G.W. Stimson, *Introduction to Airborne Radar*, El Segundo CA: Hughes Aircraft Company, 1983.

Cited References:

[7] F.J. Harris, "On the Use of Windows for Harmonic Analysis with the Discrete Fourier Transform," *Proceedings of the IEEE*, vol. 66, no. 1, pp. 51-83, January 1978.

[8] *SAW Technology*, Newbridge, Midlothian, Scotland: RACAL MESL, Ltd.

[9] F.E. Nathanson, *Radar Design Principles*, New York: McGraw-Hill, 1969.

[10] RACAL MESL, Ltd. models 510 and 511 pulse compression modules, data sheets.

[11] G.R. Welti, "Quaternary Codes for Pulsed Radar," *IRE Transactions on Information Theory*, June 1960.

[12] M.J.E. Golay, "Complementary Series," *IRE Transactions on Information Theory*, vol. IT-7, pp. 82-87, April 1961.

[13] H. Jensen, L.C. Graham, L.J. Porcello, and E.N. Leith, "Side-looking Airborne Radar," *Scientific American*, vol. 236, no. 4, Oct. 1977.

[14] D.L. Mensa and K. Vaccarro, "Two-Dimensional RCS Image Focusing," *Proceedings of the 1987 Meeting of the Antenna Measurement Techniques Association*, Seattle.

[15] Gupta et al., "A Novel Method for 2-D and 3-D Radar Imaging," Ohio State University, Electro Science Laboratory.

[16] B.D. Steinberg, D.L. Carlson, and W. Lee, "Experimental Localized Radar Cross Sections of Aircraft", *Proceedings of the IEEE*, vol. 77 no. 5, pp. 663-669, May 1989.

14

SPECIAL RADAR TOPICS

Introduction

In this chapter, we discuss several radar topics which do not fit into any of the previously encountered broad categories. These include:

- Radar coordinate systems and netting, which is transforming target data into a coordinate system understood by all and passing the data among radars and other information users
- Secondary surveillance radar (SSR), including identification friend or foe (IFF) and the air traffic control radar beacon system (ATCRBS)
- Radiation hazard to personnel
- Radomes for protecting and hiding radar antennas
- Radar electronic warfare introduction and definitions

14-1. Radar Data Processing Introduction

Radar data processing traditionally encompasses four basic functions: detection, height-finding, track-while-scan, and coordinates and conversions.

Detection: The detection function sets the detection threshold and decides whether or not the signal meets the criteria for declaration of a target's presence. It is covered in Ch. 5.

Height-finding: Height-finding determines the height of the target above some reference, usually mean sea level (MSL). It is found in Ch. 6.

Track-while-scan and tracking algorithms: These are the calculations used to smooth (interpolate) tracking data and to predict (extrapolate) a target's position at some time in the future. They are used in tracking radars as part of the track servo function and by search radars to keep track of individual targets while the radar searches for others. The principles are covered in Ch. 7.

Coordinates and netting: This function covers three topics: the coordinate systems used by radars, conversions between them, and networking radars by passing target information between them in some coordinate system understood by all members of the net. It is covered in the next section.

14-2. Coordinate Systems and Netting

Netting covers two basic, related functions: It reports to external users radar information about targets, and it ties multiple sensors together so that one may receive target positions from others. Both functions require that data be formatted in such a manner that other radars and users can understand the data from sending radars and that receiving radars can put the received information in a form usable to the system.

Radar target positions are commonly reported in one or more of six coordinate systems, listed and described below. The author is indebted to R.T. Jones, D. Christensen, and R. Benton [1] for much of the material on geodetic coordinates and on non-spherical earth corrections. Also see [2] for general information on earth-referenced coordinates and transformations.

— **Body-referenced radar-centric spherical coordinates:** In this system, the target position is described by an azimuth angle ($\theta_{AZ(B)}$), an elevation angle ($\theta_{EL(B)}$), and a slant range (R). The body's "horizontal" reference is a plane passing through the radar's center of radiation and parallel to the plane defined by the roll and pitch axes of the vehicle carrying the radar. The vertical reference is the line passing through the center of radiation and parallel to the vehicle body's yaw axis. This is the coordinate system in which the radar makes its measurements and is the system into which designation data (external data used to point the radar for target acquisition) must ultimately be translated. In land-based systems, this coordinate system is usually the same as the local earth-referenced coordinates. The origin is the center of radiation of the radar's antenna. The coordinates in this system are

$\theta_{AZ(B)}$ = the body-referenced azimuth angle from radar to target on the body's horizontal plane. Zero azimuth is parallel to the body roll axis. A positive angle is normally clockwise from the reference.

$\theta_{EL(B)}$ = the body-referenced elevation angle from radar to target above the body's horizontal plane. Up is positive (away from the landing gear or keel).

R = the straight-line distance from the radar to the target.

— **Body-referenced radar-centric Cartesian coordinates:** This system uses the same horizontal and vertical reference as the spherical coordinates above. The coordinates, described below, are X_B, Y_B, and Z_B, and the origin is the center of radiation of the radar's antenna.

X_B = the straight-line distance on the body's horizontal plane, in a direction parallel to the body's pitch axis, from the projected target position on this plane to the roll axis. Positive is to the right of the roll axis when viewed from the nominal top of the body with the nose (bow) at the top. The body X-axis is the *axis around which the body pitches*.

Y_B = the straight-line distance on the body's horizontal plane, in a direction parallel to the body's roll axis, from the projected target position on this plane to the pitch axis. Positive is toward the front (nose or bow) of the body. The Y axis is the *axis around which the body rolls*.

Z_B = the straight-line distance, in a direction perpendicular to the body's horizontal plane, from the target to the body's horizontal plane. Positive is the nominal top of the body, away from the landing gear or keel. The body Z axis is the *axis around which the body yaws*.

- **Earth-referenced radar-centric spherical coordinates, also known as local inertial spherical coordinates:** In this system, also called local spherical coordinates, the target position is described by an azimuth angle ($\theta_{AZ(L)}$), an elevation angle ($\theta_{EL(L)}$), and a slant range (R). The vertical reference is a radius from the center of the earth passing through the radar's center of radiation. The local horizontal reference is a plane passing through the radar's center of radiation perpendicular to the vertical reference. The origin is the center of radiation of the radar's antenna, and the coordinates are

    $\theta_{AZ(L)}$ = the earth-referenced azimuth angle from radar to target on the local horizontal plane. Zero azimuth is usually true north. A positive angle is clockwise from the reference (east).

    $\theta_{EL(L)}$ = the earth-referenced elevation angle from radar to target above the local horizontal plane. A positive angle is up.

    R = the straight-line distance from the radar to the target.

- **Earth-referenced radar-centric local Cartesian coordinates:** This system, also known as local inertial Cartesian coordinates, uses the same horizontal and vertical reference as the earth-referenced spherical coordinates above. The coordinates are X_L, Y_L, and Z_L, and are described below. The origin is the center of radiation of the radar's antenna.

    X_L = the straight-line distance on the local horizontal plane, in an east-west direction, from the projected target position on this plane to the north-south axis. Positive is east.

    Y_L = the straight-line distance on the local horizontal plane, in a north-south direction, from the projected target position on this plane to the east-west axis. Positive is toward north.

    Z_L = the straight-line distance, in a direction perpendicular to the local horizontal plane, from the target to the local horizontal plane. Positive is up, away from the center of the earth.

- **Geodetic, or earth-referenced geo-centric (earth-centered) spherical coordinates:** This is the most commonly used system for earth-oriented navigation. It is a universal earth-referenced system and is commonly used to pass information from one location to another. It describes a position in space by its latitude angle, longitude angle, and altitude. The coordinates are described below, with the origin at the center of the earth.

    θ_{LAT} = the latitude angle. It is the angle formed by lines from the point in question to the earth's center and from the earth's center to the equator, with both lines on a plane passing through the earth's axis of rotation. If the target is north of the equator its latitude is positive (called north latitude), and if south it is negative (south latitude).

    θ_{LON} = the longitude angle. It is the angle formed by the intersection of two planes. One plane includes the earth's axis and the point in question, and the other includes the earth's axis and a mark on the floor of the Royal Observatory in Greenwich, England. An angle to the east of the Royal Observatory is positive longitude (east longitude) and one to the west is negative (west longitude). The curve on the earth's surface formed by the surface's intersection with the last-named plane is called the *prime meridian*.

H = the height (altitude) above the reference ellipsoid. It is the distance from the point in question to the reference ellipsoid on a line from the point in question to the earth's center. The use of the reference ellipsoid allows correction for irregularities in the shape of the earth. If this corrections are not made, H is the height above mean sea level (MSL).

Latitudes and longitudes are traditionally measured in degrees, minutes (1/60 degree), and seconds (1/60 minute). In radar usage, degrees and decimal fractions are most often used. Altitude is usually in meters or feet.

— **Earth-referenced geo-centric Cartesian coordinates:** This system is used primarily as an intermediate step in the conversion of geodetic coordinates of one point in the earth's atmosphere or on the earth to another point. The references are the earth's center, its axis of rotation, and the earth's radius passing through the prime meridian. The coordinates are E, F, and G, and are defined below, with the origin at the center of the earth.

E = the axis formed by a radius of the earth from its center to the point on the equator passing through the prime meridian. The intersection of the E axis, the equator, and the mean sea-level surface is 0° latitude (the equator), 0° longitude (the prime meridian), and 0 altitude (MSL). Positive E is in the eastern hemisphere, negative is in the western. The E-axis intersects the earth's surface in the Atlantic Ocean about 1,200 km west of the African nation of Gabon.

F = the Cartesian axis from the earth's center intersecting the equator at +90° longitude. The intersection of the earth's surface with the F-axis is in the Indian Ocean approximately 1000 km southeast of the island of Sri Lanka. Positive F, viewed from above the north pole, is in the longitudes from 90° east moving clockwise to 90° west. Negative is from 90° east moving counterclockwise to 90° west.

G = the axis of the earth's rotation. Positive G is north of the equator, negative is south.

14-3. Coordinate Conversions

Conversions between body-referenced spherical and body-referenced Cartesian coordinates: These conversions form intermediate steps in preparing radar target data for transmission in geodetic coordinates and in accepting geodetic data for radar designation. The following equations show the conversions; the only symbols defined here are those not previously defined.

$$S_B = R \cos \theta_{EL(B)} \qquad (14\text{-}1)$$

S_B = the distance from the radar to the target projected onto the body horizontal plane.

$$X_B = S_B \sin \theta_{AZ(B)} \qquad (14\text{-}2)$$
$$Y_B = S_B \cos \theta_{AZ(B)} \qquad (14\text{-}3)$$
$$Z_B = R \sin \theta_{EL(B)} \qquad (14\text{-}4)$$

The inverse conversions are

$$S_B = \sqrt{X_B^2 + Y_B^2} \qquad (14\text{-}5)$$
$$R = \sqrt{X_B^2 + Y_B^2 + Z_B^2} \qquad (14\text{-}6)$$
$$\theta_{EL(B)} = \tan^{-1}(Z_B / S_B)^* \qquad (14\text{-}7)$$
$$\theta_{AZ(B)} = \tan^{-1}(X_B / Y_B)^* \qquad (14\text{-}8)$$

*The arc-tangent used in the above two equations must yield results in all four quadrants, such as does FORTRAN's ATAN2. An algorithm for the four-quadrant arc-tangent is in Ch. 6.

Conversions between local earth-referenced spherical and local earth-referenced Cartesian coordinates: The following equations give these conversions, which are often needed in local data usage and in netting.

$$S_L = R \cos \theta_{EL(L)} \qquad (14\text{-}9)$$

S_L = the distance from the radar to the target projected onto the local earth-referenced horizontal plane.

$$X_L = S_L \sin \theta_{AZ(L)} \qquad (14\text{-}10)$$
$$Y_L = S_L \cos \theta_{AZ(L)} \qquad (14\text{-}11)$$
$$Z_L = R \sin \theta_{EL(L)} \qquad (14\text{-}12)$$

The inverse conversions are

$$S_L = \sqrt{X_L^2 + Y_L^2} \qquad (14\text{-}13)$$
$$R = \sqrt{X_L^2 + Y_L^2 + Z_L^2} \qquad (14\text{-}14)$$
$$\theta_{EL(L)} = \tan^{-1}(Z_L / S_L)^* \qquad (14\text{-}15)$$
$$\theta_{AZ(L)} = \tan^{-1}(X_L / Y_L)^* \qquad (14\text{-}16)$$

*This arc-tangent must yield four-quadrant results.

Note that the range (R) in body coordinates and range in local earth-referenced coordinates is the same, since the radar is at the origin of both.

Conversion between body coordinates and earth-referenced local coordinates: These conversions prepare data from an airborne or shipboard radar for use at locations other than within the vehicle carrying the radar. The following matrices define the translation in Cartesian coordinates. If the data are in spherical coordinates, they are first converted to Cartesian.

14-3. Coordinate Conversions

$$\begin{bmatrix} X_B \\ Y_B \\ Z_B \end{bmatrix} = \mathbf{A} \begin{bmatrix} X_L \\ Y_L \\ Z_L \end{bmatrix} \tag{14-17}$$

$\mathbf{A}$ is a coordinate-rotation matrix, defined below.

$$\mathbf{A} = \begin{bmatrix} A_{11} & A_{12} & A_{13} \\ A_{21} & A_{22} & A_{23} \\ A_{31} & A_{32} & A_{33} \end{bmatrix} \tag{14-18}$$

$$A_{11} = \cos\theta_R \cos\theta_Y \tag{14-19}$$
$$A_{12} = -\cos\theta_R \sin\theta_Y \tag{14-20}$$
$$A_{13} = -\sin\theta_R \tag{14-21}$$
$$A_{21} = \cos\theta_Y \sin\theta_R \sin\theta_P + \sin\theta_Y \cos\theta_P \tag{14-22}$$
$$A_{22} = -\sin\theta_Y \sin\theta_P \sin\theta_R + \cos\theta_Y \cos\theta_P \tag{14-23}$$
$$A_{23} = \cos\theta_R \sin\theta_P \tag{14-24}$$
$$A_{31} = \cos\theta_Y \sin\theta_R \cos\theta_P - \sin\theta_Y \sin\theta_P \tag{14-25}$$
$$A_{32} = -\sin\theta_Y \sin\theta_R \cos\theta_P - \cos\theta_Y \sin\theta_P \tag{14-26}$$
$$A_{33} = \cos\theta_R \cos\theta_P \tag{14-27}$$

θ_R = the body roll angle
θ_P = the body pitch angle
θ_Y = the body yaw angle

The local inertial Cartesian coordinates are found from the body Cartesian coordinates and the body roll, pitch, and yaw angles using the following equation, where matrix $\mathbf{B}$ is the inverse of matrix $\mathbf{A}$.

$$\begin{bmatrix} X_L \\ Y_L \\ Z_L \end{bmatrix} = \mathbf{B} \begin{bmatrix} X_B \\ Y_B \\ Z_B \end{bmatrix} \tag{14-28}$$

$$\mathbf{B} = \mathbf{A}^{-1} \tag{14-29}$$

Matrix $\mathbf{B}$ is the inverse of matrix $\mathbf{A}$, and in this special case, is its transpose. Its members are defined by Eqs. 14-19 through 14-27.

$$\mathbf{B} = \begin{bmatrix} A_{11} & A_{21} & A_{31} \\ A_{12} & A_{22} & A_{32} \\ A_{13} & A_{23} & A_{33} \end{bmatrix} \tag{14-30}$$

Conversion between geodetic coordinates and E, F, and G: This conversion puts geo-centric latitude, longitude, and altitude into geo-centric Cartesian coordinates, a necessary step in finding the geodetic coordinates of a target.

$$E = (R_{E0} + H) \cos\theta_{LAT} \cos\theta_{LON} \tag{14-31}$$
$$F = (R_{E0} + H) \cos\theta_{LAT} \sin\theta_{LON} \tag{14-32}$$

$$G = (R_{E0} + H) \sin \theta_{LAT} \tag{14-33}$$

R_{E0} = the nominal radius for a spherical earth (see below about ellipsoidal and eccentric earth calculation)

H = the point of interest's height above the ellipsoid or above mean sea level (MSL).

Ellipsoidal and eccentric earth: The earth is not spherical in shape. It is flattened at the poles, assuming a roughly ellipsoidal shape. In addition, there are minor eccentricities in the ellipsoid. In very accurate radars, these factors must be accounted for in the data. One way of doing this is to use a local "earth's radius" rather than the average radius. One correction is as follows [2]:

$$R_{NU} = D_A / [1 - E_{SQ} (\sin \theta_{LAT})^2]^{1/2} \tag{14-34}$$
$$D_A = 6378135.000 \text{ meters} \tag{14-35}$$
$$F = 1 / 298.260000 \tag{14-36}$$
$$D_B = D_A (1 - F) \tag{14-37}$$
$$E_{SQ} = F (2 - F) \tag{14-38}$$

R_{NU} = the length of the normal terminated by the ellipsoidal minor axis at the location in question

D_A = the WGS-72 semi-major axis of the earth's ellipsoid (6378135. meters)

F = the earth's ellipsoid flattening factor

D_B = the WGS-72 semi-minor axis of the earth's ellipsoid

E_{SQ} = the eccentricity of the earth's ellipsoid, squared

To account for this ellipsoidality and eccentricity, the conversions from geodetic to E, F, and G coordinates (Eq. 14-31 through 14-33) become

$$E = (R_{NU} + H) \cos \theta_{LAT} \cos \theta_{LON} \tag{14-39}$$
$$F = (R_{NU} + H) \cos \theta_{LAT} \sin \theta_{LON} \tag{14-40}$$
$$G = [(1 - E_{SQ}) R_{NU} + H] \sin \theta_{LAT} \tag{14-41}$$

Equations 14-39 through 14-41 can be used for a spherical earth model by letting $R_{NU} = D_A$, $F = 0$, and $E_{SQ} = 0$.

The inverse conversions, from E, F, and G to latitude, longitude, and altitude are given below. The latitude calculation is an approximation which is claimed to have an error less than 10^{-8} [2]. The more exact ellipsoidal calculations are shown.

$$\theta_{LON} = \tan^{-1}(E / F) \tag{14-42}*$$
$$\theta_{LAT} = \tan^{-1}\{[G + (E_{SQ} / (1.000 - E_{SQ})) D_B \sin^3 U] / [P - E_{SQ} D_A \cos^3 U]\} \tag{14-43}†$$
$$U = \tan^{-1}[(G / P) (D_A / D_B)] \tag{14-44}†$$
$$P = \sqrt{E^2 + F^2} \tag{14-45}$$

14-3. Coordinate Conversions

$$H = (P / \cos\theta_{LAT}) - D_A / \sqrt{1 - E_{SQ}(\sin\theta_{LAT})^2} \qquad (14\text{-}46)$$

The spherical-earth approximations of the above conversions are

$$\theta_{LON} = \text{Tan}^{-1}(E/F) \qquad (14\text{-}47)\,^*$$
$$\theta_{LAT} = \text{Tan}^{-1}(G/\sqrt{E^2+F^2}) \qquad (14\text{-}48)\,^\dagger$$
$$H = (E^2+F^2)^{1/2} / \cos\theta_{LAT} \qquad (14\text{-}49)$$

* Four-quadrant arc-tangent
† Two-quadrant arc-tangent, defined from $-90°$ to $+90°$

Target and radar location in *E*, *F*, and *G* coordinates: Finally, we must be able to transpose the radar's coordinates in E, F, and G plus the target's coordinates in X_L, Y_L, and Z_L into target position in E, F, and G, and vice versa.

$$\begin{bmatrix} E_{TGT} \\ F_{TGT} \\ G_{TGT} \end{bmatrix} = \mathbf{C} \begin{bmatrix} X_{TGT(L)} \\ Y_{TGT(L)} \\ Z_{TGT(L)} \end{bmatrix} + \begin{bmatrix} E_{RDR} \\ F_{RDR} \\ G_{RDR} \end{bmatrix} \qquad (14\text{-}50)$$

$$C_{11} = -\sin\theta_{LON(RDR)} \qquad (14\text{-}51)$$
$$C_{12} = -\cos\theta_{LON(RDR)} \sin\theta_{LAT(RDR)} \qquad (14\text{-}52)$$
$$C_{13} = \cos\theta_{LON(RDR)} \cos\theta_{LAT(RDR)} \qquad (14\text{-}53)$$
$$C_{21} = \cos\theta_{LON(RDR)} \qquad (14\text{-}54)$$
$$C_{22} = -\sin\theta_{LON(RDR)} \sin\theta_{LAT(RDR)} \qquad (14\text{-}55)$$
$$C_{23} = \sin\theta_{LON(RDR)} \cos\theta_{LAT(RDR)} \qquad (14\text{-}56)$$
$$C_{31} = 0 \qquad (14\text{-}57)$$
$$C_{32} = \cos\theta_{LAT(RDR)} \qquad (14\text{-}58)$$
$$C_{33} = \sin\theta_{LAT(RDR)} \qquad (14\text{-}59)$$

E_{TGT}, F_{TGT}, and G_{TGT} = the target's E, F, and G dimensions, respectively

E_{RDR}, F_{RDR}, and G_{RDR} = the radar's E, F, and G dimensions, respectively

$\theta_{LON(RDR)}$ and $\theta_{LAT(RDR)}$ = the radar's longitude and latitude, respectively

$X_{TGT(L)}$, $Y_{TGT(L)}$, and $Z_{TGT(L)}$ = the target's local X, Y, and Z dimensions, respectively

The inverse of the above calculation is used to find the X_L, Y_L, and Z_L if the E, F, and G coordinates are known for both the radar and the target.

$$\begin{bmatrix} X_{TGT(L)} \\ Y_{TGT(L)} \\ Z_{TGT(L)} \end{bmatrix} = \mathbf{D} \begin{bmatrix} E_{TGT} - E_{RDR} \\ F_{TGT} - F_{RDR} \\ G_{TGT} - G_{RDR} \end{bmatrix} \qquad (14\text{-}60)$$

$$\mathbf{D} = \mathbf{C}^{-1} \quad (14\text{-}61)$$

D is the inverse of **C**, and in this case is its transpose. The members of **D** are

$$\mathbf{D} = \begin{bmatrix} C_{11} & C_{21} & C_{31} \\ C_{12} & C_{22} & C_{32} \\ C_{13} & C_{23} & C_{33} \end{bmatrix} \quad (14\text{-}62)$$

Finding target geodetic coordinates: Following are the steps necessary to convert a target's radar coordinates to geodetic coordinates and thus prepare the data for transmission in geodetic coordinates. Example 14-1 shows their use.

1. Measure the target's azimuth angle, elevation angle, and range from the radar using the radar platform's body spherical coordinate system.
2. Correct the data for atmospheric refraction (not discussed here).
3. Convert the body spherical coordinate data to body Cartesian coordinates.
4. Rotate the body Cartesian data to local Cartesian coordinates.
5. Find the E_{RDR}, F_{RDR}, and G_{RDR} coordinates of the radar.
6. Calculate the E_{TGT}, F_{TGT}, and G_{TGT} coordinates of the target.
7. Calculate the latitude, longitude, and altitude of the target.

Finding radar designation data: Radars can be designated to a point in space given the point's geodetic coordinates, the geodetic coordinates of the radar, and the body's roll, pitch, and yaw angles. Following are the steps necessary to convert a target's geodetic coordinates to radar pointing information. Please see Example 14-1. Note that target data is sometimes communicated in E, F, and G coordinates, saving one step in each conversion.

1. From the latitude, longitude, and altitude of the target, calculate its E_{TGT}, F_{TGT}, and G_{TGT}.
2. From the radar's latitude, longitude, and altitude, calculate E_{RDR}, F_{RDR}, and G_{RDR}.
3. Calculate the radar-to-target X_L, Y_L, and Z_L Cartesian coordinates.
4. Correct for refraction.
5. From the above data and the body angles (roll, pitch, and yaw), calculate the radar's body X_B, Y_B, and Z_B.
6. Calculate the body-referenced pointing data $\theta_{AZ(B)}$, $\theta_{EL(B)}$, and R.

Example 14-1: A shipboard radar tracks a target at azimuth 247.323°, elevation 5.876°, and range 242.341 km. This information is to be translated to geodetic coordinates data-linked to an aircraft, and translated to pointing information to help the aircraft's radar find the target.

The shipboard antenna's position is latitude 13.958° south, longitude 131.488° west, and is 22.0 m above mean sea level. The ship's heading (yaw angle) is 154.318° true (reference true north), its roll angle is port 6.863°, and its pitch angle is bow-down 1.744°.

The aircraft is at latitude 11.557° south, longitude 130.345° west, and is 12200 m above MSL. The aircraft's heading (yaw) is 213.122°, its roll is port 30.632° and its pitch is nose-up 11.553°.

(a) Find the latitude, longitude, and altitude of the target. Assume an ellipsoidal earth with $F=1/298.26$. Assume no refraction.

(b) Find the pointing data for the airborne radar; the body-referenced azimuth angle, body-referenced elevation angle, and range. Again, assume no refraction.

Solution: The signs of the angles are as follows: ship latitude negative, ship longitude negative, all ship radar pointing angles positive, ship yaw positive, ship pitch negative, ship roll negative. Aircraft latitude negative, aircraft longitude negative, aircraft heading positive, aircraft roll negative, aircraft pitch positive.

(a) From Eqs. 14-1 through Eq. 14-4, the target Cartesian coordinates referenced to the ship's body are $X_B = -222.431$ km, $Y_B = -92.940$ km, and $Z_B = +24.810$ km.

Equations 14-28 through 14-30 remove the ship's roll, pitch, and yaw and find the target's Cartesian coordinates with respect to local inertial space (horizontal, local vertical, and north). The target is $X_L = +162.065$ km (east of the ship), $Y_L = +180.176$ km (north of the ship), and $Z_L = +0.849$ km (above the ship).

Using Eqs. 14-31 through 14-33, the radar's E, F, and G are $E_{RDR} = -4102.131$ km, $F_{RDR} = -4638.573$ km, and $G_{RDR} = -1528.775$ km. Equations 14-50 through 14-59 give the target's earth-centered Cartesian coordinates as $E_{TGT} = -4010.066$ km, $F_{TGT} = -4779.108$ km, and $G_{TGT} = -1354.124$ km.

Equations 14-42 through 14-46 give the target's geodetic coordinates: $\theta_{LAT(TGT)} = -12.326°$ (south), $\theta_{LON(TGT)} = -129.999°$ (west), and $H_{TGT} = 6.730$ km.

(b) Equations 14-31 through 14-33 give the airborne radar's E, F, and G coordinates. They are $E_{ACF} = -4054.243$ km, $F_{ACF} = -4772.994$ km, and $G_{ACF} = -1272.044$ km.

Equations 14-60 through 14-62 give the local aircraft-centered earth-referenced Cartesian coordinates of the target: $X_{TGT(ACF)} = +37.628$ km (east of the aircraft), $Y_{TGT(ACF)} = -85.212$ km (south of the aircraft), and $Z_{TGT(ACF)} = -7.010$ km (below the aircraft). This places the target at a true bearing of 156.175° (approximately southeast) from the aircraft, 4.304° below the local horizontal at the aircraft, and 93.414 km from the aircraft (Eqs. 14-13 through 14-16).

Equations 14-17 through 14-27 are used to rotate the local earth-referenced radar Cartesian coordinates to body-referenced coordinates. Equations 14-5 through 14-7 are then used to find the antenna and range-machine pointing data. The results are: the aircraft radar's azimuth pointing angle (reference the aircraft body) is 308.627°, its elevation pointing angle is +14.184°, and its range is 93.414 km.

A good check of the above (admittedly complex) calculation is to take the target coordinates as viewed by the airborne radar and translate them back to pointing angles for the shipboard system.

14-4. Secondary Surveillance Radar (SSR)

Secondary surveillance radars use a transmitter to interrogate a transponder carried in most civil and military aircraft, using the reply from the transponder as an enhanced signal return and interpreting its reply code to obtain information about the aircraft.

There are two similar and overlapping forms of SSR: identification friend or foe (IFF) and air traffic control radar beacon system (ATCRBS). The military IFF is used to locate aircraft and to segregate friendly aircraft from hostiles, neutrals, and unknowns. ATCRBS is used to identify and report the altitude of civil and military aircraft in the air traffic control (ATC) system. IFF and ATCRBS use the same frequencies and, in many cases, the same transponders, although usually only military transponders have IFF capability.

The unit interrogating the transponder is often co-located with a primary radar and is called an interrogator. Interrogation takes place on a common frequency of 1030 MHz and consists (at present) of a three pulse code. The interrogation code tells the transponder whether or not to reply and tells it the reply mode expected. The transponder replies on a second common frequency of 1090 MHz. The echo from the interrogation (1030 MHz) is not received and no attempt is made to use it.

Reference [5] is devoted to secondary surveillance radar. See Sec. 3-17 for a listing of other types of secondary radar.

SSR modes: The replies from transponders convey several different types of information called modes. The mode in which the transponder replies is determined by (a) the capability of the transponder, (b) the transponder mode(s) selected in the aircraft, and (c) the interrogation code. At present, five modes are used in North America, listed in Table 14-1. Other modes are defined. For details, see Federal Aviation Administration (FAA) Advisory Circular 00-27 or later versions [3].

Table 14-1. IFF/ATCRBS Modes.
(Adapted from [3] and [5])

Mode	Description
Mode 1	ID mode, presently for military use
Mode 2	ID mode for military use
Mode 3A	To initiate transponder response for identification and tracking
Mode C	To initiate transponder responses for automatic pressure altitude transmission
Mode 4	For military use
Mode S	Selective interrogation and data link — future use

At present, any transponder set to the mode being interrogated and meeting the interrogation signal criteria transmits a reply, making for much congestion on the reply frequency (1.090 GHz). The new Mode S features selective interrogation and up- and down-link data transmission capability. It is described in a later part of this section.

IFF and ATCRBS system specifications: Table 14-2 gives the system specifications for SSR in the United States.

Table 14-2. IFF/ATCRBS Specifications.
(Adapted from [3])

Range coverage	1 to 200 nautical miles
Elevation coverage	0.5° to 45° above the horizontal plane
Altitude coverage	Limited only by service ceiling of aircraft
Range accuracy	± 1000 ft
Azimuth accuracy	± 1.0°
Altitude correspondence	Within ± 125 ft (95% probability) of the pressure altitude (reference the standard atmosphere of 29.92 " Hg)
Interrogation frequency	1030 MHz ± 0.2 MHz
Interrogation power	P_1 and P_3 not to exceed ERP of +52.5 dBW, P_1 within 1 dB of P_3
ISLB	Reply if P_1 is more than 9 dB above P_2, or if P_2 is missing
Reply frequency	1090 MHz ± 3 MHz
Reply power	At least +21 dBW and not greater than +27 dBW
Polarization	Vertical

Interrogation: The interrogation for the ATC modes (1, 3A, and C) and IFF Mode 2 consists of three pulses transmitted at 1030 MHz, named P_1, P_2, and P_3 (Fig. 14-1). P_1 and P_3 are usually transmitted from a high-gain directional antenna pattern (often called the sum pattern). P_2 is not required for initiating a reply, but is used to suppress replies from interrogations through sum pattern sidelobes. See the later discussion of interrogate sidelobe suppression.

The time spacing between P_1 and P_3 is switchable and is used to tell the transponder the reply mode expected. The interrogator can be set to cycle through various spacing patterns to initiate replies from all modes desired. The transponder is supposed to reply only to interrogations for the modes set into it. For example, most civilian ATC transponders are switched to reply to Mode 3/A (ATC identification) and Mode C (altitude reporting). They reply with a preset identification pulse train to interrogations with a $P_1 - P_3$ spacing of 8.0 ±0.2 µs (Mode 3/A), and with the aircraft's pressure altitude with a spacing of 21 ±0.2 µs (Mode C). P_2, when present, always occurs 2.0 ±0.15 microseconds after P_1. Table 14-3 gives the interrogation pulse characteristics.

Reply: The reply signal specifications for Modes 1, 2, 3/A, and C are given in Fig. 14-2 and Table 14-4. They consist at a minimum of 2 framing pulses and up to 12 data pulses. The presence of a pulse in a particular position indicates a binary "one" in the code; its absence is a zero. Replies are given octal values from 0000_8 to 7777_8, yielding 4096 unique values.

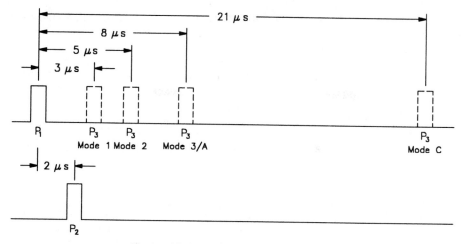

Figure 14-1. SSR Interrogation Pulses.

Table 14-3. IFF/ATCRBS Interrogation Pulse Characteristics. (From [3])

P_1, P_2, and P_3 pulse widths	0.8 ±0.1 μs	
P_1, P_2, and P_3 rise times	0.05 to 0.1 μs	
P_1, P_2, and P_3 decay times	0.05 to 0.2 μs	
P_1 to P_2 interval	2.0 ±0.15 μs	
P_1 to P_3 interval	Mode 1	3 ± 0.1 μs
	Mode 2	5 ± 0.2 μs
	Mode 3A	8 ± 0.2 μs
	Mode C	21 ± 0.2 μs

The data pulses are divided into four three-pulse groups, A, B, C, and D (Fig. 14-2). Group A encodes the most significant octal digit, while group D encodes the least significant.

Example 14-2: What reply pulse code results from a Mode 3/A interrogation with the transponder set to a value of 4573_8?

Solution: The ID digit "4" is encoded by the A (most significant) group. Thus a pulse would be present in the A_4 position and absent in the A_1 and A_2. The next digit "5" is encoded with the B group and B_1 and B_4 are present and B_2 is absent. Likewise, C_1, C_2, and C_4 are present, as are D_1 and D_2, with D_4 absent. The X-pulse is reserved for future use.

The Special Position Identification (SPI) pulse occurs 4.35 μs *after* the second framing pulse (F_2). It is present in ATCRBS replies, except Mode C, for a few seconds after the "Ident" button on the transponder is pushed. This reply is initiated at the request of air traffic controllers and causes a special display on their scopes, giving a positive identification of a particular target.

14-4. Secondary Surveillance Radar

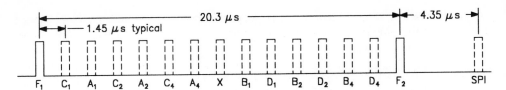

Figure 14-2. SSR Reply Pulses.

Table 14-4. Reply Pulse Specifications.
(From [3])

Pulse width	0.45 ± 0.10 μs
Rise time	0.05 to 0.10 μs
Fall time	0.05 to 0.20 μs
Framing pulses	Two pulses spaced 20.3 μs apart
Information pulses	Spacing is from first framing pulse, all pulse positions are ± 0.10 μs
C1	1.45 μs
A1	2.90 μs
C2	4.35 μs
A2	5.80 μs
C4	7.25 μs
A4	8.70 μs
X	10.15 μs
B1	11.60 μs
D1	13.05 μs
B2	14.50 μs
D2	15.95 μs
B4	17.40 μs
D4	18.85 μs
SPI	24.65 μs

Several discrete Mode 3/A reply sequences are decoded for special displays and alarms on the air traffic control consoles. A value of 7700_8 (all pulses in reply groups A and B present, with the pulses of groups C and D absent) is the emergency "Mayday" code. Code 7600_8 indicates that the aircraft has lost communication with the controllers. Code 7500_8 has another control function. Code 1200_8 indicates that the aircraft responding is not working within the air traffic control system. Most of the other code values are known as "discrete" codes and are available for assignment by air traffic controllers to identify particular aircraft. To prevent confusion, $77xx_8$ and $76xx_8$ are usually not assigned (the xx indicates that the C and D groups can have any value).

Note that the replies from Modes 1, 2, 3/A, and C have identical characteristics and cannot be discriminated from one another in the interrogator set. The only way to separate Mode 3/A and Mode 2 replies is by the interrogation code which causes that reply. The interrogator set must be able to correlate target positions and reply time of arrival to assure that only replies initiated by that interrogator are processed, and those from other interrogator sets are ignored.

Monopulse reception: Normal reception from a scanning antenna limits angle accuracy to some fraction of the beamwidth, typically 20 to 40 percent or so. By creating a sum-and-difference pattern in a manner similar to that shown in Ch. 7, a more accurate estimation of the location of the transponder reply can be made. Unlike monopulse tracking, IFF/ATCRBS monopulse is in one dimension only (azimuth). Figure 14-3 shows the principle and Fig. 14-4 shows typical SSR antenna patterns. The sum pattern provides the normal reception and a reference for the monopulse. The difference pattern provides monopulse location capability. The control pattern is used for suppression of sidelobe replies.

Monopulse improves single-signal accuracy; resolution between multiple transponder replies is determined principally by the IFF/ATCRBS antenna's beamwidth. See Ch. 7 for the principles of monopulse.

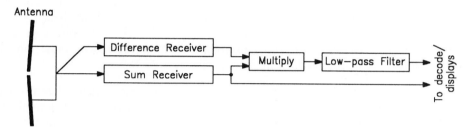

Figure 14-3. SSR Monopulse Principle.

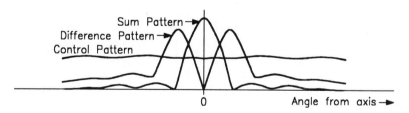

Figure 14-4. SSR Antenna Patterns.

SSR and antenna sidelobes: Since SSR is one-way communication, the interrogation antenna's sidelobes can be quite high in their response, allowing transponders close to the system to be interrogated at angles other than that of the antenna's main beam (transponder antennas are omnidirectional). The reply can also enter the system through interrogator antenna sidelobes. Without some sort of suppression of these sidelobe replies, a situation such as shown in Fig. 14-5 can develop, with the transponder's reply seen over much of the circle, causing interference with other targets and loss of azimuth data.

Two methods of suppressing sidelobe replies are used. In the first, interrogate sidelobe blanking (ISLB), prevents the transponder from replying to interrogations from other than the interrogator antenna main lobe. The second, called reply sidelobe blanking (RSLB),

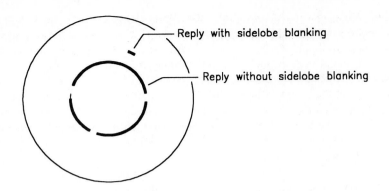

Figure 14-5. PPI Display of IFF/ATCRBS Replies
with and without Sidelobe Blanking.

identifies whether a received reply comes from the interrogator antenna's main lobe or a sidelobe. ISLB prevents excessive replies from cluttering the 1090-MHz frequency, and is the primary method of sidelobe suppression. RSLB is useful for rejecting replies caused by other interrogators, which cause "fruit" (uncorrelated clutter caused by these foreign interrogations — see defruiting, later in this section).

Interrogate sidelobe blanking: It is important that transponders not be interrogated from sidelobes of the interrogator antenna. If this were to happen, the airwaves would be crowded with emissions on 1.090 GHz (they already are, in some locations), and the IFF/ATCRBS may not be able to determine the azimuth angle of desired target. If a transponder is relatively close to the interrogator, it may be possible to detect the reply for essentially the entire scan.

Suppression of sidelobe interrogations is accomplished by using two interrogator antennas, one high gain and one omnidirectional in the horizontal plane. This latter antenna is known as the "omni" or "control" antenna. Interrogation pulses P_1 and P_3 are transmitted through the high gain antenna and P_2 is transmitted through the omni antenna. Figure 14-6 shows the relative horizontal beam patterns of the two antennas.

At the transponder, the interrogation pulses are received and their amplitudes detected. The relative amplitudes between P_1 and P_2 determine whether or not the transponder replies.

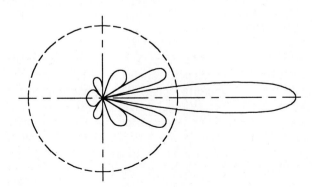

Figure 14-6. IFF/ATCRBS
Antenna Patterns.

Figure 14-7 shows the two interrogation cases. When the transponder is interrogated from the main lobe of the interrogator's high gain antenna, P_1 is of higher amplitude than P_2. This is an indication for the transponder to reply. When the P_2 amplitude is greater than P_1, the interrogation was through a sidelobe and the transponder does not reply.

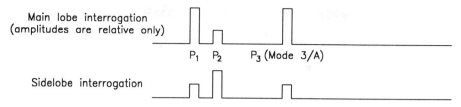

Figure 14-7. Interrogation Amplitudes at Transponder.

The implementation of an omnidirectional control antenna poses some problems. To obtain the omnidirectional pattern, it must be a separate antenna from the high-gain interrogator antenna. If it is mounted on the rotating pedestal, perhaps above the main beacon antenna, a separate waveguide rotary joint or switching is necessary to separate P_2 from P_1 and P_3. For this reason, in many installations the omni antenna is mounted on a pole alongside of the rotating pedestal. If the omni antenna is mounted lower than the high-gain antenna, the high-gain antenna blocks the omni and introduces modulation at the scan rate. If the omni is higher, it blocks the high-gain. Despite these problems, the separate, (usually) non-rotating omni antenna is commonly used.

Another interrogate sidelobe blanking method uses the sum and difference patterns of the SSR antenna. As seen in Fig. 14-4, the difference pattern has higher gain than the sum at all angles except in the sum's main beam. Thus P_1 and P_3 can be transmitted with the sum pattern and P_2 with the difference.

Many modern SSR antennas have three patterns: the sum, the azimuth monopulse difference, and a control pattern (Fig. 14-4). In these antennas, P_1 and P_3 are transmitted over the sum pattern and P_2 over the control pattern. The radar shown in Figure 1-54 has such an antenna mounted above the radar.

Reply sidelobe blanking (RSLB): Interrogate sidelobe blanking is the primary defense in IFF/ATCRBS to antenna sidelobe problems. In situations where many interrogators are present within an area, it may also be necessary to test replies to see if they were from the high gain antenna's main lobe or sidelobes. See Fig. 14-8 for the reply conditions at the interrogator.

Defruiting and processing for railing interference: One form of EMI occurs when another radar's transmit pulses are received and interfere with target echoes. When the interfering radar's PRF is different from that of the radar being interfered with, the distinctive pattern produced on a PPI is called railing.

When associated with secondary radar, this interference, called fruit, comes from transponder replies from other interrogators arriving uncorrelated in range, often through antenna sidelobes. The process of removing this interference is called defruiting and is identical to the removal of railing interference. The easiest way to remove fruit and railing is to recognize that it is uncorrelated in range and that the probability of two consecutive

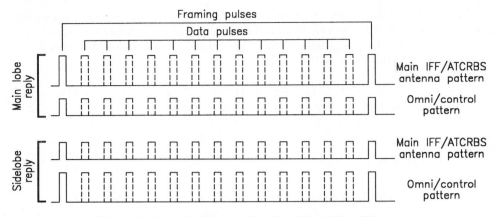

Figure 14-8. Reply Sidelobe Blanking Signal Conditions.

occurrences in the same range bin for two consecutive interrogations is low. This scheme simply requires that a target echo or transponder reply be received in the same range bin for two consecutive interrogations or transmit pulses.

Special applications of SSR: IFF is used extensively in military operations. In addition to the familiar uses on surveillance radars, many fighter aircraft carry interrogators so that beyond-visual-range (BVR) weapons will not be fired at friends. Figure 14-9 shows the IFF dipole array on the front surface of the planar slotted array antenna used in the AN/AWG-9 radar of the U.S. Navy F-14. The dipoles use the radar antenna surface as a reflector, increasing their gain and directionality. To the IFF, the radar antenna surface appears solid because the X-Band (wavelength of 3 cm) slots are much too small to be efficient radiators at the IFF wavelengths (wavelength of about 28 cm). The dipoles constitute a small aperture blockage to the radar antenna. Some tactical aircraft do not carry interrogators and instead depend on land-based, shipboard, and AWACS interrogators to identify their targets.

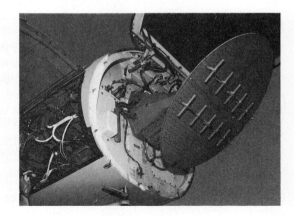

Figure 14-9. AN/AWG-9 Radar Antenna with IFF Dipole Array. (Photograph courtesy of Hughes Aircraft Company, Radar Systems Group)

The advent of highly portable anti-aircraft missiles for battlefield defense has requirement for companion IFF interrogators which can be carried in light vehicles or by people. An example is the AN/PPX-3B man-portable battery-powered IFF interrogator with Mode 4 capability.

Mode S: Mode S, selective interrogation, was developed to alleviate several problems with existing SSR. First, it will relieve much of the frequency congestion presently existing at 1.090 GHz. Second, it will allow two-way linking of data, which it is hoped will eventually remove the need for most voice-channel radio communications for air traffic control. Third, it will allow aircraft to interrogate other aircraft for collision avoidance with a system known as TCAS (terminal collision avoidance system). At present, aircraft-to-aircraft interrogation is limited by the 1.090 GHz congestion.

Mode S, when implemented, will share the frequencies of standard SSR, and thus the two systems must be totally compatible. There must be two Mode S interrogations, one which causes Mode S transponders to reply and suppresses replies from Mode 3A/C transponders, and one which induces Mode 3A/C replies and suppresses Mode S replies. Figure 14-10 shows the Mode S interrogation signal [5].

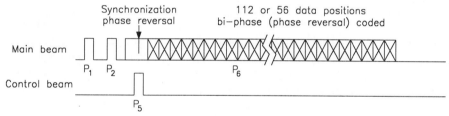

Figure 14-10. Mode S Interrogation.

Both P_1 and P_2 are transmitted over the main beam, in their normal Mode 1/2/3A/C positions and with equal amplitudes. This alone suppresses Mode 1/2/3A/C replies in that to these transponders, the equal P_1 and P_2 signals identifies the signal as from an interrogation sidelobe. A new long pulse, P_6, follows the start of P_2 by 1.5 μs. 1.25 μs after P_6 starts, the carrier wave is phase-reversed. This is the synchronization point for the following data code.

An interrogate sidelobe suppression pulse, P_5, is transmitted over the control (or omni) beam, coincident with the synchronization phase reversal being transmitted on the main beam. If the interrogation is through the main beam's main lobe, P_6 will, at the transponder, have a larger amplitude than P_5 and the transponder will recognize the synchronization phase reversal, causing a Mode S reply. If the interrogation is from a main beam sidelobe, P_5 will swamp the synchronization phase reversal and no reply is initiated. Thus the Mode S interrogation implements interrogate sidelobe blanking in a Mode S transponder and suppresses replies from Mode 1/2/3A/C.

Mode 1/2/3A/C interrogations will not cause Mode S transponders to reply because there is no synchronization phase reversal.

Mode S interrogators have a function called "all-call" which interrogates both transponder types (Mode 1/2/3A/C and S). Its interrogation waveform is shown in Fig. 14-11. There is no synchronization phase reversal, but Mode S transponders are also

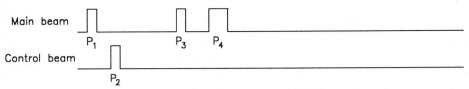

Figure 14-11. All-Call Interrogation.

interrogated by the P_4 pulse, which occurs 2 μs after P_3. Mode 1/2/3 A/C transponders ignore P_4. In all-call, Mode S sidelobe blanking is accomplished with P_1 and P_2, as in mode 1/2/3 A/C.

In all-call, the P_4 pulse can have two widths: 0.8 and 1.6 μs. The wide pulse causes the Mode S transponder to reply with its address (part of the reply data word). The short pulse elicits no reply from Mode S transponders, and is intended for airborne TCAS interrogations of Mode 1/2/3 A/C transponders.

Mode S reply: The Mode S reply is on 1.090 GHz and is as shown in Fig. 14-12. The preamble is two groups of two pulses, spaced such that no combination of Mode 1/2/3 A/C framing and data pulses within position tolerances can mimic it. This prevents the older modes from interfering with Mode S replies.

The data stream following the 8.0-μs preamble consists of either 112 or 56 bits, each of which is 1.0 μs long. Each bit is one transmitted pulse 0.5 μs long. Its value depends on its position in the 1.0-μs width. If the pulse occupies the first half of the bit space, it is a binary 1. If in the second half, it is 0 (this is *pulse position modulation* — PPM). The four data bits shown in the figure are 1101.

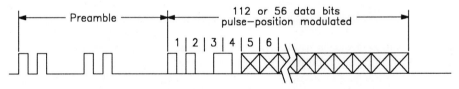

Figure 14-12. Mode S Reply.

Address and parity: In both the short-form interrogation and reply (56 data bits) and long form (112 bits), the last 24 bits are address and parity, giving both ends of the communication both the address and a measure of the quality of the signal. A 24-bit parity word is calculated from the data stream (32 or 88 bits) and is added modulo 2 to the 24-bit address of the transponder. Considering the interrogation link, the transponder decodes the data stream and calculates its own parity word, which it subtracts modulo 2 (identical to adding) from the last 24 bits. If the result is the transponder's address, it knows the data quality was good and it replies. If not, either the message was intended for another transponder or there was a transmission error; in either case there is no reply. A no-reply to an interrogation causes the interrogation to be repeated.

The transponder replies with data and a calculated parity word added modulo 2 to its own address. The interrogator recalculates parity from the data and subtracts it from the last 24 bits. If the result is the address of the transponder interrogated, the data are accepted as good. If not, the data are ignored and the interrogation is repeated.

14-5. Radiation Hazard

Personnel and equipment working on or near radio-frequency radiators, such as radars, experience some hazard because of the RF fields present. Radiation in the microwave bands is thought not to be ionizing and the principal hazard is from heating effects. This implies that average power density (average power per unit area) is the critical factor in evaluating the hazard.

This section presents the theory of calculating power densities produced by microwave transmitters and antennas, both in the antenna's far field and near field, and includes a sample calculation for a particular radar.

Radiation hazard standard for human exposure: The U.S. Air Force exposure limits [6] are given in Table 14-5. The IEEE standard for radiation exposure is given in [6a]. Note that the exposure standards are different for some other nations (the Soviet standards are considerably lower, for example). The permissible power density for short-term exposures are higher than those shown in the table. This information is provided as an example only. The reader is urged to consult the appropriate standards for further information in conducting evaluations and reviews of radiation hazard.

There are also exposure standards for ordnance and fuel. See U.S. Air Force Technical Order 31Z-10-4 [7] for further information.

Table 14-5. Permissible Exposure Levels (PEL) for Personnel (From [6])

For frequencies between 10 kHz and < 10 MHz	For frequencies between 10 MHz and 300 GHz inclusive
50 mW/cm^2 (for continuous exposures)	10 mW/cm^2 (for continuous exposures)
or	or
18,000 mW-s/cm^2 (in any 6-minute period)	3600 mW-s/cm^2 (in any 6-minute period)

Note: All exposures shall be limited to a maximum (peak) E-Field of 100 kV/m. *Standard subject to change.*

Calculation of average power density: The power density in an antenna's beam is dependent on the average power radiated from the antenna, the physical area of the antenna, the antenna's gain, and the distance from the antenna. The type of calculation used depends on whether one is in the near or far field of the antenna (Fig. 14-13).

In the near field, the cross-section of the beam is different from the geometrically expanding beam present in the far-field. The power density in the near-field follows approximately a relationship developed by Hansen [8], Eq. 14-63. This relationship is plotted in Fig. 14-14.

$$P/A = 26.1\left[1 - \frac{16}{X}\sin\left(\frac{\pi}{8X}\right) + \frac{128X^2}{\pi^2}\left(1 - \cos\left(\frac{\pi}{8X}\right)\right)\right] \quad (14\text{-}63)$$

$$X = R/(2D^2/\lambda) \quad (14\text{-}64)$$

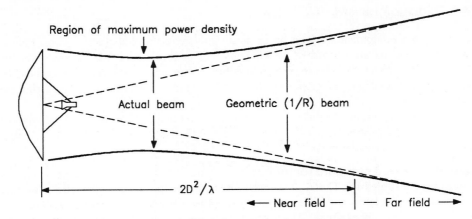

Figure 14-13. Near- and Far-Field Beam Conditions.

P/A_{N-NF} = the power density in the near-field normalized to the power density at $R = 2D^2/\lambda$ (Watts per square meter)
R = the distance to the power density being calculated (meters)
D = the antenna's physical length (meters)
λ = the wavelength (meters)

Note in Fig. 14-14 that the maximum power density occurs at approximately 1/10 the distance to the far field. *In the main beam of the antenna, this distance represents the most hazardous location so far as RF radiation is concerned.*

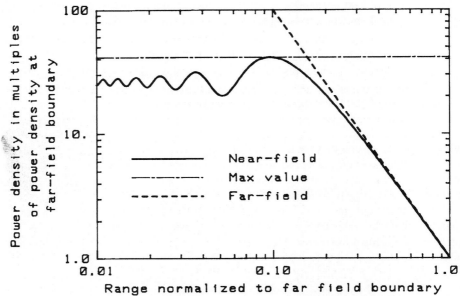

Figure 14-14. Power Density as a Function of Distance in the Near-Field.

$$R_{RFmax} \approx 0.2\, D^2 / \lambda \tag{14-65}$$

R_{RFmax} = the distance to the point of maximum power density in the main beam of the antenna (Watts per square meter)

The power density in the far-field is calculated from Eq. 14-66.

$$P/A_{FF} = \frac{P_{avg}\, G_T}{4\pi R^2} \tag{14-66}$$

P/A_{FF} = the power density in the far field of the radar's transmitting antenna (Watts per square meter)

P_{avg} = the average power emitted by the radar's antenna (Watts)

G_T = the gain of the radar's transmitting antenna

R = the distance from the transmitting antenna (meters)

The power density at the boundary of the far field ($R = 2\, D^2 / \lambda$) is

$$P/A_{FFB} = \frac{P_{avg}\, G_T\, \lambda^2}{16\pi D^4} \tag{14-67}$$

P/A_{FFB} = the power density at $R = 2\, D^2 / \lambda$ (Watts per square meter)

The power density in the near-field is thus

$$P/A_{NF} = P/A_{FFB}\, P/A_{N-NF} \tag{14-68}$$

P/A_{NF} = the power density at near-field ranges (less than $R = 2\, D^2 / \lambda$) (Watts per square meter).

Example 14-3: A single-antenna radar operating at 10.0 GHz radiates 20 kW peak power with a rectangular pulse width of 1.0 μs and a PRF of 250,000 pps. The antenna is 36 inches in diameter. Find (a) the radiation power density at 1000 m, (b) the power density at 35-m distance, (c) the range of maximum power density, (d) the maximum power density, and (e) the power density at the antenna's surface. All answers are to be found in the main beam of the antenna.

Solution: The wavelength is 0.03 m, and the antenna diameter is 0.914 m. The antenna's gain is, from Ch. 2 and assuming the effective aperture is one-half the actual aperture, 4580 (36.61 dB). The distance to the far field is 55.7 m (55-60 m). The duty cycle and average power, from Ch. 2, are 0.25 and 5000 W, respectively.

(a) 1000 m is in the antenna's far field and Eq. 14-66 gives the power density as 1.82 W/cm^2, or 0.182 mW/cm^2. This is within the AFOSH 161-0 limits for long-term exposure [6].

(b) Twenty m is within the antenna's near field and Eq. 14-68 must be used to find power density. The power density at the far-field boundary is (Eq. 14-67) 588 W/m^2 or 59 mW/cm^2. The distance in question is 0.36 times 2 D^2 / λ. Figure 14-14 gives the power density at this distance to be about 7.5 times the density at the far-field boundary, or about 440 mW/cm^2. This is many times the allowable level and personnel must avoid this area when the radar is radiating.

(c) The distance of highest power density is about 1/10 of 2 D^2 / λ, or about 6 m from the antenna.

(d) Figure 14-14 shows that the maximum power density is about 42 times that at the far-field boundary, or about 2500 mW/cm^2 (2.5 W/cm^2!). This is a very high level and should be considered quite dangerous.

(e) From Fig. 14-14, the power density at the surface of the antenna is about 28 times that at the far-field boundary, or 1650 mW/cm2.

Caution: The power density near the antenna is very much a function of the antenna's illumination taper, and the numbers calculated above must be adjusted for this fact. In addition, reflector antenna have feed radiators which are usually quite small, thus giving extremely high radiation power densities. The feeds are the most hazardous locations in reflector antennas.

Further caution: The eyes are the most sensitive organ to damage from radiation heating, although they are by no means the only tissues harmed. NEVER look at open waveguides, waveguide feeds, or any component or system where the power density exceeds the PEL. No part of the body should be subjected to levels above PEL. Effects other than heating exist. See AFOSH 161-9 or other appropriate data.

Note: Some commands, agencies, and companies use as their exposure standards lower levels than those given in AFOSH 161-9.

To prevent hazards to ground personnel, many military aircraft have a "squat switch" on the landing gear which disables the radar transmitter when weight is on the aircraft's wheels. This switch, if present, can be bypassed for maintenance, so the fact that the aircraft is on the ground or on a carrier deck is no guarantee that the radar is not emitting.

Power densities can reach hazardous levels in the sidelobes of some systems. The calculations are the same as given above, using the gain of the sidelobes at the angles in question. Remember that the sidelobe gain is the main lobe peak gain (dB) minus the measured sidelobe level (dB). The sidelobe gain of a system with a main lobe peak gain of 40-dB and 25-dB sidelobes is 15 dB.

Radiation hazard diagrams: Many systems have published diagrams showing the minimum distances from them for safe EM radiation levels. Figure 14-15 shows an example for a hypothetical tactical aircraft (the actual diagram would give distances). The area within the figures can have power densities exceeding PEL. The area showing the greatest safe distance is at those angles where the antenna main beam can be pointed. The other areas are sidelobes.

Field strength calculation: The permissible exposure level (PEL) for RF radiation includes power density and electric field strength (E-Field) standards, both of which must be met simultaneously. Unlike the power density standard, which is for average power, the permissible E-Fields are peak. The electric field strength, from Ohm's law, is

$$E_{PK} = (P/A_{PK}\, Z_0)^{1/2} \qquad (14\text{-}69)$$

E_{PK} = the RMS E-Field while the power is present (volts per meter)

P/A_{PK} = the power density during the pulse (Watts per square meter)

Z_0 = the characteristic impedance of the medium in which the power propagates ($120\pi \approx 377\Omega$ for the atmosphere)

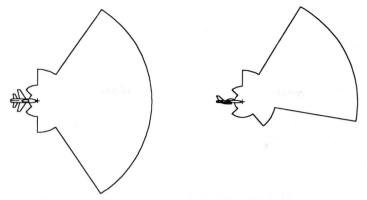

Figure 14-15. Hypothetical Radiation Hazard Diagram.

For electric fields in the atmosphere, Eq. 14-85 becomes

$$E_{PK(atm)} = (P/A_{PK}\, 377)^{1/2} \qquad \text{(if } P/A \text{ is in W/m}^2) \qquad (14\text{-}70a)$$

$$E_{PK(atm)} = (P/A_{PK}\, 3770)^{1/2} \qquad \text{(if } P/A \text{ is in mW/cm}^2) \qquad 14\text{-}70b)$$

Example 14-4: (a) Find the electric field strength in the main beam of the radar of Ex. 14-3 at a range of 1000 m. (b) Find the range at which the electric field strength in the previous example exceeds the PEL.

Solution: (a) The peak power density at 1000 m (Example 14-3a) is 0.728 mW/cm^2, which is four times the average power density (duty cycle was 0.25). By Eq. 14-70b, the electric field strength is 52 V/m. This is well under the 100 kV/m standard.

(b) By Eq. 14-70b, the power density which gives 100 kV/m electric field strength is 2.65 x 10^6 mW/cm^2. At no location does this radar's radiation cause this large a power density or E-Field level.

14-6. Radomes

A radome is a dielectric shield covering all or part of the radar's antenna. Its purpose is to protect the antenna from the elements, maintain pressurization where needed, hide the antenna from view when security requires, and in the case of airborne radars present an aerodynamically efficient shape to the airstream. An example of a radome is the air route surveillance radar in the Big Horn Mountains of Wyoming shown in Fig. 14-16. The discussions below provide only general information. Specific design details, which are beyond the scope of this book, can be found in [9], [10], and [11].

Radome classifications by how they cover the antenna: Many different types of radomes are used. They can be grouped into several classes by how they fit into the radar's design, as listed below.

Figure 14-16. Radome Covering Air Route Surveillance Radar.
(photograph by author)

— **Feed covers:** The feed cover radome is an RF-transparent window covering the feed aperture. It is most commonly used with reflector antennas. Most radars with exposed reflectors have feed covers, both to prevent contamination of the waveguides and to support waveguide pressurization. Many with radomes covering the entire antenna also require a feed cover for pressurization.

— **Aperture covers attached to the reflector:** This type radome is attached directly to the aperture of a reflector antenna and turns with it. One of its advantages is the ability to shape the radome for optimum electrical performance and minimum beam deflection. Another is that the beam always looks through the radome at the same angle, thus allowing any refraction correction to be constant with antenna pointing angle. An example is the TAS radar antenna (Ch. 9). When used with cassegrain reflector antennas, the secondary reflector is often made part of this type radome (Fig. 14-17).

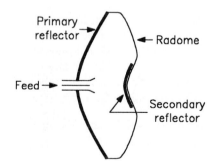

Figure 14-17. Aperture-Cover Radome with Cassegrain Secondary Reflector.

— Array element covers: One common variety of this of type radome is a dielectric sheet bonded to the front of some planar slot array antennas. Another type uses dielectric "plugs" to fill the slots in other planar slot arrays.

Radome structure: Radomes are also classified by how they are supported.

— Cloth over framework: Many radomes are composed of dielectric cloth sheets stretched over a rigid frame, which is made of either metal or dielectric structural members. Metals have in the past tended to be stronger than dielectrics, although this is no longer true. Dielectric frames usually have less effect on the electrical performance of the dome than metal. The radome shown in Fig. 14-16 is of this type.

— Rigid dielectric: Some radomes are of rigid plastic and are self-supporting. This type is commonly used with smaller domes and those which must maintain their shape regardless of airflow over them, such as those used to cover airborne radar antennas.

— Inflated (air-supported) radomes: Some radomes are of non-rigid dielectric material which is inflated with a small positive internal pressure. These domes can be of thinner materials having less electrical effect on the radar's performance, and at the same time maintain a roughly constant shape in moderate winds. They are almost invariably spherical in shape.

Radome materials: Radomes are further classified by the dielectric materials from which they are made, which can include rubberized cloth, fiberglass cloth with epoxy, other composites, ceramics, monolithic foam, metal-included dielectrics, and others.

A single-layer radome is simply one layer of a homogeneous material. Thickness is a factor in losses and refraction, with multiples of half-wavelengths in the material and at the angle of incidence of the EM wave being preferred [9]. However, many single-layer radomes are simply thin cloth approximating zero thickness. These thin radomes must be supported by a framework or be air supported (inflated).

Many thicker self-supporting radomes are made of multi-layered materials called sandwiches (see Fig. 14-18). These sandwiches provide high mechanical strengths at low weights.

An A-Sandwich is two thin skins of dense, rigid material, enclosing a thick low-density core. The skins are usually some composite, such as fiberglass cloth-reinforced epoxy. The cores are foam or honeycomb. The characteristic of the A-Sandwich is that the core has a lower dielectric constant than the skins. The skin and core thicknesses may be tuned so as to cancel reflections.

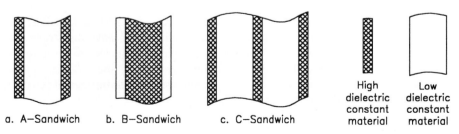

Figure 14-18. Sandwich-Type Radome Materials.

14-6. Radomes

B-Sandwiches are A-Sandwiches inside out, with a high dielectric constant core and low dielectric skins. They are heavier than A-Sandwiches with the same electrical characteristics and strength.

The C-Sandwich radome is a five-layer material with two dense outer skins, a dense center skin, and two intermediate low-density cores. It is essentially two A-Sandwiches. Its advantages are that it can be made stronger than an A-Sandwich, and that one of the cores can be hollow to circulate warm air for deicing. It also provides more design freedom in setting electrical properties. Sandwich designs can have more layers for special mechanical and electrical properties.

Some radomes have metal inclusions. In some cases the inclusions act as filters to restrict the bandpass of the radome. In others, the inclusions act as reflectors to certain signals, usually in cassegrain and polarization-twist antennas. When inclusions are selective reflectors, the wave polarization usually determines whether the signal is transmitted or reflected (see Ch. 10). In airborne radars, the inclusions are often part of the lightning-protection scheme.

Design and performance factors: Several factors affect the design and performance of radomes.

— Electrical characteristics: The dome should be transparent to the radar's RF energy, both on transmit and receive. Losses of signals caused by the radome become part of the radar's system loss.

— Beam deflection: Radomes, being composed of materials having a different index of refraction than the atmosphere, refract the radar's electromagnetic signal as it passes through them. The resulting bending of the beams can result in a degradation of the beam or a shift in the antenna's boresight direction, or both. See the discussion of radome shapes.

— Weather protection: A primary goal of the radome is to protect the antenna from weather. The radome should thus be rigid enough to withstand the winds for which it is designed, and must not leak.

— Ice protection and de-icing: Ice formation on radomes changes both their mechanical and electrical properties. Radomes which are operated in environments where ice formation is likely are usually provided with some method of ice prevention and removal. These range from pumping heated air through layers of the radome, to pneumatic breaking of the ice by flexing the dome with compressed air, to using anti-icing chemicals which prevent ice from sticking to the dome. All anti-ice and de-ice methods must allow the radome to retain its proper electrical characteristics.

— Lightning protection: Most radomes have lightning rods or other protections from penetration by lightning. This problem is particularly acute in airborne radomes, where lightning penetration, cracking, or shattering of the radome can have disastrous consequences. Airborne radomes must have some method of bleeding the electrical charge from the lightning to the airframe. This is usually done by making the radome somewhat conductive, either by using a somewhat conductive composite, or burying metallic electrodes in the plastic. Either method degrades the electrical performance of the dome.

Refraction of electromagnetic energy through radomes: Radomes which cover the entire antenna are usually either spherical with the radar's center of radiation at the center of the sphere, or aerodynamically shaped. Spherical radomes have the advantage that, regardless of the radar's pointing angle, the EM waves always look through the same shape. Spherical radomes cause the beam to spread as shown in Fig. 14-19, thus widening the beam and somewhat lowering the antenna's gain. The smaller the sphere is with respect to the antenna size, the more severe is this problem, since with smaller radomes angles at which the waves look through the dielectric vary more from antenna center to edge than they do with larger spheres. With this shape, there are generally minimal boresight errors.

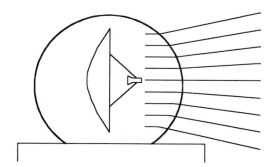

Figure 14-19. Refraction through a Spherical Radome.

Aerodynamically shaped radomes can exhibit boresight shifting as the beam passes through it (Fig. 14-20). The boresight shift depends on the direction the antenna beam is pointed. In addition, refraction usually causes the beam to spread, degrading the performance of the antenna. Also, aircraft radomes must be mechanically stronger than those used in ground-based and shipboard systems, and the thicker materials affect the EM waves more. The errors produced by these domes are usually sufficient to require correction in the radar's data processor. In some aircraft, a generic correction table is loaded; in others, the correction table may have to be customized to the particular dome in use.

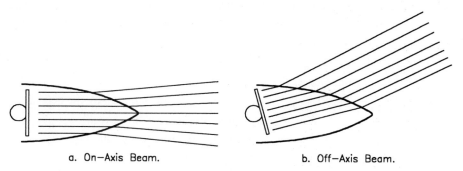

a. On−Axis Beam.　　　　　　　　　b. Off−Axis Beam.

Figure 14-20. Aerodynamically Shaped Radome.

14-7. Radar Electronic Warfare (EW) Introduction

Material on radar EW is scattered throughout this text. It is the aim of this section to summarize the facets of electronic warfare of interest to radar specialists and to provide references to the detailed information necessary for analysis and design. Several specialized texts exist which provide further information ([12] and [13], for example).

Electronic warfare is defined in JCS-Publication 1 [14] as "military action involving the use of electromagnetic energy to determine, exploit, reduce, or prevent hostile use of the electromagnetic spectrum and action which retains friendly use of the electromagnetic spectrum." Electronic warfare has several divisions, depicted in Fig. 14-21 and defined below [14].

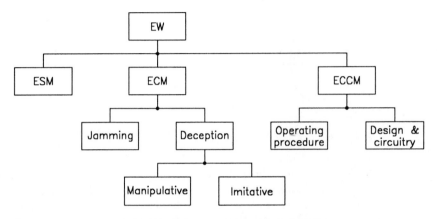

Figure 14-21. Electronic Warfare Family Tree.
(After Johnston [15])

Electronic warfare objectives: Electronic warfare systems are operated by both sides in a conflict. The two sides can be considered the "Threats" and the "Defenders." The EW objectives of the Threats and Defenders are listed below:

Threat objectives:
— Prevent detection of the presence of the threat
— Prevent the determination of the intentions of the threat
— Prevent detection of the threat's electromagnetic (EM) radiation by the defenders
— Prevent location of the threat by location of its EM signals
— Prevent recovery of the information content of the threat's EM signals
— Prevent interference with the threat's use of the EM spectrum

Defender objectives:
— Detect the presence of threats
— Determine the location and intentions of the threats using defender's EM radiation
— Detect the presence of the threat's EM radiation.
— Identify the threat's EM radiation.

— Identify the threat's intentions through analysis of the threat's EM radiation.
— Determine the threat's location through the threat's EM radiation.
— Determine the information content of the threat's EM radiation.
— Deny the use of the EM spectrum by the threat.

— *Electronic Support Measures (ESM):* ESM is "that division of electronic warfare involving actions taken to search for, intercept, locate, record, and analyze radiated electromagnetic energy for the purpose of exploiting such radiations in the support of military operations" [14]. ELINT (*EL*ectronic *INT*elligence) is sometimes classed separately from ESM. In cases where this division occurs, ELINT refers to the gathering of intelligence about the enemy's electromagnetic capability, and ESM is the tactical reception of signals to locate, identify, and analyze signals so their effects can be avoided or countered.

— *Electronic CounterMeasures (ECM):* This is "that division of electronic warfare involving actions taken to prevent or reduce an enemy's effective use of the electromagnetic spectrum" [14]. ECM itself consists of two types. The first is jamming (sometimes called denial ECM), where the use of the electromagnetic signals is denied. The second, deception, is where emitted ECM signals are meant to imitate desired signals, but with changes, or are meant to manipulate the system into reporting

— *Electronic Counter-CounterMeasures (ECCM):* ECCM "is that division of electronic warfare involving actions taken to insure friendly effective use of the electromagnetic spectrum despite the enemy's use of electronic warfare" [14].

— *Electromagnetic compatibility:* Although not part of the EW tree, Electromagnetic Compatibility (EMC) is sometimes included in discussions of EW. Johnston [15] defines it as "the ability of communications electronics equipments, subsystems, and systems to operate in their intended operational environments without suffering or causing unacceptable degradation because of unintentional electromagnetic radiation or response."

Electronic support measures: Electronic support measures, from the standpoint of radar, have three functions: interception (detection) of radar emissions, analysis of radar signals, and direction finding to the origin of radar signals. The primary use of ESM in tactical situations is in radar warning receivers (RWRs), which warn that a radar is radiating and attempt to ascertain the function of the radar (search, missile illumination, and so forth) and its direction from the RWR.

Interception: Ordinarily, the radar warning receiver in a potential target can intercept a radar's illumination signal at considerably longer range than the radar can detect the echo from the target. This is because of the one-way (R^{-2}) nature of interception and two-way (R^{-4}) nature of echo detection. In some radar designs, however, the radar's process gain to echoes from its own illumination so far exceeds the process gain of the interceptor to the same illumination signal that echo detection takes place at a longer range than interception. Radars of this type emit waveforms coded in such a way as to be difficult to detect without knowing the code, and are known as *Low Probability of Intercept (LPI)* radars. Further information on LPI can be found in [19].

The radar equation for target echoes in noise interference (from Ch. 3), assuming a monostatic radar with one antenna for transmit and receive, is

$$R_{Dmax}^4 = \frac{W_T G^2 \lambda^2 \sigma}{(4\pi)^3 L K T_0 F S/N_{min}} \quad (14\text{-}71)$$

R_{Dmax} = the maximum range for echo detection (meters)
W_T = the illumination energy for a dwell (Joules)
G = the radar antenna's gain
λ = the wavelength (meters)
σ = the target's radar cross-section square meters)
L = the total loss, including system, atmospheric, and integration losses
K = Boltzmann's constant (1.38 x 10^{-23} J/°K)
T_0 = 290°K
F = the radar's noise factor
S/N_{min} = the minimum signal-to-noise ratio for detection

To a an ESM receiver, the intercept signal-to-noise ratio is

$$S/N_I = \frac{P_T G_{TI} \lambda^2 G_{IT} G_{PI}}{(4\pi)^2 R^2 L_{INT} K T_0 B_I F_I} \quad (14\text{-}72)$$

P_T = the radar transmitted power
G_{TI} = the radar transmitter antenna's gain in the direction of the intercept antenna
G_{IT} = the intercept antenna's gain in the direction of the radar transmitter antenna
G_{PI} = the intercept signal processor's process gain to the radar's illumination signal
R_I = the distance from the interceptor to the radar
L_{INT} = the total intercept loss
B_T = the intercept receiver's bandwidth. (The most advantageous intercept receiver filter is matched to the illumination waveform. This factor in the intercept equation cannot be less than the radar's illumination bandwidth.)
F_I = the intercept receiver's noise factor.

The maximum range at which the radar's illumination is intercepted is

$$R_{Imax}^2 = \frac{P_T G_{TI} \lambda^2 G_{IT} G_{PI}}{(4\pi)^2 L_{INT} K T_0 B_I F_I S/N_{Imin}} \quad (14\text{-}73)$$

R_{Imax} = the maximum range at which interception can occur for the conditions given

S/N_{Imin} = the ESM system's minimum signal-to-noise ratio for successful interception

The probability of intercept is found in much the same way as the probability of detection from Ch. 6. An intercept probability of false alarm is chosen, and by techniques given in Ch. 6, a threshold is set. Given the intercept signal-to-noise ratio of Eq. 14-72, an intercept probability is found for one sample. Since integration is seldom possible in ESM receivers, this calculation gives the intercept probability *if the minimum conditions of Eq. 14-72 are met* (antenna gains, distances, and so forth). The overall probability of intercept is then the product of the calculated probability of intercept and the probability that the conditions of the calculation will be met (will the antenna gain assumed be present, for example).

$$P(I) = P(I)_{S/N} P(R) P(G)_{RI} P(G)_{IR} P(F) \qquad (14\text{-}74)$$

$P(I)$ = the overall probability of intercept

$P(I)_{S/N}$ = the probability of detection of a signal of the S/N calculated in Eq. 14-72 for a chosen intercept probability of false alarm

$P(R)$ = the probability that the distance from interceptor to radar will be the value used in Eq. 14-72

$P(G)_{RI}$ = the probability that the gain of the radar's antenna in the direction of the interceptor is the value used in the Eq. 14-72 calculation

$P(G)_{IR}$ = the probability that the gain of the interceptor's antenna in the direction of the radar is the value used in the Eq. 14-72 calculation

$P(F)$ = the probability that the interceptor receiver is tuned to the radar's frequency

Example 14-5: A radar has the following specifications:

Center frequency	10.0 GHz
Transmit peak power	20 kW
Pulse width	1.2 µs
PRF	22,000 pps
Antenna peak gain	38 dB
Antenna sidelobes	At least 45 dB over the circle
Waveform matched bandwidth	5.2 MHz
Losses	Neglect
Noise figure	4.3 dB
Pulses per dwell	64
Process gain to echoes	16 dB
Detection S/N	13.5 dB
Detection probability	0.70
False alarm probability	10^{-8}

The intercept receiver's specifications are:

Antenna gain	At least 9.3 dB over the circle
Instantaneous bandwidth	200 MHz

Noise figure	10.0 dB
Process gain to radar signal	0.4 (mismatched, one pulse)
Interception S/N	15 dB
Intercept probability	0.9
Intercept false alarm probability	10^{-5}

(a) Find the range at which the radar detects a 2 m² target.

(b) Find the range at which the interceptor intercepts the radar's illumination.

(c) If interception can take place in the radar antenna's main lobe, find the intercept range.

(d) If interception is in the radar antenna's main lobe and the radar's frequency agility is disconnected and the interceptor knows the radar's operating frequency, find the intercept range.

Solution: (a) The dwell energy (Ch. 3) is 1.54 J. Equation 14-71 gives the maximum echo detection range as 123 km.

(b) Equation 14-73 gives the predicted intercept range for the conditions given. If it is assumed that the intercept and radar specifications given above are simultaneously met, the intercept specifications used for the equation are:

P_T	20 kW
G_{TI}	38 − 45 = −7 dB
G_{IT}	9.3 dB
G_{PI}	0.4
L_{INT}	0 dB
B_I	200 MHz
F_I	10.0 dB
$S/N_{I\min}$	15 dB

From Eq. 14-73 with the above values, the intercept range is 17.4 km, which is considerably less than the radar's detection range. This is because of the large discrepancy in antenna gains between radar target detection and radar intercept (45 dB), because of the large process gain the radar has to echoes and which the interceptor does not have, and because of the bandwidth difference.

(c) Intercepting in the radar's main lobe modifies the G_{TI} to 38 dB. The new intercept range is 151 km.

(d) Giving the interceptor the same bandwidth as the radar (no frequency agility and good intelligence), and allowing interception from the radar's main lobe illumination makes B_I 5.2 MHz and raises the intercept range to 30,500 km.

ESM and LPI radar principles: Factors which make the radar's illumination signal more difficult to intercept are listed and described below.

− Radar transmitter and illumination waveform: It can be seen from Eqs. 14-71 and 14-72 that target detection is most affected by illumination *energy*, while interception is done on illumination *power*. It is thus to the radar's advantage to illuminate with a high energy, low power waveform. Wide pulse width, low-power transmit pulses help avoid interception, while wide illumination bandwidths (coding or chirp) maintain usable range resolutions. Coded waveforms give the radar the process gain advantage of knowing the code transmitted, allowing the use of correlation detectors. If these codes are secure, the interceptor does not have this same advantage.

- **Radar transmitter antenna:** Radar transmitting antennas have narrow beamwidths and the probability that the radar antenna's peak gain is pointed at the ESM antenna is low, unless the ESM antenna is aboard the target the radar is trying to detect. Thus many intercepts are through illumination sidelobes. The antenna design factor which most affects the intercept probability is therefore low illumination antenna sidelobes.
- **Illumination frequency:** As can be seen in Eq. 14-72, it is advantageous for the intercept receiver's bandwidth to match the illumination waveform. Thus random frequency hopping (agility) on the part of the radar will force the intercept receiver to have a wide bandwidth to accept any possible illumination frequency. If not, the factor $P(F)$ will be small.

ESM analysis: It is of great importance that the ESM systems be able to distinguish a search radar from a missile illumination radar which is "locked onto" the vehicle carrying the ESM system. The second role of ESM, then, is to analyze the intercepted signal to determine what it is and whether or not it is a threat. This involves determining the signal's frequency and frequency hopping (agility) pattern, pulse width, PRF and PRF pattern, waveform coding, antenna scan rate, tracking type, whether the radar is pointed at the ESM receiver, and many other factors. Much of the resources allocated to ESM are directed at this task, which may involve extremely complex calculations. See reference [20] for further information.

ESM direction finding: The final ESM function is to determine the angle-of-arrival (AOA) of the radar's illumination. Two fundamentally different techniques are used. One is to use a highly directional ESM antenna and rotate the beam rapidly. The advantages of directional antennas are that the ESM antenna can have high gain and can determine AOA accurately. The disadvantage is that with directional rotating beams, the probability that the ESM beam is pointed at the radar while the radar beam is pointed at the ESM antenna $[P(G)_{IR}]$ becomes quite small.

The other method is to use interferometer-type antenna systems which are omnidirectional, but which can identify the AOA of received signals. Their advantage is that they look in all directions simultaneously. Their disadvantage is relatively low gain. This type of antenna is the one most commonly used in radar warning receivers.

Electronic countermeasures: Electronic countermeasures can be either passive or active. Passive ECM is any interference technique which does not involve emitting an electromagnetic jamming signal. Examples include chaff (Ch. 4), decoys, and any technique involving tactics only, such as hiding targets in clutter. Active ECM is jamming with generated electromagnetic waves.

Active ECM (jamming) is divided into two categories: *denial* and *deceptive*. Denial jamming attempts to deny the use of the electromagnetic spectrum to the radar, usually by providing so much jamming signal that the radar's signal processor and displays are occupied almost totally with the jamming to the exclusion of targets. Deceptive jamming attempts to "fool" the radar into reporting targets at the wrong range, or wrong Doppler, or wrong angles. It also attempts to increase the number of realistic targets present, and either saturates defenses or forces resources to be expended where no threat exists.

Denial ECM: How denial jamming affects a radar depends on whether or not an effective CFAR threshold-setting circuit exists in the radar. If this circuit is not present, the false alarm rate increases to the point of hiding desired targets. If effective CFAR does exist, the manifestation of effective jamming is that the threshold is raised to prevent the jamming from causing excessive false alarms, and the targets vanish.

Several jamming signals can be used for denial. Narrow-band noise which is contained entirely within the radar's bandpass can be effective against non-agile radars. Any technique where the jammer's energy is entirely within the radar's bandpass is called spot jamming. Another useful ECM waveform is wide-band noise, used to accommodate radar frequency agility. It is called barrage jamming, and is less effective than spot noise because most of the jamming energy is rejected by the radar's receiver filters (see Ch. 3).

Very high-amplitude high-bandwidth pulses, called impulse jamming, is effective against receivers and signal processors which saturate easily (Ch. 5). Many other waveforms are used [16].

Deceptive ECM: Deception is the use of electromagnetic emissions or other techniques to produce targets which have the wrong range or position or Doppler shift, or which produce many false targets. Techniques are numerous and many of the effective ones are not published in the open literature. Van Brunt [16] describes many of them. Examples include range-gate pull-off (RGPO) for range deception, velocity-gate pull-off (VGPO) for Doppler deception, and cross-eye for angle deception. Treatment of specific methods are beyond the scope of this text.

Manipulative ECM: This form of ECM attempts to induce actions on the part of its victims which are detrimental to the victim's well-being. An example would be to set up a fake navigation radio station which would lure enemy aircraft into hostile territory (or into the side of a mountain).

Electronic counter-countermeasures: ECCM techniques are applied to several parts of the radar to minimize the effects of ECM.

Transmitter ECCM: The main factor in determining signal-to-jamming ratios is transmit energy (see Ch. 3). Energy can be obtained with high power, wide pulses, and large dwell numbers (Ch. 2).

Waveform and signal processor ECCM: Process gain to echo signals is achieved with waveform coding, whether phase coding or FM chirp. If the ECM signals do not have the same coding as the illumination and echoes, the process gain in the radar is less to the jamming signal than it is to the echoes, and thus the signal-to-jamming ratios are improved. The more coding (longer, more complex waveforms) is present and the more secure the codes, the less likely it will be that the jammer has the proper codes and can effectively interfere with the radar.

Antenna ECCM: Much denial ECM enters the radar through antenna sidelobes, making suppression of sidelobe responses a high priority for radars operating in this environment.

In Sec. 9-6, three techniques were discussed: low sidelobe designs, sidelobe blanking (SLB), and coherent sidelobe cancellation (CSLC).

Receiver ECCM: The receiver has two main roles in rejecting ECM. First, it should have effective bandpass filters so that as much broadband jamming as possible is rejected, recalling that the signal-to-jamming ratio is proportional to the ratio of jammer bandwidth to radar bandwidth (Ch. 3). The second role of the receiver is that it amplify both echoes and jamming linearly to prevent intermodulation products which the signal processor may not be able to remove.

14-8. Further Information

Coordinates and transformations: Much of the information in this section can be found in texts on guidance, autopilots, and advanced navigation.

[1]   R.T. Jones, D Christensen, and R. Benton, private communication.

[2]   G. Bomford, *Geodesy*, London: Oxford University Press, 1980.

Secondary surveillance radar: The original source of most information on SSR is from the U.S. Federal Aviation Administration in Oklahoma City. Advisory circulars such as 00-27, quoted herein, give detailed signal specifications.

[3]   FAA, *Advisory Circular 00-27*, Oklahoma City: Federal Aviation Administration, 1968.

[4]   W.H. Cole, *Understanding Radar*, London: Collins, 1985, Chs. 10-12.

[5]   M.S. Stevens, *Secondary Surveillance Radar*, Norwood MA: Artech House, 1988.

Radiation hazard: Two U.S. Air Force publications provide standards and useful design and test data for radiation hazard. Near-field power densities are described in the third source.

[6]   U.S. Air Force, *AFOSH Standard 161-9*, 10 October 1978. See also [6a].

[6a] IEEE Standard C95.1-1991, *IEEE Standard for Safety Levels with Respect to Human Exposure to Radio Frequency Electromagnetic Fields, 3 kHz to 300 GHz.*

[7]   U.S. Air Force, Technical Order 31Z-10-4.

[8]   R.C. Hansen, "Axial Power Density in the Near Field," in *The Microwave Engineers' Handbook and Buyers' Guide*, Norwood MA: Horizon House, 1961-62, p. TD-115.

Radomes: The major sources of radome data and design information are:

[9]   G.K. Huddleston and H.L. Bassett, "Radomes," Ch. 44 in R.C. Johnson and H. Jasik (eds.), *Antenna Engineering Handbook*, 2nd ed., New York: McGraw-Hill, 1984.

[10] J.D. Walton, Jr (ed.), *Radome Engineering Handbook*, New York: Marcel Dekker, 1970.

[11] V.J. DiCaudo, "Radomes," Ch. 14 in M.I. Skolnik, *Radar Handbook*, 1st ed., New York: McGraw-Hill, 1970. This topic is very much abbreviated in the second edition.

Electronic warfare: The following two books cover electronic warfare in general.

[12] D.C. Schlenher, *Electronic Warfare*, Norwood, MA: Artech House, 1986.

[13] F. Neri, *Introduction to Electronic Defense Systems*, Norwood, MA: Artech House, 1991.

Electronic warfare definitions are given in the following reference.

[14] Department of Defense, Joint Chiefs of Staff, *Dictionary of Military and Associated Terms*, JCS Pub-1, September 1974.

Much of the information on EW is not in the public domain. Four of the best unclassified sources for overall information are given below. The Van Brunt books are especially helpful in describing techniques.

[15] S.L. Johnston (ed.), *Radar Electronic-Counter Countermeasures*, Norwood MA: Artech House, 1979.

[16] L.B. Van Brunt, *Applied ECM* (two volumes), Dunn Loring VA: EW Engineering, Inc., 1978.

[17] *The International Countermeasures Handbook*, Palo Alto CA: EW Communications Inc., published annually.

[18] B. Blake, *Jane's Radar and Electronic Warfare Systems*, Coulsdon, Surrey, UK: Jane's Information Group, published biannually (2nd ed. is 1990).

Low probability-of-intercept (LPI) radar, ESM, and ELINT signal analysis are covered in the following sources:

[19] R.G. Wiley, *Electronic Intelligence: Interception of Radar Signals*, Norwood MA: Artech House, 1985.

[20] R.G. Wiley, *Electronic Intelligence: The Analysis of Radar Signals*, Norwood MA: Artech House, 1982.

A

UNIT CONVERSIONS

* Indicates an exact conversion.
† Indicates an exact conversion with a repeating fraction.

°	= degree	Kt	= Knot
cm	= centimeter	m	= meter
cm^2	= square centimeter	m^2	= square meter
cm^3	= cubic centimeter	m^3	= cubic meter
deg	= degree	m/s	= meter per second
deg^2	= square degree	mil	= mil (1/6400 circle)
fps	= foot per second	mil^2	= square mil
ft	= foot	mph	= statute mile per hour
ft^2	= square foot	nmi	= nautical mile
ft^3	= cubic foot	nmph	= nautical mile per hour (Knot)
grad	= grad (1/400 circle)	rmi	= radar mile
$grad^2$	= square grad	rmph	= radar mile per hour
in	= inch	smi	= statute mile
in^2	= square inch	smph	= statute mile per hour
in^3	= cubic inch	steradian	= solid angle in a sphere
ips	= inch per second	rad	= radian

A. Length Conversions:

1 m	= 3.28083989501 ft	1 nmi	= 6076.11548557 ft
1 m	= 39.37007874 in	1 nmi	= 1.150779448 smi
1 m	= 100 cm *	1 nmi	= 1.012685914 rmi
1 ft	= 0.3048 m *	1 smi	= 1609.344 m *
1 ft	= 12 in *	1 smi	= 5280 ft *
1 ft	= 30.48 cm *	1 smi	= 0.868976242 mni
1 in	= 0.0254 m *	1 smi	= 0.88 rmi *
1 in	= 2.54 cm *	1 rmi	= 1828.8 m *
1 cm	= 0.01 m *	1 rmi	= 6000 ft *
1 cm	= 0.0328039895 ft *	1 rmi	= 0.987473002 nmi
1 cm	= 0.393700787402 in	1 rmi	= 1.136<u>36</u> smi †
1 nmi	= 1852 m *		

B. Velocity Conversions:

1 Kt	=	1.0 nmph *
1 Kt	=	0.51444 m/s †
1 Kt	=	1.687809857 fps
1 Kt	=	1.150779448 mph
1 Kt	=	1.012685914 rmph
1 smph	=	0.44704 m/s *
1 smph	=	1.4666 fps †
1 smph	=	0.868976242 Kt
1 smph	=	0.88 rmph *
1 rmph	=	0.508 m/s *
1 rmph	=	1.666 fps †
1 rmph	=	0.987473002 Kt
1 rmph	=	1.1363̲6̲ mph †
1 m/s	=	3.280839895 fps
1 m/s	=	1.943844492 Kt
1 m/s	=	2.236936292 mph
1 m/s	=	1.968503937 rmph
Mach 1	≈	338 m/s at sea level

C. Area Conversions:

1 ft^2	=	0.092903040 m^2 *
1 ft^2	=	929.03040 cm^2 *
1 ft^2	=	144 in^2 *
1 in^2	=	0.00064516 m^2 *
1 in^2	=	6.4516 cm^2
1 in^2	=	0.0069444 ft^2 †
1 cm^2	=	0.0001 m^2 *
1 cm^2	=	0.001076391 ft^2
1 cm^2	=	0.155000310 in^2
1 m^2	=	10000 cm^2 *
1 m^2	=	10.763910417 ft^2
1 m^2	=	1550.00310001 in^2

D. Volume Conversions:

1 ft^3	=	0.028316847 m^3
1 ft^3	=	28,316.847 cm^3
1 ft^3	=	1,728 in^3 *
1 in^3	=	0.0000163870600 m^3
1 in^3	=	16.3870600 cm^3
1 in^3	=	0.000578703704 ft^3 *
1 cm^3	=	0.000,001,000 m^3 *
1 cm^3	=	0.061023744 in^3
1 cm^3	=	0.00003531466672 ft^3
1 m^3	=	1,000,000 cm^3 *
1 m^3	=	61,023.74409 in^3
1 m^3	=	35.31466672 ft^3

E. Plane Angle Conversions:

π	=	3.141592654
1 circle	=	2π rad *
1 circle	=	360° *
1 circle	=	400 grads *
1 circle	=	6,400 mils *
1 mil	=	π / 3200 rad *
1 mil	=	0.000981747704 rad
1 mil	=	0.0562500 deg *
1 mil	=	0.0.062500 grad *
1 grad	=	π / 400 rad *
1 grad	=	0.00785398164 rad
1 grad	=	0.900 deg *
1 grad	=	16.000 mils *
1 deg	=	π / 180 rad *
1 deg	=	0.01745329252 rad
1 deg	=	1.11111 grad †
1 deg	=	17.777 mils †
1 rad	=	3200 / π mils *
1 rad	=	1,018.591636 mils
1 rad	=	200 / π grads *
1 rad	=	63.66197723 grads
1 rad	=	180 / π deg *
1 rad	=	57.29577951 deg

F. Solid Angle Conversions:

1 sphere	=	4π steradians *
1 sphere	=	41,252.96125 deg^2
1 sphere	=	50,929.58179 grad^2
1 sphere	=	13,037,972.95 mil^2
1 mil^2	=	9.638285543 x 10^{-7} steradian
1 mil^2	=	0.003164062499 deg^2
1 mil^2	=	0.0039062500 grad^2 *
1 grad^2	=	0.0002467401101 steradian
1 grad^2	=	0.81000 deg^2 *
1 grad^2	=	256.000 mil^2 *
1 deg^2	=	0.0003046174198 steradian
1 deg^2	=	1.234567901 grad^2
1 deg^2	=	316.0493828 mil^2
1 steradian		3,282.806350 deg^2
1 steradian		4,052.847345 grad^2
1 steratian		1,037,528.921 mil^2

B

TRANSMITTER SAFETY

Transmitters and modulators are two of the three most dangerous parts of radars. The other is the moving antenna. Three principal hazards exist: high voltages and currents, microwave radiation, and X-rays. Note that any statements made here about transmitters apply equally to modulators and high voltage power supplies.

Electrical hazards: The voltages and currents in radar transmitters are so high that normal electrical safety principles must be augmented. These systems are totally lethal. Following are some suggestions regarding working with transmitters safely. *Always follow the system's operating manual procedures and exercise good electrical safety practice.*

— Do not work on, touch, or come close to transmitters with power applied and cabinet doors open.
— When opening transmitter cabinet doors, be sure power is off, interlocks are activated (open), and all capacitors are discharged. Some transmitters provide a shorting bar which must be applied to certain points immediately on opening the doors, and some automatically short certain points when the doors open. *Be sure to follow the procedures in the system manuals.*
— Some transmitters have openings other than the main cabinet doors which are not interlocked. Be particularly careful around them.
— Transmitter monitoring and troubleshooting can usually be done with the cabinets closed using test points provided by the manufacturer. If tests must be done with the doors open *and the manuals and owner allow such practice*, be sure the transmitter is totally dead, including discharging capacitors, before connecting the appropriate test equipment. Stand well back before activating the transmitter. *Do not* reach into the cabinet to change a meter scale or for any other purpose with the transmitter hot. Go through the shutdown procedure, including shorting capacitors, before adjusting or removing the test equipment. *Follow the manual procedures religiously.*
— Do not allow yourself to be grounded or become part of a circuit, even on a "dead" transmitter. Work with one hand. Do not depend on distance to isolate you from the transmitter — the voltages are high enough to arc over surprisingly long distances.
— *Never* work alone.
— Be sure you and all other persons working at radar sites know CPR.

Microwave radiation hazard: Microwave radiation is not normally a problem in well-designed transmitters with all transmission lines intact. Microwave radiation causes damage by heating, and thus the power density is the controlling factor. The primary hazard is in locations where power levels are high and the power is concentrated into a small area. In intact radars, the problem area is around the antennas and particularly around the antenna feeds. If waveguides or other transmission lines are opened for maintenance, that hazard is transferred to the opening. *Never operate transmitters with the transmission lines opened.* It is hard on the transmitter as well as on the personnel. If a transmitter must be tested disconnected to the antenna, use a dummy load approved by the manufacturer and procedures approved by the manufacturer and owner. See Ch. 14 for more information on radiation hazard.

X-ray Hazard: Many transmitters contain voltages high enough to cause rather hard (short wavelength, highly energetic, and penetrating) X-rays to form. This is particularly true with high-power linear-beam tubes (klystrons, TWTs, and twystrons), although cross-field tubes also generate X-rays. These systems normally have X-ray shielding around the critical parts of the tube (the collector, mainly). *NEVER operate the transmitter without the X-ray shielding in place.* Some transmitters have a portion of the X-ray shielding in the cabinets. This is one more reason for not operating transmitters with their cabinet doors open.

C

AN/ EQUIPMENT DESIGNATION

Many U.S. military systems are identified using the AN designation scheme. The AN scheme is composed of three parts. First is the prefix **AN/** which indicates that the AN designation is being used. Second is a three-letter code identifying the environment in which the system is used and its function. Third is a serial number. An example is the AN/APG-71 (the F-14D's radar). It is the 71st airborne radar guidance and control system designated by this scheme. The AN scheme is described in detail in *MIL-STD-196D*.

The first letter of the three-letter code designates the *environment* in which the system is used. The environments are

- A   Airborne (piloted aircraft)
- B   Underwater mobile (submarine)
- D   Pilotless carrier
- F   Fixed ground
- G   Ground, general
- K   Amphibious
- M   Ground mobile (in vehicle whose function is to carry the equipment)
- P   Portable (man or animal carried)
- S   Shipboard (surface ship)
- T   Ground transportable
- U   Utility (more than one class)
- V   Ground vehicular (in vehicle whose primary function is other than to carry the equipment; tank, for example)
- W   Water, surface and underwater
- Z   Airborne vehicle (piloted/pilotless) combination

The second letter tells what the equipment is. For example, P is radar, R is radio, K is telemetry, and so on.

- A   Infrared, invisible light
- C   Carrier, wire
- D   Radiac (RAdioactive Detection, Indication, And Computation)
- E   Laser

- F Photographic
- G Telegraph/Teletype
- I Interphone and public address
- J Electromechanical (not covered elsewhere)
- K Telemetry
- L Electronic countermeasures (sensing, analysis, direction finding, and jamming)
- M Meteorological
- N Sound in air
- P Radar
- Q Sound in water (sonar)
- R Radio
- S Special or combination of types
- T Telephone (wire)
- V Visible light
- W Weapon systems not otherwise covered
- X Television, facsimile
- Y Computer and data processing

The third letter indicates what the system does. For example, S is search and detection and is normally part of a radar's designation. N is navigation, which can also be a radar's function.

- A Auxiliary (part of a system, but not a complete system in itself)
- B Bombing
- C Communication
- D Direction finding and reconnaissance
- E Ejection
- G Fire control (also used in multi-mode systems where fire control is a major function)
- H Recording (graphic meteorological)
- K Computer
- M Maintenance and testing
- N Navigation
- Q Special or combination of functions
- R Receiving and passive detecting
- S Search and/or detection
- T Transmitting
- W Control, automated flight or remote
- X Identification and recognition
- Y Surveillance and control, multi-target tracking, fire control, and air control

A suffix letter indicates a major change in the system without the issuance of a new AN/ designation. For example, the AN/SPS-48E is the sixth major block of AN/SPS-48 radars (the AN/SPS-48 was the first, the AN/SPS-48A the second, and so on).

The letter V in parentheses following the serial number indicates that the system has major variants. For example, the AN/FPS-16(V) is a ground-based instrumentation tracking radar of which about 50 were built. Probably no two are exactly alike and some differ from others in major ways (klystron versus magnetron transmitters, and so forth).

Civilian radar systems, unless there is joint military procurement, do not follow the AN/ designations. Some on the nomenclature describing these systems are:

- *ASR-xx* stands for Airport Surveillance Radar, a series of medium-range (50 to 100 mile) air traffic control radars. The xx is the model number. The ASR-9, therefore, is the ninth model of the airport surveillance radar series.
- *ARSR-xx* (sometimes pronounced Are'-sur) is the Air Route Surveillance Radar series of long-range (200+ mile) air traffic control radars, with xx being the serial number. Presently, the ARSR-4 is being procured.
- *ASDE* (pronounced Az'-dee) is Airport Surface Detection Equipment, short range radars meant to locate aircraft on the ground at airports.
- *TDWR* is Terminal Doppler Weather Radar, used to identify bad weather near airports. One of its tasks is to attempt to locate and identify microbursts, extremely violent downdrafts associated with thunderstorms, which have caused the loss of several airliners.
- *WSR* is Weather Surveillance Radar, a designation used by the National Weather Service for its radars. The WSR is followed by numbers which may indicate the year of first contract or manufacture. For example, the NEXRAD (NEXt generation weather RADar) is officially the WSR-88D, for Weather Surveillance Radar model 1988 Doppler.

D

GREEK ALPHABET AND METRIC PREFIXES

Greek Alphabet:

A	α	= alpha		N	ν	= nu
B	β	= beta		Ξ	ξ	= xi
Γ	γ	= gamma		O	o	= omnicron
Δ	δ	= delta		Π	π	= pi
E	ε	= epsilon		P	ρ	= rho
Z	ζ	= zeta		Σ	σ	= sigma
H	η	= eta		T	τ	= tau
Θ	θ	= theta		Υ	υ	= upsilon
I	ι	= iota		Φ	ϕ	= phi
K	κ	= kappa		X	χ	= chi
Λ	λ	= lambda		Ψ	ψ	= psi
M	μ	= mu		Ω	ω	= omega

Metric prefixes:

Prefix name	Abbreviation	Multiplier	Prefix name	Abbreviation	Multiplier
atto	a	10^{-18}	deci	d	10^{-1}
femto	f	10^{-15}	Deka	Da	10^{1}
pico	p	10^{-12}	Hecto	H	10^{2}
nano	n	10^{-9}	kilo	k	10^{3}
micro	m	10^{-6}	Mega	M	10^{6}
milli	m	10^{-3}	Giga	G	10^{9}
centi	c	10^{-2}	Tera	T	10^{12}

E

DECIBELS

Decibels are a way of representing quantities using the properties of logarithms. The advantages of using decibel notation are that dynamic ranges are compressed; that is, very large and very small numbers are represented as convenient, easily understood quantities. Also, multiplications and divisions are carried out as additions and subtractions.

Decibels are useful in communications and radar, both in expressing and calculating widely differing signal levels and in treating these signals with very large gains and attenuations. For example, the minimum discernible echo signal (MDS) of a typical radar may be 0.000000000000015 Watt and the gain of the receiver may be 133000000000. These numbers are virtually incomprehensible, but take on much more meaning if expressed as an MDS of −108.2 dBm and a gain of 111.2 dB. The receiver's output signal level becomes −108.2 dBm + 111.2 dB = 3.0 dBm or 2 mW.

Definitions: The unit "Bel," named for Alexander Graham Bell, is simply the logarithm of the ratio of two power levels. The Bel gain of a device or process is

$$\text{Bel gain} = \text{Log}_{10}(P_o/P_i) \tag{E-1}$$

P_o = the power out of the system

P_i = the power into the system

Note that to find the Bel equivalent of a number, it must be greater than zero and must be a ratio. The logarithm of zero is negative infinity and that of a negative number is not real. The logarithm of a dimension has, in most cases, no meaning. *All decibel logarithms are base 10.*

If voltage or current ratios are given, instead of power, Bel equivalents can be found from knowing that power is proportional to the square of voltage or current.

$$\text{Bel gain} = \text{Log}[(V_o/V_i)^2] \tag{E-2}$$

V_o = the voltage out of the system

V_i = the voltage into the system

$$\text{Bel gain} = 2\,\text{Log}(V_o/V_i) \tag{E-3}$$

$$\text{Bel gain} = 2\,\text{Log}(I_o/I_i) \tag{E-4}$$

I_o = the current out of the system
I_i = the current into the system

A *decibel* (dB) is 1/10 of a Bel; the prefix *deci* indicates 10^{-1}.

$$\text{dB gain} = 10 \, \text{Log}(P_o/P_i) \tag{E-5}$$
$$\text{dB gain} = 20 \, \text{Log}(V_o/V_i) \tag{E-6}$$
$$\text{dB gain} = 20 \, \text{Log}(I_o/I_i) \tag{E-7}$$

Inverse decibels: If the decibel equivalent of a quantity is known, the quantity itself can be found using antilogarithms. The relationships are

$$P_o/P_i = 10^{(\text{dB}/10)} \tag{E-8}$$
$$V_o/V_i = 10^{(\text{dB}/20)} \tag{E-9}$$
$$I_o/I_i = 10^{(\text{dB}/20)} \tag{E-10}$$

Table E-1 gives decibel values of common ratios and vice versa. Table E-2 shows the calculator keystrokes for finding decibels from ratios and ratios from decibels using commonly available calculators.

Table E-1. Decibel Tables

Power Ratio	Power Decibels	V, I Decibels	Decibels	Power Ratio	V, I Ratio
1.00	0.00	0.00	0.0	1.00	1.00
1.25	0.97	1.94	0.5	1.12	1.06
1.50	1.76	3.52	1.0	1.26	1.12
2.00	3.01	6.02	1.5	1.41	1.19
2.50	3.98	7.96	2.0	1.58	1.26
3.00	4.77	9.54	2.5	1.78	1.33
4.00	6.02	12.04	3.0	2.00	1.41
5.00	6.99	13.98	4.0	2.51	1.58
6.00	7.78	15.56	5.0	3.16	1.78
7.00	8.45	16.90	6.0	3.98	2.00
8.00	9.03	18.06	7.0	5.01	2.24
9.00	9.54	19.08	8.0	6.31	2.51
10.0	10.0	20.0	9.0	7.94	2.82
20.0	13.0	26.0	10.0	10.0	3.16
50.0	17.0	34.0	13.0	20.0	4.47
100.	20.0	40.0	16.0	39.8	6.31
1,000.	30.0	60.0	20.0	100.	10.0
10,000.	40.0	80.0	23.0	200.	14.1
100,000.	50.0	100.0	30.0	1000.	31.6
1,000,000.	60.0	120.0	40.0	10,000.	100.
			50.0	100,000.	316.
			60.0	1,000,000.	1000.

Table E-2. Decibel Calculator Keystroke Sequences		
To Find dB:	To Find Ratio:

1. RPN (HP) Calculators
   a. Key in value as ratio
   b. Log
   c. 10 (see note)
   d. X (multiply)
	1. RPN (HP) Calculators
	   a. Key in value in dB
	   b. ENTER
	   c. 10 (see note)
	   d. / (divide)
	   e. 10^x

2. Algebraic Calculators
   a. Key in value as ratio
   b. LOG
   c. X (multiply)
   d. 10 (see note)
   e. = (equals)
	2. Algebraic Calculators
	   a. Key in value as dB
	   b. / (divide)
	   c. 10 (see note)
	   d. = (equals)
	   e. INV LOG

Note: Use 10 for power decibels or 20 for voltage/current decibels.

Decibel multiplications and divisions: Multiplying two numbers is equivalent to *adding their dB equivalents*. This is a useful way of finding some decibel equivalents and of applying gains and attenuations. For an example of the first use, consider the decibel equivalent of the power ratio 14.

$$14 = 2 \times 7 = 10^{0.301} \times 10^{0.845} = 10^{(0.301 + 0.845)} = 10^{1.146} = 11.46 \text{ dB}$$

For the second use, consider a two-amplifier cascade with power gains of 20 (13.01 dB) and 400 (26.02 dB), respectively. The total gain is

$$20 \times 400 = 10^{1.301} \times 10^{2.602} = 10^{(1.301 + 2.602)} = 10^{3.903} = 8000 = 39.03 \text{ dB}$$
$$13.01 \text{ dB} + 26.02 \text{ dB} = 39.03 \text{ dB} = 8000$$

Divisions are done by *subtracting decibel equivalents*. For example, 450 (26.53 dB) can be divided by 15 (11.76 dB) as

$$450 / 15 = 10^{2.653} / 10^{1.176} = 10^{(2.653 - 1.176)} = 10^{1.477} = 30 = 14.77 \text{ dB}$$
$$26.53 \text{ dB} - 11.76 \text{ dB} = 14.77 \text{ dB} = 30$$

Decibels of values less than one: The logarithm of a number less than one (but greater than zero) is negative. Thus decibel equivalents of ratios less than one are negative. Consider the power ratio 1/200 (0.005).

$$1 / 200 = 10^0 / 10^{2.301} = 10^{(0 - 2.301)} = 10^{-2.301} = -23.01 \text{ dB}$$

Decibels and scientific notation: Once a number is in scientific notation, its decibel equivalent is easy to find, regardless of whether the number is greater or less than one. The above example in scientific notation is:

$$1/200 = 0.005 = 5 \times 10^{-3} = 10^{0.699} \times 10^{-3} = 10^{(0.699 - 3)} = 10^{-2.301}$$
or,   +6.55 dB − 30 dB = −23.1 dB

E. Decibels

Consider the following two examples.

(a) $30{,}000 = 3 \times 10^4 = 10^{0.477} \times 10^4 = 10^{4.477} = +44.77$ dB
or $30{,}000 = 3 \times 10^4 = +4.77$ dB $+ 40$ dB $= +44.77$ dB

(b) $0.0003 = 3 \times 10^{-4} = 10^{0.477} \times 10^{-4} = 10^{-3.523} = -35.23$ dB
or $0.0003 = 3 \times 10^{-4} = +4.77$ dB $- 40$ dB $= -35.23$ dB

Gains and losses — reciprocals: As implied above, the logarithm of the reciprocal of a number is the negative of the logarithm of the number, a property useful in dealing with decibel equivalents of gains and attenuations. The reciprocal is taken by simply changing the sign of the decibel equivalent.

$$10 \text{ Log } (1/X) = 10 [\text{Log}(1) - \text{Log}(X)]$$
$$= -10 \text{ Log}(X) \tag{E-11}$$
$$\text{Loss (dB)} = -\text{Gain (dB)} \tag{E-12}$$
$$\text{Gain (dB)} = -\text{Loss (dB)} \tag{E-13}$$

If an attenuator has a loss of 1000 (30 dB), it has a gain of 1/1000 (-30 dB).

Decibels — zero and negative ratios: The logarithm of zero is undefined (negative infinity). The logarithms of negative numbers are complex. Therefore, the decibel equivalents of zero and negative numbers are undefined and complex, respectively. Since interest is almost always in the magnitude of the quantities from which decibels are taken, and since most calculators and computer library routines do not recognize the logarithm of negative numbers, ratios must have their absolute value taken and zero must be trapped prior to taking the logarithm. In many cases, it is useful to assign a value of negative infinity (or a very large negative number) to the decibel equivalent of zero. A new definition of power decibels is thus

$$\text{dB} = 10 \text{ Log } (|\text{power ratio}|) \tag{E-14}$$

Decibels and dimensioned values: Another important application of decibels is in representing dimensioned values, such as power in Watts. Since generally only the logarithms of dimensionless numbers can be taken, some reference must be used to convert the quantity to a ratio. If power in Watts is referenced to one Watt, the unit dBW results, representing power in Watts.

$$\text{dBW} = 10 \text{ Log } [P / (1 \text{ Watt})] \tag{E-15a}$$
$$\text{dBW} = 10 \text{ Log } [P \text{ (Watts)}] \tag{E-15b}$$

This is a useful way of expressing power, since in this form the input power to an amplifier (in dBW) can simply be added to the gain (in dB) of the amplifier to obtain the output power (in dBW). For example, an amplifier with a power gain of 200 (23 dB) with an input signal of 5 Watts (7 dBW) gives an output of 30 dBW (7 dBW + 23 dB), which is 1000 Watts (200 x 5 Watts).

In this example, it appears that "apples and oranges" are added when dB is added to dBW. Remember, however, that addition of decibels is really multiplication and that multiplying quantities with different dimensions is allowed.

The unit dBW is limited use in communications and radar, where most professionals prefer the unit dBm, representing power in *milli*watts.

$$\text{dBm} = 10 \text{ Log [Power (milliwatts)]} \tag{E-16}$$

The ratio of milliwatts to Watts is 1000 (30 dB). To find dBm from dBW, add 30 dB. To find dBW from dBm, subtract 30 dB. A 1,000,000-Watt (one Megawatt) radar transmitter has an output power of +60 dBW, or +90 dBm.

The unit dBm, and to a lesser extent dBW, can also be used to represent the signals associated with receivers. If the minimum discernible signal (MDS) of a radar receiver were $0.000,000,000,010$ (10^{-11}) milliwatt, the input power is more readily understood if represented as -110 dBm (or -140 dBW).

Any quantity which can be expressed as a ratio can be expressed in decibels. Radar cross-section (RCS), measured in square meters, is commonly represented in dBsm (decibels reference one square meter). Radar cross-section is a power ratio. dBsm is also used to express antenna effective aperture.

$$\text{dBsm} = 10 \text{ Log } [RCS \text{ (m}^2\text{)}] \tag{E-17}$$

$$\text{dBsm} = 10 \text{ Log } [A_E \text{ (m}^2\text{)}]$$

Another dimensioned quantity which is occasionally expressed in decibels is money. If it is in dollars, the unit is the dB$ (pronounced dee-bee-dollar). dB$ is a *power* unit (naturally), and the relationship is

$$\text{dB\$} = 10 \text{ Log [money(\$)]} \tag{E-18}$$

E. Decibels

INDEX

A

A-scope 502
A/R-scope 502
Accuracy 20. *See also* Tracking accuracy
AEGIS radar. *See* AN/SPY-1
AFC. *See* Receivers: Automatic frequency control (AFC)
AGC. *See* Receivers: Automatic gain control
Air search radar. *See* Radar: Air search
Air traffic control radar beacon system 663. *See also* Secondary surveillance radar
Airframe Doppler 27
Aliasing 280
Ambiguity function. *See* Waveform ambiguity function
AN/ designation scheme 695
AN/APG-70 3
AN/APQ-164 460
AN/APY-2 447
AN/FPS-108 458
AN/FPS-115 457
AN/FPS-118 52
AN/MPQ-53 457
AN/MPS-39 459
AN/SPG-55 3
AN/SPS-48 2
AN/SPS-10 2
AN/SPS-52 463
AN/SPY-1A, -1B 50
AN/SPY-1D 456
AN/TPQ-37 457
AN/TPS-43 50
AN/TPS-71 52
Angle-of-arrival 206
Angular position 13
Angular resolution. *See* Resolution: Cross-range
Antenna controller 4
Antenna data pick-offs 464
   Binary code 468
   Direct digital encoders 468
   Gray code 468
   Incremental digital encoders 467
   Potentiometers 464
   Synchros 465
Antenna pedestals 464

Antennas 4, 416
   Aperture blockage 419
   Aperture efficiency 101
   Array 2. *See also* Arrays
   Beam development 416
   Beamwidth and beam shape 98
   Cassegrain reflectors 438
   Circular polarization 231
   Coherent sidelobe cancellation 430
   Collimation 421
   Configurations 427
   Cosecant-squared beams 100
   Cylindrical paraboloid 437
   Effective aperture 100
   Fan beams 99
   Far field 104
   Field zones 103
   Focal point feed reflectors 432
   Fourier transform relationship 417
   Function 4
   Gain 101
   Gain defined 30
   Illumination and pattern 422
   Illumination function 418
   Isotropic 97
   Length efficiency 98
   Lens antennas 440
   Microstrip 448. *See also* Arrays: Microstrip
   Near field 104
   Offset feed reflectors 438
   Pencil beams 99
   Polarization 104
   Polarization twist cassegrain 439
   Principles, functions, and parameters 95
   Radiation pattern 96
   Reciprocity 96
   Reflector antennas 431
   Sidelobe blanking 429
   Sidelobe suppression 428
   Sidelobes and sidelobe effects 102
   Spillover 434
   Temperature 115
Anti-alias filter 586
AOA. *See* Angle-of-arrival
Apparent Doppler shift. *See* Doppler shift: Apparent. *See also* Apparent frequency
Apparent frequency 283

Apparent range 296
Archer lens 461
Area targets 138
Arrays
   Bootlace. *See* Arrays: Lens-fed
   Broadwall slot 441
   Edgewall slot 445
   Electronically phase-steered 450
   Element interactions 444
   Element spacing 444
   Frequency-steered 461
   Lens-fed 460
   Manifold 444
   Microstrip 448
   Phase errors 452
   Phase shifters 453
   Principles 422
   Standing wave 443
   Traveling wave 443
   Waveguide slot radiators 441
ARSR system designation 697
ASDE. *See* Radar missions: ASDE
ASDE system designation 697
ASR. *See* Air search radar
ASR system designation 697
ASR-9 49
ATCRBS. *See* Secondary surveillance radar: Air Traffic Control Radar Beacon System
Augmentation 139
Average power. *See* Transmitter: Average power
Azimuth angle 9

B

B-2 69
B-prime-scope 503
B-scope 502
B1-B. *See* AN/APQ-164
Backscatter 6, 35
Barker codes. *See* Waveforms: Barker codes
Beacons. *See* Radar: Secondary
Bias errors. *See* Tracking accuracy: Bias errors
Bin-centered signal 542
Bistatic radar. *See* Radar: Bistatic
Bistatic radar cross-section. *See* Radar cross-section: Bistatic
Blind Doppler shifts. *See* Doppler shift: Blind
Blind velocity. *See* Doppler shift: Blind
BMEWS 458

Boltzmann's constant 37
Burnthrough 171
Butterfly. *See* Fourier transform: FFT
Butterworth filter response 559

C

C-scope 503
Cancellers. *See* Moving target indicator: Cancellers
Cell-averaging CFAR. *See* Detection: Cell-averaging CFAR
CFAR. *See* Detection: CFAR
Chaff 198, 233
   Doppler spectrum of 233
Chain Home 65
Chirp. *See* Waveforms: Chirp
Clutter 7, 216
   Airborne clutter 230
   Angel clutter 234
   Area clutter 139, 216
   Doppler shift and spectrum 225
   Isodops 241
   Land clutter generalities 224
   Land clutter model 223
   Land clutter spectrum 228
   Models 216
   Moving clutter 237
   Sea clutter data 219
   Sea clutter generalities 223
   Sea clutter horizon 223
   Sea clutter model 218
   Sea clutter spectrum 227
   Spectrum from moving radar's antenna beam shape 245
   Spiking clutter 224
   Volume clutter 139
   Weather clutter 230
   Weather clutter spectrum 233
Clutter filters 313
Clutter notches 313, 578
Clutter statistical distributions 224
   Gaussian distribution 230
   K distribution 225
   Log-normal distribution 225
   Rayleigh distribution 224
   Weibull distribution 225
COBRA DANE. *See* AN/FPS-108
COBRA JUDY 50
Coded waveforms 42
Coherence 71
Coherent integration. *See* Signal interation
Coherent oscillator. *See* COHO
Coherent phase relationships 72

Coherent-on-receive   71
COHO   38, 72
COHO locking.   *See* Receivers: COHO locking
Collapsing.   *See* Losses: Collapsing loss
Collimation.   *See* Antennas: Collimation
Complex data
   Synthesized   555
Compression ratio   78.   *See* Pulse compression: Compression ratio
Conical scan angle tracking.   *See* Tracking: Conical scan
Continuous wave.   *See* Waveform: Continuous wave
Continuous wave radar.   *See* Radar: Continuous wave
Conversions.   *See* Unit conversions
Convolution   525
   Continuous   525
   Discrete   526
   Fast   536
   Impulses   527
   Theorum   525
Coordinate systems   654
   Conversion procedures for netting   661
   Conversions   656
   Coordinate conversions for track acquisition   661
Correlation   522
   Continuous   522
   Discrete   523
   Fast   536
   Process gain   524
Coverage diagram   48, 329
Cross-range resolution   594.   *See also* Resolution: Cross-range
Crossed-field amplifiers.   *See* Transmitters: Crossed-field amplifiers
Crystal oscillator/multiplier   486
CW.   *See* Radar: Continuous wave.   *See also* Waveforms: Continuous wave

D

D-Scope   503
Data gathering time   27
Data processing introduction   653
Data processor function   5
DBS.   *See* Synthetic aperture radar: Doppler beam-sharpening
De-stagger.   *See* Moving target indicators: De-stagger
Decibels   699

Defruiting.   *See* Secondary surveillance radar: Defruiting
Demodulators   492
   Envelope   108, 492
   I/Q   108, 493
   Synchronous   493
Depth-of-null.   *See* Tracking: Depth-of-null
Detection   6, 8
   Cell-averaging CFAR   271
   Coherent integration   258
   Coherent integration loss   259
   Definitions   252
   Errors   251
   False Alarm Number   252
   False Alarm Rate   253
   False Alarm Time   252
   Goals   251
   Guard-band CFAR   273
   Integration and target fluctuations   256, 263
   Introduction   250
   Limiting CFAR   273
   M of N   268
   Noise   253
   Non-coherent integration   264
   Probability of Detection   252
   Probability of False Alarm   252
   Probability of Noise   252
   Probability of Signal   252
   Rice curves   257
Detection errors   6
Detection probability.   *See* Detection: Probability of Detection
DFT.   *See* Fourier transform: Discrete
Diagnostic radar   29
Dicke-fix   274
Digital computer signal processing   67, 507
Digital filters   555
   Aliased response   555
   Synthesis   559
Discrete Fourier transform.   *See* Fourier transform: Discrete
Displays   502.   *See also* individual scope type
   Function   5
Doppler ambiguities.   *See* Doppler shift: Ambiguous
Doppler beam-sharpening   61.   *See also* Synthetic aperture radar: Doppler beam-sharpening
Doppler deghosting   18, 307
Doppler filters   313

Finite impulse response 556
Infinite impulse response 559
Doppler granularity 19
Doppler processing 586
   CW 586
   High PRF (Pulse Doppler) 588
   Medium PRF 588
Doppler resolution 594. *See* Resolution: Doppler
Doppler shift 7, 302
   Ambiguous 306
   Antenna sidelobes 314
   Apparent 18
   Blind 306
   Defined 13
   Derivation of exact 304
   Frequency agility 310
   Semi-active radar homing 311
Doppler shift with CW and pulse Doppler waveforms 315
Doppler spectrum. *See* Target Doppler spectrum
Doppler zones 308
Duplexer 406
   Function 4
Duty cycle. *See* Transmitter: Duty cycle
Duty cycle correction factor. *See* Transmitter: Duty cycle correction factor
Dwell-time. *See* Data gathering time
Dynamic range. *See* Receivers: Dynamic range

E

E-3A AWACS 447
E-scope 503
ECCM. *See* Electronic warfare: Electronic counter-countermeasures
Echo signal capture 35
ECM. *See* Electronic countermeasures; Electronic warfare: Electronic countermeasures
Effective aperture. *See* Antennas: Effective aperture
Effective radiated power 31
Electromagnetic compatibility 683
Electromagnetic interference 8
   Defined 8
Electronic countermeasures. *See also* Electronic warfare: Electronic counter-measures
   Cross-eye 373
   Defined 8

Range-gate pull-off 371
Velocity-gate pull-off 372
Electronic warfare 682
   Deceptive ECM 688
   Denial ECM 688
   Electronic counter-countermeasures 683, 688
   Electronic countermeasures 683, 687
   Electronic intelligence 683
   Electronic support measures 683
   ESM analysis 687
   ESM direction finding 687
   Interception 683
   Low probability of intercept 686
   Manipulative ECM 688
Elevation angle 9
ELINT. *See* Electronic warfare: Electronic intelligence
Ellipsoidal and eccentric earth 659
EMC. *See* Electromagnetic compatibility
EMCON. *See* Emission control
EMI. *See* Electromagnetic interference
Emission control 143
Enhanced cross-range resolution. *See* Synthetic aperture radar
Enhanced range resolution. *See* Pulse compression
ERP. *See* Effective radiated power
Error gradient. *See* Tracking: Error gradient
Errors
   Noise 20
ESM. *See* Electronic warfare: Electronic support measures
Euler's identities 289

F

F-scope 504
False alarm 6
False alarm probability. *See* Detection: Probability of False Alarm
Fast convolution. *See* Convolution: Fast
Fast correlation. *See* Correlation: Fast
FFT. *See* Fourier transform: FFT
FIR. *See* Signal processing: System impulse response. *See also* Doppler filters: Finite impulse response
Fourier series 528, 529
Fourier transform 528
   Continuous 528
   DFT aliasing 534
   DFT errors 539. *See also* Spectral leakage; Windows

DFT gain 535
Discrete 528, 531
FFT 536
Inverse 528
Properties 529
Spherical coordinates 536
Two-dimensional 535
FPA. *See* Transmitter: Final power amplifier
Frequency agility and Doppler shift 310
Frequency allocation treaties 53
Frequency band designation systems 53
Frequency bands and radar performance 55
Frequency bands and radar types 56
Frequency generation 72
Frequency generation, timing, and control Function 4
Frequency synthesizers 487
Frequency-steered arrays. *See* Arrays: Frequency-steered
FTC. *See* Receivers: Fast time constant

G

G-scope 504
Gain inversion ECM 340
Galactic noise 115
Gated CW pulse waveform. *See* Waveform: Gated CW pulse
Gaussian distribution. *See* Noise: Statistical distribution; Clutter statistical distributions: Gaussian distribution
Glint 203
Golay codes. *See* Waveforms: Golay codes
Graceful degradation 396
Grazing angle 216
Greek alphabet 698
Guard-band CFAR 273
Gunn device 485

H

H-scope 504
Height-finding 49. *See also* Radar height-finding
High PRF 58, 288, 325
High resolution radar 593
High-voltage power supplies. *See* Transmitters: High-voltage power supplies
History of radar. *See* Radar: History
HPRF. *See* High PRF

I

I-scope 504
I/Q demodulator. *See* Demodulator: I/Q

I/Q sampling. *See* Sampling: I/Q sampling
IAGC. *See* Receivers: Instantaneous AGC
Ice protection and de-icing 680
Identification friend or foe 663. *See also* Secondary surveillance radar
Identification, Friend or Foe. *See* Secondary surveillance radar
IFF. *See* Secondary surveillance radar
IIR. *See* Signal processing: System impulse response. *See also* Doppler filters: Infinite impulse response
Image frequency. *See* Receivers: Image frequency
Imaging 28. *See also* Synthetic aperture radar
Improvement factor. *See* Moving target indicator improvement factor
Integration loss. *See* Losses: Integration loss
Integration number 156
Interception. *See* Electronic warfare: Interception
Interfering signals 7. *See also* Clutter; Electromagnetic interference; Electronic countermeasures; Jamming; Noise; Spillover
Intermodulation 121
Interpolation. *See* Signal processing: Interpolation
Interrogator 44. *See also* Secondary surveillance radar
Inverse synthetic aperture radar. *See* Synthetic aperture radar: Inverse synthetic aperture radar
ISAR. *See* Synthetic aperture radar: Inverse synthetic aperture radar
Iso-ranges 237
Isodops 241
ITU radiolocation bands 53

J

J-scope 504
Jamming. *See* Electronic warfare: Electronic countermeasures
 Self-protection 139
 Stand-off 139
Jamming margin 126
JEM. *See* jet engine modulations
Jet engine modulations 27

K

K distribution. *See* Clutter statistical distributions: K distribution

K-scope 504
Kalman filter 68, 366
Klystrons. *See* Transmitters: Klystrons. *See also* Receivers: Local oscillators

L

L-scope 504
Leading-edge ranging. *See* Ranging: Leading edge
Leakage. *See* Pulse compression: Range leakage; Spectral leakage
Lichtenstein radar 66
Lightning protection 680
Limiting CFAR 273
Linear frequency modulation waveform. *See* Waveform: Linear frequency modulation
Linearity and MTI 123
Lobing. *See* Tracking: Lobing
Local oscillator offset for moving target indicati 580
Local oscillator reradiation. *See* Reradiation
Local oscillators. *See* Receivers: Local oscillators. *See also* STALO
Log-normal distribution: *See* Clutter statistical distributions: Log-normal
Logarithmic receivers. *See* Receivers: Logarithmic
Look energy. *See* Transmitter: Look energy
Look-time. *See* Data gathering time
Losses 141
    Antenna pattern loss 148
    Atmospheric absorption loss 151
    Collapsing loss 149
    Crossover loss 149
    Ground plane loss 141, 154
    Integration loss 157, 512, 521
    Limiting loss 149
    Non-ideal equipment loss 150
    Operator loss 150
    Plumbing loss 147
    Polarization loss 148
    Propagation medium loss 141, 150
    Pulse width loss 148
    Rain attenuation 152
    Scan loss 148
    Snow attenuation 153
    Squint loss 149
    System loss 141, 147
Low observable 34
Low PRF 20, 58, 288, 324
Low Probability of Intercept radars. *See* Electronic warfare: Low probability of intercept
Low-noise amplifiers. *See* Receivers: Radio frequency amplifiers
Low-observable 68
LPRF. *See* Low PRF
Luneburg lens. *See* Radar cross-section: Luneburg lens

M

M of N Detection. *See* Detection: M of N
M-scope 505
Magic tee 409
Magnetron development 65
Magnetrons. *See* Transmitters: Magnetrons
Matched filter 22, 106, 117, 596
MDS. *See* Receivers: Minimum discernible signal
Medium PRF 7, 58, 288, 326
Metric prefixes 698
MGC. *See* Receivers: Manual gain control
Microwave components 406
MIR 459
Mixers. *See* Receivers: Mixers
Modified Rayleigh distribution 213
Modulators 4, 77, 396
    Floating deck 398
    Function 4
    Hard tube 397
    High-level 77, 397
    Low-level 77, 396
    Pulse forming network 399
    Pulse generators 397
    Pulse transformers 397
    Soft tube 399
Monopulse tracking. *See* Tracking: Monopulse
Monostatic radar. *See* Radar: Monostatic
MOTR. *See* AN/MPS-39
Moving clutter. *See* Clutter: Moving clutter
Moving target detectors 7, 570
Moving target indicator improvement factor 42
Moving target indicators 7, 564
    Amplitude-only MTI 569
    Calibration targets 590
    Cancellers 568
    Clutter maps 586
    De-stagger 577
    Filter bank 570
    Filter types 567
    Fundamentals 564

Improvement factor   566
Improvement factor limitations   582
Local oscillator offset   580
Moving radars and clutter   578
Phase MTI   569
PRF stagger   572
Principles and methods   566
Range   570
Testing   589
Vector MTI   570
MPRF. *See* Medium PRF
MTD. *See* Moving target detectors
MTI. *See* Moving target indicators
MTI-I. *See* Moving target indicator improvement factor
MTI-I limitations. *See* Moving target indicators: Improvement factor limitations
Multi-mode radar. *See* Radar: Multi-mode

N

N-scope   505
Noise   7, 213
Statistical distribution   213
Thermal   37
Noise errors. *See* Errors: Noise
Noise factor / noise figure   113
Noise figure measurement. *See* Receivers: Noise figure measurement
Non-coherent integration. *See* Signal integration: Non-coherent
Non-linear frequency modulation waveform. *See* Waveform: Non-linear frequency modulation

O

O-scope   505
Optimal codes. *See* Waveforms: Optimal codes
OTH-B. *See* AN/FPS-118

P

P-scope. *See* Plan-position indicator
Padding. *See* Signal processing: Interpolation
Parametric amplifier. *See* Receivers: Radio frequency amplifiers
PATRIOT. *See* AN/MPQ-53
PAVE PAWS. *See* AN/FPS-115
PD. *See* Radar: Pulse Doppler
Peak power. *See* Transmitter: Peak power

Phase   39
Phase shifters   453
Diode   454
Ferrite latching   454
Meander line   454
Reggia-Spencer   454
Varactor   454
YIG   454
Phase-coded waveforms. *See* Waveforms: Phase-coded
Phased arrays. *See* Arrays: Electronically phase-steered
Phasor notation   39
Piezoelectric effect   607
Plan-position indicator   505
Plunge, elevation   369
Point targets   136
Polarization. *See* Antennas: Polarization
Polarization scattering matrix   211
Power density   32
PPI. *See* Plan-position indicator
PRF. *See* Pulse repetition frequency
PRF agility   79
PRF classes   20, 58, 287. *See also* High PRF; Low PRF; Medium PRF
PRF classes evaluation   323
PRF stagger   572
Equivalent PRF   575
PRF selection   574
PRI. *See* Pulse repetition interval
Primary radar   44. *See also* Radar: Primary
Probability of detection. *See* Detection probability. *See also* Detection: Probability of Detection
Probability of false alarm. *See* False alarm probability. *See also* Detection: Probability of False Alarm
Process gain   42, 126
Propagation data   32
PRT. *See* Pulse repetition time
Pseudorandom codes. *See* Waveforms: Pseudorandom codes
Pulse compression   23, 67
Analog   607
Compression ratio   158
Digital   611
Fundamentals   595
Matched filter   596
Range leakage   609
Waveforms   604
Windows   609
Pulse Doppler radar. *See* Radar: Pulse Doppler

Pulse energy. *See* Transmitter: Pulse energy
Pulse period. *See* Pulse repetition interval
Pulse repetition frequency   78
Pulse repetition interval   79
Pulse repetition time. *See* Pulse repetition interval
Pulse width   78

Q

Quasi-coherent. *See* Coherent-on-receive

R

R-scope   502
RACON. *See* Radar: Secondary
Radar
   Air search   2, 47
   Bistatic   45
   Defined   1
   Missions   1
   Monostatic   45
   Multi-mode   3, 63
   Primary   44
   Pulse Doppler. *See* High PRF
   Pulsed   4
   Search   47
   Secondary   44
   Stability requirements   129
   Surface search   47
   Three-dimensional search   48
   Two-dimensional search   47
Radar Compared to Other Electromagnetic Sensors   64
Radar coordinate systems. *See* Coordinate systems
Radar cross-section   27, 33
   Birds   234
   Bistatic   208
   Complex objects   201
   Corner reflector   199
   Cylinder   197
   Definition   184
   Dipole   198
   Doppler spectra   209
   Flat plate   198
   Glint   203
   Insects   234
   Luneburg lens   199
   Median and mean   268
   Median RCS   203
   Model   33
   Polarization scattering matrix   211
   Scintillation   203
   Sphere   194
   T-33A   202
Radar data processing. *See* Data processing
Radar displays. *See* Displays
Radar electronic warfare. *See* Electronic warfare
Radar equation
   Area clutter   164
   Augmentation   175
   Beacons and transponders   178
   Bistatic radars   176
   Burnthrough   171
   Crossover range   171
   CW radars   162
   Detection range in self-protection jamming   171
   Detection range in stand-off jamming   174
   Detection range prediction   144
   Introduction   29, 135
   Look energy and average power   142
   Missile illumination   176
   Monostatic radars with separate antennas   144
   Point targets in area clutter   167
   Point targets in noise   140
   Point targets in volume clutter   168
   Pulse compression   157
   Pulse Doppler radars   162
   Search radars   159
   Self-protection jamming   170
   Signal processing role   155
   Signal-to-noise ratio   142
   Stand-off jamming   173
   Summary   136
   Tracking radars   161
   Volume clutter   168
Radar frequency bands   52
Radar height-finding   326
Radar horizon   329
Radar missions
   Air search   59
   Air-to-air missions   60
   Air-to-surface missions   61
   Airborne weather radar   61
   ASDE   59
   Doppler Beam-Sharpening mapping   61
   Doppler update   63
   Early warning   59
   Fire control   60
   Ground moving target search   62
   Gun director   61
   Illumination   61

Metric instrumentation 60
Over-The-Horizon (OTH) 60
Radar altimeter 63
Raid discrimination 61
Range-while-search 60
Real-beam ground mapping 61
Sea surface search 62
Spacecraft detecting and cataloging 60
Surface search 59
Surface-to-air missions 59
Surface-to-surface 58
Synthetic Aperture Radar mapping 61
Terrain following and terrain avoidance 62
Track-while-scan 60
Tracking 61
Velocity search 60
Weather radar 60
Radar principles 3-8
Radar signal processing. *See* Signal processing
Radar signature 27
Radar testing. *See* specific system (transmitter, and so forth)
Radar types 44
Radar waveforms 57
   Complex waveform pulse 58
   Gated CW pulse 57
   Modulated CW 57
   Unmodulated CW 57
Radial velocity 14, 302
Radiation hazard 673, 694
Radomes 677
   Refraction 681
Railing interference 669
Range 9
   Apparent 296
   Blind 301
   Minimum 301
Range ambiguity 11
Range ambiguity and range zones 295
Range deghosting 296, 302
Range granularity and quantization 11
Range resolution. *See* Resolution: Range
Range tracking. *See* Tracking: Range tracking
Range with CW and pulse Doppler waveforms 315
Range zones 295
Range-height indicator 505
Ranging 10, 293
   CW estimation 320
   Effective PRF 301

   Leading edge 10
   Pulse Doppler 320
   Semi-active radar homing 321
Ranging conditions 294
RAR. *See* Real aperture radar
Rayleigh distribution 213. *See also* Noise: Statistical distribution; Clutter statistical distributions: Rayleigh distribution
Raytheon R-20 450
RCS. *See* Radar cross-section
Real aperture radar 25
Receiver examples 494
Receivers 5, 471
   Automatic frequency control (AFC) 488
   Automatic gain control 110
   COHO locking 488
   Crystal video 472
   Demodulator 108
   Description and functions 105
   Dynamic range 119
   Equivalent noise bandwidth 112
   Equivalent temperature 115
   Fast time constant 111
   Function 5
   Gain controls 109
   Homodyne 472
   Image frequency 107, 473
   Instantaneous AGC 111
   Intermediate frequency amplifiers 108, 489
   Linearity 119
   Local oscillators 107, 484
   Log IF amplifiers 491
   Logarithmic 122
   Manual gain control 109
   Minimum discernible signal 113
   Mixers 107, 481
      Configurations 483
      Specifications 482
   Noise 112
   Noise factor / noise figure 113
   Noise figure measurement 501
   Radio frequency amplifiers 477
   RF attenuators 477
   RF filters 477
   RF processor 106, 475
   Sensitivity, noise, and temperature 112
   Sensitivity time control 110
   Superheterodyne 472, 473
      Introduction 106
   System temperature 115
   Tangential sensitivity 113, 500
   Types 471

Reflex klystron 484
Reradiation 483
Resolution 21, 593
   And signal processing 43
   Cross-range 24
   Doppler 27
   Range 21, 593
Resolution and radar waveforms 598
Resolution and windows. *See* Window resolution
RHI. *See* Range-height indicator
Ricean distribution 213
Rieke diagram. *See* Transmitters: Rieke diagram
Rihaczek plot 599
ROTHR. *See* AN/TPS-71
Rotman-Turner lens 461

S

Sampling 278
   Direct sampling 289
   I/Q sampling 289
   Information recovery 283
   Nyquist criterion 278
   Nyquist sampling 565
   Radar information rates 285
SAR. *See* Synthetic aperture radar
SARH. *See* Semi-active radar homing
SAW. *See* Surface acoustic wave devices
Scalloping loss 260
Scintillation 203
SCR-268 66
SCR-270 65
SCR-584 67
Sea states 218
Search radar. *See* Radar: Search
Secondary radar. *See* Radar: Secondary
Secondary surveillance radar 2, 663
   Air traffic control radar beacon system 663
   Defruiting 669
   Identification, friend or foe 3, 663
   Interrogate sidelobe blanking 667
   Interrogation 664
   Mode S 671
   Modes 663
   Monopulse reception 667
   Reply 664
   Reply sidelobe blanking 669
Self-protection jamming. *See* Jamming: Self-protection

Semi-active radar homing 3, 46
   Doppler shift 311
Sidelobe Doppler 314
Sidelobe suppression. *See* Antennas: Sidelobe suppression
Sidelooking SAR. *See* Synthetic aperture radar: Sidelooking SAR
Signal amplitude 38
Signal composites 39
Signal integration 510
   Coherent 514
   Integration loss. *See* Losses: Integration loss
   Motion compensation 519
   Non-coherent 512
Signal phase 39
Signal processing 38. *See also* individual topics
   A/D converter function 509
   Block diagrams 508
   Causality 510
   CFAR function 509
   Clutter interference 41
   Coherent integration 156
   Convolution function 508
   Correlation function 507
   Digital filters. *See* Digital filters
   Effectiveness 42
   Filtering function 507
   Functions and parameters 123
   Fundamentals 40
   Information and signal bandwidths 125
   Information capacity 125
   Integration loss. *See* Losses: Integration loss
   Integration number 156
   Interpolation 552
   Introduction and background 507
   Jamming margin 126
   Linearity 510
   Matched filter function 509
   Moving target indicator improvement factor 127
   Moving targets 42
   MTI. *See* Moving target indicators
   Noise interference 41
   Non-coherent integration 156
   Objectives 507
   Parameters 42
   Process gain 126
   Process noise 126
   Recursiveness 510

Sample-and-hold function 508
Signal filtering function 509
Signal integration. *See* Signal integration
Signal integration function 507
Signal storage function 509
Spectrum analysis function 507, 509
Sub-clutter visibility 127
System fundamentals 510
System impulse response 510
Time invariance 510
Windowing function 508
Signal processor 5
   Block diagram 124
   Function 5
Signal properties 38
Signal-to-interference ratio 37
Signal-to-noise ratio 37
Signal-to-noise ratio examples 144
Signature. *See* Radar signature
Single-target track. *See* Tracking: Single target
Sinuous feed 461
Sinusoidal FM CW 315
Slow-wave structure 381
SMO. *See* System master oscillator
Solar noise 115
Space loss 145
Spectral leakage 540
Spectrum analysis 528. *See also* Fourier transform
Spherical coordinates 9
Spillover interference 8, 162. *See also* Antennas: Spillover
SSR. *See* Secondary surveillance radar
Stability requirements. *See* Radar: Stability requirements
Stable local oscillator. *See* STALO
Stagger. *See* PRF stagger
STALO 72, 106
Stand-off jamming. *See* Jamming: Stand-off
STC. *See* Receivers: Sensitivity time control
Stealth. *See* Low observable
Sub-clutter visibility 566. *See also* Signal processing: Sub-clutter visibility
Superheterodyne receivers. *See* Receivers: Superheterodyne
Surface acoustic wave devices 607
Surface search radar. *See* Radar: Surface search
SWAP utility 532. *See also* Fourier transform: Discrete
Swerling cases. *See* Targets: Swerling cases

Synchros. *See* Antenna data pick-offs: Synchros
Synthetic aperture radar 61, 616
   Defined 25
   Doppler beam-sharpening 616, 618
   Focused SAR 633
   Focusing and rotating objects 645
   Inverse Doppler beam-sharpening 626
   Inverse synthetic aperture radar 617
   Resolution of rotating objects 642
   Rotational imaging of tactical targets 650
   Sampling criterion for rotating targets 643
   Sidelooking SAR 616, 627
   Sidelooking SAR PRF 636
   Sidelooking SAR resolution 628
   Synthetic aperture with rotating objects 640
   Unfocused ISAR 645
   Unfocused SAR 631
   Windowing 644
System master oscillator 73
System temperature 115

T

Tangential sensitivity. *See* Receivers: Tangential sensitivity
Target Doppler spectra 209
Target Doppler spectrum 27
Target fluctuations 186
Target Information Extraction 8
Target position locating 9
Target reflection 33
Target velocity. *See* Doppler shift
Targets 184
   Augmentation effects on fluctuation 193
   Field zones 185
   Fluctuation mechanisms 186
   Fluctuation models 191
   Fluctuation speeds 186
   Marcum model 191
   Multipath 190
   Nyquist sampling 193
   Swerling cases 191
TDWR system designation 697
TE_{10}. *See* Waveguide
Thermal noise. *See* Noise: Thermal
Three-dimensional search radar. *See* Radar: Three-dimensional search
Time-on-target 160
TR tubes 408
Track acquisition 369

Track-while-scan 51. *See also* Tracking: Track-while-scan
  Maneuvering gates 368
Tracking 49
  Amplitude-comparison monopulse 343
  Antenna patterns 333
  Conical scan 335
  Depth-of-null 334
  Designation coordinates 661
  Elevation plunge 369
  Error gradient 334
  Lobing 341
  Monopulse 51
  Monopulse with arrays 346
  Multi-target track 49, 331
  Phase-comparison monopulse 348
  Range tracking 352
  Secant compensation 369
  Single target track 331
  Spotlight track 331
  Track-while-scan 332, 351
  Velocity tracking 355
  Wideband monopulse 350
Tracking accuracy 356
  Angles 360
  Bias errors 356
  Multipath errors 362
  Noise errors 357
  Propagation medium errors 356
  Range 357
  Servo lag errors 356
  Target-induced errors 357
Tracking anomalies 370
Tracking servos 364
  α-β filter 367
  Kalman filter 367
Transmission line temperature 115
Transmitter 4
  Average power 80
  Duty cycle 79
  Duty cycle correction factor 80
  Efficiency 81
  Exciter/Driver 77
  Final power amplifier 77
  Function 4
  Look energy 80
  Peak power 79
  Pulse energy 80
  Waveform generator 77
Transmitter cooling 402
Transmitter functions and parameters 77
Transmitter safety 693
Transmitter vacuum 402

Transmitters
  CFA modes 390
  CFA types 391
  Coaxial magnetrons 393
  Crossed-field amplifiers 388
  High-voltage power supplies 401
  Introduction and functions 376
  Klystrons 378
  Magnetrons 391
  Monitoring and testing 403
  Rieke diagram 393
  Solid-state 394
  Traveling wave tube 381
  TWT attenuators and severs 384
  TWT control and shadow grids 386
  TWT depressed collectors 388
  TWT types 385
  Twystrons 388
  Types 377
Transponder 44. *See* Secondary surveillance radar
Traveling wave tube. *See* Transmitters: Traveling wave tube
Triangular FM CW 316
TSS. *See* Receivers: Tangential sensitivity
Two-dimensional search radar. *See* Radar: Two-dimensional search

U

Uncertainty. *See* Waveform uncertainty
Unit conversions 691

V

V-FM. *See* Waveform: V-FM
Velocity measurement and discrimination 13
Volume targets 138

W

Waveform ambiguity function 598
Waveform evaluation 598
Waveform spectra and bandwidths 81
Waveform uncertainty 600
Waveforms
  Barker codes 89, 611
  Biphase coding 88
  Chirp 604, 607
  Continuous wave 90
  Gated CW pulse 82
  General 82
  Golay codes 613
  Linear frequency modulation 84

Maximal length sequences 613
  Non-linear frequency modulation 85
  Optimal codes 614
  Phase coded 88
  Phase-coded 611
  Phase-coded FM CW 320
  Pseudorandom codes 613
  Pseudorandom sequences 89
  V-FM 87
  Welti codes 613
Waveguide 410
Waveguide pressurization 413
Waveguide rotary joints 410
Waveguide table 413
Wavelength 52
Weibull distribution. *See* Clutter statistical distributions: Weibull distribution
Welti codes. *See* Waveforms: Welti codes
White noise 118
Window resolution 549
  Clipped (non-linear) 552
  Dolph-Chebyshev 551
  Hamming 551
  Hann 551
  Kaiser-Bessel 551
  Leakage 549
  Rectangular 551
  Taylor 552
Windows 539, 541
  Clipped (non-linear) 547
  Cosine family 543
  Dolph-Chebyshev 546
  Hamming 544
  Hann 544
  Kaiser-Bessel 545
  Parameter summary 548
  Rectangular 542
  Scalloping 548
  Taylor 546
WSR system designation 697

X

X-rays 694
XAF 65

Z

Z-transform 556